AF443447

Amorphous Insulators and Semiconductors

NATO ASI Series

Advanced Science Institutes Series

A Series presenting the results of activities sponsored by the NATO Science Committee, which aims at the dissemination of advanced scientific and technological knowledge, with a view to strengthening links between scientific communities.

The Series is published by an international board of publishers in conjunction with the NATO Scientific Affairs Division

A	**Life Sciences**	Plenum Publishing Corporation
B	**Physics**	London and New York
C	**Mathematical and Physical Sciences**	Kluwer Academic Publishers
D	**Behavioural and Social Sciences**	Dordrecht, Boston and London
E	**Applied Sciences**	
F	**Computer and Systems Sciences**	Springer-Verlag
G	**Ecological Sciences**	Berlin, Heidelberg, New York, London,
H	**Cell Biology**	Paris and Tokyo
I	**Global Environmental Change**	

PARTNERSHIP SUB-SERIES

1.	**Disarmament Technologies**	Kluwer Academic Publishers
2.	**Environment**	Springer-Verlag / Kluwer Academic Publishers
3.	**High Technology**	Kluwer Academic Publishers
4.	**Science and Technology Policy**	Kluwer Academic Publishers
5.	**Computer Networking**	Kluwer Academic Publishers

The Partnership Sub-Series incorporates activities undertaken in collaboration with NATO's Cooperation Partners, the countries of the CIS and Central and Eastern Europe, in Priority Areas of concern to those countries.

NATO-PCO-DATA BASE

The electronic index to the NATO ASI Series provides full bibliographical references (with keywords and/or abstracts) to more than 50000 contributions from international scientists published in all sections of the NATO ASI Series.
Access to the NATO-PCO-DATA BASE is possible in two ways:

– via online FILE 128 (NATO-PCO-DATA BASE) hosted by ESRIN,
Via Galileo Galilei, I-00044 Frascati, Italy.

– via CD-ROM "NATO-PCO-DATA BASE" with user-friendly retrieval software in English, French and German (© WTV GmbH and DATAWARE Technologies Inc. 1989).

The CD-ROM can be ordered through any member of the Board of Publishers or through NATO-PCO, Overijse, Belgium.

3. High Technology – Vol. 23

Amorphous Insulators and Semiconductors

edited by

M. F. Thorpe

Department of Physics and Astronomy,
Michigan State University,
East Lansing, Michigan, U.S.A.

and

M. I. Mitkova

Central Laboratory of Electrochemical Power Sources,
Bulgarian Academy of Sciences,
Sofia, Bulgaria

Kluwer Academic Publishers

Dordrecht / Boston / London

Published in cooperation with NATO Scientific Affairs Division

Proceedings of the NATO Advanced Study Institute on
Amorphous Insulators and Semiconductors
Sozopol, Bulgaria
May 26–June 8, 1996

A C.I.P. Catalogue record for this book is available from the Library of Congress

ISBN 0-7923-4404-9

Published by Kluwer Academic Publishers,
P.O. Box 17, 3300 AA Dordrecht, The Netherlands.

Kluwer Academic Publishers incorporates the publishing programmes of
D. Reidel, Martinus Nijhoff, Dr W. Junk and MTP Press.

Sold and distributed in the U.S.A. and Canada
by Kluwer Academic Publishers,
101 Philip Drive, Norwell, MA 02061, U.S.A.

In all other countries, sold and distributed
by Kluwer Academic Publishers Group,
P.O. Box 322, 3300 AH Dordrecht, The Netherlands.

Printed on acid-free paper

Table of Contents

1. Glass Formation

Strong and Fragile Liquids; Glass Transitions and Polyamorphic Transitions
in Covalently Bonded Glassformers 1
C.A. Angell

Kinetics of Glass Formation 21
I. Gutzow and A. Dobreva

Chalcogenide Glasses 45
M. T. Mora

Synthesis, Structure and Some Modes of Application of New Chalcogenide
and Chalco-Halide Glasses of Silver 71
M. Mitkova

2. Structure

X-Ray and Neutron Diffraction: Experimental Techniques and Data Analysis 83
A. C. Wright

Computer Modeling of Glasses and Glassy Alloys 133
N. Mousseau

Molecular Dynamics Methods and Large-Scale Simulations
of Amorphous Materials 151
P. Vashishta, R.K. Kalia, A. Nakano, W. Li, and I. Ebbsjö

Reverse Monte Carlo Simulations for Determining Disordered Structures:
Basics and Application for Amorphous Semiconductors 215
L. Pusztai

Reverse Monte Carlo Simulations of Glasses – Practical Aspects 225
N. Zotov

vi

A Thermodynamic Model for the Calculation of the Physical Properties
and Structural Characteristics of Glasses and Melts 235
N. M. Vedishcheva, B. A. Shakhmatkin, and A. C. Wright

High Resolution and Multidimensional Nuclear Magnetic Resonance Probes
of Glass Structure 245
J.W. Zwanziger, K.K. Olsen, S.L. Tagg, and R.E. Youngman

Porous Silica: Model Fractal Materials, Their Structures and Their Vibrations 255
E. Courtens and R. Vacher

3. Vibrational States and Elasticity

The Structure and Mechanical Properties of Networks 289
M.F. Thorpe, B.R. Djordjevic, and D.J. Jacobs

The Elastic Moduli of Random Networks: Calculations 329
A. R. Day

One-Fold Coordinated Atoms, Constraint Theory and
Nano-Indentation Hardness 339
P. Boolchand, M. Zhang, B. Goodman

Silicates and Soft Modes 349
M.T. Dove

Vibrational Dynamics in Glasses 385
C. Levelut, F. Terki, Y. Scheyer, and J. Pelous

Optical Studies of Nanophase Titania 395
R.J. Gonzalez and R. Zallen

4. Electronic Properties and Defect

Electronic Structure Methods with Applications
to Amorphous Semiconductors 405
D.A. Drabold

Amorphous Silicon: Defects and Disorder 437
G. J. Adriaenssens

Tunneling States..469
    S. Hunklinger

Electronic Structure of Amorphous Insulators and Semiconductors
    by X-Ray Photoelectron and Soft X-Ray Spectroscopies...............497
    C. Sénémaud

Photograph..508

Participants 511

Index..519

Preface

The aim of this NATO ASI has been to present an up-to-date overview of current areas of interest in amorphous materials. In order to limit the material to a manageable amount, the meeting was concerned exclusively with insulating and semiconducting materials. The lectures and seminars fill the gap between graduate courses and research seminars. The lecturers and seminar speakers were chosen as experts in their respective areas and the lectures and seminars that were given are presented in this volume. During the first week of the meeting, an emphasis was placed on introductory lectures, mainly associated with questions relating to the glass-formation and the structure of glasses. The second week focused more on research seminars. Each day of the meeting, about four posters were presented during the coffee breaks, and these formed an important focus for discussions. The posters are not reproduced in this volume as the editors wanted to have only larger contributions to make this volume more coherent.

This volume is organized into four sections, starting with general considerations of the glass forming ability and techniques for the preparation of different kinds of glasses. There has been activity on the nature of the glass transition in recent years, but it remains a difficult and poorly understood subject. The second section involves the various techniques used to investigate the structure of the glasses, and includes extensive notes on computer simulation techniques. The third section discusses vibrational states, with special emphasis on the interaction between theory and experiment. In the final section, the electronic structure of the glasses is reviewed and the relationship to their electrical properties which involves an understanding of defects. Also included in this section is a review of the current thinking concerning tunneling states in glasses.

This NATO ASI was held in Sozopol, Bulgaria. Legend takes Sozopol back to 610BC when Greek settlers founded the city of Apollonia. Situated on a small rocky peninsula on the southern Black Sea coast, south east of the bustling port of Bourgas, Sozopol today is a beautiful old seaside town. It is justly proud of its past, and keeps its old traditions very much alive. Artists and photographers love its narrow cobbled streets, historic churches, weather-beaten blackened eaves, bay windows and wooden facings on its old houses, many dating from the 18th and 19th centuries. Sozopol ranks as one of Bulgaria's most enchanting and historic coastal towns.

The meeting lasted for 10 working days with a day off for a trip to the ancient coastal town of Nessebar, some distance north of Sozopol. Each day had either morning and evening sessions [with afternoons free to visit the beach or wander in the town] or morning and afternoon sessions [with the evenings free to drink coffee, wine or rakia on the cliffs overlooking the Black Sea]. On most of the free evenings, entertainment was provided which included a fashion show, a magic show, folk dancing and a disco which was the most popular.

Organizing a conference in Bulgaria is not always easy due to persistent problems with currency exchange, communications, etc. This was especially true for this meeting as it was decided to hold the meeting in the remote but beautiful small town of Sozopol. Nevertheless, this meeting demonstrated that it was possible to maintain a high scientific level and for the participants to have a comfortable and interesting stay with plenty of good food and drink. We hope that this meeting will be a model for future scientific meetings on the Black Sea coast of Bulgaria.

Some anecdotes from the meeting were provided by Professor Dick Zallen at the end of the conference, some of which are provided here without comment!

"If it doesn't close, it's not a ring."

"He is as Swedish as I am Spanish."

"Glasses form when crystals don't."

"The weakest structure former gets the glass."

"Excess entropy is the lifeblood of the liquid state."

"The nature of the glass transition is not a private question."

"The two–banana problem"

"Forget experiments, we will simulate it."

Finally, we should like to thank the NATO Science Committee for providing financial support for this meeting. We would especially like to thank Ms. Janet King for her invaluable assistance from the planning through to the report stage of this meeting. Most of the participants corresponded with her and got to know her well via e-mail.

M.F. Thorpe
M.I. Mitkova
East Lansing, July 1996

Organizing Committee
G.J. Adriaenssens
M. Balkanski
S. Hunklinger
A.C. Wright

Co–Directors
M.F. Thorpe
M.I. Mitkova

STRONG AND FRAGILE LIQUIDS; GLASS TRANSITIONS AND POLYAMORPHIC TRANSITIONS IN COVALENTLY BONDED GLASSFORMERS

C. A. ANGELL
Department of Chemistry, Arizona State University
Tempe, AZ 85287-1604

1. Introduction

In this article, we review the origin of glassforming ability in certain liquids and liquid mixtures, i.e. the reasons crystals do not form in these cases, and then discuss the details of glass formation by the continuous viscous slowdown of non-crystallizing liquid phases. We give special attention to the phenomenon of fragility in covalently bonding systems and relate decreasing fragility to the increasing crosslinking with elimination of "floppy modes," as the Phillips-Thorpe rigidity percolation threshold is approached. Discussing overconstrained systems, we encounter the phenomenon of liquid-liquid phase transition in tetrahedral phases and consequent liquid-liquid unmixing in multi-component systems. Finally, we turn to the solid state physics aspects of the glass transition phenomenon and relate the onset of diffusivity and relaxation in the amorphous phase on heating to the anharmonic terms in the interparticle potential identifying a sort of Lindemann criterion for the glass transition.

While we will focus attention, in this article, on the process of vitrification from the liquid state, it must always be borne in mind that essentially the same glass can be made in a number of quite different ways. It is because of this that we will start by defining the term "glass" somewhat more broadly than in the ASTM index in order to include cases made by unconventional routes:

"A glass is an amorphous solid which is capable of passing continuously into the viscous liquid state, usually, but not necessarily, accompanied by abrupt increase in heat capacity."

A broader definition which includes the cases of analog behavior now being encountered all across physics [1] (spin-glasses, vortex glass, dipole glasses, etc.) would be the following:

"A glass is a condensed state of matter which has become non-ergodic by virtue of the continuous slow-down of one or more of its degrees of freedom."

The study of ergodicity-breaking by slow-down of responses (or of the corresponding intrinsic fluctuation time scales) to values beyond any measurement time scale, is currently arousing much scientific curiosity. The problem is one of understanding how the coupling of these processes to the very fast microscopic

1

M. F. Thorpe and M. I. Mitkova (eds.), Amorphous Insulators and Semiconductors, 1–20.
© *1997 Kluwer Academic Publishers. Printed in the Netherlands.*

2

fluctuations associated with particle vibrations can become so weak that the relaxation time can show such tendencies to diverge.

In the case of liquids, this refers to the manner in which, in a narrow range of temperature, a great many liquids more or less suddenly lose their fluidities and become brittle solids. The particles become "localized" . Alternatively viewed from the solid, the amorphous solid collection of particles, with increasing temperature above a particular value, rapidly becomes diffusive.

The liquid is interesting in proportion to how sudden this onset of diffusivity proves to be. We use the word "fragility" to describe the rate per K of the collapse of the rigid glassy structure into the mobile liquid. Very fragile liquids are those which become fluid in a very short temperature interval. And ultimately we try to understand how this happens in terms of fundamental aspects of particle motion – after all in the glass we have, overwhelmingly, just vibrational motions – anharmonic maybe, but all particles are caged. Then quite suddenly, above a particular temperature for each substance, particles are slipping out of their cages or reorganizing their cages cooperatively and diffusing amongst each other – what happened in that narrow temperature range to make this possible?

We will see that, in some cases, it can even happen via a first order phase transition; and the best documented case in question is that of the substance which more than any other is at the center of interest of this conference – viz. amorphous silicon. We'll get to this in the fourth section of this article in which we will make bold to draw phenomenological analogies between Si and its binary alloys, and the two great solution systems of the natural world, viz.,

(i)   water + second components, (of hydrology, and life) and

(ii) SiO_2 + second components (of geochemistry).

2.   Origin of Glassforming Ability

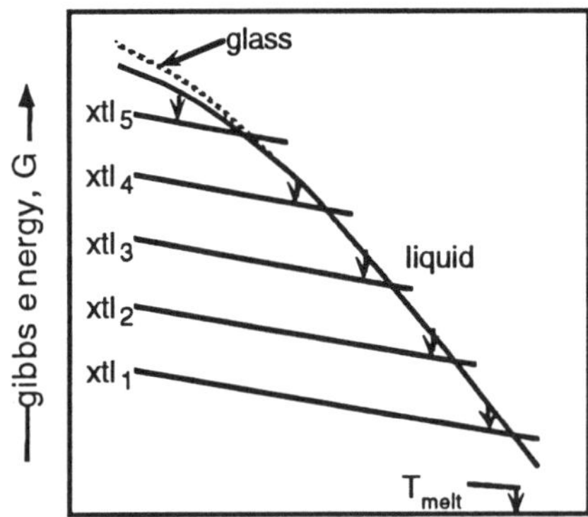

Figure 1. Schematic variation of Gibbs free energy (G) with temperature, showing access to different metastable crystals from the supercooled liquid. Slower cooling rates move the glass transition to lower temperature, permitting access to more metastable crystalline forms.

To start, however, I want to spend some time on why glasses exist and how you form them if you want to study them. After all, they shouldn't exist. In almost every case, they are *metastable* with respect to the crystal state of the same composition (if it is a one-component system) or a combination of crystals (if it is a multi-component system). This means they only exist because the system didn't have time to reach the lowest available free energy surface. "Not having time" means that crystals don't have time to nucleate and grow during the process of cooling the liquid to below the glass transition temperature. For crystals to nucleate and grow, fluctuations in order must occur, and these will occur more slowly the more viscous the liquid. Liquids which vitrify easily are almost always rather viscous at their freezing temperatures.

Let us try to understand how this can happen, so we can know how to *make* it happen if we want to.

Figure 1 shows the free energy vs. temperature relations for a hypothetical system which has a variety of possible crystal states of different lattice energy (the lattice energy is G at T = OK when the reference state is the dilute gas state). For simplicity, we assume the same heat capacity-temperature relations, hence the same curvature of G. The liquid phase, which has a higher entropy, is drawn to cut across each of the crystal curves. The intersection of G(liquid) with each G(crystal) determines the melting point each would have if transitions between crystal polymorphs were excluded. Clearly the polymorph with the poorest crystal packing (smallest lattice energy) has the lowest melting point, and would accordingly melt to the liquid of highest viscosity.

From this "thought" example, we would conclude that systems which vitrify easily should be those which lack any large lattice energy crystal packing arrangements. We see, by this argument, that the ability of a substance to vitrify is determined in the crystalline state, *not* in the liquid state as is often supposed.

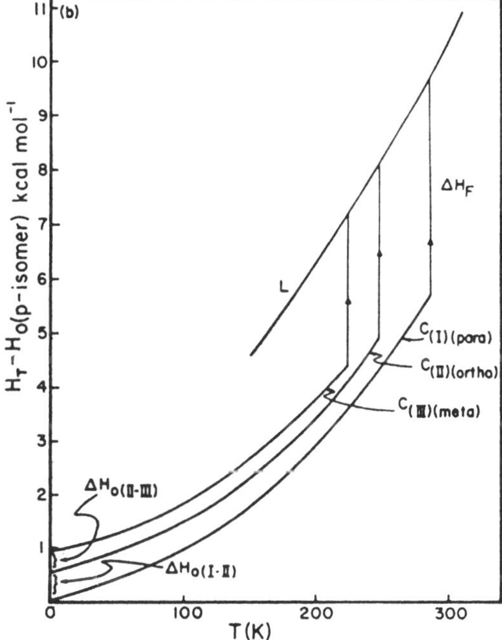

Figure 2: Enthalpy vs. temperature relations relative to enthalpy of p-xylene at O K for crystal and liquid isomers of xylene, showing how smaller lattice energy (less negative H_O) due to packing problems in case of m-isomer leads to low fusion temperature, hence greater glassforming ability for this isomer.[3]

To support this line of thought, we consider the case of the three isomers of the disubstituted benzene xylene. The liquids all boil within a few degrees of each other, hence we know that the attractive forces between the molecules are roughly the same, and consistent with this, the viscosities of the three isomers measured at the same temperature are very similar. However, only the meta-isomer is observed to form a glass on fast cooling [2,3]. At a simple level, the explanation follows immediately on examination of the melting points of the three isomers – the meta-isomer has quite the lowest fusion point (185K compared with 211 K for the o-isomer and 216 K for the p-isomer). This means that, consistent with our earlier generalization, the meta-isomer has the largest viscosity at the temperature at which it becomes thermodynamically meta-stable. Our problem, therefore, becomes one of explaining why the meta-isomer should have the lowest melting point of the three.

The explanation according to our argument above should lie in the difficulty of packing the meta-isomer molecules in long range three-dimensional order, and in this case, we can prove it by calculating that there is a smaller crystal lattice energy for the meta-isomer than for either of the others. This is done in Figure 2[3] using precise

4

thermochemical data available for these substances [4]. The difference in lattice energy responsible for the ~70 K difference in melting points of m- and p-isomers is only ~1 kcalmol^{-1}.

In summary, therefore, we attribute the stability of certain materials in the amorphous state to the failure of nature to find suitable solutions to the three-dimensional long-range order-packing problem in the case of these substances.

The probability of glass formation in such cases will clearly be enhanced by the formation of liquid mixtures, in which the free energy of the liquid phase is decreased, while that of the phase which is to crystallize remains unchanged in the general case in which the crystallizing solid is pure. The more strongly the components in the mixture interact, the lower the activity coefficient of the solvent, hence the more rapidly the freezing point of the solvent will be depressed. The more the freezing point is depressed, the more viscous becomes the liquid at the point (the liquidus) where thermodynamic instability is reached, hence the slower the growth of nucleogenic fluctuations, and the less likely the crystallization.

However, if attractive interactions become too great, a new crystal structure will become favored. The more stable the new compound formed, the higher in temperature one must heat it before the increased entropy of the liquid state leads to free energy crossover, hence fusion. The higher the fusion temperature, the less viscous the liquid at the liquidus temperature and the more probable the crystallization of the *compound* on cooling. The most favorable situation for formation of glasses on cooling of binary solutions will therefore be the intermediate case where the solvent-solute interactions are strong enough to depress the liquidus much more than calculated for ideal solutions, but not strong enough to generate a new and competitively crystallizing compound.

It is usually found that in the glassforming region of such systems, there are one or more crystalline compounds with very low melting points, implying poor structures. The optimum situation for glass formation is when the exothermic mixing is strong enough to produce a compound, but only just. This maximizes the melting point lowering – thus we name this approach to finding glassforming compositions as the "barely stable compound" approach. This implies that, rather than multicomponents having some nice, stable amorphous arrangement of particles (to be evaluated by scattering techniques, and appreciated for its stability, as is the common approach to interpretation of glass structure), our line of thought would suggest the opposite. Glasses form as a weak, disordered reflection of a three-dimensional crystal structure which is so energetically incompetent that it can barely compete with the disordered form at 0 K.

It is quite consistent with the above line of thought that, in metallic systems, those which vitrify most easily are those which do not show the existence of any stable binary compounds, but which have strong solvent-solute interactions, hence deep eutectic temperatures. A good example is the system Au-Si [5]. It is furthermore consistent that, during warm-up of glasses formed during rapid quenching of the binary liquids, crystalline *compounds* can ge generated which are metastable, have no thermodynamically stable range, and later (at higher T) decompose to a combination of the components (or of one component plus a binary solution).

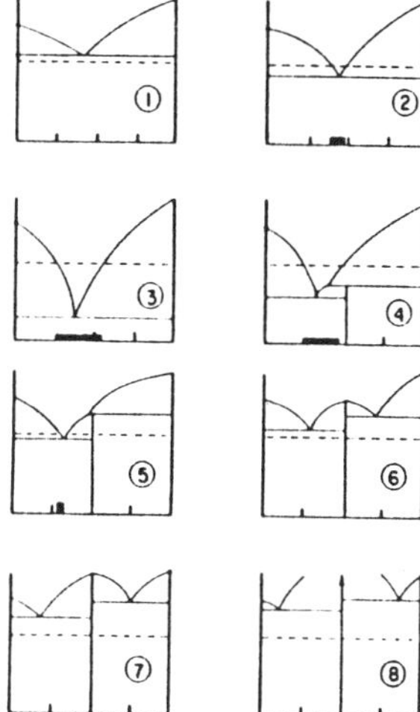

Figure 3: Illustration of how effects of increases in solute-solvent interaction may influence liquidus temperatures in binary systems, hence glassforming composition ranges (hatched bars) and glass stability within that range.

We illustrate the above argument by showing in Figure 3 a representative series of possible phase diagrams for the system A + B in which the strength of the A - B interaction increases systematically in the sequence 1-8. In the unlikely event that the viscosity were to be independent of composition in all these cases, hence a function only of temperature, then the glassforming composition ranges in the above systems could be related straightforwardly to the position of the liquidus curves, and for a fixed cooling rate would be those indicated in the diagram. Although the glassforming range would be similar in cases 3 and 4, system 3 would have the most stable glass composition (near the deep eutectic). A dependence of viscosity on composition would modify these considerations quantitatively. Nevertheless the basic ideas suffice to qualitatively explain the glassforming ranges observed in many binary system families of limited glassforming ability, e.g. the $BiCl_3$-alkali chloride glasses described by Topol et al [6,7]; nitrate glasses [8], and calcium alumina glass [9] in which melting point lowerings of ~1000 K are involved.

To the extent that the above line of reasoning is valid, theories for stable glassforming ranges in multi-component systems can be regarded as theories for free-energy lowering of given solid phases on dissolution of solutes, as a function of solute character. Even the famous Zachariasen rules[8] for glassforming substances may be regarded as rules for predicting structures which have lattice energies which are weakly competitive with their amorphous equivalents, hence which have low melting points relative to the forces acting between the particles. There are now known a great many inorganic glassforming systems , usually mixtures satisfying the above "barely stable compound" principle, in which the "glassformers," e.g. ZrF_4 [10] does not conform to the Zachariasen rules.

The understanding of glassforming ability which we are promoting here can be summed up by the banality, "glasses form when crystals don't."

3. Behavior of Liquids Approaching Vitrification: Strong and Fragile Liquids

There has been much discussion in recent years of the behavior of glassforming liquids and polymers in terms of the property known as fragility, referred to in the introduction. To reiterate, fragility is a qualitative concept related to the rapidity with which a liquid structure, which was arrested at the glass transition during cooling, becomes disrupted on reheating. The disruption is monitored by the structural relaxation time of the liquid which decreases, concomitantly, from the values of T_g $(10^2 s)$ toward the picosecond values characteristic of highly fluid systems. The relaxation time for shear stress perturbations is simply related to the viscosity by a Maxwell relation

$$\tau_\eta = \eta_s/G_{s,\infty} \qquad (1)$$

6

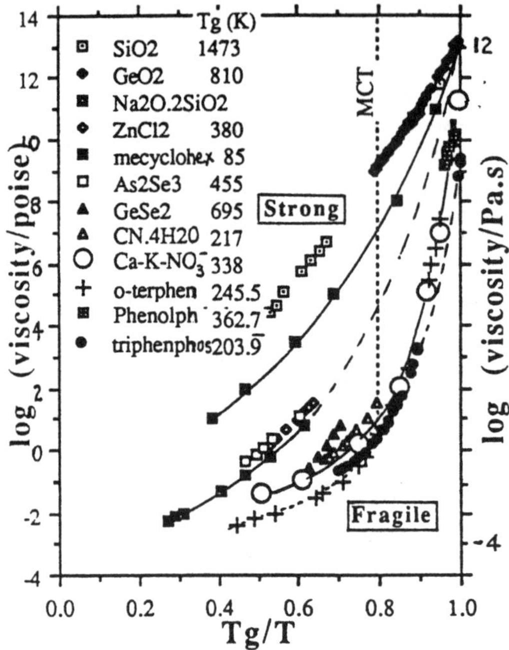

Figure 4. T_g-scaled Arrhenius plot for viscosity of various liquids including recently studied highly fragile case, tri-phenyl phosphite and convenient (T_g < ambient) molecular system, phenolphthalein.[13]

The behavior of a selection of liquids is shown in Figure 4. The data are presented in a scaled Arrhenius form [11,12,13] using, as scaling parameter, the temperature at which the glass transition is manifested during heating at 10 K/min. This corresponds to the temperature at which the enthalpy relaxation time is ~100 sec. The diagram[13] shows that the more fragile liquids have viscosities which are well below the traditional 10^{13} poise at the T_g defined from scanning calorimetry as above, but would be well above 10^{13} p if T_g were defined to be 0.9 $T(\tau_H = 100s)$. An intermediate choice $T\tau_H \approx 10^3$ s would consolidate the viscosities at ~$10^{13.5}$ p. While this would serve to maintain, roughly, the traditional view of the glass transition s as an isoviscous phenomenon, it would be achieved at some expense. Variations of G_∞ between different liquids would then mean the shear relaxation times, which are of more fundamental interest (but are usually not available), would be different at T_g.

The liquids selected in Figure 4 include the chalcogenides $GeSe_2$, As_2Se_3 and Se. Se, studied very near T_g, appears more fragile than in Figure 4, possibly due to the additionally influence of a ring-chain equilibrium [14] which sets in near T_g. The position of Se in the large scale pattern is more like that of other inorganic chain polymers such as $NaPO_3$ which is found to be intermediate in the Figure 3 pattern.

The latter case raises the question of how the fragility is best defined. It has become popular [15,16,17] to use the slope at $T_g/T_O = 1$ in a Figure 4 type plot of a relaxation time or a viscosity. This is, m = $dlog\tau/d(T_g/T)$, which can vary between 16 (Arrhenius behavior) and 200. This would be consistent with a definition based on the position in the Figure 3 pattern if the latter could be accurately represented by the Vogel-Fulcher equation inthe form

$$\tau = \tau_O \exp [T_O/F(T - T_O)] \qquad (2)$$
$$= \tau_O \exp - (F\varepsilon)$$

in which T_O/F replaces the usual B parameter and $\varepsilon = (T/T_O - 1)$. Then F, which varies between 0 (for Arrhenius behavior) and 1 (for a quasi-first order transition from fluid to glass), would be the fragility and would be related straightforwardly to m defined above.

Unfortunately, a single set of parameters rarely suffices, so some confusion between fragilities defined in different temperature domains can be expected. Since much of the usefulness of the fragility concept lies in the way its value seems correlated with other properties of materials interest, some of which are manifested below T_g (e.g. physical aging of polymers), a definition made near T_g might seem most appropriate. On the other hand, for these interested in liquid state behavior, for instance, the testing of mode coupling theory, a definition based on overall behavior such as a best fit value of F could be more useful.

3.1   THE GLASS TRANSITION IN COVALENT SYSTEMS

At this ASI, attention is directed particularly at systems in which more or less abrupt changes in behavior, due to percolation phenomena, can ensue as a composition variable is explored. For this reason, we concentrate attention from this point forward, on the behavior of glasses and liquids in the chalcogenide system Ge-As-Se. This is representative of the general class of covalent bonded glasses in which individual elements express different capacities for bonding to neighbors according to the 8 - n rule, where n is the periodic table group number.

Chalcogenide systems have been recognized as remarkable for their strongly glassforming characteristics in certain composition domains for a long time [18]. Interest in these systems ran very high in the 1960s because of their potential as IR transmitting media and reversible switching materials [19,20]. The interest content of chalcogenide systems was raised to a new level by the work of Phillips [21,22] and Thorpe [22,23] who saw that their covalent bonding character made them model systems for testing of constraint theory predictions [21-23]. Thorpe in particular predicted that a rapid increase of mechanical rigidity would set in when the bond density passed a critical value [23]. The bond density, also called the average coordination number (meaning coordination of covalently bonded, as opposed to merely spacefilling, neighbors), was defined as

$$<r> = \Sigma n_i X_i \tag{3}$$

where n is the number of covalent bonds made by species i according to the (8-n) valence electron rule and X is its mole fraction in the multi-component solution. The argument [21-23] given in detail elsewhere in this volume counts off the total degrees of freedom for a mole of atoms against the total of bond length and bond angle constraints imposed by the intact covalent bonds and predicts that rigidity should percolate through the structure for any value of <r> greater than 2.4.

An ideal system for testing these ideas is the ternary system Ge-As-Se in which all components are of similar mass (neighbors on the periodic table).

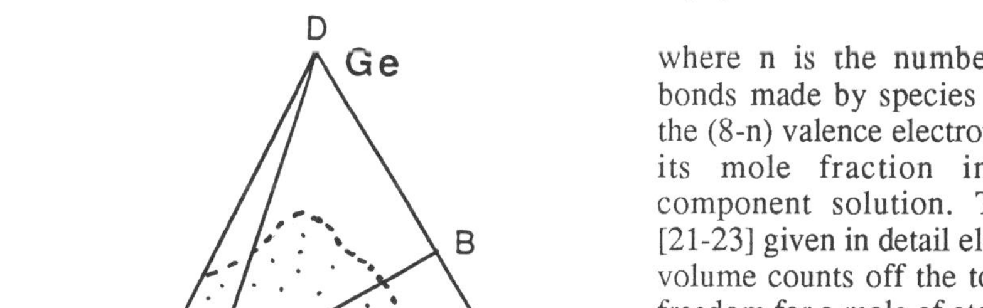

Figure 5. The model covalent glass system Ge-As-Se, showing the domain of easy glass formation and the condition for rigidity percolation, <r> = 2.4, within this domain. The ternary composition lines on which interest is focused in this paper are AB, and CD.

Because of the different valences, 4, 3, and 2 respectively, the condition <r> = 2.4 can be satisfied for a range of different ternary mixtures [24] as shown by the line marked <r> = 2.4 in Figure 5. Figure 5 also shows the glassforming region for moderate rate cooling of the liquid alloys. Although the predictions of breaks in the mechanical properties at <r> = 2.4 has met with only moderate success [24] because of neglect of non-bonding interactions and other factors, it was found by Tatsumisago et. al. [25] that properties characteristic of the viscous liquid and the glass transition showed special behavior near this <r> value.

We will give the next part of this paper to a review of the latter findings for these systems, which suggest that they may provide an excellent testing ground for general theories of ergodicity breaking and the overall problem of the glass transition. The following part will be devoted to the broader aim of establishing a phenomenological relationship between silicon- (or germanium-) based covalent bonded alloy solutions and two other major classes of solutions which together control much of what happens in the liquid state of the natural world. These are the aqueous solutions, typified by the well studied system H_2O + LiCl [26], and the siliceous solutions typified by the even more thoroughly studied system SiO_2 + Na_2O [27]. We will reaffirm the idea that much that is interesting and unusual about the latter two systems is due to the intrinsic structural incompatibility, at low temperatures, of the tetrahedral network structure of the parent components and the structures of homogeneous solutions which form at higher second component contents when the network has been sufficiently disrupted. Since binary solution data in the Si- (or corresponding Ge-) based systems at high Si contents are sparse, most of this part of the paper will be devoted to arguing, from experimental and computer simulation studies, that liquid silicon and germanium have the same special characteristics possessed by H_2O and SiO_2 which dominate their binary solution behavior, so their binary solutions should be phenomenological analogs.

The first finding of interest in the Ge-As-Se

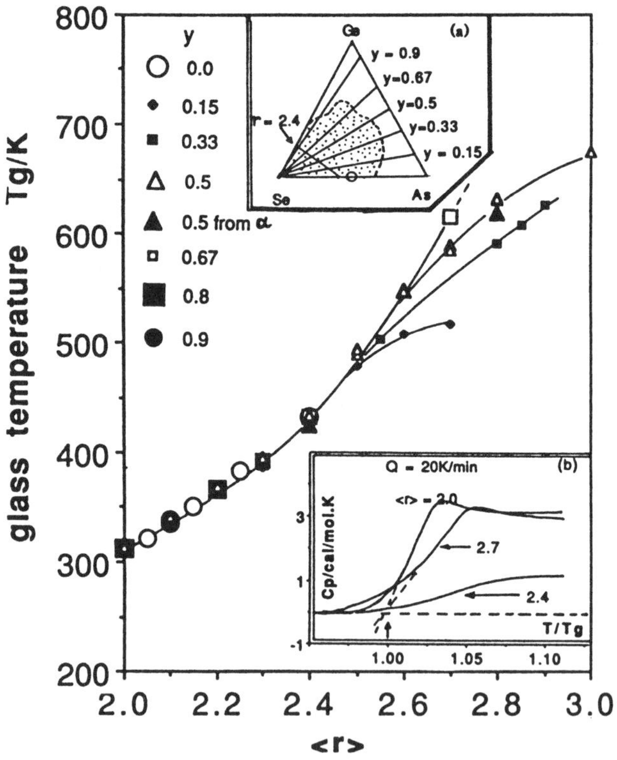

Figure 6. <r> dependence of T_g from DSC and expansivity, for Ge-As-Se glasses along various pseudobinary cuts $Se_{(1-x)}(Ge_ySe_{1-y})_x$ as indicated in inset (a). Inset (b) shows behavior of the heat capacity C_p through the glass transition for <r> values at, and on either side of, the rigidity-percolation threshold along the y = 0.5 [25].

system is that for glasses in the composition range below and a little above $\langle r \rangle = 2.4$ have T_g values which depend only on $\langle r \rangle$. This is shown in Figure 6 [25]. It was noted that the glass transition near $\langle r \rangle = 2.4$ was very "smeared out" (see Figure 6 insert). To investigate this further, we study behavior along a single line, AB, in the ternary Figure 5 along which G:As = 1. Figure 7 shows how both the activation energy for viscosity and the change of heat capacity at T_g go through minimum values close to $\langle r \rangle = 2.4$. A recent paper by Senapati and Varshneya [28] shows similar behavior for the related system Ge-Sb-Se and shows that a minimum also occurs in the change in expansion coefficient.

The correlation seen in Figure 7 is predictable from the C_p minimum via the Adam-Gibbs equation for relaxation processes [29]. For instance, the shear relaxation time, τ_s, whose temperature dependence dominates that of the viscosity, is written according to Adam and Gibbs as

$$\tau_s = \eta/G_\infty = \tau_0 \exp(C/TS_c) \qquad (4)$$

where η is the viscosity, G_∞ is the shear modulus at high frequencies, $\omega\tau \gg 1$. Because ΔC_p controls the temperature dependence of S_c in Eq. (4), it also controls the magnitude of deviations from the Arrhenius equation, hence also the slope of the Arrhenius plot of η measured at T_g.

The reason for ΔC_p having a minimum at $\langle r \rangle = 2.4$ is less obvious but must have to do with the bond distribution being optimal at this composition hence the drive to thermal disruption via increased degeneracy, a minimum. In terms of potential energy hypersurfaces [30], $\langle r \rangle = 2.4$ must correspond to a hypersurface with a relatively small number of minima, presumably because of a more or less unique bond distribution. This would correspond to a small entropy and a small increase in entropy per unit increase in energy.

It has been shown elsewhere [25] how the viscosity behavior at $\langle r \rangle = 2.4$ is that of a "strong" liquid – even stronger than B_2O_3 and certainly stronger than the tetrahedral network compound $GeSe_2$, seen in Figure 4. The system Se-(Ge:As) thus exhibits the full range of strong to fragile behavior previously only seen by reference to a variety of other

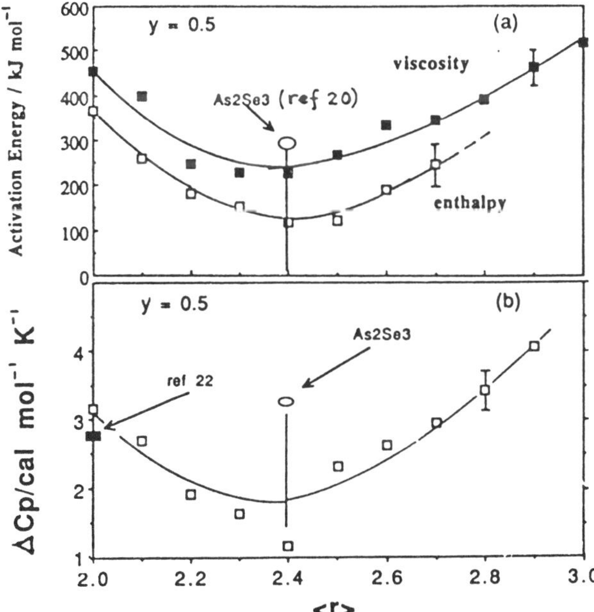

Figure 7. Excess molar heat capacities, and activation energies for viscosity and enthalpy at T_g, along the line AB of Figure 5 (adapted from Ref. 25).

liquids, hence this system is rich in the phenomenology of the liquid-to-glass transition. Being simply constituted (no internal degrees of freedom) it is therefore to be regarded as a model system. Addition of iodine, which would break up chain structures, might be expected to lead to even more fragile behavior, but this is not strongly supported by the available data [31].

Consistent with strong liquid behavior at r = 2.4 is the relation between T_g and the Kauzmann temperature T_K at which extrapolations of the excess of liquid entropy over crystal entropy (which is ΔS_f at the fusion temperature) show the excess vanishing. The large value of T_g/T_K for As_2Se_3, 1.90, is contrasted with the small value, 1.23, for the relatively fragile Se. Detailed discussions of this Kauzmann vanishing excess entropy conundrum are given elsewhere [32,33,34,35].

Two further correlations of liquid behavior with fragility may be made using data from the Ge-As-Se systems.

Firstly it is found that manner in which the liquids approach equilibrium after a perturbation, i.e. the relaxation function, is controlled by the fragility. An example of linear relaxation in the time domain is shown in Figure 8 for a chalcogenide alloy [36]. If the relaxation function is represented by the popular "stretched exponential" function

$$\theta(t) = \exp\text{-}([t/\tau_0]^\beta), \quad 0 < \beta < 1.0 \tag{5}$$

which fits the data very well except at very short times, then β (the "non-exponentiality of relaxation" is correlated with the fragility m (defined from the temperature dependence of τ_0 as explained earlier). This is demonstrated in Figure 9, in which complimentary data from studies on chain polymers [37] are included.

Secondly, if the relaxation is monitored after dropping the temperature below T_g, so that the system is "out of equilibrium," i.e. non-ergodic, then the extent to which therelaxation time depends on the departure from equilibrium seems to depend on the fragility. This "non-linearity of relaxation" is shown in Figure 10 for the extreme case of

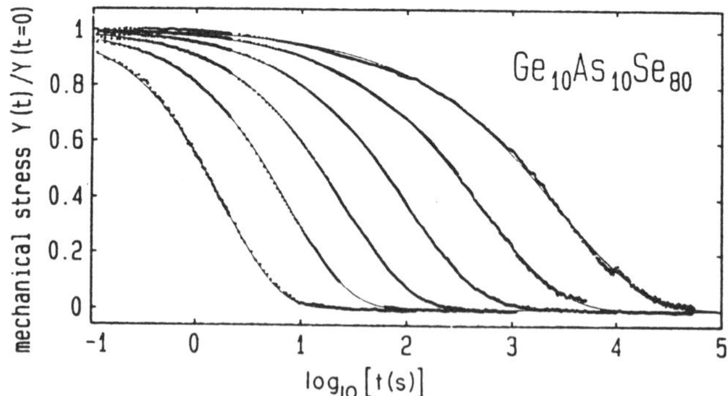

Figure 8. Normalized stress relaxation of a Ge-As-Se supercooled liquid as measured at temperatures (from left to right) 433.4, 422.6, 413.3, 403.7, 394.5, and 385.3 K. The *solid lines* are fits with Eq. (5). The stretching exponent *b* is temperature dependent. The *dashed line* represents monoexponential relaxation. Adapted from Ref. 36

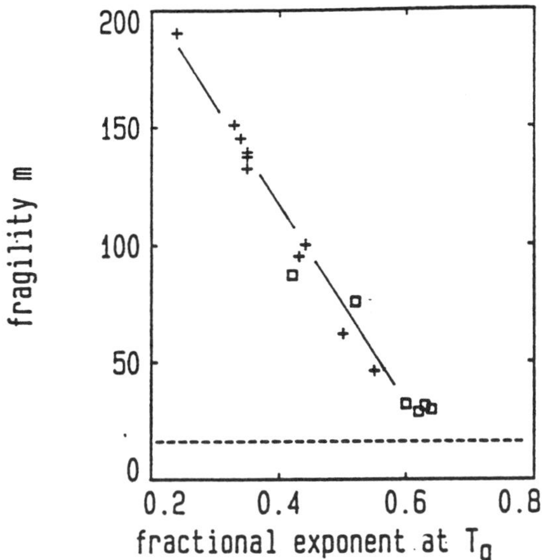

Figure 9. Fragility m versus fractional exponent b at the glass transition for polymers (circles, taken from Plazek and Ngai (Ref. 37?) and chalcogenides (squares, this work). The broken line indicates the minimum fragility.

selenium in which we focus on the part of a complex relaxation function which is believed to be the visco-elastic part (there is a longer time part assigned to a quasi-chemical ring-chain motif equilibration, which is described elsewhere[14].

The diagram shows that the relaxation function depends on waiting time, and that it requires some 60 hours of waiting at T = 300 K before the structure responsible for the viscoelastic relaxation, seen in this time window, fully equilibrates. For compositions at <r> = 2.4, no waiting time dependence could be observed [38]. This has the important consequence that glasses of this composition should be dimensionally stable over long periods of time near and below T_g.

In summary, a collection of measurements confirm that the rigidity percolation threshold of <r> ~2.4 is of significance in covalently bonded glasses.

4.   Overconstraining, and Liquid-Liquid Phase Transitions

Solutions with <r> > 2.4 are more than optimally constrained. The term "over-constrained" has been used [23]. With an interest in the consequences of excessive overconstraint, we turn attention to the line CD in Figure 5 which contains compositions on either side of <r> = 2.4 but avoids passing through any compound, e.g. $GeSe_2$ along the line Ge-S. By analogy with SiO_2-based and H_2O-based system behavior, we expect that, as the ultimately overconstrained composition, pure Ge, is approached, the alloy will become unstable with respect to two separate liquid or amorphous phases, one of which is the random network of a-Ge itself. While this remains to be documented experimentally, and will be difficult to establish except for a small range near the edge of the glassforming edge (because of the strong tendency of a-Ge to crystallize at T > 800K), we present reasons for expecting it, as follows.

Firstly, a-Ge has already been observed to form exothermically from a binary liquid system studied near its glass transition temperature: solutions of Ge in GeO_2 quite near the composition GeO_2 (hence at ~37 at % Ge in the system Ge-O [c.f. Ge-Se]) were observed by de Neufville and Turnbull [39] to precipitate a-Ge from the glass, with release of heat, on reheating above the T_g of 633K. One of their suggested interpretations is very close to our line of thought given above.

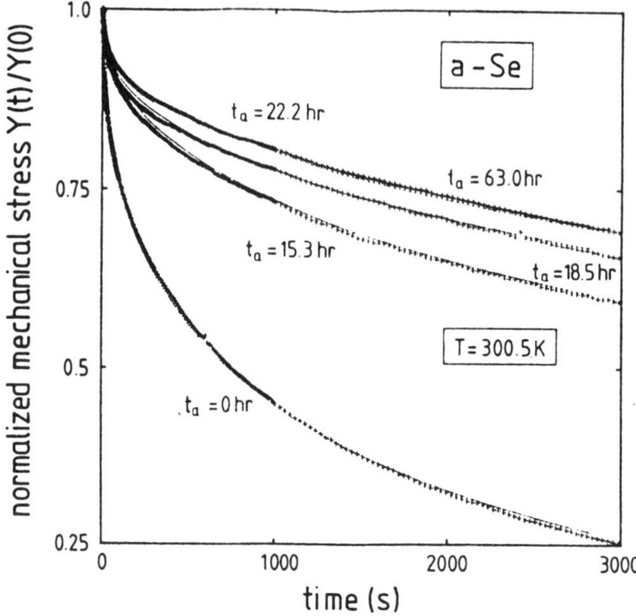

Figure 10. Time dependence of normalized tensile stress Y(t)/Y(t = 0) of amorphous selenium, which has previously been quenched from an equilbrium state obtained by annealing at 335K, well above the glass transition temperature. The labels indicates the times at which the measurements have been started relative to the time of commencement of the first run. The solid lines are calculated using Eq. (5) with β = 0.47 [14] .

Secondly, according to computer simulation studies [40] discussed below and reported in more detail elsewhere [41], Si shows all the anomalies exhibited by SiO_2 and H_2O, both of whose binary solutions are known to exhibit the phenomenon of splitting out of a second liquid (or amorphous) phase at compositions rich in the pure "tetrahedral" liquid. There is no reason to expect liquid Ge to behave differently. Hence we propose that Ge-Se and, more so, Ge-Te binary solutions will behave like SiO_2-based and H_2O-based solutions and split out an amorphous Ge phase from quenched binary glasses in restricted composition ranges as identified in the caption to the Figure 11.

Thirdly, the metallic-liquid-to-covalent-amorphous-network phase transition suggested as a possibility by deNeufville and Turnbull[39] has actually been observed by Thompson et al in laser melting and quenching experiments in the analogous case of silicon [42]. It even occurs at the same reduced temperature ($T_{\ell\text{-}a}/T_m$ = 0.88 where T_m is the melting point) suggested for a first order liquid-to-amorphous network (fragile-to-strong liquid-liquid transition) in the case of water [43,44].

The relevant binary phase diagrams are shown in Figure 11, and are annotated in the captions to highlight the common oddities. Note that the two-liquid region at high temperatures in the Ge-Se system [45] is a consequence of the incompatibility of the metallic germanium liquid and the semiconducting binary solution. This two-liquid region disappears in the corresponding Ge-Te system in which the whole liquid range is metallic, so the Ge-Te phase diagram [45] looks more like the others, Figs. 3(a) and (b). Unfortunately the glassforming composition range found between the compositions Ge-Se and $GeSe_2$ of the Ge-Se system also disappears. Thus we focus on the Ge-Se system and treat the two-liquid (metallic and semiconducting) range as a distraction to be ignored because our interest lies in the incompatibility of the *semiconducting* binary liquid with the *semiconducting*-tetrahedral phase which appears below a particular (metallic-liquid-to-semiconductor) transition temperature. The existence of the two-liquid region does, however, imply that the semiconducting solution – semiconducting tetrahedral phase unmixing line – will have a more complex shape than for the other systems (a) and (b) though we postpone comment on this to our "concluding remarks."

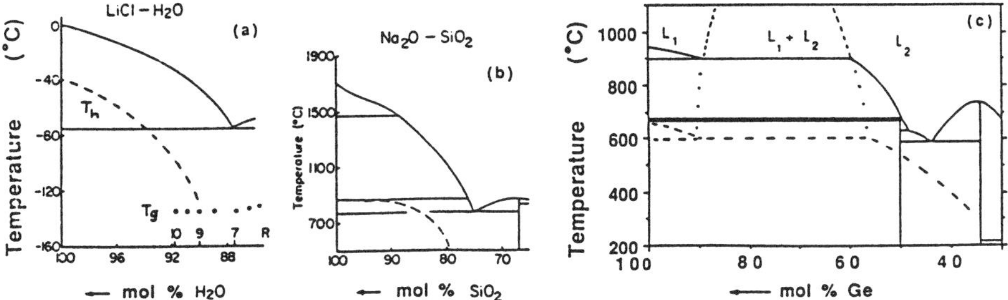

Figure 11. Comparison of phase diagrams of systems in which tetrahedral structure of parent component bestows anomalous behavior, including density maxima and liquid-liquid phase separation, on the supercooled binary solutions. (a) SiO_2 + Na_2O; (b) H_2O + LiCl; (c) Ge + Se. The dashed line in the water-LiCl system lying below the liquidus line is the line of homogeneous nucleation temperatures obtained from emulsion studies, and now interpreted as a close reflection of the line of homogeneous solution-to-(solution + amorphous) transitions at which crystallization of ice Ic from the amorphous phase becomes extremely rapid. The dashed line in the SiO_2-Na_2O system is the known boundary of the homogeneous solutions before unmixing occurs. The dashed line below the liquidus in the Ge-Se system is our conjectured line of liquid solution-to-(solution + amorphous) transitions. The arrow on the water-LiCl diagram indicates the composition where evidence for liquid-liquid separation without ice formation has been reported [26].[41]

Unfortunately we do not yet have simulation data on Ge nor binary solution experiments on Ge-GeS quenched solutions. Thus in the ensuing section of this contribution, we will support our case for the solution analogies by emphasizing the phenomenological analogies between water and liquid silicon (and germanium) both above and approaching the liquid-to-amorphous transition temperature. For this purpose, we will use both laboratory and computer simulation data.

We commence with a comparison of the behavior of the densities in the liquid states as a function of pressure. Figure 12(a) shows the effect of temperature on the density of silicon simulated using the Stillinger-Weber pair potential [46]. This potential, although intrinsically unable to reproduce the effect of the electron delocalization which occurs at the crystal-liquid transition in real silicon (hence also at the liquid-amorphous transition), has been shown to give a good qualitative account of the major increase in coordination and density on fusion, and also reproduces the melting point of the diamond cubic crystal exactly. At very high pressures it also shows the crystallization of liquid into the BC-8 phase of real silicon. Earlier studies of this potential by Lüdke and Landman [47] showed many features which are reproduced, in greater detail, in the present study. Somewhat more accurate results are obtained by Jank and Hafner [48] using effective pair potentials with volume dependence derived from pseudopotential- and linear-response theories [49]. The only way to cope fully with the metal-to-non-metal problem in simulation studies is to have the electrons treated explicitly, as is achieved in the Car-Parrinello approach [50] which has now been applied to Si [51] and Ge [52] for very small systems. In these, however, amorphous-amorphous phase transitions are obscured by the small system sizes. It is the large system sizes coupled with long runs of the present study which makes possible our success in observing the phenomena of interest to this paper, viz., the well defined density maxima seen in Figure 12a and the liquid amorphous phase change as well as the liquid-vapor tensile stability limit discussed below.

14

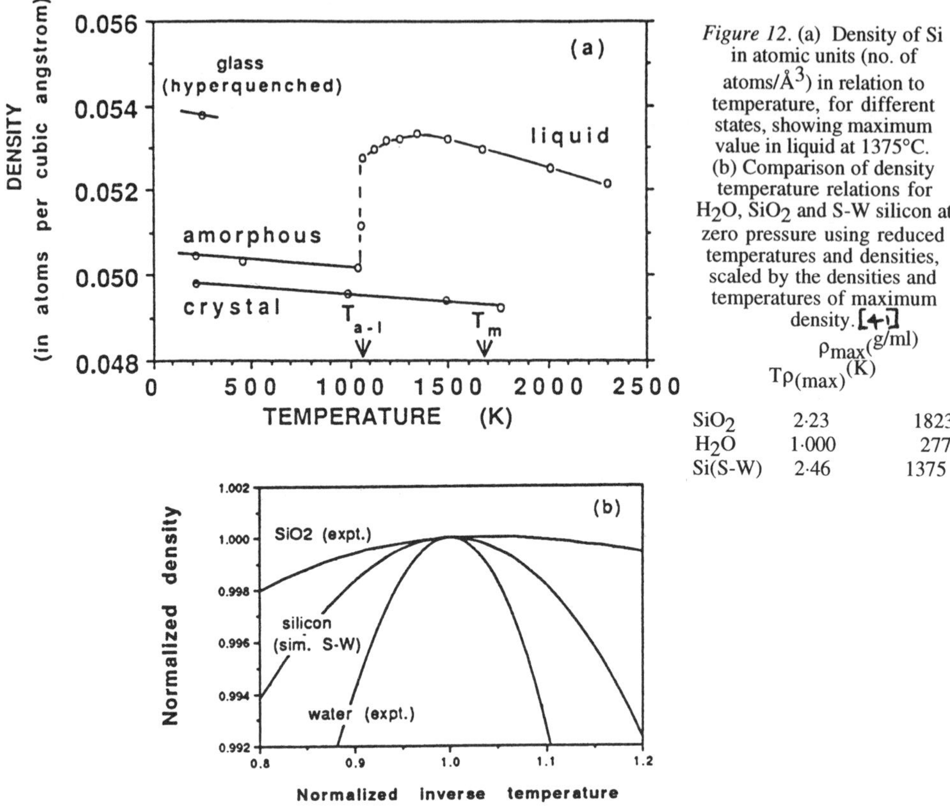

Figure 12. (a) Density of Si in atomic units (no. of atoms/$Å^3$) in relation to temperature, for different states, showing maximum value in liquid at 1375°C. (b) Comparison of density temperature relations for H_2O, SiO_2 and S-W silicon at zero pressure using reduced temperatures and densities, scaled by the densities and temperatures of maximum density.[47]

	ρ_{max}(g/ml)	$T_{\rho(max)}$(K)
SiO$_2$	2·23	1823
H$_2$O	1·000	277
Si(S-W)	2·46	1375

The behavior of the density of S-W liquid silicon is compared with that of experimental water and experimental silicon dioxide in Figure 12(b) using the same scaling scheme used in early comparisons of the latter two substances [53]. It appears that, relative to SiO_2, the density maximum in silicon is quite sharply defined as in water.

The effect of pressure (or tension) on the silicon density maximum is illustrated by the behavior of the isochoric cooling curves shown in Fig 13a. This behavior, which is highlighted by the dashed line drawn through the points of maximum density (isochoric pressure minima or tension maxima), may be compared with that known for laboratory water shown in Figure 13b [29] . It appears that the density maximum in Si "washes out" at high enough pressure though the possibility that this is due to equilibration problems probably cannot be excluded.

Also included in Figure 13b is the line of homogeneous nucleation temperatures, T_h, obtained from emulsion measurements and the line of singular points, T_s, obtained from data-fitting of divergent thermodynamic [53] and transport data (in particular the spin lattice relaxation times of Lang and Lüdemann [54]).

The latter singularities are to be compared with the most spectacular feature of the liquid silicon simulations, a convincing example of which was first reported by Lüdke and Landman [47]. These are the sudden occurrences, at temperatures which depend on the density of the silicon being simulated, of well-defined transitions from quite highly

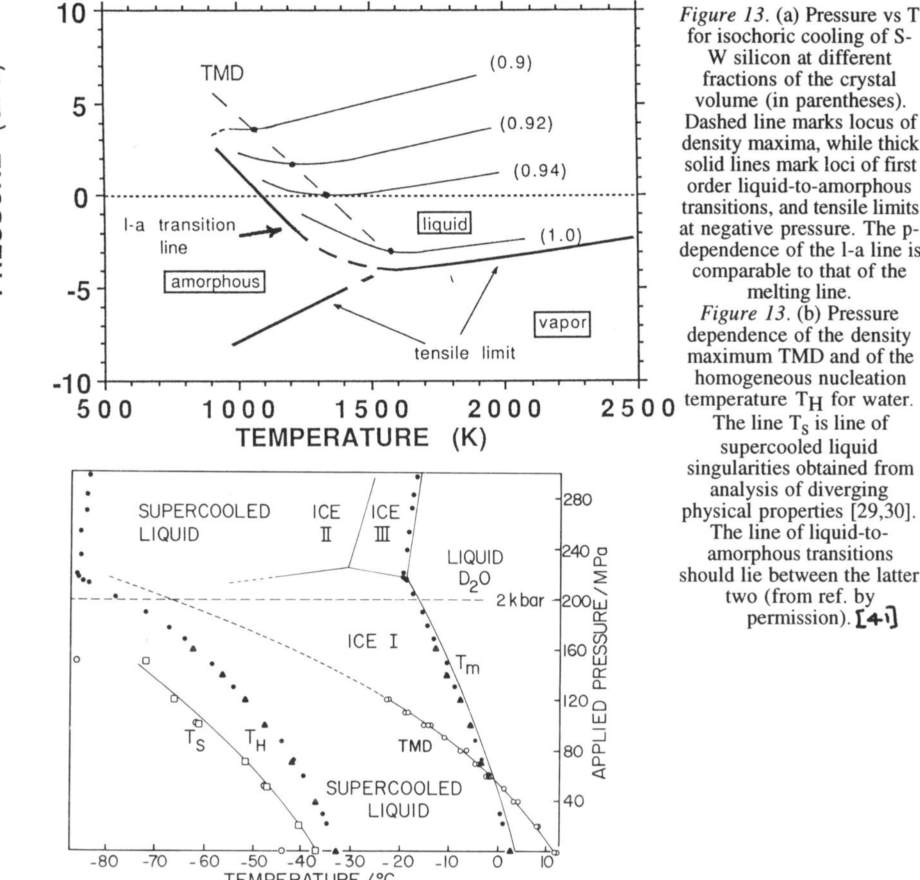

Figure 13. (a) Pressure vs T for isochoric cooling of S-W silicon at different fractions of the crystal volume (in parentheses). Dashed line marks locus of density maxima, while thick solid lines mark loci of first order liquid-to-amorphous transitions, and tensile limits at negative pressure. The p-dependence of the l-a line is comparable to that of the melting line.

Figure 13. (b) Pressure dependence of the density maximum TMD and of the homogeneous nucleation temperature T_H for water. The line T_S is line of supercooled liquid singularities obtained from analysis of diverging physical properties [29,30]. The line of liquid-to-amorphous transitions should lie between the latter two (from ref. by permission). [41]

diffusive ($D_{Si} = \sim 10^{-5}$ cm^{2}s^{-1}) liquid states (whose coordination number changes with decreasing temperature), to essentially non-diffusive ("amorphous") phases with coordination number fixed at 4.1 independent of temperature. The transition to the amorphous phase occurs by nucleation at various points in the sample, and it is at the surface of these microdroplets that the crystallization of the system to the diamond cubic polymorph may be observed to initiate after a relatively short delay.

At zero pressure, the transition was observed {See Figure 12(a)} at 1060K, well below the value 1480K suggested by experiment [42] but close to the "amorphization temperature" reported by Stich et al [51] from ab initio simulations. Crystallization of Si was never observed at temperatures above this transition, despite the greater mobility of the "metallic" liquid phase. This is entirely consistent with the recent observation by Filipponi et al [55] that levitated liquid Ge (metallic) can be supercooled to 680°C (~260K of supercooling). This temperature corresponds closely with the metallic-to-semiconducting liquid-liquid transition temperature calculated by Ponyatovsky and Barkalov [56] from their two-liquid model for silicon-like liquids.

The temperature at which the transition occurs is depressed by increase of pressure as would be expected from the sign of the volume change according to the

Clapeyron equation. The observed transition temperatures are plotted in Figure 13(a). The pressure dependence is comparable to, but smaller than, that of the melting transition.

The occurrence during supercooling of a first order transition to a new amorphous phase which is of much higher viscosity, is exactly the interpretation that was recently given to the behavior of water during hyperquenching [42], and is also exactly what is predicted by the Haar-Gallagher-Kell equation of state for water [57] which was developed to best represent the precisely determined behavior of laboratory water in its stable domain. On the other hand, the occurrence of a first order transition in supercooling water at ambient temperature has been called into question by the results of computer simulations, using several different pair potentials, by Poole et al [58] who find that such first-order transitions would only occur above a critical pressure. Whether or not the use of more realistic cooperative potentials will yield different results remains to be seen. According to the bond-modified van der Waals model for liquids with tetrahedral

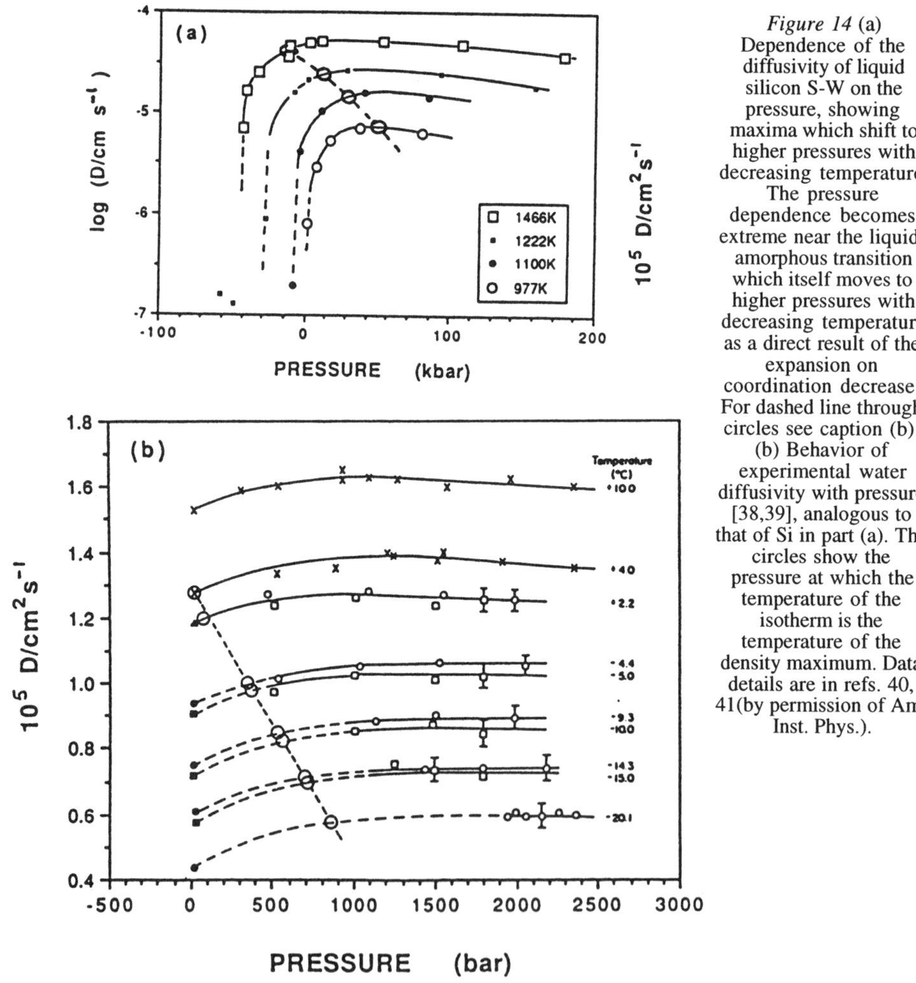

Figure 14 (a) Dependence of the diffusivity of liquid silicon S-W on the pressure, showing maxima which shift to higher pressures with decreasing temperature. The pressure dependence becomes extreme near the liquid-amorphous transition which itself moves to higher pressures with decreasing temperature as a direct result of the expansion on coordination decrease. For dashed line through circles see caption (b). (b) Behavior of experimental water diffusivity with pressure [38,39], analogous to that of Si in part (a). The circles show the pressure at which the temperature of the isotherm is the temperature of the density maximum. Data details are in refs. 40, 41(by permission of Am. Inst. Phys.).

bonding [59] the two scenarios are closely related. It appears from the present simulations that in S-W silicon, in contrast to the case of simulated water [58], the transition can be observed directly in simulations of reasonable length, and that it also can occur at ambient pressure.

To summarize the phase behavior of liquid Si, heavy lines marking the loci of observed liquid-to-amorphous transitions, liquid-to-cavitated-liquid transitions, and amorphous-to-cavitated-amorphous transitions have been added to Figure 13(a). The latter lines fall at negative pressures and mark the tensile limits of the condensed phrases [60]. The tensile strength of the amorphous phase of Si seen in Figure 13(a) is comparable to that of rigid ion SiO_2 [61].

In a binary system, the temperature of any liquid-amorphous transition will be depressed by addition of second components for the same reason that the melting point is depressed. Therefore, a subsidiary feature of any phase diagram for a system containing silicon, or water, plus some second component which is soluble in the *normal* liquid phase, will be the separation of a viscous liquid or amorphous solid phase from the binary solution at some temperature below that of the pure component phase transition - as shown in two of the phase diagrams of Figure 11 [62]. In the Si- (or Ge-) based and the water-based systems, this amorphous phase separation will, however, be followed immediately by a crystallization of the amorphous phase unless the quenching is extremely rapid, or unless the solution composition is chosen so that the unmixing occurs very close to the T_g (i.e. higher Se contents).

Finally in this demonstration of analogies we show in Figure 14 the similar behavior of the transport coefficients in liquid S-W Si and laboratory water [63,64]. Such similarity is in fact expected, in view of the negative expansivity regime common to each, from the Adam-Gibbs equation for the relaxation time, Eq. (4). A negative expansivity requires, by a Maxwell relation, a positive pressure dependence of the entropy and hence, according to Eq. (4), a relaxation time which decreases with increasing pressure [65]. The parallels in behavior seen in Figure6 are impressive. Comparable results are obtained also for the temperature dependence of the diffusivity at constant pressure, each system showing extremely non-Arrhenius behavior [41].

Of course, future large sample *ab initio* simulations of liquid Si are sure to show quantitative differences from the behavior displayed here. However, it seems likely to us that the qualitative trends we discuss will hold true and that the analogies drawn in Figure 11 will therefore be validated.

5.   View from the Solid

While it is natural, in view of the way glasses are usually formed, to view the glass transition as an extension of the liquid state problem, it is also possible to treat it as a problem in solid state physics. The problem is to account for the transition from more or less harmonic behavior, well below T_g, to behavior so anharmonic at T_g that irreversible displacement of particles from their original sites occur and stresses can be relaxed by deformation on time scales of 100s of seconds – something which does not happen in defect-containing crystals unless the crystals are very small.

Some sort of Lindemann criterion seems to apply because both neutron scattering [66,67,68] and computer simulation [69] studies show that the experimental glass transition tends to occur when the mean square vibrational amplitude reaches a value of about 1/5 of the atomic diameter. In the case of selenium where values at both glass

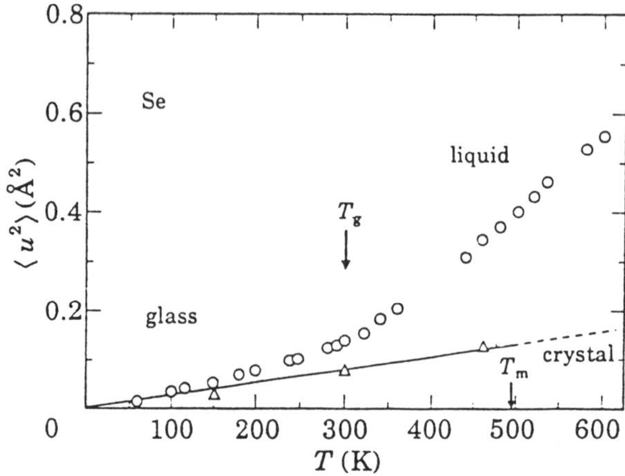

Figure 15: Temperature effect on the mean square displacement of particles in selenium glass, crystal, and liquid (from ref. 66).

transition and melting are known [66], they are coincident, as seen in Figure 15. Clearly the rate of increase in $\langle r^2 \rangle$ per unit of thermal energy increase has to be much greater in the glass than in the crystal, which must be accountable in terms of the more random potentials in which the vibrations occur.

Whereas the limiting value of the mean square displacement in a crystal leads to the complete collapse of the structure in favor of the liquid, the consequence in the glass is only that very occasional local rearrangements of particles occur. Indeed the diffusivity of molecules in the crystal via defect diffusion near T_m is usually greater than that in the glass at the same $\langle r^2 \rangle$. However, the situation changes rapidly above T_g due largely to a change in structure which is usually (but not necessarily) accompanied by expansion, and is always accompanied by an increase in heat absorption per unit temperature increase. Evidently when the $\langle r^2 \rangle$ reaches $\langle 0.66 r_{Se} \rangle^2$, configurationally excited states (we suggested the name "configurons" in an earlier paper [70]) of the system can form at the expense of what would otherwise be excessive oscillatory excursions. The configuron states cannot be very extensive in space because microemulsion droplets of a glassformer (which exist at the bulk chemical potential) exhibit the same behavior (at least with respect to T_g) as the bulk [71].

The configurons must be coupled to the phonons by the anharmonic terms in the interparticle potential. The more harmonic the effective potential, the weaker the coupling, hence the less probable the excitations, i.e. the greater the energy barrier to the phonon-configuron exchange, and the "stronger" the resulting liquid. The enhanced harmonicity in these cases is reflected in the overshoot in $\langle r^2 \rangle$ vs t seen in computer simulations [69], which persists above T_g. This is the undamped remnant of harmonic "ringing" seen in the simulation of crystals before dephasing destroys the correlations [72]. It is expected that, in network glasses, with longer intermediate range order, this damping/dephasing will be less marked so networks, intermediate range order, persistent boson peaks, and vibrational harmonicity, should all be linked together with "strong liquid" behavior [69,71].

Spin glasses, while showing a Vogel-Fulcher-like temperature dependence, do not depend on an anharmonicity coupling mechanism to establish equilibrium between phonon bath and spin system, so the basis for the spreading of systems in a strong vs. fragile pattern does not exist. This is presumably why a T_g/scaled presentation of spin glass behavior shows that most examples of the phenomenon are excessively fragile by comparison with structural glasses [73]. It is the high fragility which is the reason that most of the experimentation in spin glasses is carried out in the non-ergodic (non-linear) regime below T_g – the inverse of the situation with "structural glasses," which are mostly explored in their ergodic or viscous liquid states. Failure to recognize this essential difference is often a source of confusion when comparisons between the two phenomenologies are made.

6. Concluding Remarks

The formation of glasses from supercooled liquids is a branch of physical science with a rich phenomenology and many unsolved problems. It may be expected that it will command increasing attention from theoreticians and experimentalists in the coming years.

7. References

1. See Proc. 14th Citges Conference on Statistical Pysics (1996), in press.
2. Alba, C., Busse, L.E. and Angell, C.A. (1990)*J. Chem. Phys.* **92**, 617-624.
3. Wong, J. and Angell, C.A. (1976) *Glass: Structure by Spectroscopy,* Marcel Dekker, New York (Chapter 1).
4. Pitzer, R.S. and Scott, D.W. (1943) *J. Amer. Chem. Soc.* **65**, 803.
5. Chen, H.S. and Turnbull, D. (1968) *J. Chem. Phys.* **48**, 2560; (1967) *J. Appl. Phys.* **38**, 3646; (1969) *Acta. Metall.* **17,** 1021.
6. Topol, L.E., Mayer, S.W. and Ransom, L.D. (1960) *J. Phys. Chem.* **64**, 862.
7. Angell, C.A. and Ziegler, D.C. (1981) *Mat. Res. Bull.* **16**, 269.
8. Thilo, E., Wieker, C. and Wieker, W. (1964) *Silic. Tech.* 15, 109.
9. Phase Diagrams for Ceramicists, Ed Levin E.M.,Robbins,C.R. McMurdie H. F., American Ceramic Society, Diagram No 391
10. Poulain, M., Chantanasingh, M.,Lucas J., (1977) Mater. Res. Bull., 12, 151-156
11. Angell, C.A. (1991) *J. Non-Cryst. Sol.* **131-133**, 13-31.
12. Angell, C.A. (1985) in K. Ngai and G.B. Wright (eds.), *Relaxations in Complex Systems,* National Technical Information Service, U.S. Department of Commerce, Springfield, VA 22161, pg. 1.
13. Angell, C.A., Alba, C., Arzimanoglou, A., Böhmer, R., Fan, J., Lu, Q., Sanchez, E., Senapati, H. and Tatsumisago, M. (1992) *Am. Inst. Phys. Conference Proceedings* No. **256**, 3-19.
14. Böhmer, R. and Angell, C. A. (1993) *Physical Review B* **48**, 5857-5864.
15. Bohmer, R., Ngai, K.L., Angell, C.A. and Plazek, D.J. (1993) *J. Chem. Phys.* **99**, 4201-4209.
16. Ngai, K.L., Cramer, C., Saatkamp, T. and Funke, F. (in press) in Proc. of Workshop on Non-Equilibrium Phenomena in Supercooled Fluids, Glasses, and Amorphous Materials, Sept. 25-29, 1995.
17. Fujimori, H. and Oguni, M. (1995) Solid State Commun. **94**, 157-162.
18. Rawson, H. (1967) *Inorganic Glassforming Systems*, Academic Press, London, pp. 216-218.
19. Zallen, R. (1983) *The Physics of Amorphous Solids*, John Wiley & Sons, New York.
20. Ovshinsky, S.R. (1968) *Phys. Rev. Lett.* **21**, 1450.
21. Phillips, J.C. (1979) *J. Non-Cryst. Solids* **34**, 153.
22. Phillips, J.C. and Thorpe, M.F. (1986) *Solid State Commun.* **53**, 847.
23. Thorpe, M.F. (1983) *J. Non-Cryst. Solids* **57**, 355.
24. Halfpap, B.L. and Lindsay, S.M. (1986) *Phys. Rev. Lett.* **57**, 847.
25. Tatsumisago, M., Halfpap, B.L., Green, J.L., Lindsay, S.M., and Angell, C.A. (1990) *Phys Rev Lett.* **64**, 1549.
26. (a) Angell, C.A. and Sare, E.J. (1968) *J. Chem. Phys.* **49**, 4713; (1970) **52**, 1058.
    (b) Kanno, H. (1987) *J. Phys. Chem.* **91**, 1967.
    (c) MacFarlane, D.R.(1985) *Cryoletters* **6**, 33 ; (1986) **23**, 230.
27. Kreidl, N.J. (1983) in *Glass: Science and Technology Vol. 1*, Academic Press, New York, pp. 105-299.
28. Senapati, U. and Varshneya, A.K. (1995) *J. Non-Cryst. Sol.* **185**, 289.
29. Adam, G. and Gibbs, J.H. (1965) *J. Chem. Phys.* **43**, 139.
30. Angell, C.A., (1991) *J. Non-Cryst. Sol.* **131-133**, 13.
31. Lucas, J., Ma, H.L., Zhang, X.H., Senapati, H., Böhmer, R. and Angell, C.A. (1992) *J. Sol. State Chem.* **96,** 181.
32. Kauzmann, W. (1948) *Chem. Rev.* **43**, 219.
33. Angell, C. A., (1970) J. Chem. Ed., 47, 583.
34. Gibbs, J.H. (1960) in McKenzie,J.D. (edl), *Modern Aspects of the Vitreous State*, Butterworths, London, ch. 7.
35. Angell, C.A. and Sichina, W. (1976) *Ann. N.Y. Acad. Sci.* **279**, (Proc. Workshop on the Glass Transition and the Nature of the Glassy State) 53.
36. Böhmer, R. and Angell, C.A. (1992) *Phys. Rev. B.* **45**, 10091-10094.
37. Plazek, D. J., and Ngai, K.-L., (1991) Macromolecules, 24, 1222.

39. de Neufville, J.P. and Turnbull, D. (1971) *Faraday Society Discussions* **327**, 0000.
40. Grabow, M.H. and Gilmer, G.H. (1989) *MRS Symposia Proceedings* **141**, 341.
41. Grabow, M.H., Borick, S.S. and Angell, C.A. (1996) *J. Non-Cryst. Sol.* (in press).
42. Thompson, M.O., Galvin, G.J. and Mayer, J.W. (1984) *Phys. Rev. Lett.* **52,** 2360.
43. Angell, C.A. (1993) *J. Phys. Chem.* **97**, 6339.
44. While computer simulations using different potentials, ST2, TIP4P, SPC, deny this for normal pressure, they confirm it for somewhat "higher" pressures. On the other hand, all the best equations of state require it for normal pressures, albeit in a region of considerable extrapolation, hence unreliability [Kisilev, S.B., Levelt-Sengers, J.M.H., and Zheng, Q. (1995) Proc. Water and Steam Conf., in press].
45. (a)   Borisova, Z.U. (1981) *Glassy Semiconductors*, Plenum, New York.
    (b)   Ipser, H., Gambino, M. and Schuster, W. (1982) *Monatsh. Chem.* **113**, 389.
    (c)   Binary Alloy Phase Diagrams, Eds.
46. Stillinger, F.H. and Weber, T. (1985) *Phys. Rev. B* **31**, 5262.
47. Lüdke, W.D. and Landman, U. (1988) *Phys. Rev. B* **37**, 4656; (1989) **40**, 1164.
48. (a)   Jank, W. and Hafner, J. (1990) *Phys. Rev. B.* **41,** 1497.
    (b)   Jank, W. and Hafner, J. (1989) *J. Phys. Condensed Mater.* **1**, 4235.
49. Hafner, J. (1987) *From Hamiltonians to Phase Diagrams,* Springer, Berlin.
50. Car, R. and Parrinello, M. (1985) *Phys. Rev. Lett.* **55**, 2471.
51. Stich, I., Car, R., and Parrinello, M. (1989) *Phys. Rev. Lett.* **63**, 2240; (1991) *Phys. Rev. B,* **44,** 4262.
52. Kresse, G. and Hafner, J. (1994) *Phys. Rev B* **49**, 14251.
53. Angell, C.A. and Kanno, H. (1976) *Science* **193**, 1121; (1980) *J. Chem. Phys.* **73**, 1940.
54. Lang, E.W. and Lüdemann, H.-D. (1977) *J. Chem. Phys.* **67**, 718; (1982) *Angew. Chem. Int. Ed. Engl.* **21**, 315.
55. Filipponi, A. and Dicicco, A. (1995) *Phys. Rev. B* **51**, 12322.
56. Ponyatovsky, E.G. and Barkalov, O.I. (1992) *Mater. Sci. Rep.* **8**, 147.
57. Haar, L., Gallagher, J. and Kell, G.S. (1985) *National Bureau of Standards—National Research Council Steam Tables*, McGraw Hill.
58. Poole, P.H., Sciortino, F., Essmann, U. and Stanley, H.E. (1992) *Nature* (London), **360**, 324; (1993) *Phys. Rev. E* **48**, 3799; (1993) *Phys. Rev. E.* **48**, 4605.
59. Poole, P.H., Sciortino, F., Grande, T., Stanley, H.E., and Angell, C.A. (1994) *Phys. Rev. Lett.* **73**, 1632.
60. It is noteworthy that, if the system is held near the tensile limit of the metallic phase where it is also at the limit of its stability with respect to the tetrahedral (or "amorphous") phase, then it will crystallize into an icosahedral phase (Grabow, M. unpublished work). In the laboratory this same phase can be made by heating NaSi in vacuum.
61. Kieffer, J. and Angell, C.A. (1988) *J. Non-Cryst. Solids* **106**, 336.
62. The extrapolation of the immiscibility boundary in the SiO_2-Na_2O system raises the question of whether for SiO_2 there could be a "liquid-amorphous" (actually "glass-glass") transition to a cristobalite-like glass phase at ~850°C on very long time scales.
63. Angell, C.A., Finch, E.D., Woolf, L.A. and Back, P. (1976) *J. Chem. Phys.* **65**, 3063.
64. Woolf, L.A. (1975) *J. Chem. Soc. Faraday Trans. I* **71**, 784; (1976) **72**, 1267.
65. Gibbs, J.H. private communication.
66. Buchenau, U and Zorn, R. (1992), *Europhys. Lett.* **18**, 523.
67. Frick, B., and Richter, D., (1993) Phys. Rev., B 47, 14795
68. Frick, B., Richtter, D., Petry, W., Buchenau, U., (1988), Z. Phys., B70, 73.
69. C. A. Angell, P. H. Poole, and J. Shao, *Nuovo Cimento*, **16D**, 993 (1994).
70. Angell, C. A. (1968), J. Am. Ceram. Soc., 51, 117-124
71. Dubochet, J., Adrian, M., Teixeira, J Kadiyala, R. K., Alba, C. M., McFarlane, D. R. and Angell, C. A.,(1984) J. Phys. Chem., 88, 6727.
72. We are indebted to Sir Roger Elliott at this meeting for pointing out the relation to crystalline state behavior. Simulations of crystals show thsat the corresponding ringing phenomenon persists for a very long time.
73. Souletie, J.,(1990), J. Phys. (Paris), 51, 883.

KINETICS OF GLASS FORMATION

I. GUTZOW, A. DOBREVA

*Institute of Physical Chemistry, Bulgarian Academy of
Sciences, Sofia 1113, Bulgaria*

1. Introduction

Following the traditions and the structural concepts of the 18th and 19th century, present day molecular physics still divides matter into three states of aggregation: gaseous, liquid and solid. Sometimes in analogy to the beautiful schemes from Hellenistic times where four elemental forces (air, water, earth and fire) constituting the Universe are combined, a forth state of aggregation - the plasma state - is introduced to denote the existence of partially or totally ionized state of matter.

Classical thermodynamics operates with thermodynamic phases i.e. with the different equilibrium forms of appearance of the same substance [1]. From the view point of thermodynamics the states of aggregation are also equilibrium states of a system. Thermodynamic phases may coexist (as different states of aggregation or as various phases in the same state of aggregation) in mutual contact, their coexistence in equilibrium being governed by the Gibbs phase rule. By changing the values of the external parameters phase transformations are initiated.

The first thermodynamic classification of possible types of phase transformations is due to Ehrenfest [2].

The unusual properties of glasses, their intermediate position between crystalline solids and amorphous liquids led in the early 1930ies to the characterization of glasses as the forth state of aggregation of matter (in addition to gases, liquids, and solids) (see evidence and literature cited in [3]). This proposal did not (and could not) resolve the problems connected with the definition of glasses and the vitreous state as a very particular physical state.

According to knowledge already accumulated in the beginning of 1920ies and mainly due to the investigations done by Simon [4,5] it is known that glasses are non equilibrium frozen-in systems. This description means that glasses are excluded from any thermodynamic classification either as a state of aggregation or as a thermodynamic phase, at least in the sense that Gibbs has given to the latter notion.

It turns out that the thermodynamic treatment of the frozen-in states is only possible in the framework of a phenomenological science developed in the 1930ies - thermodynamics of irreversible processes i.e. in this generalization which De Donder [6] and Prigorgine [7] have given to classical thermodynamics. It can be even stated

21

M. F. Thorpe and M. I. Mitkova (eds.), Amorphous Insulators and Semiconductors, 21–43.
© 1997 *Kluwer Academic Publishers. Printed in the Netherlands.*

that this further development and extension of the thermodynamics was to a great extend initiated by the problems of the quantitative description of the properties of glasses as the most striking examples of frozen-in states. Frozen-in states become defined when at least one additional frozen-in parameter of state is introduced.

The nature of the process by which the undercooled melt becomes a glass or more generally , the kinetics of glass formation has been a source of even more debates than the discussions connected with the vitreous state itself. In this respect the numerous attempts to treat glass formation in terms of existing thermodynamic classifications of phase transformations should be mentioned. Keeping in mind that glasses are frozen-in, non equilibrium systems, it is evident that such descriptions are logistically controversial from the very beginning. Nevertheless, as discussed in section 3, the process of vitrification of a melt was considered as a first, as a second and even as a third order thermodynamic phase transition. Now it seems more or less established convention that undercooled melts form glasses in a process of kinetic transition of state caused by structural arrest. Because this process takes place in a broader temperature interval sometimes it is also called diffuse phase transition [8]. The quantitative analysis of this process brings difficulties and it is considered as one of the greatest challenges of present day condensed matter physics [9-11].

There are different ways to describe this process. The approach used in a number of investigations initiated by very general theoretical concepts is to formulate a universal microscopic theory (e.g. in the framework of the so called mode coupling theory [9], or using bond fluctuation models [10,11] etc.) predicting the possibility (or even the necessity) of vitrification when the melt is undercooled beyond some critical value. However, most of these general and complicated models although having the great advantage of *ab initio* formulation, at present still lead to contradictory conclusions [11] or only confirm in a more universal form results already known.

The other basic approach in the physics of the vitreous state is the phenomenological analysis of glasses and of the processes leading to their formation in terms of the thermodynamics of irreversible processes. The thermodynamic properties of glasses [7, 12], the thermodynamics of vitrification [13, 14], the nature of vapour pressure and solubility of glasses [15] have been treated by using this approach. Moreover, the phenomenological analysis based on the thermodynamics of irreversible processes leads to a simple but general description of both the freezing-in process of glass forming liquids (i.e. the kinetics of vitrification) and the evolution of the frozen-in state of the glass to the equilibrium state of the corresponding undercooled liquid (i.e. the kinetics of stabilization) [12, 16, 17].

The classical kinetic treatment of vitrification is based on semi-empirical models (e.g. free volume model of viscous flow [18, 19] leading to the WLF viscosity formula, the two level energetic models formulated by Bartenev [20], Ritland [21] and Volkenstein and Ptyzin [22] etc.). The combination of a phenomenological thermodynamic approach and an appropriate *ad hoc* microscopic model is usually termed kinetic (or relaxational) theory of vitrification [16, 22]. Most of the results of present day glass science - the derivation of the Bartenev-Ritland equation [3, 16, 23], the derivation of dependences connecting the properties of glasses with cooling rate

[14], the introduction of memory effects in glass formation [24] etc.- are based on such a dual structure.

In the present contribution also a dual approach is adopted: the combination of the thermodynamics of irreversible processes (in its linear approximation as well as in its non linear extension [3, 16]) with the activated state complex approach [25, 26] gives a more universal outline of the problems connected with the vitreous state and the kinetics of its formation. Thus, as shown in the subsequent sections, adopting Simon's idea of vitrification as a process of structural arrest and introducing at least one internal de Donder-Prigorgine parameter, ξ, we can derive expressions describing:

i/ the temperature behaviour of the thermodynamic functions and of the thermodynamic coefficients at $T=T_g$;

ii/ the dependence of the glass transition temperature and thermodynamic properties of the glass on cooling rate, q;

iii/ the rate of evolution of the frozen-in glass towards equilibrium and

iv/ the non symmetric, structure dependent properties of the kinetic coefficients of glass forming melts.

No other theoretical approach has given so abundant results using so simple premises.

2. Nature of Vitrification: Experimental Evidence

Experimental data from thermodynamic investigations of approximately one hundred glass forming substances are summarized and critically evaluated in our publications [14, 27]. Here we would like to discuss only the results obtained for one typical glass former - glycerol. The temperature dependence of the specific heats, c_p, and of the dielectric constant, ε, of crystalline, liquid and vitreous glycerol are given on figures 1a and 1b, respectively. It is seen that in the vicinity of the glass transition temperature, T_g, a sigmoidal change - a jump - of $c_p(T)$, and $\varepsilon(T)$ dependences occurs.

The temperature dependence of density, ρ, for the same substance is presented on figure 1c. At $T=T_g$ a break point of the $\rho(T)$ dependence is observed. The differentiation of the $\rho(T)$ curve gives the temperature dependence of the thermal expansion coefficient, α, for vitrifying glycerol (see figure 1d). Similar to the $c_p(T)$ and $\varepsilon(T)$ dependences a jump of the $\alpha(T)$ curve for vitreous glycerol appears at $T=T_g$.

The viscosity, η, of vitrified undercooled liquid dramatically changes in the vicinity of the glass transition temperature. At $T=T_g$ a break point of the $\eta(T)$ dependence is observed. In the transformation range, the viscosity of undercooled melts changes from 10^{10} to 10^{14} dPa.s and at T_g, η reaches values of about $10^{14.5}$ dPa.s.

Knowing the $c_p(T)$ dependence for glycerol and its enthalpy of melting, ΔH_m, the temperature course of the difference of the thermodynamic functions of liquid and crystalline glycerol can be constructed by using simple thermodynamic formalism (see [3]). This is done on figure 2, following Simon's construction of the thermodynamic properties of vitrifying glycerol.

24

For the entropy difference $\Delta S(T)=S_f(T)-S_c(T)$ (the subscripts f and c being used to denote the liquid and the crystal, respectively) we have to write

$$\Delta S(T) = \Delta S_m - \int_T^{T_m} \left(\Delta c_p / T\right) dT \tag{1}$$

where ΔS_m and T_m are the entropy of melting and the temperature of melting, respectively. Assuming as indicated by the results given on figure 1a, that for $T<T_g$, $\Delta c_p(T)=c_{p,g}(T)-c_{p,c}(T)\approx0$, (the subscript g refers to the glass) it follows that the entropy value given by the expression

$$\Delta S_g \equiv \Delta S(T)_{T<T_g} \approx \Delta S_m - \int_{T_g}^{T_m} \left(\Delta c_p / T\right) dT \tag{2}$$

is frozen-in below T_g.

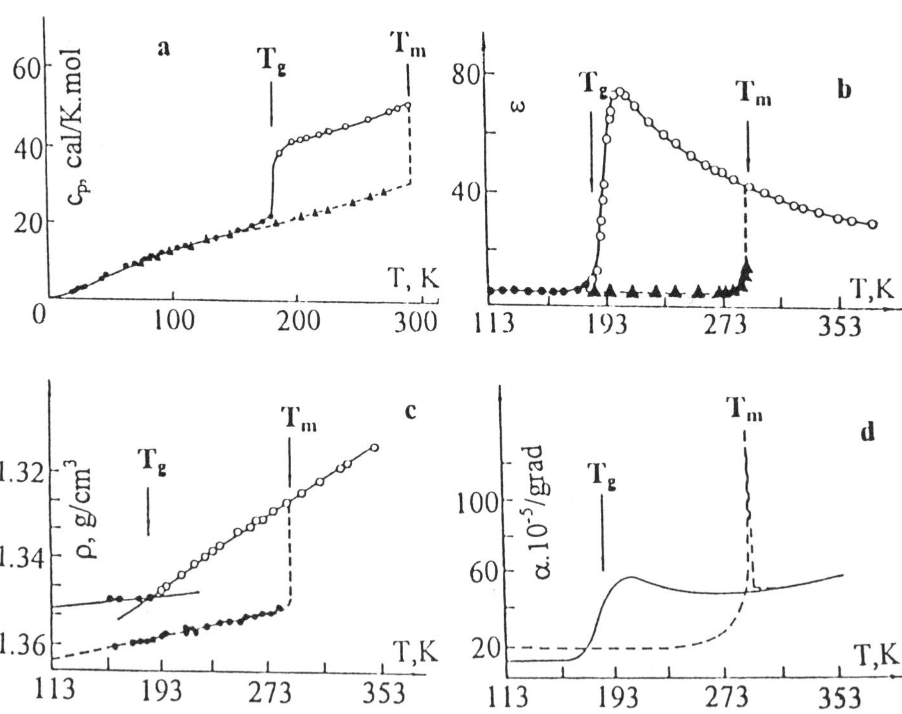

Figure 1: Temperature dependence of thermodynamic properties of glycerol. a/ specific heats of liquid (o), crystalline (▲) and vitreous (•) glycerol [28, 29]. b/ Dielectric constant of liquid (o), crystalline (▲) and vitreous (•) glycerol [30]. c/ Density of glycerol [31]. Open and closed circles of the upper curve refer to liquid and vitreous glycerol, respectively while the closed circles of the lower curve are experimental data for crystalline glycerol. d/ Thermal expansion coefficient, α, for glycerol. The bold line corresponds to the liquid and the vitreous state while the dashed line refers to the crystal.

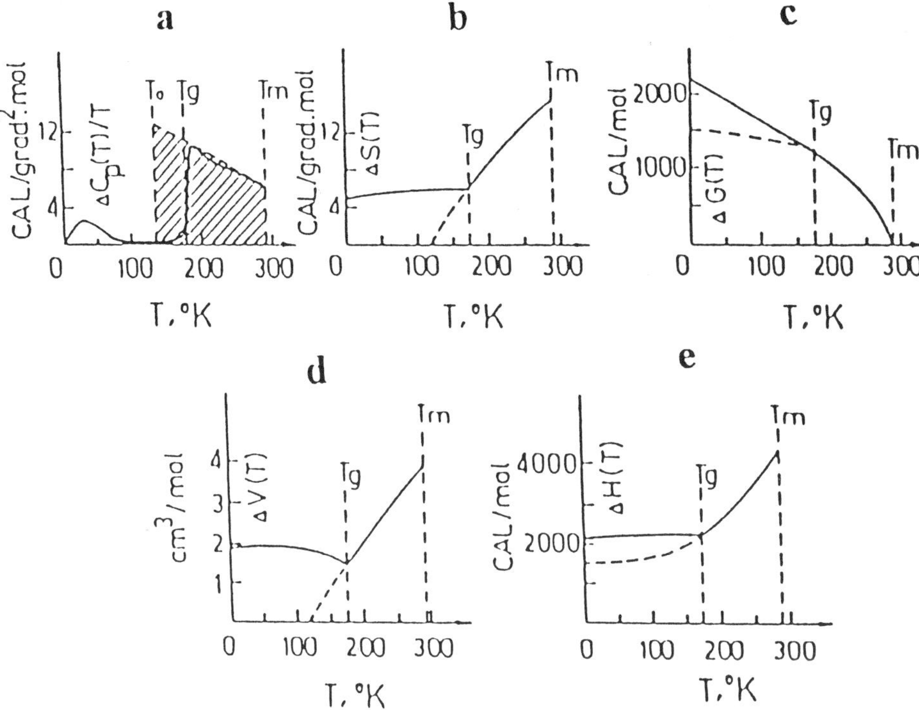

Figure 2: Temperature dependence of the thermodynamic functions of vitrifying glycerol as constructed by Simon [5]. The respective calculations are based on the measurements of the specific heats (figure 1a) of the liquid, vitreous and crystalline states of this substance. a/ Reduced specific heat difference (the shaded area from T_m to T_o gives the entropy of melting of glycerol). b/ Entropy. c/ Gibbs free energy. d/ Molar free volume (the curve is based on density measurements). e/ Enthalpy. By a dashed line the extrapolations of the thermodynamic functions of the undercooled melt below T_g are given. The curves for the undercooled melt and the glass are specified by full lines.

The result given with eq. (2) contradicts one of the basic laws of thermodynamics - the Third Principle - because it indicates that the glass approaches the absolute zero temperature with an entropy value higher than zero. This striking result was recognized for the first time by Simon [4]. In ref. [3] the remarkable story of the verification of the zero point entropy of glasses and its thermodynamic and structural implications are given in detail. It is interesting to note here that it was Einstein [32] who first foretold the possibility of existing of systems of frozen-in disorder (including glasses) with non zero values of the zero point entropy many years ago before the experiments were performed. The possible limits of ΔS_g and its structural dependence are analyzed in [33].

Following the same line of argumentation, it turns out that the enthalpy difference

$$\Delta H_g \equiv \Delta H(T)\Big|_{T<T_g} \approx \Delta H_m - \int_{T_g}^{T_m} \Delta c_p(T)dT \tag{3}$$

is also frozen-in below T_g.

Introducing constant values of ΔH_g and ΔS_g into another well known thermodynamic dependence

$$\Delta G = \Delta H - T\Delta S \tag{4}$$

a second violation of the Third law of thermodynamics becomes evident: for $T<T_g$ we have to write

$$\Delta G_g = \Delta H_g - T\Delta S_g \tag{5}$$

i.e. the thermodynamic potential of the glass approaches the zero point temperature with a constant slope

$$d\Delta G_g / dT\Big|_{T<T_g} = -\Delta S_g \tag{6a}$$

and not as required by the Third Principle

$$d\Delta G_g / dT\Big|_{T\to 0} = 0 \tag{6b}$$

The general thermodynamic significance of eq. (5) is that at $T=T_g$, the $\Delta G_g(T)$ dependence of the frozen-in system, the glass, turns out to be a tangent to the $\Delta G(T)$ course of the equilibrium system, the metastable undercooled liquid [34].

Further measurements confirmed above thermodynamic results and it was established for a great number of substances using different experimental techniques that:

i/ The first derivatives of the thermodynamic potential of the undercooled liquid (i.e. the thermodynamic functions S(T), H(T), V(T) etc.) display a break point at $T=T_g$.

ii/ The second derivatives of the thermodynamic potential of the undercooled liquid (i.e. the thermodynamic coefficients c_p, α, ε) display a jump at $T=T_g$.

iii/ As far as the kinetic properties of the melt (e.g. viscosity, η, time of molecular relaxation, τ, etc.) are determined by the thermodynamic functions of the melt the break point in the temperature dependence of the thermodynamic functions determines also a break point of the $\eta(T)$ and $\tau(T)$ courses at $T=T_g$ (see section 5).

iv/ The glass transition temperature, T_g, depends on cooling rate, $q=-dT/dt$, following a course established for the first time by Bartenev [20] and Ritland [21]

$$1/T_g = c_1 - c_2 \log q \tag{7}$$

where c_1 and c_2 are constants.

Upon heating the glass a kinetic overshoot at T_g leads to the formation of a particular "nose" in the $c_p(T)$ dependence. At cooling run experiments this "nose" has never been observed. Thus T_g has not the character of a thermodynamic equilibrium point but is a kinetically determined quantity.

Before deriving the basic dependences of the theory of glass formation we would like to mention some empirical rules obtained in examining the behaviour of glass forming melts.

The first generalized description of the thermodynamic properties of glass forming melts was given in 1948 by Kauzmann [35]. Summarizing the experimental evidence, available then, on the temperature dependence of thermodynamic functions of about 15 glass formers, he has found that:

i/ The temperature dependence of entropy in reduced coordinates (e.g. $\Delta S(T)/\Delta S_m$ vs. T/T_m) is nearly the same for all substances analyzed.

ii/ The glass transition temperature obtained at normal cooling rates (e.g. at $q=10^1$-10^2 K/s) is

$$T_g = (2/3)T_m \tag{8}$$

This expression now is known as the 2/3 Kauzmann-Beaman rule (see also [3, 16, 36].

iii/ The linear extrapolation of the $\Delta S(T)/\Delta S_m$ dependence below T_g indicates that the configurational entropy of the fictive undercooled melt in internal equilibrium approaches zero values, i.e. $\Delta S(T)/\Delta S_m \to 0$, at temperatures considerably above the absolute zero - approximately at $T_o \approx 0.5 T_m$. Thus the linear extrapolation of $\Delta S(T)/\Delta S_m$ below T_o leads to negative values (the so called Kauzmann paradox).

A further analysis [37] of the experimental data has shown that

$$\Delta c_p(T_g)/\Delta S_m \approx 3/2 \tag{9}$$

Two other simple dependences

$$\Delta S_g \approx \Delta S_m / 3 \tag{10}$$

$$\Delta H_g \approx \Delta S_m T_m / 2 \tag{11}$$

have been established in [14, 27].

The significance of above empirical rules is that they give the possible deviations from equilibrium of a glass frozen-in at normal cooling rates. Considering eqs. (8, 10, 11) we obtain

$$\left. \frac{\Delta G_g(T)}{\Delta S_m T_m} \right|_{T \to 0} = \left. \left| \frac{1}{2} - \frac{1}{3}\frac{T}{T_m} \right| \right|_{T \to 0} \approx \frac{1}{2} \tag{12}$$

3. Thermodynamics of Phase Transitions and Glass Formation Processes: a Geometric Approach

A critical reappraisal of the thermodynamic classification of phase transformations given by Ehrenfest (see also [38]) as well as an elaborated analysis of another phenomenological approach - the Landau order parameter theory- may be found in [39]. Here we would like only to compare the classical classification of phase transitions [2, 38] with the results of a simple analysis of the change of the configurational part of the thermodynamic functions in systems in which a freezing-in process takes place.

According to Ehrenfest [2] the order of the thermodynamic phase transition is determined by the order of the derivative of the thermodynamic potential of the respective phase which shows a singularity (i.e. a jump) at the equilibrium temperature

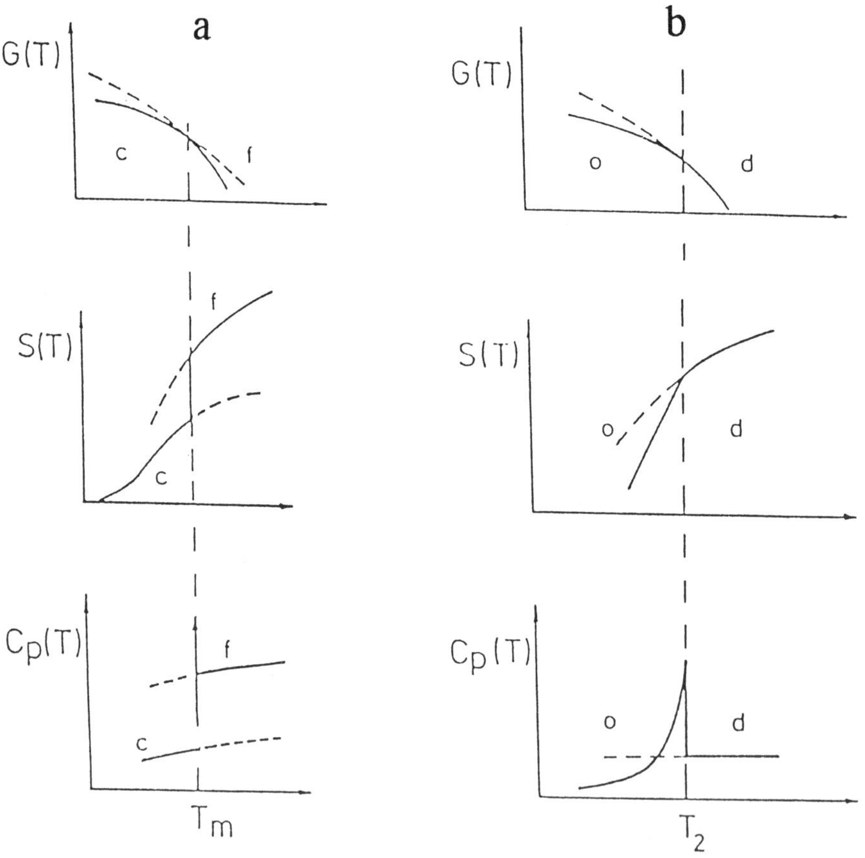

Figure 3: Temperature dependence of the thermodynamic potential, G, the entropy, S, and the specific heats, c_p, for system undergoing first (a) and second (b) order phase transitions, respectively. As an example for first order phase transitions the transition crystallization (c)/melting (f) is chosen, while as an example for second order phase transformations the transition order (o)/disorder(d) is presented .

T_e. The temperature dependence of the thermodynamic potential G(T), of the entropy, S(T), and of the specific heats, c_p(T), of a system in which a first or second order phase transition takes place is illustrated on figure 3. At the corresponding equilibrium temperature T_e (e.g. T_e=T_m for crystallization/ melting transition and T_e=T_2 for order/disorder transition) the thermodynamic potentials of both phases are equal i.e. $\Delta G(T_m)$=0 and $\Delta G(T_2)$=0.

From a geometric point of view there are three possibilities for the G(T) curves of the two equilibrium phases having a common point when they are in the same plain i.e. at p=const: to intersect, to coalesce or to tangent at T=T_e. The distinction between intersecting, coalescent and tangenting, curves can be done considering the sign of $\Delta G(T)$ for both ΔT=(T_e-T)>0 and ΔT=(T_e-T)<0. Thus for intersecting curves $\Delta G(T)$ changes its sign through passing the equilibrium point.

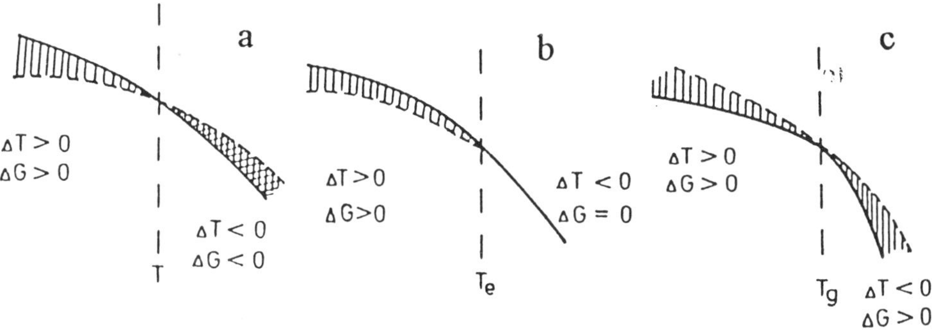

Figure 4: G(T) curves: a/ intersecting, b/ coalescent, c/ tangenting.

For coalescent curves $\Delta G(T)$ goes from $\Delta G(T)$>0 to ΔG=0 when ΔT changes from ΔT>0 to ΔT<0 and for tangenting curves $\Delta G(T)$>0 in the whole temperature interval (below and above T_g) (see figure 4).

According to above mentioned thermodynamic dependences the temperature course of $\Delta G(T)$ is determined by $\Delta S(T)$. The temperature dependence of entropy for first order phase transitions determines intersecting G(T) curves (see figure 3a). For second order phase transitions where $\Delta S(T_2)$=0 the only possibility guaranteeing the coexistence of the two phases in equilibrium is the G(T) curves of the ordered and the disordered phase to coalesce for T>T_2 [38].

In constructing the temperature dependence of the thermodynamic potential of the undercooled metastable melt and of the frozen-in glass using the corresponding S(T) and c_p dependences it turns out that the $G_g(T)$ curve, given with eq. (5), is a tangent to the $G_f(T)$ dependence at T=T_g i.e. the glass/undercooled melt difference of the thermodynamic potentials is positive below and above the glass transition temperature, T_g. This result, obtained for the first time in studying the vapour pressure and solubility of glasses [15, 34] (see also [17, 40]) gives a *thermodynamic* distinction between thermodynamic phase transitions and freezing-in processes in a glass. It shows that there is no temperature region of absolute or even relative stability of the glass. The

30

glass is always in a non equilibrium state having a higher thermodynamic potential than the corresponding metastable liquid or the thermodynamically stable crystal for both $T<T_g$ and $T>T_g$ (c.f. figure 5).

The dependence of the glass transition temperature, T_g, on cooling rate, q, (eq. 7) gives the *kinetic* distinction between thermodynamic phase transition and vitrification [17]. The dependence of $\Delta G_g(T)$ on cooling rate is evident from eq. (5) where both ΔH_g and ΔS_g are functions of q. The change of $\Delta G_g(T)$ for different cooling rates is shown on figure 6. It is seen that at higher cooling rates more unstable glasses are produced.

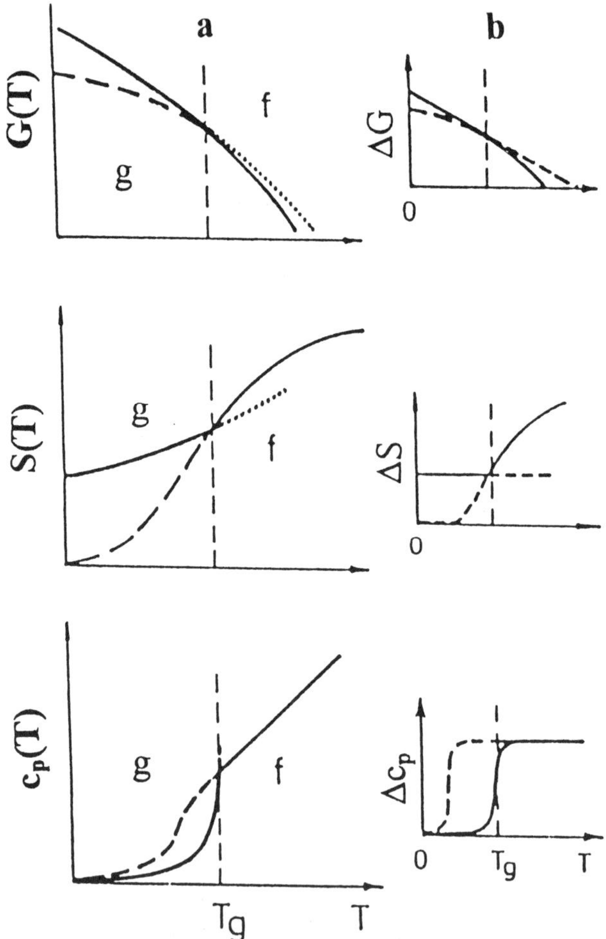

Figure 5: Temperature dependence of the thermodynamic potential, G, the entropy, S, and the specific heats, c_p, of a system undergoing vitrification. b/ Temperature dependence of the melt/crystal difference in thermodynamic potential, ΔG(T), the entropy ΔS(T), and the specific heats, Δc_p, of a vitrifying system.

The comparison of figure 3 and figure 5 shows that the temperature dependence of the thermodynamic functions of a freezing-in process resemble to some extend that of a second order phase transition. These apparent similarities were the reason that years ago Boyer and Spencer [41] considered vitrification as second order phase transition.

There have been unsuccessful attempts to classify vitrification as a first order phase transition. Uberreiter and Burns [42] gave such a suggestion misinterpreting measurements of the vapour pressure of vitrifying organic polymer melts. Other attempts in this respect are critically considered in [43, 44].

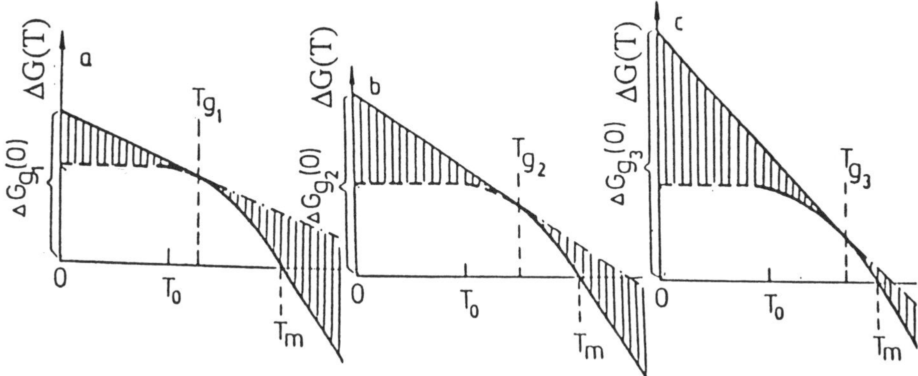

Figure 6: $\Delta G_g(T)$ diagram of the same melt, vitrified at three different cooling rates, q. For $q_1 < q_2 < q_3$ we have to expect that $T_{g1} < T_{g2} < T_{g3}$ and $\Delta G_{g1}(0) < \Delta G_{g2}(0) < \Delta G_{g3}(0)$.

4. Kinetics of Vitrification

Introducing as done in the thermodynamics of irreversible processes [3, 6, 7] an additional structural parameter, ξ, the conditions for thermodynamic equilibrium at constant pressure, p, and constant temperature, T, are

$$\left[\partial \, \Delta G(\xi) \, / \, \partial \xi \right]_{T,p} = 0 \qquad (13a)$$

and

$$\left[\partial^2 \Delta G(\xi) \, / \, \partial \xi^2 \right]_{T,p} > 0 \qquad (13b)$$

The structural parameter, ξ, having the physical significance of a structural reaction coordinate can be defined in such way that $\xi=0$ corresponds to the case of absolute configurational order (as it is in the crystal) and $\xi=1$ in a completely

32

disoredered system [14, 17, 27]. At equilibrium the structural parameter, ξ, is a single valued function of the state variables and at p=const

$$\xi = \xi(T)_p \tag{14a}$$

However, when the state of the system is abruptly changed (e.g. by rapid quenching) its structure described in a generalized way by the parameter, ξ, can not follow the alterations of temperature immedeately. In such cases at lower temperatures the melt remains frozen-in a state of disorder, characterized by the value of $\tilde{\xi}$ corresponding to a higher temperature $\tilde{T}$ i.e.

$$\tilde{\xi} = \xi(\tilde{T})_p \tag{14b}$$

Here $\tilde{T}$ is the frozen-in structural temperature which following a proposal by Tool is called also fictive temperature of the system (see [3] and [16]).

For the non equilibrium system thus obtained we have

$$\left[\partial \Delta G(\tilde{\xi}) / \partial \tilde{\xi}\right]_{T,p} \neq 0 \tag{15}$$

and as always $\tilde{\xi} > \xi$ the sign of the derivative in above inequality is positive.

According to the thermodynamics of irreversible processes the rate $d\tilde{\xi}/dt$, of the evolution of a non equilibrium system to equilibrium is given by [3, 17]

$$d\tilde{\xi} / dt = f(A) \tag{16a}$$

where

$$A = -\left[\partial \Delta G(T,\tilde{\xi}) / \partial \tilde{\xi}\right]_{T,p}$$

is the affinity of the system. The function f(A) satisfies the condition

$$f(A = 0) \equiv f(0) = 0 \tag{16b}$$

Developing the f(A) function in eq. (16a) in a Taylor series for A$\rightarrow$0 (i.e. at $\tilde{\xi} \rightarrow \xi$) and accounting for eq.(16b) we obtain

$$\frac{d\tilde{\xi}}{dt} = -L\left[1 + \frac{1}{L}\sum_n \frac{1}{n!} L_{n-1} A^{n-1}\right]\left[\frac{\partial \Delta G(T,\tilde{\xi})}{\partial \tilde{\xi}}\right]_{T,p} \tag{17}$$

where

$$L = \left[\partial f(A) / \partial A\right]_{T,p}\big|_{\widetilde{\xi} \to \xi}$$

has the significance of a kinetic coefficient and $A^{(n-1)}$ denotes the higher ordered derivatives of $f(A)$. In the linear approximation of the thermodynamics of irreversible processes it is assumed that

$$L^{-1}\sum_n (1/n!)L_{n-1}A^{n-1} \ll 1 \qquad (18)$$

Thus follows the de Donder equation

$$d\widetilde{\xi} / dt = -L\left[\partial\, \Delta G(T,\widetilde{\xi}) / \partial\widetilde{\xi}\right]_{T,p} \qquad (19)$$

Expanding the derivative in above equation in a Taylor series for $\widetilde{\xi} \to \xi$ (i.e. for small deviations from equilibrium) and accounting for eq. (13a) we obtain in a first approximation

$$d\widetilde{\xi} / dt = -(\widetilde{\xi} - \xi) / \tau \qquad (20)$$

where

$$\tau = -\left\{ L\left[\partial^2 \Delta G(T,\widetilde{\xi}) / \partial\widetilde{\xi}^2\right]\big|_{\widetilde{\xi} \to \xi}\right\}^{-1} \qquad (21)$$

is the time scale of the considered process of structural evolution to metastable equilibrium (in the case of glass stabilization, the phenomenological coefficient is determined by the fluidity of the system at temperature T).

In most cases the approximate solution (eq. 20) with a constant value of τ (eq. 21) successfully describes the kinetics of stabilization. Moreover, as shown below, it can be used to define the vitreous state and to derive kinetic dependences for both vitrification of undercooled melt and the stabilization of the corresponding glass.

Equation (20) with a constant τ value (i.e. independent of the prehistory of the system) defines a symmetric $d\widetilde{\xi}/dt$ curve in approaching equilibrium from above and from below (see figure 7). However, at greater deviations from equilibrium the non symmetric course of the $d\widetilde{\xi}/dt$ curve becomes evident (figure 8) and the above approximation does not hold any more. In this case non linear effects should be introduced i.e. higher ordered terms in eq. (19) should be considered, as proposed in [17].

34

In general it is expected that τ depends on temperature exponentially e.g.

$$\tau\,(T) = \tau_o \exp\!\big(U(T)\,/\,kT\big) \tag{22}$$

where τ_o and k denote as usual the time of eigen vibrations of the system and the Boltzman constant, respectively. The activation energy U(T), for undercooled liquids increases with descreasing temperature i.e.

$$dU(T)\,/\,dT < 0 \tag{23}$$

and thus τ(T) becomes a function changing even steeper than the exponential one [3, 17, 46].

According to eq. (20) a frozen-in system at T<T_g is defined as

$$d\widetilde{\xi}\,/\,dt\,\Big|_{T<T_g} = 0 \tag{24a}$$

Accounting for the exponential τ(T) dependence (eq. 22) it is seen that the above requirement is fulfilled even at temperatures a few degrees lower than T_g where we have to expect τ(T)$\to\infty$. As far as $\widetilde{\xi}$ = const for T<T_g, eq. (24a) can be also written as

$$d\widetilde{\xi}\,/\,dT\,\Big|_{T<T_g} = 0 \tag{24b}$$

Thus, according to the thermodynamics of irreversible processes glass is defined as a frozen-in non equilibrium system by the simultaneous implementation of eqs. (24).

According to a fundamental kinetic criterion at the temperature of vitrification the time of molecular relaxation, τ, of the undercooled liquid has to be of the same order of magnitude as the characteristic observation or stay time, Δt, for the considered process i.e.

$$\tau_{T \to T_g} = \Delta t \tag{25}$$

Differentiating above equation with respect to temperature and introducing the cooling rate, q=-dT/dt, we obtain

$$\big[d\tau(T)/dT\big]\,q = -1 \tag{26}$$

When the activation energy U(T) in eq.(22) is considered approximately as a constant (i.e. U(T)≈const=U_o) eq. (26) gives for T=T_g

$$q\tau(T_g) = RT_g^2 / U_o = c_o \qquad (27)$$

Above dependence is known as the Frenkel -Kobeko equation. It was found by Bartenev that for typical glass formers $c_o \approx$ (1-5) K. Taking logarithm from both sides of eq.(27) we obtain

$$1 / T_g = (2.3R / U_o)\log(c_o / \tau_o) - (2.3R / U_o)\log q \qquad (28)$$

Denoting $c_1=c_2\log(c_o/\tau_o)$ and $c_2=(2.3R/U_o)$ the Bartenev-Ritland equation (7) follows immediately.

In [14, 27] we used more complicated τ(T) dependences to derive more accurate T_g(q) functions. It is seen that eq. (28) gives for q→0 the value of T_g(0)→0. By using the VFT equation, as done in [14, 27] for q→0 the value $T_g(0)\approx T_o=T_m/2$ is obtained. This result is of significance in connection with the Kauzmann paradox: it turns out that the Kauzmann temperature, T_o, is reached at q=0.

If the dependence of T_g on q and the temperature course of the thermodynamic functions of undercooled melt are known the thermodynamic functions of the corresponding glasses frozen-in at different cooling rates can be constructed. In [14, 27] we solved this problem by using analytical expressions for the thermodynamic functions following either from particular microscopic considerations or assumed in the framework of phenomenological models.

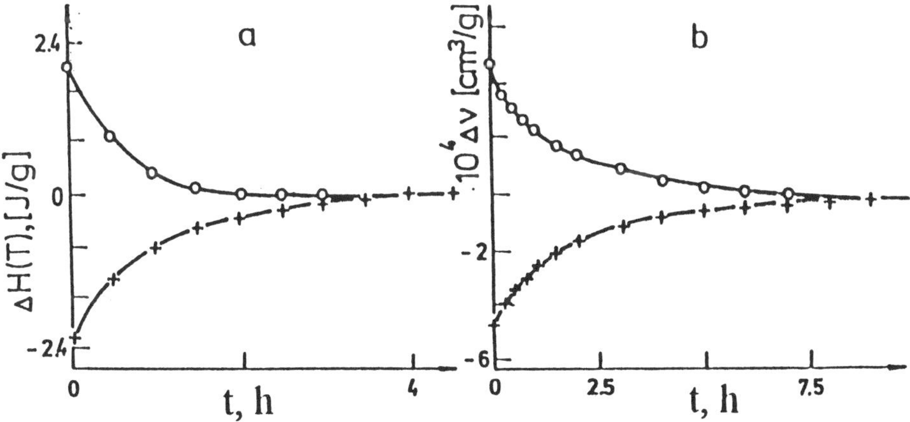

Figure 7: Experimental curves for the relaxation of vitrified glass forming melts in the vicinity of Tg [12]. a/ relaxation of the enthalpy of glycerol at temperature T=185K. b/ Relaxation of the volume of glucose at a temperature T=304K. + approach to equilibrium from below; o approach to equilibrium from above.

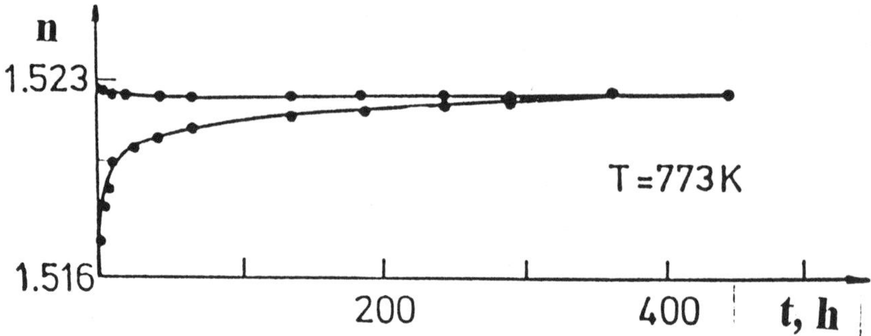

Figure 8: Relaxation of the coefficient of refraction, n, of a borosilicate glass according to measurements of Winter [45]: example of a non symmetric approach to equilibrium.

The processes of vitrification and   of subsequent stabilization have to be reconsidered taking into account that the kinetic coefficients in eq. (19) depend on ξ and thus on the previous history of the system. An elaborated discussion in this respect may be found in [3, 17] where a non symmetric course of the kinetics of stabilization is predicted when equilibrium is approached from above and from below.

By using the structural parameter, ξ, the jump of the temperature dependence of the specific heats (and thus of all other thermodynamic functions) can be also readily explained.

The specific heats at constant pressure are defined as

$$c_p = (dH / dT)_p \tag{29}$$

Introducing the structural parameter, ξ, the enthalpy, H_f, of the undercooled liquid becomes

$$H_f = H(T,\xi)_p \tag{30}$$

and its total differential at constant pressure can be written as

$$dH_f(T,\xi) = \left(\partial H_f / \partial T\right)_{p,\xi} dT + \left(\partial H_f / \partial \xi\right)_{T,p} d\xi \tag{31}$$

Thus with eq. (29) the specific heats of the melt are

$$c_{p,f} = \left(\partial H_f / \partial T\right)_{p,\xi} + \left(\partial H_f / \partial \xi\right)_{T,p} \left(d\xi / d T\right)_p \tag{32}$$

From eq. (32) it is seen that the specific heats of undercooled liquids are constituted of two parts: a phonon one

$$c_{p,ph} = \left(\partial H_f / \partial T \right)_{p,\xi} \tag{32a}$$

which is practically equal to the phonon specific heats of the crystal and a configurational one

$$c_{p,conf} = \left(\partial H_f / \partial \xi \right)_{T,p} \left(d\xi / dT \right)_p \tag{32b}$$

For $T<T_g$, according to eq. (24b) we obtain

$$c_{p,conf} \Big|_{T<T_g} = 0 \tag{33a}$$

and thus

$$c_{p,f} \Big|_{T<T_g} \equiv c_{p,g} \approx c_{p,c} \tag{33b}$$

as in fact is observed in vitrification (see figure 1a). Above derivation has been made by Prigorgine and Deday [7]. A more elaborated discussion, including also the derivation of the Prigorgine -Defay ratio and the introduction of more than one structural parameter, ξ_i, can be found in [3].

5. Temperature Dependence of the Thermodynamic Functions of Undercooled Melts: Strong and Fragile Liquids.

In general the experimentally observed configurational entropy- temperature functions, $\Delta S(T)$, of glass forming liquids can be divided into two categories: those for which the $\Delta S(T)$ curves in the experimentally accessible temperature region from T_m to T_g are concave and those for which the $\Delta S(T)$ curves are convex in $\Delta S(T)$ vs. T coordinates (see figure 9) [46, 47]. Model computations in the framework of MFA and MC simulations [47] have shown that simple (in the sense of non associating) liquids constituted of equal structural units show concave $\Delta S(T)$ curves. On the other hand glass forming melts with considerably increased configurational and computational disorder are convex in the indicated sense. Concave $\Delta S(T)$ behaviour corresponds also to a concave $\Delta c_p(T)$ dependence while convex $\Delta S(T)$ curves correspond either Δc_p=const or Δc_p increases as the temperature falls (see figure 10). In this respect the temperature dependence of the configurational specific heats can be used to make more or less distinct structural conclusions.

38

In order to achieve a unified description, any $\Delta c_p(t)$ dependence may be approximated with a constant, as proposed in [14, 48] i.e.

$$\Delta S_m = \int_0^{T_m} \left(\Delta c_p(T)/T\right)dT \approx \int_{T_o}^{T_m} \left(\Delta c_p(T)/T\right) \approx \Delta c_p(T_g)\ln(T_m/T_o) \qquad (34)$$

In this way, a dimensionless number, the thermodynamic structural factor, $a_o=\Delta c_p(T_g)/\Delta S_m$, introduced in [14, 27] can be used to determine the Kauzmann temperature, T_o, i.e. the temperature where $\Delta S(T)=0$, as

$$T_o/T_m = \exp\left(-1/a_o\right) \qquad (35)$$

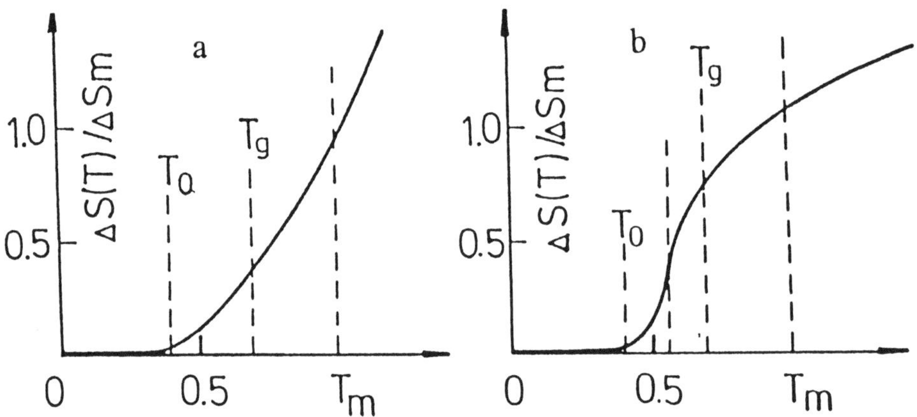

Figure 9: Temperature dependence of the configurational entropy of undercooled liquids. a/ A concave ΔS(T) dependence of simple liquids. b/ A convex ΔS(T) dependence for complex melts with increased configurational disorder.

The value of a_o, as discussed in [14, 27] varies from 2 (for organic polymer liquids) to 1 for simple non associating systems. For metallic liquids even $a_o<1$ can be observed. Thus, eq. (34) with different a_o values can be used to describe the $\Delta S(T)$ dependence of both simple and polymeric liquids.

The temperature dependence of the viscosity, $\eta(T)$, can be determined by the temperature dependence of $\Delta S(T)$ according to an equation analogous to the VFT free volume equation or to the Adams -Gibbs equation as [49, 50]

$$\eta\ (T) = \eta_o \exp\left(U(T)/RT\right)\exp\left(b_o/\Delta S(T)\right) \qquad (36)$$

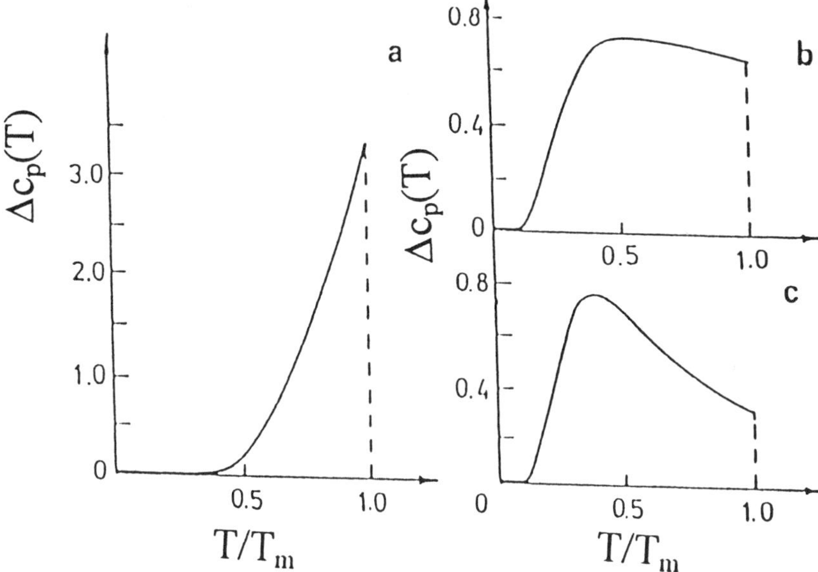

Figure 10: Temperature dependence of the configurational part of the specific heats obtained in a MFA computation [46, 47]. a/ Simple non associating liquids. b/ Undercooled liquids with medium degree of polimerization. c/ liquids with considerable degree of polymerization.

Here η_o is a constant and b_o is a dimensionless number we called kinetic structural factor [14, 27]. As discussed in [51] for simple liquids b_o approaches unity. However, for complex liquids, where flow and processes of self diffusion require the co-operative movement of many molecules, b_o reaches values up to 5, 6 or even 8 [51]. In this sense the interplay of both structural factors - the thermodynamic one, a_o, and the kinetic one, b_o, determines the temperature dependence of the viscosity and eq. (36) can be written as

$$\eta \approx \eta_o \exp(\gamma_o T_m / T) \exp(b_o / a_o \ln(T_m / T)) \tag{37}$$

Above equation follows assuming that $U(T) \approx U_o = \gamma_o RT_m$. The construction of viscosity data in $\log \eta(T)$ vs. T_m/T co-ordinates would give curves with slopes depending on both γ_o and the ratio of the kinetic and thermodynamic structural parameters b_o/a_o.

Angell [52, 53] introduced in glass science an interesting concept. Based on the temperature dependence of the viscosity, glass forming melts can be divided into two categories: strong liquids for which $\log \eta$ vs. T_g/T gives a straight line and fragile liquids for which a changing $(d\log\eta)/(dT^{-1})$ slope of the $\log \eta(T_g/T)$ dependence is obtained.

According to the 2/3 Beaman-Kauzmann rule (eq. 8) it is evident that the $\log \eta(T)$ vs. T_m/T and $\log \eta(T)$ vs. T_g/T dependences should be equivalent, at least for glasses obtained at normal cooling rates. However, from eq. (37) it is evident that the normalization factor should be T_m/T. It is also to be noted that from a thermodynamic point of view and in terms of the more general formulation of the theory of physicochemical similarity [54] it is necessary to introduce in such considerations as a

40

normalizing factor an equilibrium temperature (either the critical temperature, T_{cr}, the boiling temperature, T_b, or as indicated by eq. (37) the melting temperature, T_m) depending on the configurational part of the thermodynamic functions of the liquid.

In terms of above MFA models and eqs. (36, 37) Angell's strong liquids are systems with smaller configurational possibilities for which $\gamma_o T_m > b_o/a_o \ln(T/T_m)$ and b_o/a_o is a small number while fragile liquids are those with increased configurational possibilities.

It is also obvious that upon vitrification $\Delta S(T)$ in eq. (36) becomes a constant and thus a break point in the log η vs. $1/T$ dependence should be observed as indicated in section 2.

6. Kinetics of Structural Evolution and Kinetics of Relaxation. A Non Linear Approach

Equation (20) describing the rate of evolution of the frozen-in structure towards equilibrium can be written in the form

$$d\Delta\xi \, / \, dt = -\Delta\xi \, / \, \tau \tag{38}$$

where $\Delta\xi = \tilde{\xi} - \xi$. The integration of eq. (38) with the boundary condition

$$\Delta\xi = \Delta\xi_o = \left(\tilde{\xi}_o - \xi\right) \qquad at \qquad t \to 0 \tag{38a}$$

gives

$$\Delta\xi = \Delta\xi_o \exp(-t \, / \, \tau) \tag{39}$$

It is readily seen that eq. (39) corresponds to the Maxwell formula giving the time dependence of relaxation.

However, existing experimental evidence shows (see section 4) that the linear approach (eqs. 38, 39) in most cases can not describe neither stress relaxation, nor stabilization of glasses.

Two solutions of this problem have been proposed:
i/ To introduce structure dependent τ values i.e.

$$\tau\,(\xi) = \tau \, / \, \psi(\Delta\xi) \tag{40}$$

where $\psi(\Delta\xi)$ is a dimensionless function, and eq. (38) is transformed to

$$d\Delta\xi \, / \, dt = -\psi(\Delta\xi)\Delta\xi \, / \, \tau \tag{41a}$$

Appropriate $\tau(\Delta\xi)$ (or $\psi(\Delta\xi)$) functions can be obtained using either empirical dependences or molecular models. Thus in terms of the Prandtl-Eyring potential barrier model the respective $\psi(\Delta\xi)$ function reads [55]

$$\psi(\Delta\xi) = \sinh(a\Delta\xi) / a\Delta\xi \tag{42}$$

where a is the so called viscous volume.

ii/ The second approach is to introduce a time dependent dimensionless function, $\varphi(t)$, and thus

$$d\Delta\xi / dt = -\varphi(t)\Delta\xi / \tau \tag{41b}$$

It is interesting to note that eq. (41b) which is more general than Maxwell's formula (38) was introduced in science by both Kohlrauschs (father and son) [56, 57] in the course of prolonged investigation of dielectric relaxation and inelastic after-effects in different materials before Maxwell's dependences were established. For

$$\varphi(t) = A_k / t^P \tag{39a}$$

where A_k and p are empirical constants the integration of eq. (41b) under the boundary conditions given with eq. (38a) leads to the Kohlrausch stretched exponent formula [56, 57]

$$\Delta\xi = \Delta\xi_o \exp\left(-t / \tau_k\right)^b \tag{39b}$$

where b=1-p<1 and $\tau_k=(b\tau/A_k)^{1/b}$. In [58] these two approaches have been united and the expression connecting the $\varphi(t)$ and $\psi(\Delta\xi)$ functions has the form

$$\varphi(t) = -\tau \frac{d}{dt}\left\{\ln\left[F^{-1}\left[F(\Delta\xi(t))\right]\right]\right\} \tag{40}$$

where F^{-1}, the inverse function of

$$F\left(\Delta\xi(t)\right) = F\left(\Delta\xi_o\right)\exp(-t / \tau) \tag{40a}$$

is determined by the law of change of $\Delta\xi$ i.e. by the $\psi(\Delta\xi)$ function. A detailed derivation of eq. (40) and the respective physical argumentation can be found in the original contribution [58].

7. References

1. Gibbs, J. W. (1875-78) On the equilibrium of heterogeneous substances, *Transaction Connecticut Academy of Sciences* **3**, 108.

2. Ehrenfest, P. (1933) Phase transitions in the usual and in the generalized sense, classified in accordance with the singularities of the thermodynamic potential (in German), *Proc. Amsterdam Acad.* **36**, 153.

3. Gutzow, I. and Schmelzer, J. (1995) *The Vitreous State: Thermodynamics, Structure, Rheology and Crystallization*, Springer, Berlin.

4. Simon, F. (1931) On the state of undercooled liquids and glasses (in German), Z. *anorganische und allgemeine Chemie* **203**, 219.

5. Simon, F. (1956) The third low of thermodynamics. An historical survey. 40-th Gutrie Lecture, *Year Book Phys. Soc. London* , 1.

6. de Donder, Th. and Rysselberghe, P. (1936) *Thermodynamic Theory of Affinity*, Stanford University Press, Stanford.

7. Prigorgine, I. and Defay, R. (1954) *Chemical Thermodynamics*, Longmans, London.

8. Zallen, R. (1983) *The Physics of Amorphous Solids*, J. Wiley, New York.

9. Elliot, S. R. (1990) *Physics of Amorphous Materials*, J. Wiley, New York.

10. Binder, K. (1994) Monte Carlo simulation of the glass transition of polymer melts, *Progr. Colloid Polym. Sci.* **96**, 7.

11. Binder, K. (1996) Theoretical concepts on the glass transition on polymers and their test by computer simulation, Z. *Bunsenges. Phys. Chem.*, in press.

12. Davies, R. O. and Jones, G. O. (1953) Thermodynamic and kinetic properties of glasses, *Adv. Phys.* **2**, 370.

13. Cooper, A. (1977) Internal parameters, ordering parameters, history, and the glass transition, in G. H. Frischat (ed.) J. *Non Cryst. Solids*, Trans. Tech Publications, Aedermansdorf, p. 384.

14. Gutzow, I. and Dobreva, A. (1991) Structure, thermodynamic properties and cooling rate of glasses, *J. Non Cryst. Solids* **129**, 266.

15. Grantcharova, E. and Gutzow, I. (1986) Vapour pressure, solubility and affinity of undercooled melts and glasses, *J. Non Cryst. Solids* **84**, 99.

16. Mazurin, O. V. (1986) *Vitrification* (in Russ), Nauka Publ., Leningrad.

17. Gutzow, I., Dobreva, A. and Pye, L. D. (1995) A geometric and analytic approach to the process of vitrification, *J. Non Cryst. Solids* **180**, 107, 117.

18. Bueche, F. (1956) Derivation of the WLF equation for the mobility of molecules in molten glasses, *J. Chem. Phys.* **24**, 418.

19. Cohen, M. H. and Turnbull, D. (1959) Molecular transport in liquids and glasses, *J. Chem. Phys.* **31**, 1164.

20. Bartenev, G. M. (1966) *Structure and Mechanical Properties of Glass* (in Russ.), Buil. Mat. Press, Moscow.

21. Ritland, H. N. (1954) Density phenomena in the transformation range of a borosilicate crown glass, *J. Am. Ceram. Soc.* **37**, 370.

22. Vol'kenstein, M. V. and Ptizyn, O. B. (1956) Relaxational theory of vitrification (in Russ.), *J. Theor. Phys.* **26**, 2204.

23. Cooper, R. A. and Gupta, P. K. (1982) A dimensionless parameter to characterize the glass transition, *Phys. Chem. Glasses* **23**, 44.

24. Avramov, I. and Milchev, A. (1984) Thermodynamic and relaxation properties of glass forming melts, *Soviet Journal Glass Phys. and Chem.* **10**, 44.

25. Laidler, K. J. (1963) *Reaction Kinetics. II. Reactions in Solution*, Pergamon Press, London.

26. Glasstone, S., Laidler, K. J. and Eyring, H. (1941) *The Theory of Rate Processes,* Princeton University, New York.

27. Gutzow, I. and Dobreva, A. (1992) Thermodynamic functions of glass forming systems and their dependence on cooling rate, *Polymer* **33**, 451.

28. Simon, F. and Lange, F. (1926) On the question of the entropy of amorphous substances (in German), *Z. Physik* **38**, 227.
29. Gibson, G. E. and Giauque, W. F. (1923) The third law of thermodynamics. Evidence from the specific heats of glycerol that the entropy of glasses exceeds that of the crystal at the absolute zero, *J. Am. Ceram. Soc.* **45**, 93.
30. Ubbelohde, A. R. (1965) Melting and Crystal Structure, Clarendon Press, Oxford.
31. Schultz, A. K. (1954) Density of glycerol as an undercooled liquid, as a glass and as a crystal, *J. Chem. Phys. Biol.* **51**, 324.
32. Einstein, A. (1914) Contributions to the quantum theory (in German), *Verh. der Deut. Physik. Ges.* **16**, 820.
33. Gutzow, I. (1962) On the zero point entropy of glasses (in Germann), *Z. Phys. Chem.* **221**, 152.
34. Gutzow, I. (1972) On the vapour pressure and solubility of undercooled melts and glasses, *Z. Phys. Chem. (N.F.)* **81**, 195.
35. Kauzmann, W (1948) The nature of the glassy state and the behaviour of liquids at low temperatures, *Chem. Rev.* **43**, 219.
36. Gutzow, I. (1972) The thermodynamics of supercooled glass forming liquids and the temperature dependence of their viscosity, in B. Ellis, R. Douglas (eds.) *Amorphous Materials*, J. Wiley, New York, p. 159.
37. Wunderlich, B. (1960) Study of the change in specific heats of monomeric and polymeric glasses during glass transition, *J. Phys. Chem.* **68**, 1052.
38. Justi, E. and von Laue, M. (1934) Third order phase equilibria, *Phys. Z.* **35**, 945.
39. Binder, K. (1995) Phase transition at surfaces, in D. R. Pettifor (ed.), *Cohesion and Structure of Surfaces*, Elsevier, Amsterdam, p. 121.
40. Nemilov, S. V. (1988) The thermodynamics of the Prigorgine-Defay relation and the structural difference glass/liquid, *The Vitreous State* (in Russ.), Nauka Publ., Leningrad, p. 15.
41. Boyer, R. F. and Spencer, R. S. (1944), Glass transition in high polymers, *J. Appl. Phys.* **15**, 398.
42. Uberreiter, K. and Bruns, W. (1966) On the nature of vitrification (in German) *Ber. Bunsenges. Phys. Chem.* **68**, **70**, 541, 17.
43. Filipovich, V. N. (1992) Discussion remarks, *Phys. Chem. Glasses (in Russ.)* **18**, 125.
44. Mazurin O. V. (1992) On the general conclusions for a criticism of the relaxational theory of vitrification, *Phys. Chem Glasses (in Russ.).* **18**, 128.
45. Winter, A. (1943) Transformation region of glasses, *J. Am. Ceram. Soc.* **26**, 189.
46. Gutzow, I. (1979) Structure and thermodynamics of glass forming systems (in German) *Wiss. Zs. Fr. Schiller Univ.* **28**, 243.
47. Petroff, B., Milchev, A. and Gutzow, I. (1966) Thermodynamic functions of both simple and polymeric melts: MFA and MCS, *J. Macrom. Sci. (Phys.)* **B35**, 763.
48. Gutzow, I. (1981) On the electrochemical behaviour of undercooled melts and glasses, *J. Non Cryst. Solids* **85**, 203.
49. , I. Avramov, A. and Kastner, K. (1990) Glass formation and crystallization, *J Non Cryst. Solids* **123**, 97.
50. Gutzow, I., Kashchiev, D. and Avramov, I. (1985) Nucleation and crystallization in glass forming melts: old problems and new questions, *J. Non Cryst. Solids* **73**, 477.
51. Angell, C. A. (1985) Spectroscopy simulation and scattering and the medium range order problem in glass, *J. Non Cryst. Solids* **73**, 1.
52. Macedo, B. Litovitz, T. A. (1965) On the relative roles of free volume and activation energy in the viscosity of liquids, *J. Chem. Phys.* **42**, 245.

CHALCOGENIDE GLASSES

MARÍA TERESA MORA
Grup de Física de Materials I, Departament de Física,
Universitat Autònoma de Barcelona, 08193-Bellaterra, Spain

1. Introduction

The modern technology allows to obtain different kinds of glassy materials. Most of them show relaxation effects, glass transition and crystallization under heat treatment, with very similar phenomenological peculiarities. Chalcogenide glasses contain elements from Group VIb of the Periodic Table, namely S, Se, Te. Oxides are excluded in this category mainly form historic as well as scientific reasons. Traditionally, the oldest-known glass-forming systems are oxide materials, while chalcogenide ones were discovered rather recently. Oxides have typically a more significant ionic contribution to the chemical bonding than chalcogenides, which are essentially covalent materials.

A number of reviews have been devoted to specific aspects of chalcogenide glasses [1-5] The general properties of these materials are discussed in the book of Borisova [6]. This paper is based on these previous reviews, and will treat the following topics:

Glass forming ability: structural considerations and treatment of short range order in continuous networks.

Glass transition: experimental data on heat capacity, enthalpy and residual entropy and relaxation.

Ideal glass in thermodynamic terms: Gibbs free energy of formation.

Thermodynamic criteria of glass formation based on the transformation temperatures and the energetics of the process.

Non-equilibrium crystallization kinetics: empirical evaluation and modeling of predictive transformation diagrams..

2. Glass forming ability

Besides crystalline solids, glass formation is not limited by stoichiometry rules but only by the ability to cool the molten alloy enough rapidly to inhibit crystallization. Another requirement is to force homogenization of the melt previous to quench.

The glass forming ability, as well as the physical and chemical properties of the glasses, are determined by the character of the chemical bond between the atoms that make up the glass. Glass forming ability is enhanced in compound forming systems.

M. F. Thorpe and M. I. Mitkova (eds.), Amorphous Insulators and Semiconductors, 45–69.
© 1997 *Kluwer Academic Publishers. Printed in the Netherlands.*

46

That is, when there is a strong negative interaction between unlike constituent atoms, in the liquid state, as indicated by a negative heat of mixing and a low-lying eutectic. This behaviour suggests that in the liquid some of the atoms (or ions) are free while other atoms associate into small groups or clusters (bound chemically) with a composition about that of the minimum of the heat of mixing. The number of atoms associated in a cluster is not large, that is, the cluster may be considered as a molecule or complex of composition A_pB_q. In differs, however, from a real molecule because it is likely to have a flexible structure in which the interatomic distance between unlike atoms is kept constant, whereas the bonding angle is variable over a wide range. Strong interactions are evidenced by a marked peak of the liquidus temperature near the melting point of the compounds participating in the relatively low-eutectic reactions.

Quenching of the melt, splat cooling or any other rapid quenching method is supposed to freeze the liquid structure giving rise, in the glass, to a characteristic short-range order described by molecular structural units, formed by an atom and its nearest neighbours. In chalcogenide glasses structural units are covalently bonded.

Most important glass forming binary chalcogenide (and related compounds) from which ternary or higher order component alloys glasses are produced are:
   (a) Pnictogen-Chalcogen (V-VI): As-S, As-Se, P-Se
   (b) Tetragen-Chalcogen (IV-VI): Ge-S, Ge-Se, Ge-Te, Si-Se
   (c) Halogen-Chalcogen: Te-Cl

2.1. STRUCTURAL CONSIDERATIONS

The bulk glasses are believed to be formed into a fully cross-linked random network connecting the covalently bonded structural units. Pure chalcogen glasses, like Se, and chalcogen alloys as SeTe mixtures up to 70 at. % Te show in the liquid state a random chain structure which is conserved in the glassy state. Radial distribution and spectroscopic studies on As_xX_{1-x} and Ge_xX_{1-x} ($X \equiv S, Se$) alloy glasses have confirmed that in these materials the Ge, As and X atoms are four-, three- and two-fold coordinated respectively in chalcogen-rich compositions. There are AsX_3 pyramidal units for the As-X glasses and GeX_4 tetrahedral units for the Ge-X glasses. Glass forming ability has been extensively investigated from the pioneering studies of Kolomiets [7]. Glass forming regions for air quenched glasses in ternary systems based in pnictogen-chalcogen glasses extend in a wide composition range when adding Ge or Si whereas tetragen-chalcogen based glasses may easily incorporate another IV or VI element. As there is no absolute criterion for glass formation, glass forming ability is often related to the critical cooling rate needed to avoid crystallization on quenching from the melt.

2.1.1. Structural criteria
The chemical ordering in covalent systems follows the 8-N rule [8]. That rule allows accommodation of several elements in a random network with almost no dangling bond. However, for elements other than Groups IV to VII of the periodic table charge transfer effects have to be included in the evaluation of N (= number of valence

electrons) [9]. Therefore, short range order (SRO) is often described in terms of coordination polyhedra, as shown in Fig. 1. The quantities Z_j, r_{ij} and ϑ_{ij}, representing the coordination number of atom j, the i and j nearest-neighbour bond length and bond angle respectively, are often used to describe the coordination polyhedra geometry.

Topological disorder is obtained from small distortions in both the nearest-neighbour bond length and bond angle in the following coordination polyhedra of the continuous network.

Fig. 2 shows the comparison between the chain structure in trigonal and amorphous Se. In alloy glasses the chemical short range order is related to the nature of the nearest-neighbour elements. The schematic representation of some elemental coordination polyhedra and their structural unit formula is shown in Fig. 3. Fig. 4 shows a schematic planar view of topologically disordered but chemically ordered As_2S_3 glass.

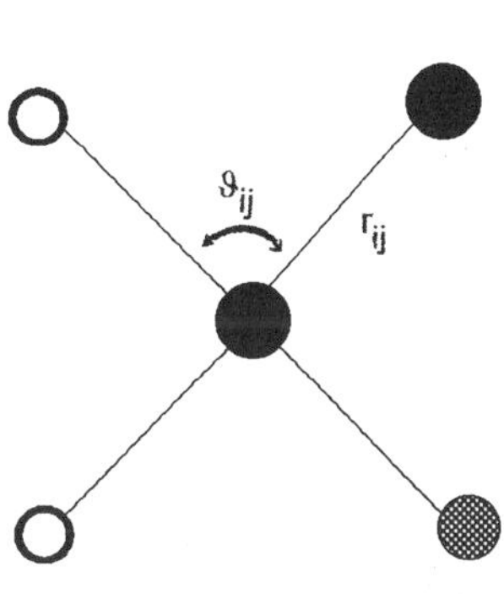

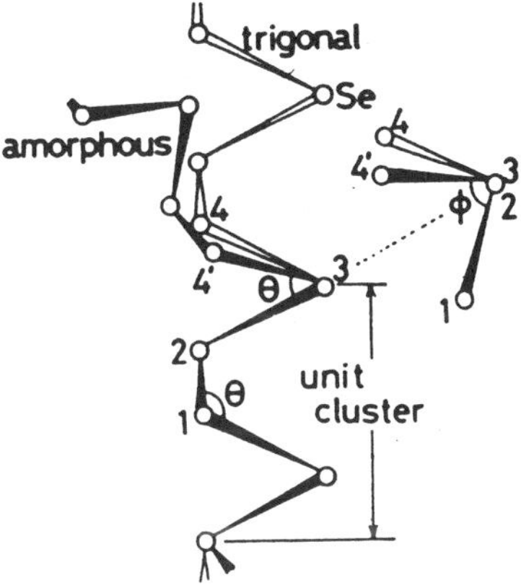

Figure 1. Coordination polyhedra

Figure 2. Helicoidal chains in amorphous Se

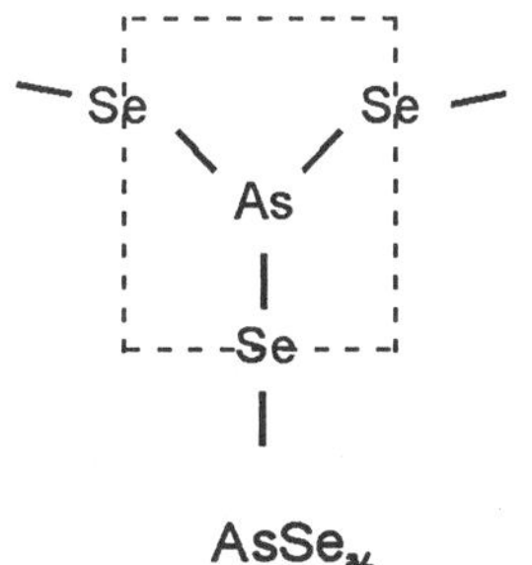

$AsSe_{3/2}$

$GeSe_{4/2}$

Figure 3. $AsSe_{3/2}$ and $GeSe_{4/2}$ structural units

48

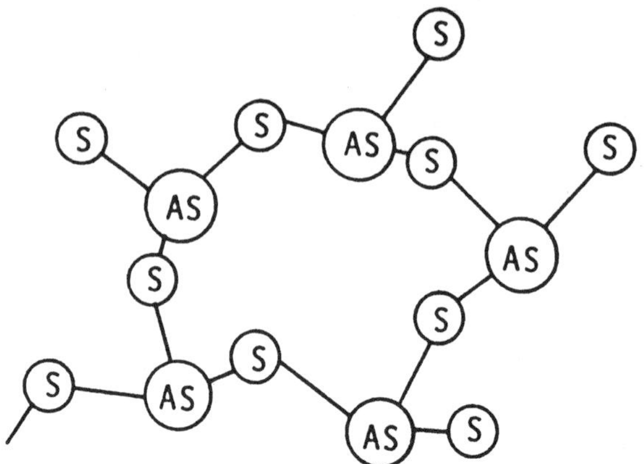

Figure 4. As$_2$Se$_3$ chemically ordered glass structure

Other structural quantities used to describe SRO in a multicomponent alloy glass are the average coordination number

$$m = \sum_c x_c Z_c \tag{1}$$

where x_c is the atomic fraction of element c, and the average number of covalent bonds is

$$N_c = \frac{m}{2} \tag{2}$$

Topological criterion. The general idea of the topological approach is that one can optimize the glass forming ability by matching the number of constraints per atom, N_c, with the number of degrees of freedom per atom, which is simply N_d, the dimensionality of the space in which the solid is embedded [10]. In order to evaluate N_c a particular network glass is considered to be subject to interatomic forces acting on an atom in terms of the *valence-force-field* model in which the potential energy U_s is expressed as a sum of contributions from bond-stretching and bond-bending forces as

$$U_S = \tfrac{1}{2}\,\alpha_S\,\Delta r^2 + \tfrac{1}{2}\,\beta\,\Delta \vartheta^2 \tag{3}$$

where α_s and β are the bond stretching and bending force constants, respectively. There are one constrain per atom associated with the first term of Eq. (3) and one other constraint associated with the second term for a two-fold coordinated atom. Adding each additional bond gives two more constraints because the angles between two existing bonds must be specified [11]. Further the fact that N_d +1 bonded angles (in N_d-dimensional space) are not linearly independent must be taken into account [12]. This yields [4]

$$N_c = m/2 + m\,(m-1)/2 = m^2/2 \qquad \text{if } m \leq N_d - 1$$
$$N_c = m/2 + (N_d - 1)(2m - N_d)/2 \qquad \text{if } m \geq N_d - 1 \tag{4}$$

The topological criterion for covalent materials is

$$N_c = N_d \tag{5}$$

If $N_c > N_d$ the glass is overstrained and it will be difficult to form by conventional methods: On the contrary, if $N_c < N_d$ it is an understrained glass, *i.e.*, easy to form.

The optimum average coordination number becomes

$$
\begin{aligned}
m_c &= \sqrt{2}\,N_d & \textit{when } m &\leq N_d - 1 \\
m_c &= N_d\,(N_d + 1)/(2N_d - 1) & \textit{when } m &\geq N_d - 1
\end{aligned}
\tag{6}
$$

Values of m_c in 1 to 3 dimensional networks are shown in Table 1 whereas the compositions at which Eq. (5) is satisfied in specific binary alloys is given in Table 2.

TABLE 1. Values of m_c for an ideal glass

N_d	1	2	3
m_c	2	2	2.4

TABLE 2. Ideal glass compositions in some binary systems

$N_d = 3$	$(Ge$ or $Si)_x(S$ or $Se)_{1-x}$	$x_c = 0.16$
$N_d = 3$	$Ge_{1-x-y}As_xSe_y$	$x_c = 1.43 - 1.71 y_c$
$N_d = 2$	Polyacetylene $(CH)_n$	$x_c = 0.5$

The values predicted in Table 2 agree with experimental observation of glass formation in these systems and may be easily applied to ternary or higher order systems. However, as already stated [11] the topological criterion must not be used indiscriminately. For instance in Si_xO_{1-x} the Si-O-Si bond angle is known to take a range of values, therefore it is reasonable not to count the angular constraint at the oxygen atoms. this leads to the boundary of glass formation at x = 0.33 which is more reasonable as SiO_2 is a good bulk glass.

Empirical theory of glass-formation. According to Dembovskii [13] the extend of the glass forming region may be evaluated by

$$G = (A + E)(EVE - K)/2 \tag{7}$$

where A is the number of atoms of each species in the alloy, E the number of different kinds of structural units coexisting in the network, EVE the external valence electrons and K the coordination of the melt. The glass forming ability depends on the value of G. If $G > 4.0 \pm 1.0$ then the alloy glass is easy to form; if $G < 3$ no glass is formed even by sudden quenching. The application of this criterion is compared to experimental glass forming ranges in Table 3. Improvement of the criterion has been developed to account for dependence on cooling rate and for analysis of the changes in coordination during melting [14]. In particular, it has been suggested that changes in the mean coordination of the melt at the boundary of the glass-formation region occur.

50

TABLE 3. Extents of regions of glass formation in chalcogenide systems

System	Glass forming region, at.%	
	Experimental	Theoretical
Se-Te	0-50 Te	0-33 Te
Te-In	10-20 In	4-24 In
Se-Ge	0-45 Ge	0-43 Ge
Te-Ge	10-25 Ge	3-35 Ge
S-As	0-70 As	0-67 As
Se-As	0-70 As	0-68 As
Te-As	20-65 Te	20-57 Te
$GeSe_2$-Sb	0-33 Sb	0-42 Sb
$GeSe_2$-Sb_2Se_3	0-90 Sb_2Se_3	0-95 Sb_2Se_3
As_2S_3-Sb	0-28 Sb	0-30 Sb
Te-As_2Se_3	0-53 Te	0-55 Te

Glass-formation in ternary systems. When considering the insertion of a third element in a binary glassy network, Suchet [15] considered the presence of additional structural units formed by using mostly the p-orbitals. For instance, in the As-Se-Ge, apart from the structural units $SeSe_{2/2}$, $GeSe_{4/2}$ and $AsSe_{3/2}$, the consideration of the additional $As_2Se_{4/2}$ one (see Fig. 5) gives a rather good account of the limits of glass formation in that system.

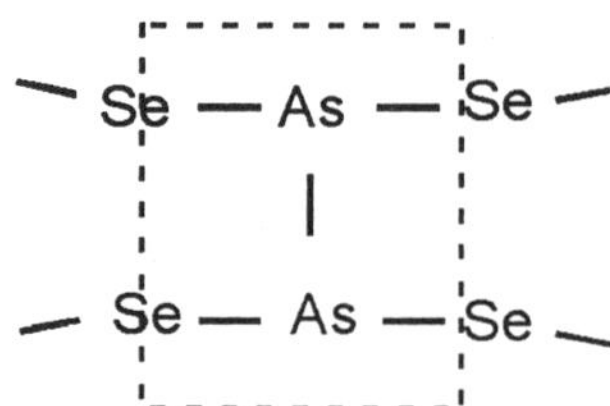

Figure 5. "Foreign" structural unit in the As-Se-Ge system

2.2. SHORT RANGE ORDER IN CONTINUOUS NETWORKS

As a consequence of the previous structural considerations, the description of local chemical order may be presented in terms of the number of bonds i-j per atom, N_{i-j} by taking into account the atomic coordination, the distribution of bonds between atoms and the "molecular" structure of the network forming groups. There are three models based on the extent to which chemical SRO develops in the network alloy glass. These are the random covalent network (RCN), the chemically-ordered network (CON) and the strongly ordered network (SON) models.

In the RCN model the relative differences between i-j bond energies are neglected. Therefore, the statistical distribution of bonds completely determined by the local atomic coordinations Z_i and Z_j. The assumptions of the model are equivalent to consider an ideal solution in thermodynamic terms. In a binary $A_{x_1}B_{x_2}$ the resulting i-j bonds per atom are:

$$N_{A\text{-}A} = \frac{(Z_A x_1)^2}{2(Z_A x_1 + Z_B x_2)}$$

$$N_{B\text{-}B} = \frac{(Z_B x_2)^2}{2(Z_A x_1 + Z_B x_2)} \tag{8}$$

$$N_{A\text{-}B} = \frac{2\, Z_A Z_B x_1 x_2}{2(Z_A x_1 + Z_B x_2)}$$

The CON model considers that heteropolar bonds are favored at all compositions and that chemically ordered phases have only A-B bonds. When there is only one binary structural unit, the axis composition is divided in two parts by its stoichiometry ($x_2 = Z_A/(Z_A+Z_B)$). The respective numbers of i-j bonds in each of them are

1) A-rich alloys: $0 < x_2 < Z_A/(Z_A+Z_B)$

$$N_{A\text{-}A} = N_e - N_{A\text{-}B} = [(Z_A+Z_B)x_1 - Z_B]/2$$
$$N_{B\text{-}B} = 0 \tag{9}$$
$$N_{A\text{-}B} = Z_B x_2$$

2) B-rich alloys: $Z_A/(Z_A+Z_B) < x_2 < 1$

$$N_{A\text{-}A} = 0$$
$$N_{B\text{-}B} = N_e - N_{A\text{-}B} = [Z_B - (Z_A+Z_B)x_1]/2 \tag{10}$$
$$N_{A\text{-}B} = Z_A x_1$$

where $N_e = (Z_A x_1 + Z_B x_2)/2$ is the overall number of bonds per atom.

When there are two structural units the composition range is divided in three parts. For instance in $As_x Se_{1-x}$ glasses, if the structural units are (i) $AsSe_{3/2}$ ($x = 0.4$) and (ii) $As_6 Se_{8/2}$ ($x \approx 0.6$), in the CON model for the composition range $0.4 \le x \le 0.6$ there are only As-Se bonds -mixture of structural units (i) and (ii). Spectroscopic techniques may differentiate between heteropolar bonds of each of both local order structures. In Fig. 6 are represented the expected fractions of each structural unit in this composition range, as well as the Raman scattering data [16]. The low value obtained for unit (ii) comes from the large number of atoms involved in that unit (expanded network former). The dashed line in Fig. 6 was constructed assuming some chemical disorder occurs, namely some statistical probability of As atoms in the Se *sites* of the (i) unit.

The SON model assumes a statistical distribution between unlike "molecular" units $A_p B_q$ ($p/Z_A = q/Z_B$) and "free" atoms A, B whose relative amounts follow a mass action law according to the dissociation reaction [17]

52

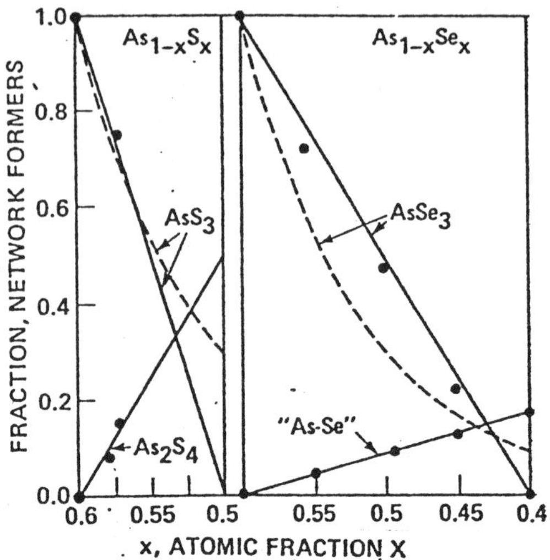

Figure 6. Molecular structures in (S, Se)-X systems as calculated (dashed and solid lines) and Raman scattering data (points).

$$A_pB_q \Leftrightarrow p\,A + q\,B \tag{11}$$

When there is only one binary structural unit, the respective numbers of i-j bonds in each of them are

$$\begin{aligned}
N_{A\text{-}A} &= n_{A\text{-}A}\, x_A \\
N_{B\text{-}B} &= n_{B\text{-}B}\, x_B \\
N_{A\text{-}B} &= n_{A\text{-}B}\, x_{A_pB_q}
\end{aligned} \tag{12}$$

where $n_{i\text{-}j}$ is the number of i-j bonds in the molecular unit α and their fractions x_α are given by mass balance. The assumptions of the model are equivalent to consider an strongly associated solution in thermodynamic terms [18]. Fig. 7 shows some example calculations in the S-As system (solid lines). Dashed lines reproduce the results of the CON model whereas thin lines correspond a total demixing between A and B atoms.

3. Glass transition

From several thermodynamic measurements, it is possible to estimate the thermodynamic quantities, such as the Gibbs free energy, enthalpy and entropy of the liquid, stable and metastable crystalline phases over a wide temperature interval. However, as firstly stated by Kauzmann [19], there is a low temperature limit at which the properties of the supercooled liquid can be extrapolated which is called the isentropic temperature, T_S. That is, a supercooled liquid whose entropy falls below that of the crystalline phase must undergo massive freezing to a glass. Applying an *inverse* Kauzmann argument to the problem of melting, an entropy catastrophe has been predicted when the entropy of the superheated crystal exceeds that of the liquid phase.

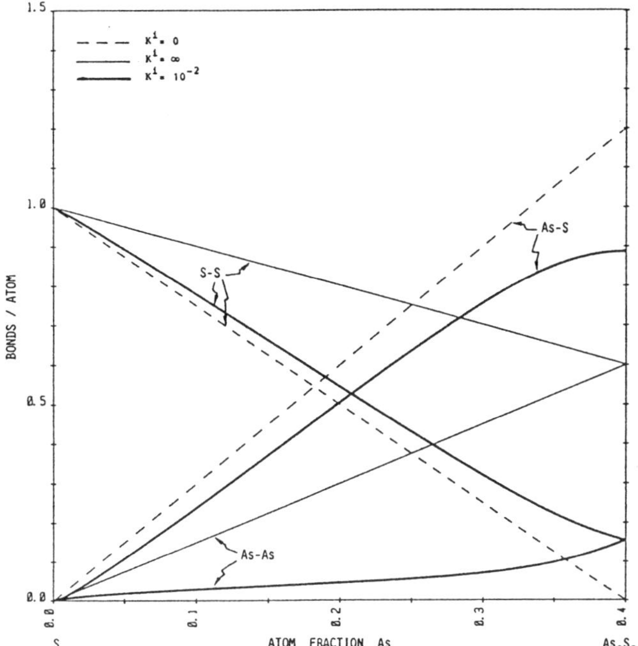

Figure 7. Bonds per atom calculated in the As-S system with SON (solid lines) CON (dashed lines) and demixing of the elements (thin lines).

Real glasses exhibit glass transition at the glass transition temperature, T_g, normally above T_S and, consequently, they retain a residual entropy relative to the crystal.

3.1. EXPERIMENTAL DATA

The nature of the glass transition phenomenon is a question that remains open. However it may be easily observed calorimetrically in most glasses. This is primarily due to the fact that on heating (cooling) the heat capacity, C_p, of a glass increases (decreases) suddenly by at least about half its original value in a very small temperature interval as the glass changes to a liquid (the liquid changes to a glass). Usually T_g is defined at the inflection point in the C_p versus temperature curve. It is also seen as rounded discontinuities of other quantities like compressibility and thermal expansion. The physical origin of these discontinuities is the "freezing" of the liquid diffusive atomic motion. In particular, the self diffusion coefficient of a molecule approaches an essential singularity with

$$D \sim exp \left\{ -\frac{A}{T - T_o} \right\} \tag{13}$$

variously known as the Vogel-Fulcher or Dolittle law, until right at the transition a power law finally prevails. (see Fig. 8).

54

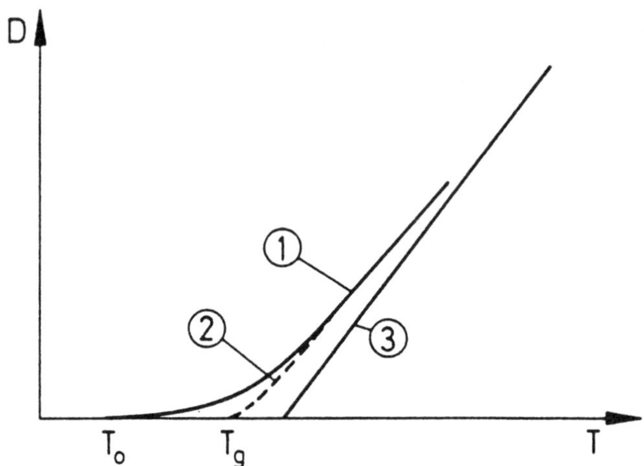

Figure 8. The behaviour of the diffusion coefficient near the glass transition. T_o is the Vogel-Fulcher temperature. 1: Vogel-Fulcher law; 2: power law , and 3: mean field theory. From Ref. [22]

3.1.1. Measurements of heat capacity, enthalpy and Gibbs free energy difference between liquid and crystalline phases.

In chalcogenide systems typically the heat capacity difference between the undercooled liquid and the glass, ΔC_p , ranges between R and 3R while the heat capacity of the glass is very close to the corresponding one for the stable crystal at T_g. Various models have been proposed with the underlying idea that glass transition is a thermodynamic phase transition. Among them is the theory developed by Gibbs and coworkers [20,21], whereby the configurational entropy of linear molecular chains was related to the viscosity. A particularly simple model of a liquid of long rod molecules was presented [22] that permits a simple picture of the glass transition and can be translated into a theoretical model for molecular diffusivity.

Also, Rao [23] developed a cluster model of glasses in which glass transition is recognized as corresponding to the melting of clusters. In another approach Goldstein [24,25] has considered three main contributions to the to the excess "configurational" specific heat of liquids above glass transition, namely, molecular rearrangements and changes in both lattice frequencies and in anharmonicity due to these rearrangements. Another approach to glass transition is the free volume theory [26-28] which emphasizes the concomitant decrease in volume and fluidity of glass forming melts in the undercooled region. The main origin of the glass transition in this theory is the contraction of the free volume down to a critical value, when using percolation theory.

Knowing the heat capacity of the supercooled liquid, at least in a certain temperature interval, its enthalpy, entropy and Gibbs free energy may be evaluated. But, there are kinetic and thermodynamic contributions to the heat capacity values in the glass transformation range, because above T_g the supercooled liquid is allowed to explore all the structures or configurations at equilibrium while below T_g the glass is trapped kinetically in one of such configurations and can only relax towards that corresponding to a local minimum of the potential energy of the system. Then the existence of potential energy barriers which are large compared to the thermal energy are intrinsic to the occurrence of the glassy state. As a consequence, the heat capacity, entropy or any other thermodynamic quantity is dependent on the thermal treatment the system received while passing through the transformation region. This is the reason why the *ideal glass* concept has been introduced.

3.1.2. *Applicability of thermodynamics to glasses*
A liquid which has been undercooled to a temperature between the melting point and glass transition is in a metastable state of equilibrium. In most experiments this state is found to be independent of previous history, so that it is well defined in terms of the usual thermodynamic state variables. The problem faced with glass transition is that of irreversibility. Below glass transition the system can still be considered in a metastable state. But in the intermediate region there is an entropy production. The change in entropy is split into two terms

$$dS = d_m S + d_i S \tag{14}$$

where $d_m S = (C_{p,app}/T)\, dT$ is the measured entropy change due to exchange of heat with the surroundings, and $d_i S\ (\geq 0)$ is the entropy created not provided by external exchange.

Ideal glass. To rid the apparent heat capacity of its irreversible contribution is assumed that the liquid alloy exists in a metastable state down to the isentropic temperature T_S at which it becomes an *ideal* glass with a heat capacity equal to that of the mixture of the stable crystalline phases.

$$C_{p,app} = C_{p,\ell}(T) \text{ if } T > T_S$$
$$C_{p,app} = C_{p,g}(T) \text{ if } T < T_S$$

The isentropic temperature T_S is defined by:

$$S^{\ell}(T_S) = S^{c}(T_S) \tag{15}$$

If $\Delta C_p \equiv C_{p,\ell} - C_{p,g}$ is constant in the temperature interval $T_S < T < T_m$, the value of T_S is:

$$T_S = T_m\, exp(-\Delta S_m/\Delta C_p) \tag{16}$$

where ΔS_m is the melting entropy.

3.2. KINETIC TREATMENT OF A GLASS

The thermodynamic treatment of glassy materials is complicated by the fact that the glassy state is not always a metastable but sometimes an unstable state. According to Angell [29], glasses have supercooled liquid viscosity behaviour in between two extreme cases labeled strong and fragile. These labels refer to the stability of the intermediate range order in the liquid as temperature increases from glass transition towards melting. Strong glasses have an Arrhenius dependence of the supercooled liquid with temperature while for fragile glasses a Vogel-Fulcher expression is needed.

Two kinds of structural instability of glassy alloys will be considered. The first one is the instability against relaxation within the amorphous state itself, with a typical relaxation time τ_{in}. The second one is the instability against relaxation out of the amorphous state into the crystalline state, with a typical relaxation time τ_{out}. In most glasses, τ_{out}, which is no more than the characteristic nucleation time, is much longer than τ_{in}. However, there are important classes of glass, for instance most metallic glasses, in which $\tau_{out} << \tau_{in}$, a fact that is responsible for the inability to observe the glass transition in this class of glasses.

3.2.1. *Relaxation* The relaxation process within the amorphous phase has several features that currently lack a complete explanation. The reason is threefold: i) in most cases the relaxation process is non-Arrhenius in its temperature dependence, ii) it is non-exponential in its time dependence and iii) it is non-linear in its structural state dependence [30].

One manifestation of the relaxation processes comes from experiments in which the temperature is changed continuously at a constant rate such as those related with the measure of the apparent heat capacity by differential scanning calorimetry. On heating and cooling the glass in the glass transition region, a marked hysteresis is sometimes observed, as shown in Fig. 9. Sometimes a large peak in $C_{p,app}$ is observed on heating that is absent on cooling. DSC is a very appropriate technique to study the relaxation process because it allows one to perform successively both the heat treatment and the experimental determination of the thermodynamic quantities such as the glass transition temperature and the evolution of $C_{p,app}$. Both quantities reflect in general the relaxation effects caused by the previous annealing [31-34]. The relaxation enthalpy induced in the glass by a previous annealing at a temperature close to glass transition may be measured by the area under the curve of the excess heat capacity versus temperature on further heating up to the undercooled liquid state. This relaxation enthalpy is equal to the enthalpy that was released by the sample during the annealing and has to be re-absorbed by the sample to reach the undercooled metastable liquid state. Fig. 10 shows schematically the differences in enthalpy of two glasses, g_1 and g_2. If on cooling at a certain rate the sample through the glass transition region it reaches temperature T_a with the state g_1, during annealing at that temperature for a given time it may evolve towards state g_2. Now, on heating up to temperature T_b it has to recover. Therefore ,the difference between ΔH_1 and ΔH_2 is the relaxation enthalpy.

Normally, the relaxation enthalpy is evaluated as a function of the annealing time/temperature and the concept of a spectrum of activation energies is assumed to describe the non-exponential behaviour of the relaxation enthalpy.

3.2.2. *Crystallization* Stability of the glass against crystallization is highly related to stability of the supercooled liquid against nucleation and crystal growth. Typically, in glass forming systems the temperature range in which nucleation is important is overlapped with the temperature range for which crystal growth is not negligible. Therefore, glass stability is linked to glass forming ability.

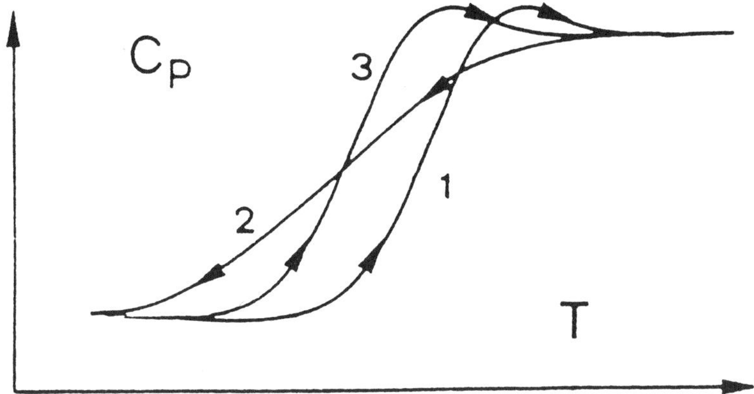

Figure 9. Hysteresis form of the apparent heat capacity on cycling the system through the glass transition region. Values of $C_{p,app}$ obtained on: heating an as-quenched glass (curve 1); cooling (curve 2); and, re-heating (curve 3).

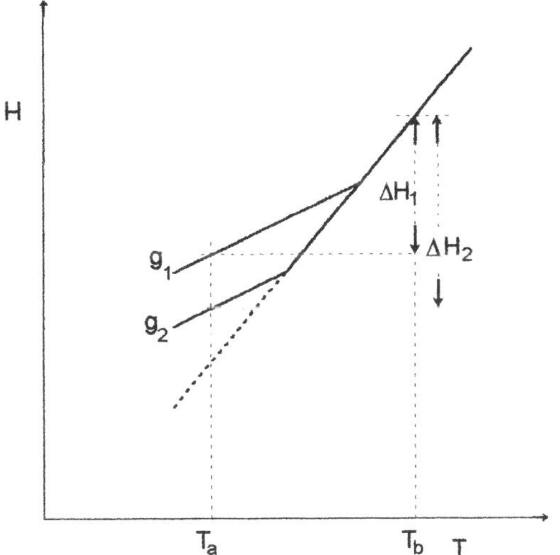

Figure 10. Enthalpy vs. temperature of quenched and relaxed glasses.

4. Thermodynamic criteria of glass formation

The ability of a molten alloy to form metastable non-crystalline structures depends on both structural and thermodynamic criteria. As there is no absolute criterion for glass formation, empirical parameters are extensively used for quantitative characterization of the glass forming ability of molten alloys. In particular, as already mentioned, both glass transition temperature and onset of crystallization are often easy to measure for chalcogenide glasses. Further, the liquidus temperature remains within values directly accessible for current techniques to be obtained. For those reasons, we will consider thermodynamic criteria of glass formation for these systems in detail. That term will include criteria or empirical parameters related to both the transformation temperatures and the energetics of the transformation process.

4.1. CRITERIA BASED ON TRANSFORMATION TEMPERATURES

Some parameters relate glass formation to the phase diagram. One of them is the reduced glass temperature

$$T_{rg} = T_g/T_m \tag{17}$$

where T_m is the liquidus temperature. T_{gl} takes a value of about 2/3 for a large number of glassy substances [35]. Another parameter is the Hruby parameter , K_{gl} , defined as [36]

$$K_{gl} = \frac{T_r - T_g}{T_m - T_r} \tag{18}$$

where T_r is the temperature of the onset of the crystallization peak. Fig. 11 shown an example of application of this parameter to glass forming ability of Ge-S and Ge-Si-S glasses.

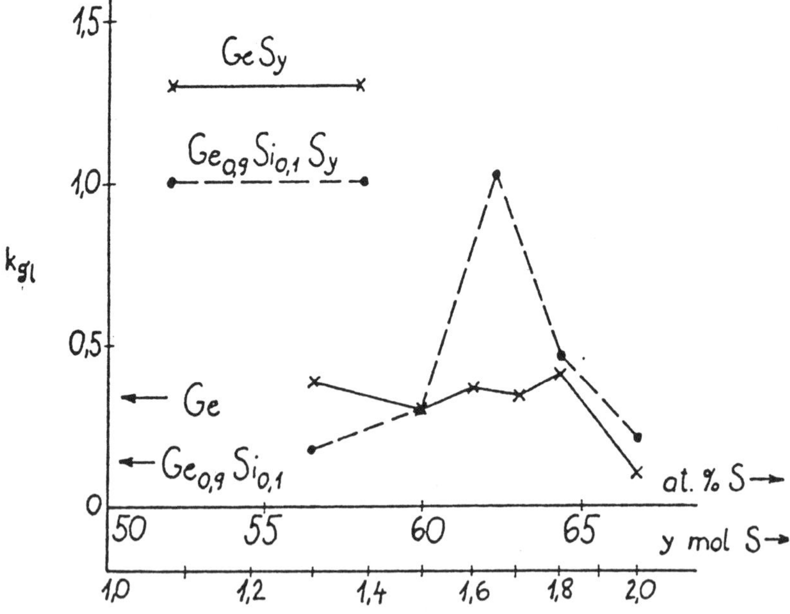

Figure 11. K_{gl} vs. composition for two chalcogenide glasses.

4.2. CRITERIA BASED ON THE ENERGETICS OF THE TRANSFORMATION

One of the important thermodynamic quantities appearing in the thermodynamic prediction of glass formation is the Gibbs free energy difference between the supercooled liquid and the crystalline phase or phases at that temperature, also called the driving force, ΔG. At the eutectic temperature the minimum value of ΔG occurs at the eutectic composition. This is the basic thermodynamic argument backing the experimental observation that eutectic compositions are in general good glass-formers when rapid solidification techniques are used. A complementary kinetic argument is that for eutectic compositions the equilibrium solidification temperature interval reduces to a single temperature. As an example, in Fig. 12a is presented the phase diagram and the values of the glass transition temperature T_g of glasses in the system Sb_2Se_3-$GeSe_2$. The values calculated for ΔG at various temperatures are shown in Fig. 12b as a function of composition as well as the minimum quenching rate (MQR) needed to form a glass in this system [37]. There is a good correlation between the minimum of ΔG (indicated by arrows in Fig. 12b and the glass forming ability obtained experimentally by the maximum of MQR.

A better description of the glass forming ability, in thermodynamic terms, comes from the consideration of the Gibbs free energy of formation, G_f, of a glass from the stable elemental components. In an ideal monocomponent glass, it is equal to the value of ΔG at the isentropic temperature, T_S

$$G_f(T_S) = G^\ell - G^c = \Delta G(T_S) \tag{19}$$

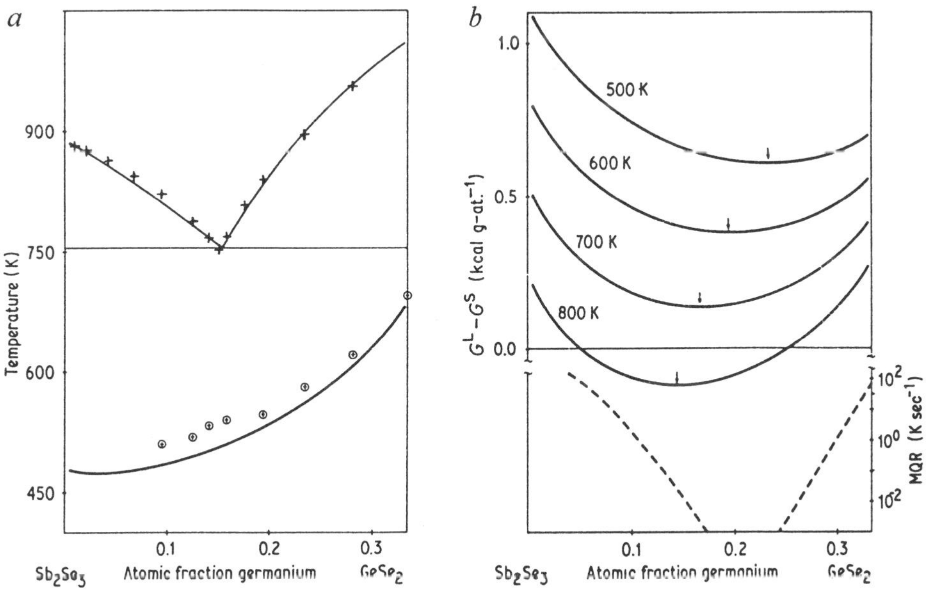

Figure 12. Sb_2Se_3-$GeSe_2$ system. a) Phase diagram (+) and glass transition temperatures (⊕); b) Values of ΔG (continuous curves) and of minimal quenching rate (dashed curve)

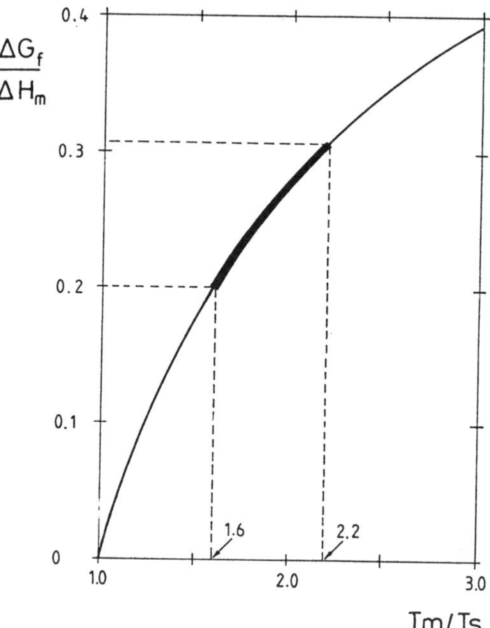

Figure 13. ΔG_f vs. T_m/T_S for a monocomponent system

The Gibbs free energy of formation of a monocomponent ideal glass is presented, as a function of the ratio T_m/T_S in Fig. 13 [38].

The glass forming regions predicted in multicomponent systems from thermodynamic calculations of G_f are normally based on the previous modeling of the liquid solution phase. In the evaluation of glass forming ability, associated solution models are being currently considered.

5. Non-equilibrium crystallization

We will consider that crystallization (on heating a glass) occurs above glass transition. Then, crystallization, as well as solidification, deals with the transformation undercooled liquid $\Rightarrow$ crystal. The empirical description makes emphasis in the fact that crystallization is driven by thermally activated processes while modeling considers the mechanisms which limit the process. Both approaches establish equations between the measured kinetic quantities. In general terms these are the crystallized fraction, x, or its time derivative, dx/dt, (the reaction rate) either as a function of time, under isothermal regime, or as a function of temperature (under constant heating rate, β) conditions.

5.1. EMPIRICAL DESCRIPTION

As crystallization proceeds it is assumed that the rate of reaction may be written as

$$dx/dt = K(T)\,f(x) \tag{20}$$

In the temperature range in which crystallization can be investigated, the rate constant, $K(T)$, is generally assumed to follow the Arrhenius equation

$$K(T) = K_o \, exp(-E/RT) \qquad (21)$$

where K_o is the pre-exponential factor and E the apparent activation energy. There are various differential and integral methods to calculate these quantities from continuous heating data. In particular Kissinger method [39] assumes that at a given heating rate β

$$\frac{dln(\beta/T^2_p)}{d(1/T_p)} = -E/R \qquad (22)$$

where T_p is the peak maximum temperature, $(d^2x / dt^2)_{T=T_p} = 0$.

Ozawa method [40] assumes that

$$\frac{dln\beta}{d(1/T_p)} = -E/R \qquad (23)$$

Both methods are currently used to get E and the values obtained are very similar.

If the assumption of a constant apparent activation energy is satisfied, a linear behaviour of $ln(dx/dt)$ against $1/T$, for a fixed value of x, is expected irrespective of the experimental procedure that was used (isothermal or continuous heating regime). The slope of the straight line gives also E.

The product $K_o \cdot f(x)$ can be evaluated from the experimental dx/dt data and the apparent activation energy by taking [41]

$$K_o \cdot f(x) = dx/dt \, exp(E/RT) \qquad (24)$$

5.1.1. *Empirical construction of the transformation diagrams*

An important tool in the study of non-equilibrium crystallization processes is the use of transformation diagrams. The time-temperature or T-T-T curves are constructed from the explicit dependence of the crystallized fraction on annealing temperature and annealing time. That diagram uses the coordinates of temperature versus time for a fixed value of the crystallized fraction. By the same procedure the Temperature-Heating (Cooling) rate or T-HR-T (T-CR-T) curves are constructed as the temperature versus heating (cooling) rate for a fixed value of the crystallized fraction.

The empirical construction of the T-T-T and T-HR-T transformation diagrams is grounded on the recognition that Eq. (20) when applied to isothermal and continuous heating regimes, gives, in differential form, the functions $x = x(T,t)$ and $x = x(T,\beta)$, respectively [42].

In general terms, under isothermal conditions (at temperature T), the integrated form of Eq. (20) is

$$g(x)/K_o = t \, exp(-E/RT) \qquad (25)$$

with

62

$$g(x) = \int\limits_{0}^{x} dx / f(x) \tag{26}$$

Equation (25) can be written in the form $x = x(T,t)$. This equation, when represented as the temperature versus time needed to crystallize a fixed fraction x of the material gives the T-T-T curve.

By the same procedure, integration of Eq. (20) under a constant scan rate β gives

$$g(x) / K_o = \beta^{-1} \int\limits_{0}^{T} exp(-E / RT) \, dT \tag{27}$$

That is, the extent of crystallization is only a function of temperature and scan rate, i.e., $x=x(T,\beta)$. Assuming a fixed value of x, the plot of the temperature against the heating rate produces a T-HR-T diagram.

5.1.2. *Accuracy in the determination of the reaction rate in differential scanning calorimetry (DSC) measurements*

As very sensitive DSC equipments become available, they are widely used as a tool to study kinetics of crystallization. Nevertheless, there are limits to the accuracy on the determination of dx/dt, the most important coming from the assumption that the rate of heat generated during the crystallization process, $\delta\dot{Q}$, is proportional to the transformation rate. This assumption, valid for eutectic- and polymorphous-type transformations, is taken even for a primary crystallization process due to the difficulty of finding precisely the dependence of the rate of enthalpy generated on the crystallization rate when both the crystalline nuclei and the amorphous matrix composition may change during the process.

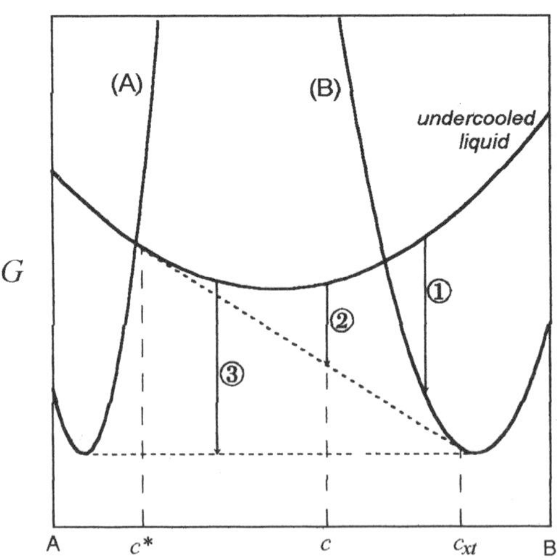

Figure 14. Crystallization processes possible in a binary undercooled liquid:
① polymorphous; ② primary; and ③ eutectic transformations.

In Fig. 14 are plotted, schematically, the Gibbs free energy of several phases which compete at a given temperature, T. Starting from the undercooled liquid, and for various values of its composition, the arrows indicate three possible crystallization processes.

Polymorphic (eutectic) crystallization occurs without partition (long range diffusion). In primary crystallization a concentration gradient build up in the direction of growth. In the following this situation will be analyzed in more detail.

Let c^*, c_{xt} and c be, respectively, the interface, crystal and initial undercooled liquid concentration of B element. There is an initial transient in the growth in which the local equilibrium in between the emerging crystal and the liquid ahead of the crystal-liquid interface is established (see Fig. 15a). Once the stationary regime is achieved, it stays until the diffusion fields between neighbouring crystalline grains overlap significantly (see Fig. 15b). The assumption $\delta\dot{Q} \propto dx/dt$ is only valid at this stage. In the final stages of the primary crystallization process the overall concentration of the remaining undercooled liquid attains the value c^* (see Fig, 15c).

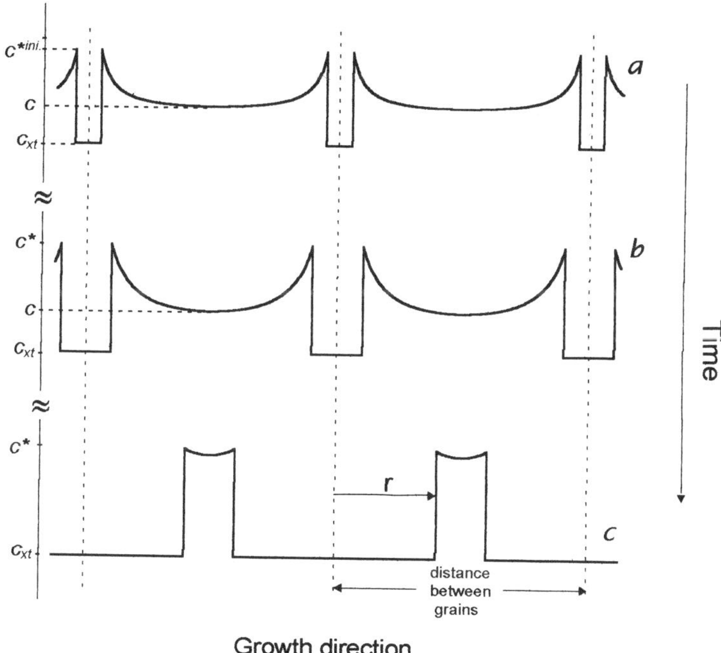

Figure 15. Concentration profiles ahead of the interface between crystalline grains: a) Initial transient; $b)$ stationary regime; and c) final transient:.

5.2. MODELING

In general the modeling of the crystallization of the undercooled **liquid** treats separately the processes of nucleation and crystal growth [43-48].

5.2.1. Nucleation

The classical theories of nucleation and growth have been developed independently by several authors. Assuming homogeneous nucleation, the nucleation frequency is given by

$$I = I_o\ exp\ [\ -\ 16\pi\sigma^3/3RT(\Delta G)^2] \tag{28}$$

where $16\pi\sigma^3/3(\Delta G)^2$ is the energy barrier for the formation of a nucleus of critical size $r^o = 2\sigma\Omega/\Delta G$ (with Ω^3 the mean volume per atom) and

$$I_o = N_v kT/3\pi a_o^3 \eta \tag{29}$$

Here N_v is the mean fraction of atoms in the supercooled liquid, a_o is the mean atomic diameter, $\eta = A\ exp[B/(T-T_o)]$ is the viscosity and σ is the molar crystal-liquid interfacial energy.

Both kinetic and thermodynamic factors determine the value of I. The kinetic factors are taken into account by the viscosity while three thermodynamic factors have to be considered. These are:

- the Turnbull factors [49] $\alpha=\sigma_g/\Delta H_m$ and $\beta=\Delta H_m/RT_m$
- the reduced Gibbs free energy difference between the supercooled and the crystal, $\Delta G_r = \Delta G/\Delta H_m$, given by

$$\Delta G_r = (\Delta T_r)(1-\gamma)-\gamma T_r\ lnT_r\ \ \text{with}\ \gamma=\Delta C_p/\Delta S_m\ ,\ T_r=T/T_m\ \text{and}\ \Delta T_r = 1-T_r\ .$$

Among all three parameters, β and γ may be accessible from experimental measurement or by modeling whilst α is difficult to evaluate and, normally, is only estimated.

Heterogeneous nucleation results in a reduction of the energy barrier of formation of a nucleus of critical size by a factor $f(\theta)$, whose dependence on the wetting angle θ is [50]

$$f(\theta) = \frac{(2+cos\theta)(1-cos\theta)^2}{4} \tag{30}$$

5.2.2. Crystal growth

Interface controlled crystal growth as it occurs on polymorphous crystallization has a crystal growth rate, u, which depends on two thermodynamic factors β and γ. It is given by

$$u = u_o\ [1 - exp\{\ -\ (\Delta G/RT)\}] \tag{31}$$

or

$$u = u_o\ [1 - exp\{-(\Delta G_r\beta/T_r)\}] \tag{32}$$

with

$$u_o = f\,kT/3\pi a_o^2\eta \tag{33}$$

Here f is the fraction of crystal sites on the nucleus surface where atoms are preferentially added.

When primary crystallization occurs the growth mechanism may follow the sequence:

i) the initial steps of growth are controlled by the interface between the nanocrystals and the surrounding matrix. The radius at time t of a spherical nucleus formed at time τ is

$$r(T,t,\tau) = u \cdot (t-\tau) + r^o \qquad (34)$$

ii) subsequent growth is limited by diffusion. The rate of growth under stationary conditions is

$$\frac{dr}{dt} = D \frac{c^* - c}{c^* - c_{xt}} \frac{1}{r} \qquad (35)$$

Here c^*, c_{xt} and c are, respectively, the interface, crystal and initial amorphous concentration and D the diffusion coefficient of the slowest diffusing element [51].

5.3. MODELING OF THE TRANSFORMATION DIAGRAMS

The crystallization kinetics modeling is based on Avrami's equation [45-47]

$$x = 1 - exp(-x_{ex}) \qquad (36)$$

where x_{ex} is the extended volume fraction, evaluated ignoring impingement between crystalline grains.

Time-Temperature-Transformation curves. Under isothermal regime, assuming that both nucleation frequency and growth rate are time-independent and further neglecting the initial critical size of the nuclei, Eq. (36) may be written in the general Johnson-Mehl-Avrami-Kolmogorov (JMAK) form

$$x = 1 - exp[-(Kt)^n] \qquad (37)$$

with t the time and K and n, respectively, the rate constant, **and the** kinetic exponent which depend on the mechanism of crystallization. Table 4 lists the respective values of K and n for various crystallization processes corresponding either to nucleation/growth ($N_I = 0$) or to growth of pre-existing nuclei ($I = 0$), with N_I the density of pre-existing nuclei.

In a more general situation in which $I \neq 0$ and $N_I \neq 0$, x_{ex} may be evaluated by

$$x_{ex}(T,t) = \int_0^t I(T,\tau)V(T,\tau,t)d\tau \\ + \sum_n N_n(T)V_n(T,t) \qquad (38)$$

Here $I(T,\tau)$ is the nucleation frequency at time τ, $V(T,\tau,t)$ the volume at t of nuclei formed at τ, $N_n(T)$ the density of pre-existing nuclei at T and $V_n(T,t)$ the volume at t of these pre-existing nuclei. The T-T-T curves are constructed by the representation of the function $T = T(t)$ for a fixed value of x.

Temperature-Heating Rate-Transformation curves. A very similar expression can be written for the extended fraction crystallized under continuous heating of the supercooled liquid up to a temperature T at a rate β. This is

$$x_{ex}(T,\beta) = \int_0^T I(T')V(T,T')dT / \beta + \sum_n N_n V_n(T) \tag{39}$$

TABLE 4. Avrami rate constant, K, and kinetic exponent, n, for various crystallization processes

Morphology of the crystalline grains	Nucleation $N_I = 0$		Pre-existing nuclei $I = 0$	
	K	n	K	n
One-dimensional growth: fibers of section S	$\left(\dfrac{SIu}{2}\right)^{1/2}$	2	$SN_I u$	1
Two-dimensional growth: rods of thickness L	$\left(\dfrac{\pi LIu^2}{3}\right)^{1/3}$	3	$(\pi N_I L)^{1/2} u$	2
Three-dimensional growth: spheres	$\left(\dfrac{\pi Iu^3}{3}\right)^{1/4}$	4	$\left(\dfrac{4\pi N_I}{3}\right)^{1/3} u$	3

Here $I(T')$ is the nucleation frequency at the nucleation temperature T', $V(T,T')$ the volume at T of nuclei formed at T', N_n the density of pre-existing nuclei and $V_n(T)$ the volume at T of these pre-existing nuclei. Therefore, the crystallization behaviour under continuous heating can be graphically represented as T-HR-T curves. These are obtained by plotting, in a temperature-heating rate diagram, the function $T = T(\beta)$ for fixed values of x. Typically, it is assumed that the onset of detectable crystallization corresponds to $x = 10^{-6}$. The coupling of experimental data of both T-T-T and T-HR-T curves has recently proven to get reliable values of both I and u for the first stages of primary crystallization [52].

Temperature-Cooling Rate-Transformation curves. This type of transformation diagram applies under continuous cooling of the liquid alloy and gives the temperature at which a fixed value of the crystalline fraction is obtained as a function of the cooling rate, β_c, used in the experiment.

6. Conclusions

The description of structural, topological and chemical SRO in chalcogenide glasses has been related to the glass forming regions observed in multicomponent systems. Glass transition and stability against relaxation and/or crystallization was performed by introducing thermodynamic quantities. For that purpose we use the concept of *ideal glass*. This concept is related to the fact that the entropy of any liquid cannot become lower than that of the stable crystalline state. Therefore, this *ideal glass* is the one with the smallest entropy, at any temperature, among those that can be prepared from the liquid with the same composition, irrespective of the specific technological preparation method (including or not the liquid phase in the preparation route).

The kinetic treatment of the crystallization under non-equilibrium conditions of a supercooled molten alloy has been analyzed on the basis of nucleation and crystal growth. Some of the thermodynamic factors which are of relevance on the process, like the enthalpy and Gibbs free energy of fusion, may be obtained from explicit knowledge of the thermodynamic properties of all the competing phases. These include the initial supercooled liquid, the equilibrium or metastable stoichiometric compounds and the equilibrium crystalline mixtures. Other thermodynamic factors, in particular up to our knowledge, the molar interfacial Gibbs free energy between the supercooled liquid and each of the competing crystalline phases, are estimated from indirect measurements.

7. Acknowledgments

Support form the "Comissió Interdepartamental de Recerca i Tecnologia (CIRIT)" by Project No. 1995SGR-00514 and from the "Comisión Interministerial de Ciencia y Tecnología (CICYT)" by Project No. MAT96-0769 is acknowledged.

8. References

1. H. Rawson (1967), *Inorganic glass forming Systems*, London, New York, Academic press
2. N.J. Kreild (1983) in *Glass Sicence and Technology* Vol. 1, D.R.Uhlmann, N.J. Kreild Eds., New York, AcademicPress pp 231-259
3. R. Zallen (1983), *The Physics of Amorphous Solids*, New YorK Wiley
4. S.R. Elliot (1984), *Physics of Amorphous Materials* London Longman, 1990
5. S.R. Elliot in *Materials Science and Technology* Vol. pp. 375-454
6. Z.U. Borisova (1981), *Glassy Semiconductors*, New York, Plenum Press
7. B:T: Kolomiets, *phys. stat. sol.*, *7* (1964) 713
8. N.F. Mott, *Adv. Phys.*, *16* (1967) 49
9. J.Z. Liu and P.C. Taylor, *Sol. State Comm.*, *70* (1989) 81
10. J.C. Phillips, *J. Non-Cryst. Solids 34* (1979) 153
11. G.H. Döhler *et al.*, *J. Non-Cryst. Solids 42* (1980) 87
12. M.F. Thorpe, *J. Non-Cryst. Solids 57* (1983) 355

13. S.A. Dembovskii, *Zh. Neorg. Khim.*, 22 (1977) 3187

14. S.A. Dembovskii and L.M. Ilizarov, *Izv. Akad. Nauk. SSSR, Neorg. Mater.*, *14* (1978) 1997

15. J.P. Suchet, *Mat. Res. Bull.*, *6* (1971) 491

16. G. Lucovsky, F.L. Galeener, R.H. Geils and R.C. Keezer (1977), in P.H. Gaskell (ed.) *The Structure of Non-Crystalline Materials*, Taylor & Francis Ltd., London, p 127-130.

17. M.T. Clavaguera-Mora, N. Clavaguera and J. Onrubia, *CALPHAD*, *8* (1984) 163

18. M.T. Clavaguera-Mora and N. Clavaguera, *J. Phys. Chem. Solids 43* (1982) 963

19. W. Kauzmann, *Chem. Rev.*, *43* (1948) 305

20. S:F: Edwards and Th. Vilgis, *Physica Scripta, T13* (1986) 7

21. J.H. Gibbs and E.A. DiMarzio, *J. Chem. Phys.*, *28* (1958) 373, 807

22. G. Adam and J.H. Gibbs, *J. Chem. Phys.*, *43* (1965) 139

23. K.J. Rao, *Proc. Indian Acad. Sci. (Chem. Sci.).*, *93* (1984) 389

24. M. Goldstein, *J. Chem. Phys.,51* (1969) 3728

25. M. Goldstein, *J. Chem. Phys.*, *64* (1976) 4767

26. M.H. Cohen and D. Turnbull, *J. Chem. Phys.*, *31* (1959) 1164

27. D. Turnbull and M.H. Cohen, *J. Chem. Phys.*, *34* (1961) 120

28. G.S. Grest and M.H. Cohen, *Adv. Chem. Phys.*, *48* (1981) 455

29. C.A. Angell, *J. Non-Cryst. Solids, 102* (1988) 205

30. I.M. Hodge, *J. Non-Cryst. Solids, 169* (1994) 211

31. C.T. Moynihan, A.J. Easteal, J. Winder and J. Tucker, *J. Chem. Phys.*, *78* (1974) 2673

32. M. Lasocka, *Mater. Sci. Eng.*, *23* (1978) 5655

33. C. Bergman *et al.*, *J, Non-Cryst. Solids, 70* (1984) 367

34. M.T. Clavaguera-Mora, *Thermochim. Acta, 148* (1989) 261

35. S. Sakka & J.D. Mackenzie, *J. Non-Cryst. Solids, 6* (1971) 145

36. Hruby, *Izv. Akad. Nauk. URSS, Ser. Mater. 3* (1937) 355

37. N. Clavaguera, M.T. Clavaguera-Mora and J. Onrubia, *J. Mater. Sci.*, *20* (1985) 3917

38. M.T. Clavaguera-Mora and N. Clavaguera, *J. Mater. Res.*, *4* (1989) 906

39. H.E. Kissinger, *Anal. Chem. 29* (1957) 1702

40. T. Ozawa, *Bull. Chem. Soc. Japan 38* (1965) 1881

41. S. Suriñach, M.D. Baró, M.T. Clavaguera-Mora and N. Clavaguera, *J. Non-Cryst. Solids, 58* (1983) 209

42. S. Suriñach, M.D. Baro, J.A. Diego, N. Clavaguera and M.T. Clavaguera-Mora, *Acta metall. & mater., 40* (1992) 37

43. A.N. Kolmogorov, *Izv. Akad. Nauk. URSS, Ser. Mater. 3* (1937) 355

44. W.A. Johnson and K.F. Mehl, *Trans. Am. Inst. Mining Met. Eng.*, *135* (1939) 416

45. M. Avrami, *J. Chem. Phys.*, *7* (1939) 1103

46. M. Avrami, *J. Chem. Phys. 8* (1940) 212

47. M. Avrami, *J. Chem. Phys. 9* (1941) 177

48. B.V. Yerofeev and N.I. Mitzkevitch, in *"Reactivity of Solids"*, Elsevier, Amsterdam, 1961, p. 273

49. D. Turnbull, *J. Appl. Phys. 21* (1950) 1022

50. W. Kurz and D.J. Fisher (1986), *Fundamentals of solidification*, Switzerland, Trans Tech Publications
51. N. Clavaguera and M. T. Clavaguera-Mora, *MRS 1995 Fall Meeting Proc., 398* (1996) (in press)
52. N.Clavaguera and J.A.Diego, *Intermetallics, 1* (1993) 187

SYNTHESIS, STRUCTURE AND SOME MODES OF APPLICATION OF NEW CHALCOGENIDE AND CHALCO-HALIDE GLASSES OF SILVER

M. MITKOVA,
Central Laboratory of Electrochemical Power Sources,
Bulgarian Academy of Sciences,
1113 Sofia, Bulgaria,
E-mail: Selen@Bgearn.Acad.Bg

1. Introduction

The amorphous chalcogenide semiconductors were discovered by Kolomiets [1,2] in the early 1950's. Actually this was the first class of disordered semiconductors to be found. The investigations of Kolomiets showed that a great number of chalcogenide glasses with widely varying stoichiometry in binary A_xB_{100-x} and more complicated compounds can be synthesised and that they behave like intrinsic semiconductors and their electrical conductivity cannot be affected by adding dopants. Most of these additives, specially added to the melt, occupy the lowest-energy configuration (i.e. satisfy the 8-N rule). Thus they hardly perturb the equilibrium between acceptor and donor defects and hence they do not have major effects on the physical properties. In 1957 Spear [3] reported the first drift mobility measurements in vitreous selenium and Tauc [4] reported the first studies of amorphous germanium in the early 1960's.

Especially fascinating for a large number of solid state scientists in the late 60's were the works of Ovshinsky [5-6] - the discovery of switching and memory effects in chalcogenide glasses and creation of the so called "ovonic" devices that could combine the work of complicated and difficult for fabrication integrated circuits. Later the optical memory effects, imaging, photodoping and the reversible photostructural changes were discovered which suggest possibilities for wide application of the non-crystalline materials and particularly mainly the chalcogenide glasses. These phenomena demonstrate that there is a large field of material science which is not explored completely, giving good possibilities for both fundamental and applied research. The fundamentals of the properties of these materials are given by Zallen [7].

M. F. Thorpe and M. I. Mitkova (eds.), Amorphous Insulators and Semiconductors, 71–82.
© 1997 *Kluwer Academic Publishers. Printed in the Netherlands.*

2. Glassformation and structure of the chalcogenide systems

2.1.GENERAL CONSIDERATIONS

The first interesting and important results were the reason for very extensive investigations on a number of chalcogenide glassy systems and many new glasses have been synthesised. I should like to point out that when I am speaking about "glassformation" in a concrete system I will have in mind glassformation occurring under "standard" conditions (cooling rate of about 10^2 deg.min^{-1}). At higher cooling rates, for example, by a splat cooling method or by deposition onto a cool substrate one can obtain glasses even from metals and alloys with a great coordination number, but if we consider such kind of glasses the results in terms of glassformation will be not comparable.

The problem for the vitrification is widely discussed and the general considerations concerning glasssformation are given by Gutzow [8]. Fig. 1 shows a part of the Periodic system of the elements where the chalcogenide and other glasses are placed. In the

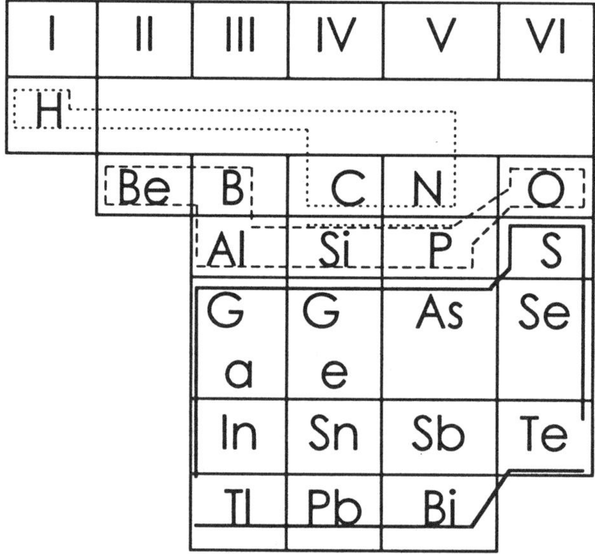

Figure 1. Position of the elements in the Periodic system taking part in the chalcogenide glasses.

bicomponent systems only a limited number of p-elements are important for the glassformation and especially these that are adjacent in the Periodic system of the elements (in the same period). This follows from the principles of their chemical structure in connection with the criteria for the chemical bonding of the covalent compositions with saturated valence requirements in which the coordination numbers are equal to the valence and correspond to the group numbers. So the role of the elements is clearly defined. For the three-component systems, again the p-elements play the basic role for the vitrification but it is also possible that s-elements like silver or quicksilver which behave more like s- rather than p-elements to take part in the glasses. Lately, special interest is also drawn to the chalcogenide glasses containing halides which, like the chalcogenide elements, contain a great number of lone pair electrons. These glasses are widely investigated by Lucas [9,10], McKenzie [11] and others. The glassformation in some chalcogenide systems containing silver halides [12-15] are investigated by us. Cheng *et al.* directed their interest towards the As-Ge-Ag-Se-Te-In system and new lead-halides-based glassforming systems [16,17].

The physico-chemical factors for the glassformation in the chalcogenide systems are as follows:

♦ Presence of glassforming composition in the chalcogenide system - this is especially important for the polycomponent systems where the identification of these compositions is more difficult.

♦ Existence of solid solutions in wide regions - especially clearly expressed in bichalcogenide systems which tend to form copolymer chains.

♦ Presence of eutectics in the system. This leads to decrease of crystallisation ability and leads sometimes to the formation of a second glass forming region.

2.2. STRUCTURE OF THE CHALCOGENIDE GLASSES

The structure of the chalcogenide atoms is defined from the covalent bonds of the chalcogenide atoms to the other elements. The simplest chalcogenide glass, for example selenium, has normally one-dimensional structure represented by a layered chain structure. The atomic structure is different in the bulk material and in the thin selenium films: in thin films the selenium atom chains are shorter, while in the bulk material the chains are long enough to form spherolids. The coordination radius r_c and coordination number N_c for different selenium phases vary by $\pm 10\%$ [18]. The resistivity of the selenium depends also strongly on the atomic structure: amorphous selenium has resistivity of 10^{12} Ohm.cm and crystalline rhombic selenium has a much lower resistivity of about - 10^6 Ohm.cm .

The binary chalcogenides from the As_2S_3 type have two-dimensional nonperiodic network structure with ground elements nonplanar triangles. The molecular structural unit is tetrahedral AsX_3. Every As atom has five valence electrons ($4s^2 4p^3$) from which 3 electrons ($4p^3$) can form chemical bonds with the chalcogen element. In the chalcogenide atoms X (S - $3s^2 3p^4$; Se - $4s^2 4p^4$; Te - $5s^2 5p^4$) two of the p electrons form chemical bonds and the other two form non-bonding lone pair [19]. As a result, every As atom is covalently bonded to three chalcogen atoms and each chalcogen to two As atoms. A definite short range order is imposed as each atom fulfils its chemical valence requirement according to the 8-n rule. Stronger heteropolar bonds (As-S) are favoured over homopolar (S-S and As-As) and the bond lengths are within 1% the same as those found in crystals. There are even off membered rings of different sizes and the structure seems as if formed by ground elements nonplanar triangles.

When an element from Group 4 of the Periodic system is included in the chalcogenide glasses, for example Ge, Si, etc., a three dimensional network is formed in which the elementary cell is one tetrahedron.

Another major element of randomness is the variation of bond angles and the fact that the flexibility of covalent bond angles is largest for the twofold coordinated group

VI elements and less for the tetrahedrally coordinated group IV elements. The reason for this is the greater variety of admixture from the other atomic orbitals to the covalent bond when the coordination number is less than the number of valence electrons.

In the chalcogenide glasses containing halide elements the structure variety is limited because bonding between the tetrahedrons, triangles or chains is impossible by means of monovalent halide elements. One result of this is for example the low transformation temperature of these glasses. The steric effect of the lone pare electrons on the chains and the fact that they form no π bonds are very favourable for the free rotation of the Te-Te σ bond. This leads very easily to structural disorder without the need of chemical bonds modification. The fact that the elongation of the chemical bonds normally is one-dimensional favours also the easy defect formation in the chains and the loss of periodicity. This plays also a very important role for the glassformation during freezing of the melts although the coordination is very low. There is no information about the chain lengths in the solid state but it seems trustworthy that they are quite short because the monovalent iodine forms a lot of mono bonds and serves as a chain terminator.

3. Glassformation and structure of the Silver-chalcogenide systems

3.1. GLASSFORMATION IN THE BICHALCOGENIDE SILVER SYSTEMS

The glassformation in the silver-containing chalcogenide and chalcohalide glasses follows the main rules of glassformation mentioned above. The glassforming region in the system Ag-Se-S includes a large area containing up to 20 at.% silver and 50 at.% sulphur - Fig.2. This is perhaps a consequence of the fact that,

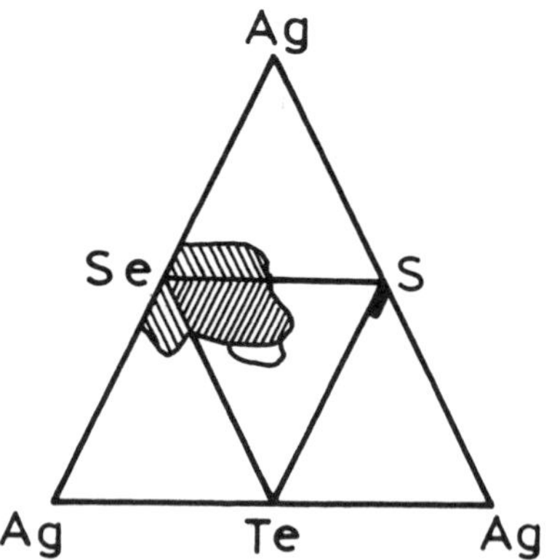

Figure 2. Glassforming regions in the silver chalcogenide systems.

when in a liquid state, selenium and sulphur build mixed chains which favour glassformation. The Ag-Se-Te system [12]has much smaller dimensions - up to 17 at.% silver and up to 20 at.% tellurium. It must be pointed out that substituting selenium with tellurium causes a decrease of the glassforming region concerning the content of tellurium in comparison with sulphur. This is in agreement with the generally accepted ideas that the strongly directed covalent bonds play an essential role in glassformation. In the literature tellurium is mentioned as a poor glass-former. This seems to be determined from its tendency to octahedral coordination similar to the distorted cubic lattice type NaCl. In the latter the ratio $r_2/r_1 = 1.199$, while in hexagonal selenium this ratio is $r_2/r_1 = 1.433$. The coordination of the vitreous tellurium is assumed to be approximately equal to 4 which leads to a great overconstrains in the structure and prevents glassformation. The Ag-S-Te system has a very small glassforming region - the glasses contain up to 10 at.% Ag. The alloys containing more than 15 at.% tellurium are microheterogeneous. The droplike structure of the heterogeneities testifies in favour of their liquation character. There may be a tendency to formation of structural units close to these of the solid solutions characteristic for the two-component systems. There is an excellent agreement between the glassy regions in the chalcogenide systems of silver and the system of the three chalcogen system, investigated by Suvorova *et al.* [21] - Fig.2.

3.2. GLASSFORMATION IN THE Se - Ag - I SYSTEM

Including a halogen element to the silver chalcogenides we investigated the glassformation in the systems Ag-Se-I [13] and Ag-Te-I. [14] There is very scarce information about the system Ag-I in which the composition AgI is identified. In it the ratio of the covalent part of the bond strength to melting temperature $E_{b.cov.}/T_m$ is 0,017 kcal/mol.K which characterises it as a poor glass progenitor. However, Gutzow *et al.* report that they could not synthesise AgI in a glassy state even at extreme cooling rates $(10^4 - 10^5$ K/s). We experimented glassformation in a more complicated system with Se as a glassformer and obtained the region shown in Fig. 3a. The samples glassforming

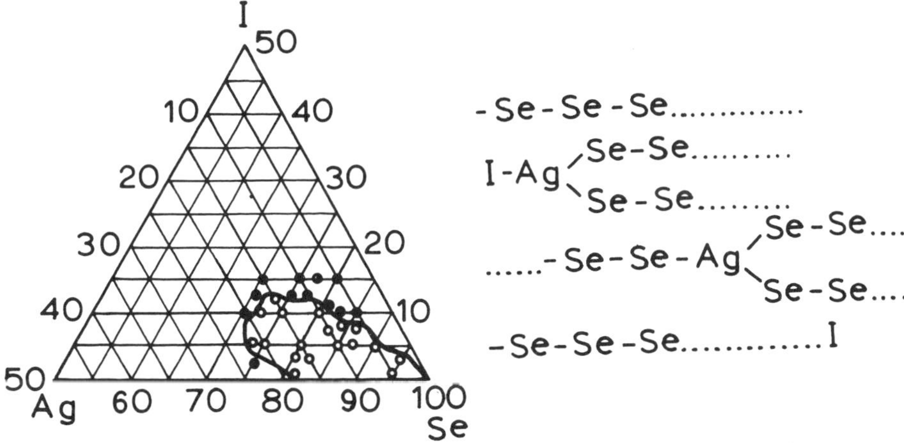

Figure 3a. Glassforming region for the system Se-Ag-I.

Figure 3b. Structural model of the glasses from the system Se-Ag-I region have marked crystallisation peaks, the highest being those of crystalline selenium and in some cases - of crystalline AgI. This violates Gibbs' phase rule and shows that rapid quenching prevents reaching an equilibrium. The electron microscope investigations proved that the selenium-rich samples are homogeneous. With increase in the silver and iodine content the glasses display a tendency to liquation. The phase separation in the melt and later in the solid phases is also affected by the cooling conditions because of the non-equilibrium distribution of the initial phases.

The Raman spectra of the glasses and the X-ray investigations of crystallised phases give grounds to suggest that in the general case, the glasses have the structure, depicted in Fig.3b. The number of selenium rings is strongly reduced due to the high temperature of the synthesis and selenium exists primarily as hexagonal chains. In the literature silver is considered to be the linking element between the individual selenium chains. It is plausible that it finds a place there since weak Van der Waals forces act between the chains. Silver has one electron shell, consisting in the last two levels of $4s^2 4p^6 4d^6 5s^1$ electrons. So it can take part in configurations with a ternary coordination in which one 5s electron and two 4d electrons form common electron pairs with the electrons of two selenium atoms from two selenium chains, creating dative bonds and the third electron links to one 5p electron of an iodine atom. In this way a trigonal pyramid with planar character is formed that separates the selenium chains and involves the basic physicochemical properties of the glasses as microhardness and density. The excess of Ag relative to iodine forms three-fold coordination, bridging three selenium chains, and the excess of iodine with respect to silver may be positioned at the end of the selenium chains, since it is known that it may serve as a chain terminator and can reduce their length and package density of the atoms.

3.3 GLASSFORMATION IN THE Te - Ag - I SYSTEM

The glassforming region in the system Te-Ag-I is shown on Fig. 4a. Its dimensions are comparable to those of the glassforming region in the system Se-Ag-I the region being shifted to the compositions richer in iodine. The isomorphic substitution of selenium with tellurium as discussed earlier should result in restricting the possibilities for glassformation in the system but in the case of interest, probably the fact that the three elements belong to one and the same period of the periodic system plays an important role. The inclusion of large quantities of iodine in the glassy phases proves its contribution to glassformation in the system under the particular conditions.

Tellurium exists mainly in the form of hexagonal chains in which the free rotation of the Te-Te bonds results in a loss of the periodicity without modification of the chemical bond energy. These chains are the basic matrix, dictating the structure of the investigated glasses. The X-ray investigations and the Raman spectroscopy indicate the presence of AgI and Ag_2Te in the crystallised phases. This means that most probably silver takes part in a donor bond with two lone pair electrons from two adjacent tellurium chains and at the same time it links ionically with one iodine atom. In the case of iodine excess with respect to silver, the remaining quantity of iodine is probably distributed along the tellurium chain. The two lone pair electrons of the tellurium atoms are bonded covalently with the two monovalent iodine atoms, and in some cases it is possible that iodine breaks the tellurium chains, forming single bonds. This structure is presented on Fig. 4b.

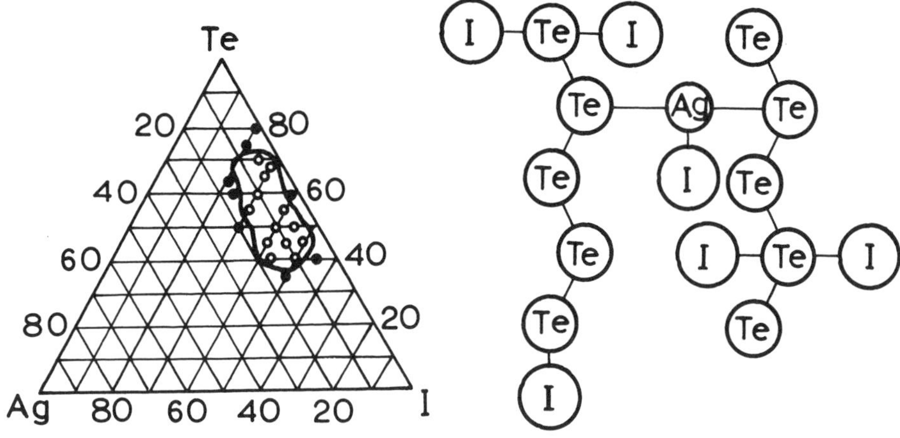

Figure 4a. Glassforming region
for the system Te-Ag-I.

Figure 4b. Structural model of the glasses from
the system Te-Ag-I.

4. Electronic structure and defects in the chalcogenide glasses

The electronic structure of the chalcogenide glasses depends largely on the degree of disorder. Therefore, it is difficult to determine specific values for the band gap E_g and the electronic structure of defects for a particular species. The extended electronic states can be seen in the broadness of absorption spectra of chalcogenides. Empirical criteria

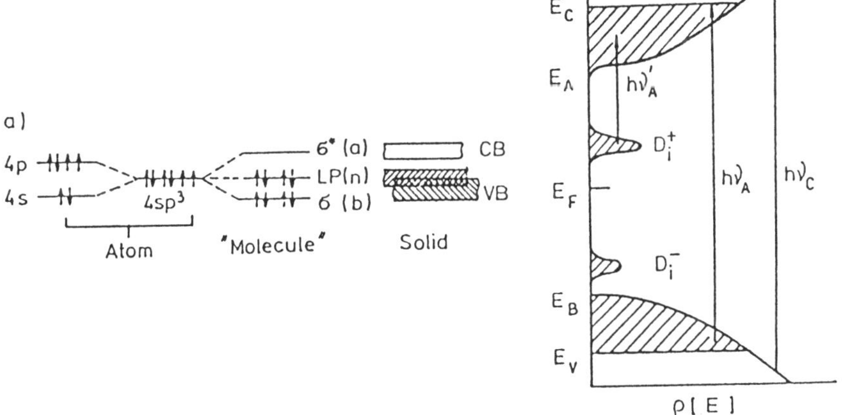

Figure 5. Molecular orbital schemes for the electronic configurations and band structure for Se-like semiconductors.

are often used to determine the band gap energy region E_g (or wavelength λ_g) where $k_a(hv_g) \sim 10^4$ cm^{-1} and measurements of the electron conductivity can also be used. The strong covalent bonding leads to a high energy gap, E_g [19,21]. Fig. 5 shows the molecular orbital scheme for the electronic configuration and band structure for Se-like semiconductors.

There are three kinds of defects in amorphous chalcogenides. The first model to describe them was created by Mott, Davis and Street (the so-called MDS model) [22]. It has been proved successful in explaining a variety of externally induced phenomena in chalcogenides such as photoinduced processes, thermal processes and luminescence. According to MDS, defects have three charged states - D^0, D^+ and D^- depending on the local coordination. D^0 is a neutral atom, with an unpaired dangling electron, because these atoms are singly coordinated, situated at the end of the Se-chain. D^- are also singly coordinated but they have an extra lone pair. D^+ are triply coordinated sites. D^0 centres are unstable and react exothermically in the charge transfer process shown in Fig. 6.

$$2D^0 \rightarrow D^+ + D^-, \tag{1}$$

Note that this is not a simple two-centre, single-step electron hop but rather a two step process, where the second electron transfer involves creating a D^+ out of the previously defect-free site. Fig.5 shows one typical energy level diagram for chalcogenide semiconductors, including the energies of the D^- and D^+ defects. Note that defect energies lie within the energy gap.

A similar model was proposed by Kastner, Adler and Fritzsche (KAF) [23]. This model is somewhat more flexible since it includes more possible combinations of charge states and degrees of coordination. In general, all atoms are designated by C^k_i where k is the charge on the atom and I the coordination number. Thus, a normal Se atom, i.e. neutral and doubly coordinated, is designated C^0_2. Of course, one can easily transform between the MDS and KAF models. Fig. 6 shows the designation for reaction (1) in both schemes. However, in binary or more complex amorphous semiconductors, the

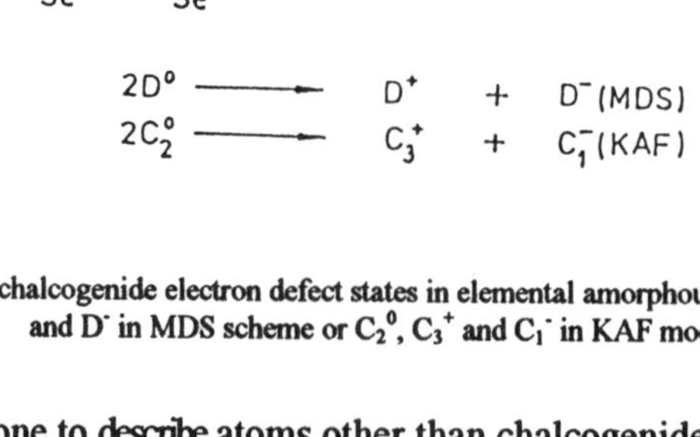

$$2D^0 \longrightarrow D^+ + D^- \, (MDS)$$
$$2C^0_2 \longrightarrow C^+_3 + C^-_1 \, (KAF)$$

Figure 6. Formation of chalcogenide electron defect states in elemental amorphous Se, corresponding to D^0, D^+ and D^- in MDS scheme or $C_2{}^0$, $C_3{}^+$ and $C_1{}^-$ in KAF model.

KAF model allows one to describe atoms other than chalcogenides.

5. Photoinduced changes - nature and some modes of application

5.1 NATURE OF THE PHOINDUCED CHANGES

Photoinduced processes are of a great interest in the context of the investigation of the chalcogenide glasses as they have unique properties that attract the researchers interest. They are connected with the lone pair electrons and the low coordination number of the chalcogenide atoms which is the reason for a strong electron-phonon coupling in the system of the chalcogenide glass and gives a great mobility of a number atomic local configurations. Tanaka [24] gave an exhaustive review of these phenomena and Elliott [25] offered a unified model for the photostructural changes in the chalcogenide glasses. Using EXAFS technique to analyse reversible photostructural changes in annealed a-As$_2$S$_3$, he found a stronger change in the second maximum of the radial distribution function during illumination. This demonstrates a change in the medium range order caused by the formation of an interlayered bond. Thus, the concentration of homopolar bonds (As-As, S-S) decreases and the concentration of heteropolar bonds (As-S) increases. Lakshmikumar [26] created a new model for the photodiffusion of silver in the amorphous chalcogenide materials. Tikhomirov and Lyubin [27] reported on new photoinduced effects in the chalcogenide glasses.

Photoinduced changes can be both direct, via electronic excitation which promotes atoms to nonbonding states, in bond breaking and charge transfer, or indirect, when the excited atoms relax by converting the excessive electronic energy to vibrational energy (phonons) thereby inducing thermal activated structural changes (short and medium range order). The latter is similar to thermal processes which also affect nuclear motion. The first can occur already at low light intensity, the second usually requires higher light intensities.

These two types of processes in thin films can be easily separated from each other since the rate of the thermal process will depend on the size of the irradiated sample, or rather on the radius of the laser spot whereas the rate of a direct photoinduced electronic process is size independent.

All these effects affect the optical (α,n), mechanical (d, H$_\mu$) and chemical properties of the glasses. I will here review the change of the chemical properties, which leads to photoinduced selective solubility due to a difference in the dissolution rates of the exposed and unexposed areas, which may be applied for creation of structures on thin films.

5.2 SELECTIVE SOLUBILITY IN THE SILVER BICHALCIGENIDE SYSTEMS

Part of our investigations in this respect have been devoted to the glasses from the silver bichalcogenide systems Ag-Se-Te and Ag-S-Se. The first evidence of the magnitude of the photoinduced changes created in them comes from the absorption spectra which show a shift towards the longer wavelengths after illumination with light with a wavelength within the absorption edge. So photodarkening occurs which is better expressed for the Se-S-Ag system, probably due to the smaller ion radii of S atoms which make possible their structural transformations. Besides, it is well known that in the mixed Se-S alloys there exist many mixed and monochalcogenide 8-atom rings which may be broken after light illumination.

The same tendency to a greater shift of the absorption after irradiation is manifested in the samples containing more silver for both systems. We assume that after irradiation Ag changes its coordination creating a new bridging bonds between the chalcogenide chains. Structural changes in these films are confirmed also by electron microscope diffraction patterns which show that after illumination ordering of higher degree occurs in the films .

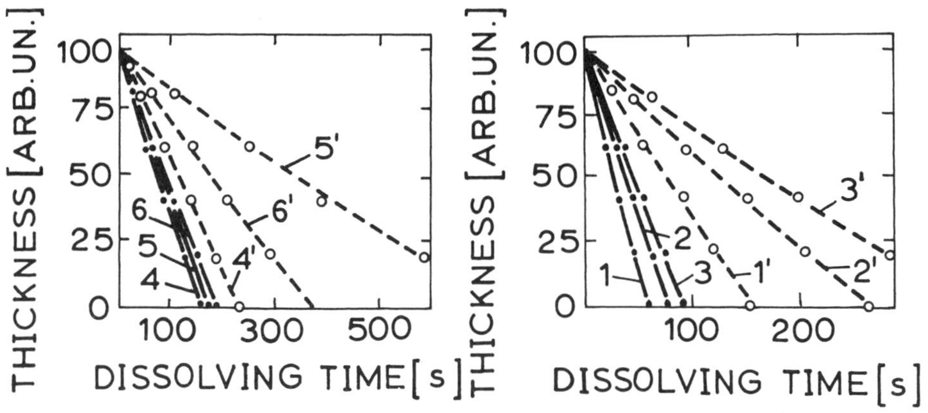

Figure 7a. Selective dissolution curves for the glasses from the system Se-Te-Ag in aqueous solutions.

Figure 7b. Selective dissolution curves for the glasses from the system Se-S-Ag in aqueous solutions.

The selective solubility shows one negative effect, i.e. the non-irradiated part of the films dissolves more quickly than the irradiated one-Fig.7. One can assume that after irradiation, due to higher ordering in the systems, the number of defects and free ends of the chains decreases as shown also in the theoretical results of Rajagopalan *et al.* [28] for other chalcogenide systems so that interaction with the dissolving agent is more difficult. This effect is better expressed for the samples with higher content of silver and in this case higher concentrations of the dissolving agent are also necessary which is not acceptable from a technological point of view. The dissolving agent for the system with Se and S was H_2SO_4 and for the system containing Te and Se - a mixture from $K_2Cr_2O_7$ and H_2SO_4.

One may conclude that in general the chalcogenide elements are essential for the dissolution process which has an oxy-reduction nature. The difference in the behaviour of both systems is caused by the fact that the oxidation of the sulphur proceeds to the end . For sulphur the six-valent condition is most stable, and for tellurium - the four-valent one. Furthermore, tellurium has a lower oxidation ability and needs a stronger oxidiser [29].

Applying these effects, the structure shown on Fig. 8 can be created which can be prepared with extremely high resolution and deep depth due to the small structural units and high chemical durability of the glasses.

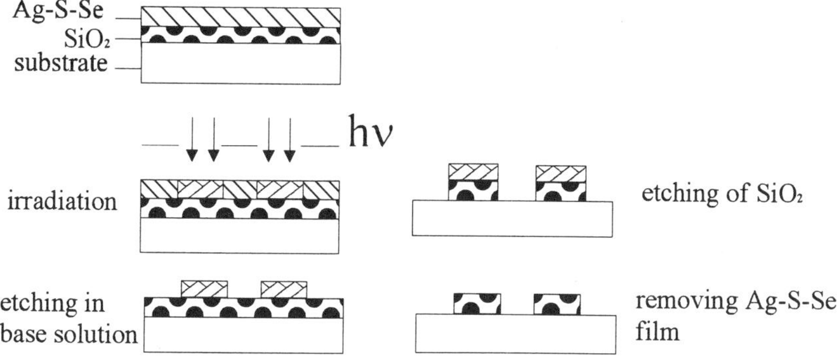

Figure 8. Photolithographic technology for obtaining of relief structure in chalcogenide glasses.

In conclusion, the main characteristics of the chalcogenide glasses are: relatively good glass-formation ability caused mainly by the presence of the lone pair electrons which contribute to the great mobility in the structures and are also the cause for considerable photoinduced changes in them.

Acknowledgements: It is my pleasure to thank to Dr. T. Petkova and to Mr. V. Boev for the preparation of the glasses and providing of some of the investigations of their properties. This work was done with the financial support of the Bulgarian NSF - grants Nr., Nr. X226 and X558 and EC grand Nr. ERBCIPA 940170.

References

1. Kolomiets, B.T. (1964) Vitreous semiconductors (I), *Phys. Stat. Sol.* **7**, 359-372.
2. Kolomiets, B.T. (1964) Vitreous semiconductors (II), *Phys. Stat. Sol.* **7**, 713-731.
3. Spear, W.E. (1957) Transit time measurement of charge carriers in amorphous selenium films, *Proc. Phys. Soc.* **B 70**, 669-675; Spear, W.E. (1960) The hole mobility in selenium, *Phys. Stat. Sol.* **76**, 826-832.
4. Tauc, J.,Grigorovici, R. and Vancu, A. (1966) Optical properties and electronic structure of amorphous germanium, *Phys. Stat. Sol.* **15**, 627-637.
5. Ovshinsky, S.R. (1968) Reversible electrical switching phenomena in disordered structures, *Phis. Rev. Lett.* **21**, N 20, 1450-1453.
6. Ovshinsky, S.R. (1992) Opticaly induced phase changes in amorphous materials, *J. of Non - Cristalline Solids* **141**, 200-203.
7. Zallen, R. (1983) *The Physics of Amorphous Solids*, John Wiley & Sons, New York.
8. Gutzow, I. and Schmelzer, J. (1995) *The vitreous state, thermodynamics, structure, rheology and crystallization*, Springer-Verlag, Berlin-Heidelberg.
9. Lucas, J. (1990) Recent progress in halide and chalcogenide glasses, *S. St. Jonics* **39**, 105-112.

10. Lucas, J., Ma Hong Li and Zwang Xiang Hua (1991) High Tg tellurium selenium halide glasses, *J. of Non - Crystalline Solids* **135**, 49-54.

11. Mc Kenzie, J.D. and Jong Heo (1989) Chalcochalide glasses, *J. of Non - Crystalline Solids* **113**, 246-252.

12. Mitkova, M. and Boncheva - Mladenova, Z. (1989) Glass forming region and some properties of the glasses from the system Se-Te-Ag, *Monatshefte für Chemie* **120**, 643-650.

13. Petkova, T. and Mitkova, M. (1991) Glass forming in the Se-Ag-I system, *Materials Chem. and Physics* **30**, 55-59.

14. Petkova, T. and Mitkva M. (1993) Glass - forming region and some properties of the glasses from the Te-Ag-I system, *Materials Chem. and Physics* **33**, 233-238.

15. Boev, V. and Mitkova, M. (1994) Glass forming in the Se-Ag-Br and Se-Ag-Cl systems, *Materials letters* **20**, 195-201.

16. Cheng, J., Chen, W. and Ye Dapeng (1995) Novel chalcohalide glasses in the As-Ge-Ag-Se-Te-I system, *J. of Non - Crystalline Solids* **184**, 124-127.

17. Cheng, J. and Jiu Zhenwu (1995) New lead - halide - based glasses - forming systems, *J. of Non - Crystalline Solids* **184**, 213-217.

18. Elliott, S.R. (1990) Amorphous materials, *Longman Scientific & Technical, London.*

19. Kastner, M. and Fritzsche, H. (1978) Defect chemistry of lone - pair semiconductors, *Phil. magazine B* **37** (2), 199-215.

20. Suvorova, A.N., Borisova, Z.U., Orlova, G.M. (1974) Glasses in the system S-Se-Te, Izv. AS USSR-Inorganic materials **10** (3), 441-445.(in Russian)

21. Owen, A.E. (1984) *Electron transport in chalcogenide glasses, coherence and energy transfer in glasses*, Plenum, New York, ed by P.A. Flenry, B. Golding, 243-278.

22. Mott, N.F., Davis, A.E.and Street, R.A. (1975) Conduction in non - crystalline systems XI. The on - state of the threshold switch, *Phil. Magazine.* **32**, 961-996.

23. Kastner, M., Adler, D. and Fritzsche, H. (1976) Valence - alternational model for localised gap states in lone - piar semiconductors, *Phys. Rev. Lett.* **37**(22), 1504-1507.

24. Tanaka, K. (1980) Reversible photostructural change: Mechanisms, properties and applications, *J. of Non - Crystalline Solids* **35 & 36**, 1023-1034.

25. Elliot, S.R. (1986) A Unfied model for reversible photostructural effects in chalcogenide glasses, *J. of Non - Crystalline Solids* **81**, 71-98.

26. Lakshmikumar, S.T. (1986) A new model for photodiffusion of silver in amorphous chalcogenides, *J. of Non - Crystalline Solids* **88**, 196-205.

27. Lyubin, V.M. and Tikhomirov, V.K. (1991) Novel photoinduced effects in chalcogenide glasses, *J. of Non - Crystalline Solids* **135**, 37-48.

28. Rajagopalan, S. (1982) Photo - optical changes in Ge - chalcogenide films, *J. of Non - Crystalline Solids* **50**, 29-38.

29. Mitkova, M. and Boncheva - Mladenova, Z. (1987) Selective solubility of some silver - chalcogenide glasses, *J. of Non - Crystalline Solids* **90**, 589-592.

X-RAY AND NEUTRON DIFFRACTION:

Experimental Techniques and Data Analysis

Adrian C. WRIGHT

*J.J. Thomson Physical Laboratory, University of Reading
Whiteknights, Reading, RG6 6AF, U.K.*

> *"L'ordre
> Est le plaisir de la raison
> Mais le désordre
> Est le délice de l'imagination."*
> Paul Claudel.

1. Introduction

As with any material, a knowledge of the structure of (inorganic) amorphous insulators and semiconductors is an important prerequisite for a full understanding of their physical properties at a microscopic level. Amorphous insulators and semiconductors both have predominantly directional bonding with significant covalent character and hence have network structures, which are traditionally described in terms of Zachariasen's random network theory [1]. This chapter will describe the general principles involved in studying the structure of amorphous network solids by X-ray and neutron diffraction and give examples of the use of structural modelling in the interpretation of diffraction data. A much more detailed account of the techniques involved and further examples, together with a comprehensive list of references, can be found in [2] and two earlier reviews by the present author [3,4].

Amorphous network solids not only comprise glasses, obtained by melt quenching*, but may be prepared by a variety of techniques from each of the three main states of matter, as summarised in Table 1. The lack of a periodic structure, combined with the fact that amorphous solids are in a state of metastable thermodynamic equilibrium, means that the structure of any given amorphous solid depends on its detailed preparation. This has important consequences in the study of amorphous solids in that any sample must be

---

*In this chapter the term *glass* will be reserved for materials covered by the ASTM definition of a glass as "an inorganic product of fusion which has cooled to a rigid condition without crystallising."

M. F. Thorpe and M. I. Mitkova (eds.), Amorphous Insulators and Semiconductors, 83–131.
© 1997 *Kluwer Academic Publishers. Printed in the Netherlands.*

84

TABLE 1. Preparation of amorphous network solids.

Precursor phase	Preparation technique	Examples
Gas	Thermal evaporation	Se, $As_{1-x}S_x$, Si, $Si_{1-x}O_x$
	Sputtering	Ge
	Vapour-phase hydrolysis	SiO_2
	Vapour-phase pyrolysis	SiO_2
	Glow-discharge decomposition	$Si_{1-x}H_x$
Liquid	Melt quenching	Se, SiO_2, B_2O_3, P_2O_5
	Gel desiccation (Sol-gel techniques)	SiO_2
	Precipitation	As_2S_3, $Ca_3(PO_4)_2$
	Electrolytic deposition	Ge
Solid	Photon irradiation	As_4Se_4
	Particle irradiation	Ge, SiO_2
	Pressure amorphisation	$AlPO_4$, Fe_2SiO_4, H_2O, ZnP_2
	Shock wave amorphisation	SiO_2
	Oxidation	SiO_2

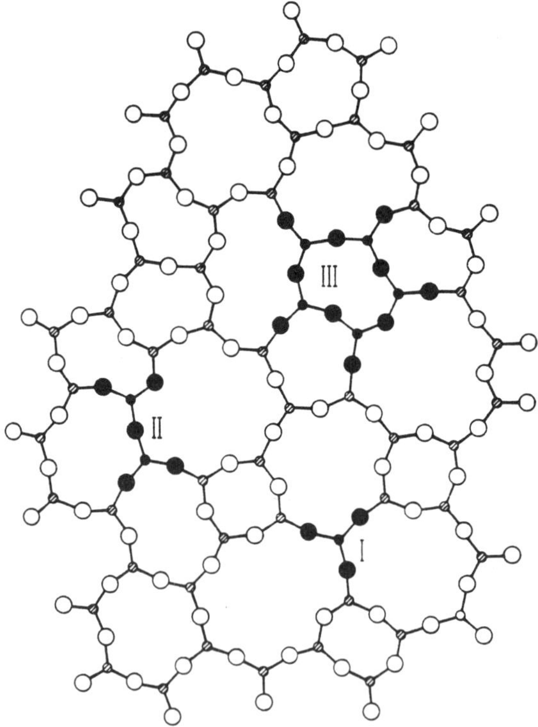

Fig. 1. Zachariasen's original random network diagram (retraced) for a glass of composition A_2O_3 [1], with structural units shaded to indicate range I, II and III order. Small atoms, A and large atoms, O.

carefully prepared and adequately characterised for the results obtained to be of any significant use. Samples of identical chemical composition may have very different structures, depending on their method of preparation, and, conversely, the preparation method may yield important clues concerning structure.

An amorphous solid is one with a structure which lacks periodicity, extended symmetry and long range order and it is the absence of the last of these that leads to a diffraction pattern which is a continuous and relatively slowly varying function of the magnitude of the scattering vector, Q, in contrast to the sharp Bragg peaks of a crystalline material. As with the latter, X-ray and neutron diffraction are the major direct structural probes but, due to the inherent differences between the two classes of materials, the problems encountered in diffraction studies of amorphous solids are somewhat different from those experienced in crystallography. The existence of a periodic structure means that, given good diffraction data over a reasonable region of reciprocal space, it is in practice possible to determine the structure of simple crystalline solids absolutely. The same is not true for amorphous solids. Because they are normally isotropic on a macroscopic scale, the maximum that can be obtained from a diffraction experiment on an amorphous solid is a one-dimensional correlation function, from which the regeneration of the underlying three-dimensional structure can never be unique. It is for this reason that modelling plays such an important role in structural studies of amorphous solids and why the choice between possible models involves a wide range of experimental techniques and not just X-ray and/or neutron diffraction.

1.1. THE RANDOM NETWORK THEORY

The random network theory was first introduced by Zachariasen [1] to describe the structure of oxide glasses and is illustrated schematically in two dimensions, for a pure glass forming oxide A_2O_3, in Fig. 1. As pointed out by Zachariasen, the atoms in a conventional melt-quenched glass are linked together by forces essentially the same as those in the corresponding crystalline materials and, if the internal energy of the glass is to be comparable to that of the crystal, then it is necessary that the oxygen polyhedra in the vitreous and crystalline states are similar. As in the latter, extended three dimensional networks are formed and the atoms vibrate about definite equilibrium positions, but in the glass the structural units (AO_3 triangles in Fig. 1) are linked together randomly to form a non-periodic structure which lacks long range order. It is clear from Zachariasen's original diagram that he expected the structural units (s.u.) in a network glass to be as regular as those in the crystal and that, for a three dimensional structure, disorder would be introduced via a distribution of the torsion angles α_1 and α_2 and the bond angle β_O, as defined for vitreous silica in Fig. 2, a view supported by modern diffraction data.

In order to obtain a random network which is relatively distortion free, Zachariasen proposed his well-known criteria for glass formation [1], which are essentially topological in nature and allow the necessary degrees of freedom for glass formation with very little excess energy resulting from defects and/or network strain. The violation of one of Zachariasen's criteria does not necessarily mean that a glass cannot be formed, merely that

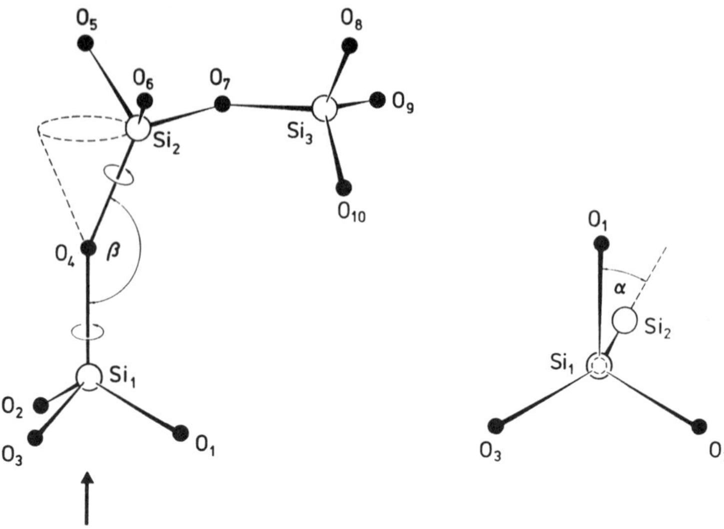

Fig. 2. Definition of the bond angle β_0 and the torsion angles α_1 and α_2 for vitreous silica.

it is energetically less favourable than one for which all the criteria are fulfilled. It should also be noted that a three dimensional network is not an essential prerequisite, even for oxide glass formation, as demonstrated by the existence of *invert* glasses.

Two important points should be noted about the random network theory. First, the perfect (continuous) random network plays the same role in amorphography as does the idealised crystal structure, represented by the unit cell, in crystallography. In real materials there will be "defects," but these will in general be undetectable in an X-ray or neutron diffraction study. Second, the random network theory was formulated to describe conventional oxide glasses quenched from the supercooled melt at relatively slow cooling rates. Thus, although the random network theory has been extended to cover other amorphous network solids, such as amorphous semiconductors (e.g. amorphous germanium and hydrogenated amorphous silicon), Zachariasen's arguments, based on internal energy considerations, concerning the relationship between the structural units found in the crystalline and vitreous states are not necessarily applicable to materials which have been prepared by techniques involving highly non-equilibrium processes such as vapour deposition or glow discharge. For example, the tetrahedral bonding in evaporated amorphous Si and Ge (cf. Section 5.1) is very much more distorted than that in the ambient crystalline modification and both materials contain a much greater fraction of broken bonds than do the corresponding oxide glasses SiO_2 and GeO_2.

1.2. CHALCOGENIDE SYSTEMS

Chalcogenide materials are, in general, structurally much more complicated than their oxide counterparts, in that even two element systems, such as As-S and Ge-Se, exist as amorphous solids over a wide range of composition and contain homopolar bonds. Models for the numbers of each type of bond present range from the random covalent model to complete chemical ordering. The former places no restriction on the bonds which may be formed, the number of each type of bond being determined solely from the 8-N rule and the concentration of each atomic species. A chemically ordered system, on the other hand, contains the maximum number of bonds between unlike atoms. In the vitreous state, the relevant bond strengths are a useful guide to the relative number of each type of bond likely to be formed but, for example, in vapour deposited materials the balance may be significantly affected by the molecules present in the vapour phase (cf. Section 7.4).

1.3. MULTICOMPONENT GLASSES

In multicomponent oxide glasses, the individual constituents are usually classified as network formers or network modifiers, although some intermediate materials can act in either capacity and this distinction is only applicable for glasses in which the network modifying cation bonding is significantly more ionic than that involving the network forming cations. Hence this terminology is not appropriate, for example, in the case of heavy metal fluoride glasses [5]. The addition of a further network former to a single component glass will result in a mixed network of two different structural units, whereas the conventional view of a network modifier is that it leads to the formation of negatively-charged non-bridging oxygen atoms. However, the addition of a network modifier may also result in an increase in the network-forming cation co-ordination number, $n_{A(O)}$, such as occurs preferentially in alkali borate glasses at low alkali contents, where $n_{B(O)}$ increases from 3 to 4.

An extremely controversial aspect of the stuctures of glasses containing network modifiers concerns the spatial distribution of the network modifying cations. Traditionally, the network modifying cations have been envisaged as occupying holes in the random network close to the non-bridging oxygen atoms, but modern diffraction and EXAFS data strongly suggest that the network-modifying cations adapt their local environment to obtain a more desirable co-ordination polyhedron, similar to that found in related crystalline materials. This does not necessarily imply the introduction of regions of crystalline order, or crystallites, but is simply the consequence of radius ratio effects along the lines discussed in [5] for ionic systems. In addition, it might be expected that the negatively-charged non-bridging oxygen atoms will tend to cluster, since they will be attracted to the positively-charged network modifying cations and vice versa, which has led to a modified random network theory [6] in which the network-modifying cations are concentrated in channels between regions of silica-rich network. This model appears to be supported by molecular graphics representations of the structures generated by modern molecular dynamics simulations of alkali silicate glasses, such as that shown later in Fig. 23, but a more detailed analysis suggests that this may not necessarily be the case [7].

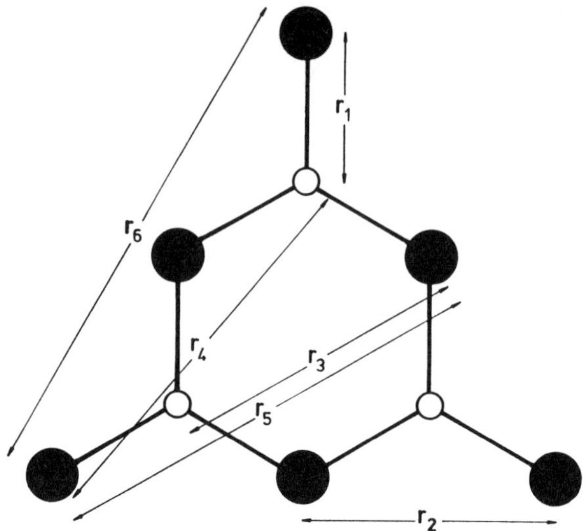

Fig. 3. Interatomic distances within a B_3O_6 boroxol group. O, B and ●, O.

1.4. SUPERSTRUCTURAL UNITS

Crystalline borate systems contain a rich variety of so-called **superstructural units**, which comprise well defined arrangements of the basic BO_3 and BO_4 structural units (cf. Fig. 1 of [8]), an example being the boroxol group in Fig. 3, which consists of a planar 3-membered ring of BO_3 triangles stabilised by delocalised π-bonding. There is increasing evidence that the same units are also present in the corresponding glasses, as revealed by sharp lines in their Raman spectra, which has led Bray [9] to suggest a structural model in which it is the superstructural units that are linked together randomly to form the vitreous network, in addition to the basic BO_3 and/or BO_4 units.

2. Quantification of Amorphous Solid Structures

Although the concept of a random network is easy to describe qualitatively, there is no way to completely define the resulting structures quantitatively, except by specifying the co-ordinates of every atom present, which is clearly impossible for any real material. Traditionally the structural order within an amorphous solid has been divided into three ranges: short, intermediate (medium) and long. However, for systems with well defined directional bonding, it is much more convenient to discuss the analysis of diffraction data using four ranges [2], the first three of which are indicated schematically in Fig. 1: I, the structural unit; II, the interconnection of adjacent structural units and III, the network topology. Typical structural parameters associated with each of the four ranges are summarised in Table 2.

TABLE 2. Ranges of order for an amorphous solid.

Range	Order	Characteristic parameters	Symbol
I	Structural unit (s.u.)	Identity	AX_n
		Internal co-ordination number	$n_{A(X)}$
		Bond length	r_{A-X}
		Bond angle	β_A
		Superstructural unit definition	-
II	Interconnection / relative orientation of adjacent units	Connection mode	C
		Connectivity	c
		Bond angle	β_X
		Bond torsion angle	α
III	Network topology / order beyond adjacent unit	Shortest path ring size	m
		Network dimensionality	D
		Topological cluster type	-
IV	Longer range fluctuations	Type (composition or density)	-
		Morphology (droplet, etc.)	-
		Radius of gyration	R_g
		Inter-region separation	R

2.1. RANGE I: THE STRUCTURAL UNIT / SUPERSTRUCTURAL UNITS

The most important fact to establish for an amorphous network solid is the identity of the structural unit(s) present. The structural parameters in range I specify the detailed geometry of the structural unit(s), including the distribution of bond lengths and angles. Superstructural units lead to well defined interatomic distances which are larger than those in the basic structural unit(s), as shown for the boroxol group in Fig. 3.

2.2. RANGE II: THE INTERCONNECTION / RELATIVE ORIENTATION OF ADJACENT STRUCTURAL UNITS

The range II order involves the relative orientation and, where appropriate, the interconnection of adjacent structural units. The number of parameters required depends on the number (if any) of shared atoms. For example the interconnection of two corner-sharing SiO_4 tetrahedra requires one (Si–O–Si) bond angle and two torsion angles as defined in Fig. 2.

2.3. RANGE III: THE NETWORK TOPOLOGY / ORDER BEYOND THE ADJACENT STRUCTURAL UNIT

A useful concept in discussing the so-called intermediate range order for an amorphous network solid is that of an underlying topological network, which can be decorated using various atomic motifs to represent different amorphous solids [2]. As discussed in [2], the topology of the network can only be fully specified by a connectivity matrix or a near neighbour table, which again is impossible for a real material, and hence it is usual to characterise a network in terms of the ring statistics by shortest path analysis.

2.4. RANGE IV: LONG RANGE DENSITY FLUCTUATIONS

Although amorphous solids lack long range order, there may nevertheless be longer range fluctuations in density and/or composition arising from phase separation, etc. The characteristic distances involved in such fluctuations are such that they must be studied using small angle scattering techniques, which are beyond the scope of the present chapter.

2.5. EXPERIMENTAL PAPAMETERS

In general the parameters just outlined are not those which are obtained directly from experiment, diffraction or otherwise. The isotropic nature of most amorphous solids means that in real space the result of a single diffraction experiment is a weighted sum of component correlation functions,

$$t_{ij}(r) = 4\pi r \rho_{ij}(r) = d_{ij}(r) + t_j^{\circ}(r), \tag{1}$$

$$d_{ij}(r) = 4\pi r [\rho_{ij}(r) - \rho_j^{\circ}], \tag{2}$$

$$t_j^{\circ}(r) = 4\pi r \rho_j^{\circ}, \tag{3}$$

in which $\rho_{ij}(r)$ is on average the number density of atoms of type j (usually elements, but it may be important to distinguish between chemically distinct atoms of the same element) a distance r from the ith atom in the composition unit (c.u.) and ρ_j° is the average number density of j atoms. For a sample containing n elements/atom types, there are n(n+1)/2 independent component correlation functions.

3. Theoretical Outline

In this section, only a very brief outline of the background theory of X-ray and neutron diffraction by amorphous solids will be given. A more detailed account, using the same formalism, can be found in [2-4] and in a series of papers discussing neutron diffraction studies of vitreous silica [10-13].

For an amorphous solid, the scattered intensity, I(Q), is a function only of the magnitude of the scattering vector,

$$Q = (4\pi/\lambda) \sin \theta, \tag{4}$$

λ being the incident wavelength and 2θ the scattering angle. This arises because any interatomic vector $\mathbf{r}_{pq}$ has equivalent vectors elsewhere in the sample in all possible orientations with respect to $\mathbf{Q}$. Hence for such materials it is possible to perform an isotropic average to yield the Debye equation

$$I_N(Q) = \sum_p \overline{a_p^2(Q)} + \sum_p \sum_q \overline{a_p(Q)a_q(Q)} \sin(r_{pq}Q)/(r_{pq}Q), \tag{5}$$

in which $\overline{a(Q)}$ represents either the X-ray form factor, $f(Q)$, or the neutron scattering length, $\overline{b}$, and the p and q summations are taken over all the atoms in the sample. The Debye equation may be used to calculate the scattered intensity if all the interatomic vectors are known or can be calculated, as in the case of a structural model (cf. Section 6.1).

For a real sample, on the other hand, the interatomic vectors are unknown and hence it is necessary to work in terms of the component correlation functions, $t_{ij}(r)$. The double summation over all atoms in Eq. (5) may thus be replaced by an integration over the component correlation functions to give

$$I(Q) = \sum_i \overline{a_i^2(Q)} + \sum_i \sum_j \overline{a_i(Q)a_j(Q)} \int_0^\infty t_{ij}(r) \sin(Qr)/Q \, dr, \tag{6}$$

the intensity, $I(Q)$, now being normalised to one composition unit. The i summation is taken over the atoms in one composition unit, while that for j is over atom types/elements. The final corrected X-ray or neutron diffraction pattern takes the general form

$$I(Q) = I^\circ(Q) + I^S(Q) + i(Q). \tag{7}$$

$I^\circ(Q)$ is the experimentally inaccessible scattering at $Q{\sim}0$, due to the average sample density, $I^S(Q)$ is known as the self (independent) scattering and $i(Q)$ as the distinct (interference) scattering. The required structural information contained within $i(Q)$ may be extracted by means of a Fourier sine transformation of the interference function $Qi(Q)$ as outlined for neutrons in Fig. 4. The experimental total neutron correlation function is given by

$$T^N(r) = T^\circ(r) + D^N(r), \tag{8}$$

in which $D^N(r)$ is the corresponding differential correlation function

$$D^N(r) = (2/\pi) \int_0^\infty Qi^N(Q)M(Q) \sin(rQ) \, dQ, \tag{9}$$

and

T(r) from a Neutron Diffraction Experiment

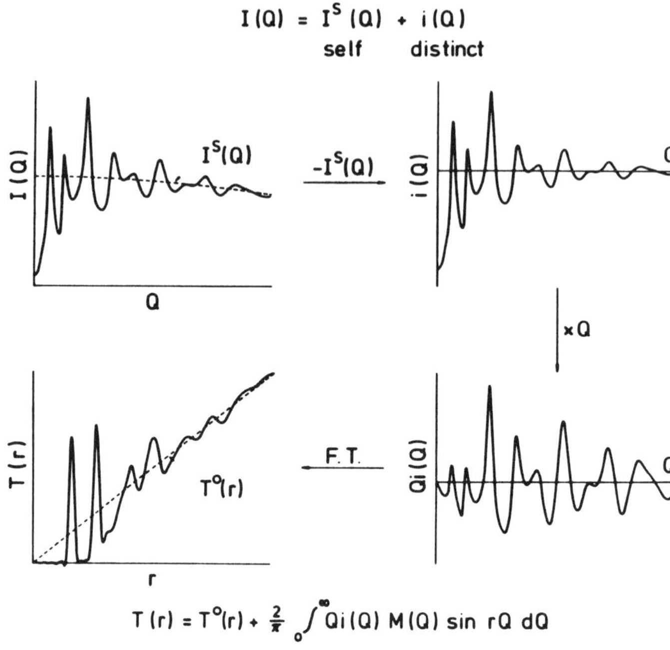

$$T(r) = T°(r) + \frac{2}{\pi} \int_0^\infty Qi(Q)\, M(Q)\, \sin rQ\, dQ$$

Fig. 4. The relationship between the corrected, normalised diffraction pattern I(Q) and the real space correlation function T(r). The data are for vitreous silica.

$$T°(r) = 4\pi r \rho° \left(\Sigma_i \overline{b_i} \right)^2. \tag{10}$$

$\rho°$ is the average composition unit number density and $M(Q)$ is a modification function to allow for the fact that $Qi^N(Q)$ can only be determined for Q less than or equal to some maximum value Q_{max} and is zero for $Q>Q_{max}$.

A simple Fourier sine transformation of the X-ray interference function along the lines of Eq. (9) leads to the electronic correlation function of Finbak [14], which suffers from poor real space resolution since the peaks reflect the finite size of the electron clouds surrounding each atom/ion. For this reason it is conventional to divide $Qi^X(Q)$ by a *sharpening function* before Fourier transformation. For systems containing more than one atomic species, the usual sharpening function is $f_e^2(Q)$, $f_e(Q)$ being the average form factor per electron for the sample in question. This yields the X-ray differential correlation function

$$D^X(r) = (2/\pi) \int_0^\infty [Qi^X(Q)/f_e^2(Q)]\, M(Q)\, \sin(rQ)\, dQ. \tag{11}$$

The expression for $T^X(r)$ is analogous to Eq. (8) for $T^N(r)$ and, in $T^o(r)$ {Eq. (10)}, the atomic number Z_i replaces $\bar{b_i}$.

The fact that data can only be obtained for $Q \leq Q_{max}$ means that the relationship between $T^N(r)$ or $T^X(r)$ and the component correlation functions $t_{ij}(r)$ is one of convolution,

$$T(r) = \sum_i \sum_j t_{ij}'(r), \tag{12}$$

where

$$t_{ij}'(r) = \int_0^\infty t_{ij}(r') \, [P_{ij}'(r\text{-}r') - P_{ij}'(r\text{+}r')] \, dr', \tag{13}$$

r' is a dummy convolution variable and the prime indicates N or X, for neutrons or X-rays respectively. The corresponding peak functions,

$$P_{ij}^N(r) = (b_i b_j/\pi) \int_0^\infty M(Q) \cos(rQ) \, dQ \tag{14}$$

and

$$P_{ij}^X(r) = (1/\pi) \int_0^\infty [f_i(Q) f_j(Q)/f_e^2(Q)] \, M(Q) \cos(rQ) \, dQ, \tag{15}$$

define the experimental resolution in real space, that for neutrons being given for the common modification functions in Fig. 5. Note that the experimental real space resolution (width of the central maximum) is inversely proportional to Q_{max} and that, for X-rays, the different Q dependence of $f(Q)$ for each element means that the shape of $P_{ij}^X(r)$ changes for each independent component. (See, for example, Fig. 14 of [2].)

In principle, $Qi(Q)$ and $T(r)$ contain precisely the same information, albeit expressed in a different form, although in practice the information content of $T(r)$ is slightly reduced by the use of a modification function other than a step function. The advantage of $T(r)$ is that the information for a particular interatomic distance is concentrated around the appropriate value of r, whereas it is spread throughout reciprocal space. Both functions are one-dimensional representations of a three-dimensional structure and are a gross average over the whole irradiated volume. In addition, as shown by Eq. (12) and (13), the experimental correlation function is broadened by $P_{ij}'(r)$ and a single diffraction experiment on a multi-element sample yields only a weighted sum of the individual components, $t_{ij}'(r)$.

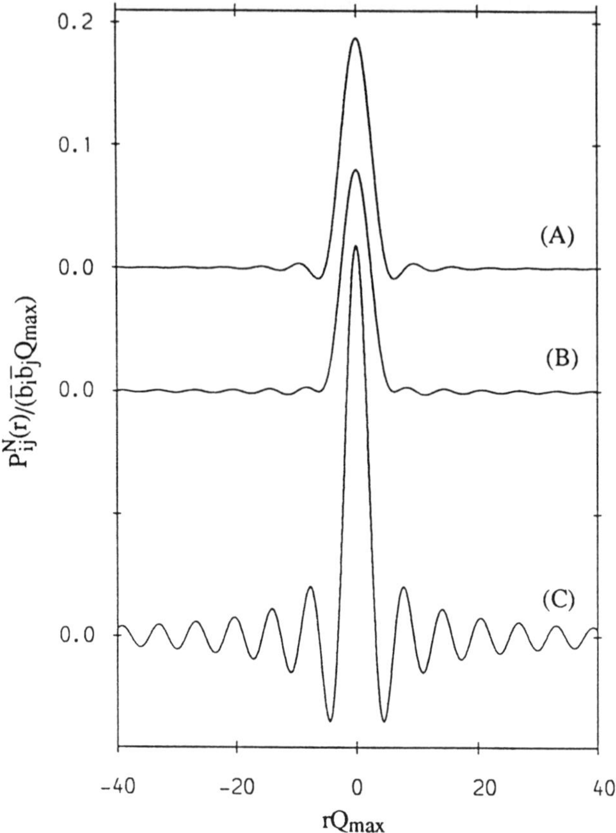

Fig. 5. Reduced neutron peak functions: (A) Lorch function [15], $M(Q) = \sin(\Delta rQ)/(\Delta rQ)$, with $\Delta r = \pi/Q_{max}$; (B) artificial temperature factor, $M(Q) = \exp(-BQ^2)$, with $B = \ln 10/Q_{max}$, and (C) Step function, $M(Q) = 1$.

4. Experimental Techniques

As indicated above, an X-ray or neutron diffraction experiment involves a determination of the scattered intensity as a function of the magnitude of the scattering vector Q. In order to achieve the required variation in Q, it is possible to scan the scattering angle, 2θ, at fixed incident wavelength, λ, (conventional technique) or to make measurements as a function of λ at fixed 2θ.

4.1. NEUTRONS

The two techniques are compared for neutrons in Fig. 6 and examples of the conventional and time-of-flight diffractometers used for amorphous solids are given in [2]. In the conventional steady state reactor twin-axis experiment (e.g. the D4b diffractometer at the

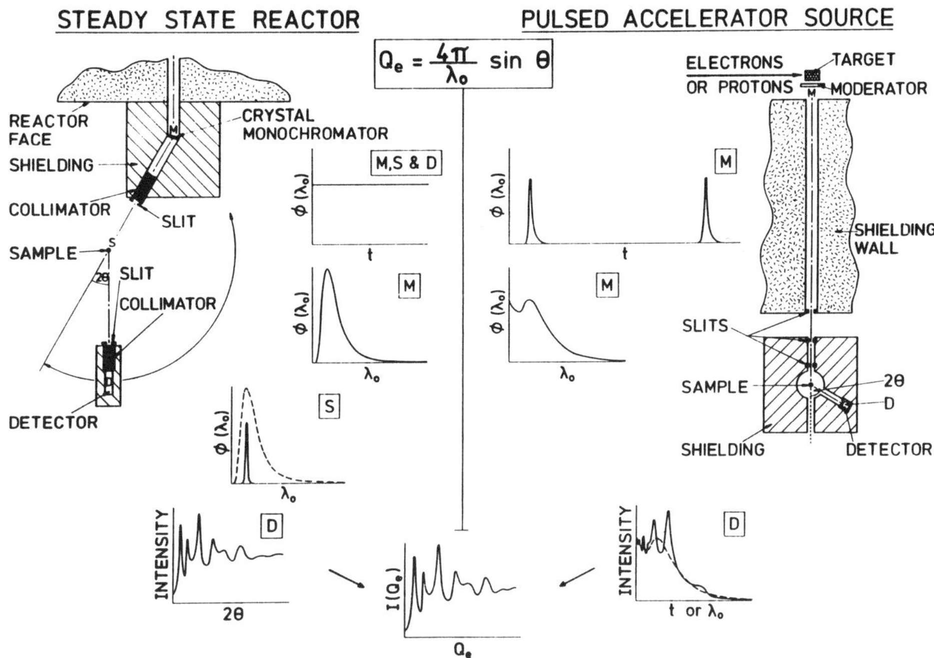

Fig. 6. A schematic comparison of reactor and pulsed source techniques for measuring the neutron diffraction pattern from an amorphous sample.

Institut Laue-Langevin), the flux, $\Phi(\lambda)$, of thermal neutrons extracted from the moderator/reflector of a steady-state reactor has a Maxwellian distribution of velocities and is time independent. A beam of wavelength λ, selected by the monochromator crystal, is incident on the sample and scattered into the detector through a variable angle 2θ. The variable λ time-of-flight technique is usually employed with a pulsed accelerator source. In a spallation neutron source, such as ISIS at the Rutherford Appleton Laboratory, protons strike a heavy metal target producing pulses of fast neutrons which are partially moderated before being incident on the sample and scattered into a detector situated at a fixed angle 2θ. The detector records the scattered intensity, as a function of the time-of-flight for the distance from the moderator via the sample to the detector. For any neutron, the time-of-flight is simply related to λ through the velocity and De Broglie's relationship and the diffraction pattern, I(Q), may be extracted by dividing the measured intensity by the incident neutron spectrum shape. The main difference between the real instruments and the schematic diagrams of Fig. 6 is that the former have multiple counters to increase the counting efficiency and, for the time-of-flight spectrometer, to provide coverage of the full range of Q. The great advantage of a pulsed accelerator source over a

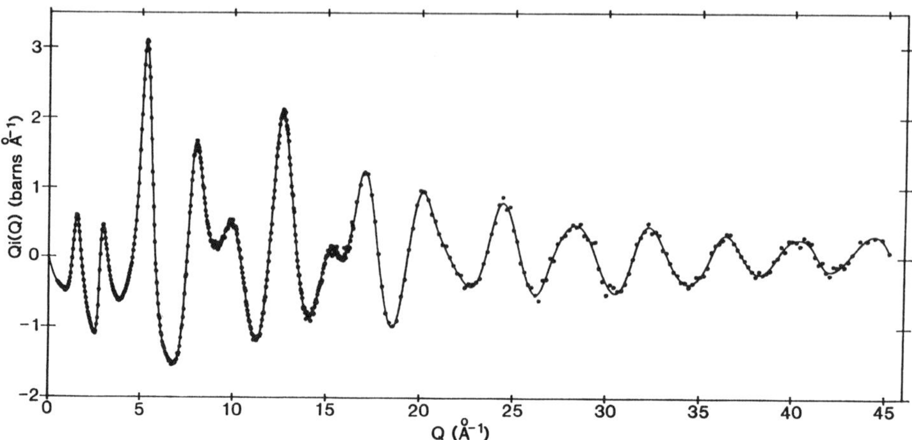

Fig. 7. The interference function for vitreous silica obtained from time-of-flight data (•, experimental points and ——, cubic spline fit).

steady-state reactor is that the former is undermoderated, which gives rise to a strong epithermal component in the incident spectrum, $\Phi(\lambda)$ (cf. Fig. 6). These short wavelength neutrons allow data to be obtained to much higher values of Q, resulting in a corresponding increase in real space resolution, and, as an example, a time-of-flight interference function for vitreous silica is shown in Fig. 7.

4.2. X-RAYS

In the case of X-rays, the situation for amorphous solids is complicated by the presence of incoherent Compton scattering, which at high values of Q can be of much higher intensity than the required coherent contribution. Hence, for accurate quantitative work on amorphous solids, a very much more sophisticated instrument is required, which employs both an incident beam monochromator and an analyser in the diffracted beam. The former is required to ensure monochromatic radiation since, unlike a polycrystalline powder, an amorphous solid does not separate the elastic scattering into clearly defined Bragg peaks. The analyser is required because neither the wavelength distribution nor the integrated intensity of the Compton scattering from a given sample can be satisfactorily calculated and therefore it is necessary either to remove this contribution or to measure the energy distribution of the scattered radiation at each angle and then extract the desired coherent elastic scattering using peak fitting techniques, e.g. by the use of a solid state detector with pulse height (X-ray energy) analysis or an analysing crystal in conjunction with a position sensitive detector.

For laboratory systems, many authors have used only a fixed analyser in the diffracted beam, tuned to the incident K_α wavelength, but the problem with this arrangement is that white background radiation from the X-ray tube can be Compton scattered into the monochromator envelope. A much better arrangement is that due to Warren and Mavel [16], shown in Fig. 8, which employs a conventional curved crystal monochromator in the incident beam (Ag $K_{\alpha 1}$: $\lambda=0.5594$ Å), but the analyser in the diffracted beam comprises a foil with an absorption edge at a wavelength slightly longer than the characteristic line of the X-ray tube (Ru K edge: $\lambda=0.5605$ Å) such that the coherent intensity will excite fluorescence whereas the Compton scattering will not. The fluorescent radiation is then recorded by the detector. Despite the obvious superiority of this technique, however, it is very rarely used with the result that X-ray data quality is frequently not up to that of its neutron counterpart. Mozzi and Warren's X-ray data for vitreous silica [17] are shown in Fig. 9.

The increasing use of synchrotron radiation sources in X-ray diffraction studies of amorphous solids is one of the most important recent developments. Provided X-rays of sufficiently short wavelength (high Q_{max}) can be generated using a wiggler, this type of source has several advantages over a standard laboratory X-ray generator and when used with an appropriate diffractometer, optimised for amorphous materials, is capable of yielding high quality data. The continuous spectrum of a synchrotron radiation source allows the selection of the optimum wavelength for anomalous dispersion experiments (cf. Section 8.2) and the highly collimated beams produced make it possible to perform grazing incidence experiments to investigate very thin films or surface layers. As with laboratory systems, a major factor limiting accuracy is the elimination of the Compton scattering and a relatively recent development has been the adaption of the Warren and

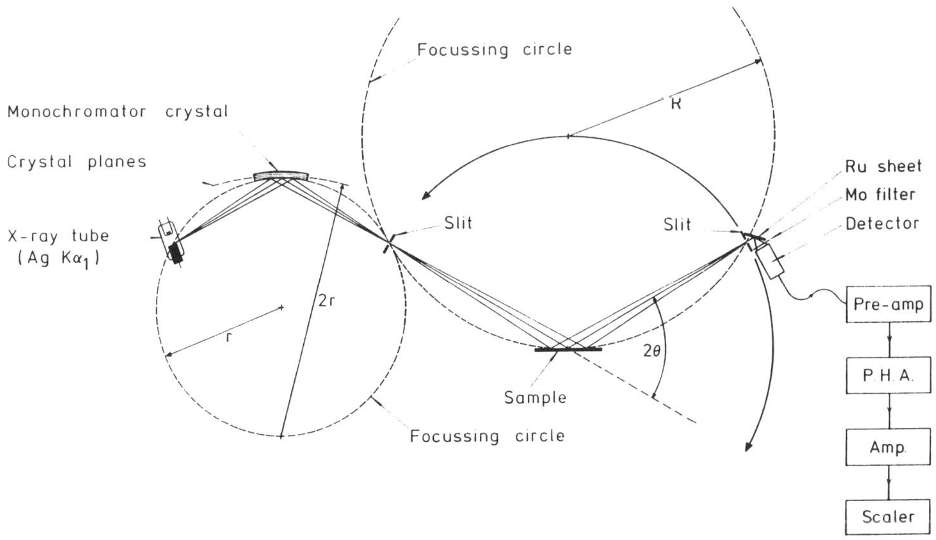

Fig. 8. The fluorescent foil X-ray technique [16].

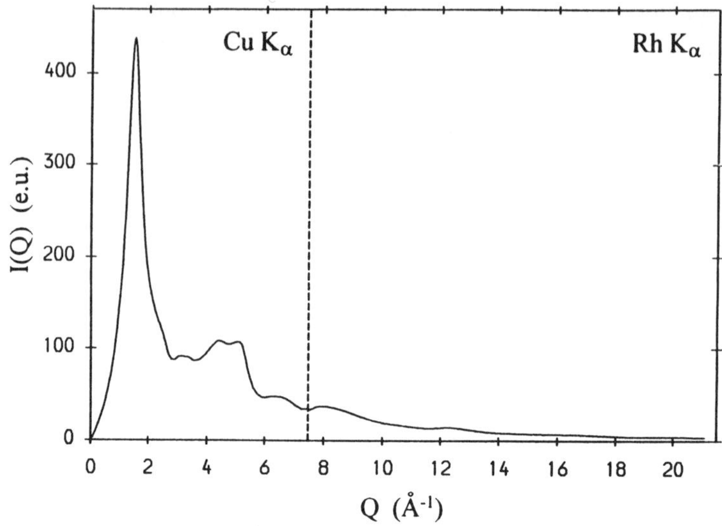

Fig. 9. Mozzi and Warren's X-ray data for vitreous silica [17], which was obtained with Cu K$_\alpha$ and Rh K$_\alpha$ radiation. The vertical dashed line indicates the change from Cu K$_\alpha$ to Rh K$_\alpha$ radiation.

Mavel technique for use on amorphous materials diffractometers at synchrotron sources [18].

The dispersion technique is also possible with X-rays, using a semiconductor detector. The advantage of this method is that, for complex sample environments (such as cryostats, furnaces and high-pressure cells), only two or three small windows are needed, a fixed angle apart. Similarly, all values of Q are examined simultaneously, making the technique ideally suited to following phase changes or studying kinetic effects. Outside such special applications, however, X-ray variable-wavelength experiments are unlikely to prove important for work on amorphous materials, owing to the difficulty of accurately determining the incident spectrum and of making corrections for absorption and Compton scattering.

4.3. DATA CORRECTION

Any measurement involves a certain number of corrections to the raw data, but the good experiment minimises these corrections or puts them in a form in which they are easily handled. A much more detailed account of the corrections applied to both X-ray and neutron data from amorphous solids can be found in [2,3,11,13] and only a very brief summary will be given here. Normally corrections are included for:

4.3.1. *Counter Paralysis Time*
This correction is becoming increasingly important with modern high intensity X-ray and neutron sources.

4.3.2. *Instrumental Background and Sample Container Scattering*
A correction is required both for the empty spectrometer background and also for the scattering from any sample container used.

4.3.3. *Absorption, Self-Shielding and Multiple Scattering*
These corrections are closely coupled and can only be treated independently if one or other of them is small.

4.3.4. *X-ray Corrections*
Corrections specific to X-ray diffraction include those for polarisation and for the residual Compton scattering not removed by the analyser in the diffracted beam.

4.3.5. *Neutron Static Approximation Distortions (Placzek Corrections)*
In a normal amorphous diffraction experiment, the detector records both the elastic scattering and the inelastic scattering arising from the thermal motion of the atoms and hence yields the instantaneous (or static) space-time correlation function $G(r,0)$ [3,10-13]. In the case of neutron diffraction, the incident energy is of a similar magnitude to that associated with thermal vibrations and hence a (Placzek) correction is required for the fact that the detector energy integration is performed at constant 2θ rather than at constant Q.

4.4. NORMALISATION

Following correction, it is necessary to normalise the data to absolute units. This may be achieved either by the use of self consistent (Krogh-Moe [19] - Norman [20]) integration techniques or, for neutrons, by measuring the (incoherent) scattering from a standard vanadium sample.

4.5. FOURIER TRANSFORMATION

One of the most controversial aspects of the analysis of X-ray and neutron diffraction data for amorphous solids is the Fourier transformation of the interference function to give the real space correlation function, mainly because the Fourier transform is a clear indication of data quality. The Fourier transformation may be performed using either Filon's quadrature [21] or a fast Fourier transformation algorithm [22], the latter being more economical on computer time, but much less flexible. The data may also be smoothed before Fourier transformation but this has very little effect on the resulting transform since the frequency range of the noise removed mainly corresponds to distances in excess of those of interest [22].

4.6. ASSESSMENT OF ACCURACY

The statistical accuracy of diffraction data in reciprocal space can be simply calculated from the total number of counts at each intensity point, but a much better indication of the overall accuracy, including systematic errors, is given by the structure in the correlation

100

function at low r below the first true peak. If the transform is well behaved in this region, then the data are of reasonable quality. Great care is needed in making this assessment, however, as some authors either plot the radial distribution function rT(r), which has the effect of reducing the relative amplitude of the error ripples at low r, or use these false oscillations as a criterion for "massaging" their data before publication and do not include the original unadulterated transform. In the absence of other information, such massaged data must be treated with the utmost suspicion. Similar techniques are sometimes used to "remove" the effects of terminating the data at finite Q_{max}. Information theory, however, indicates that it is impossible to replace the unmeasured data without making some assumption about the material under investigation so that the resulting correlation function merely becomes one possible model which "fits" (or not, as the case may be) the results, rather than an unbiased Fourier transform. Similar objections can be raised concerning the maximum entropy technique, which also has the additional problem that it is impossible to know what to do to the correlation function from a structural model in order to be able to directly compare it with the experimental maximum entropy version.

An alternative method of assessing data quality is to perform two independent sets of measurements on the same material and the results of such a study of vitreous silica are illustrated in Fig. 10. The solid line comprises twin-axis data recorded in the 1970s [11] and time-of-flight data from the Total Scattering Spectrometer (TSS; HELIOS, Harwell)

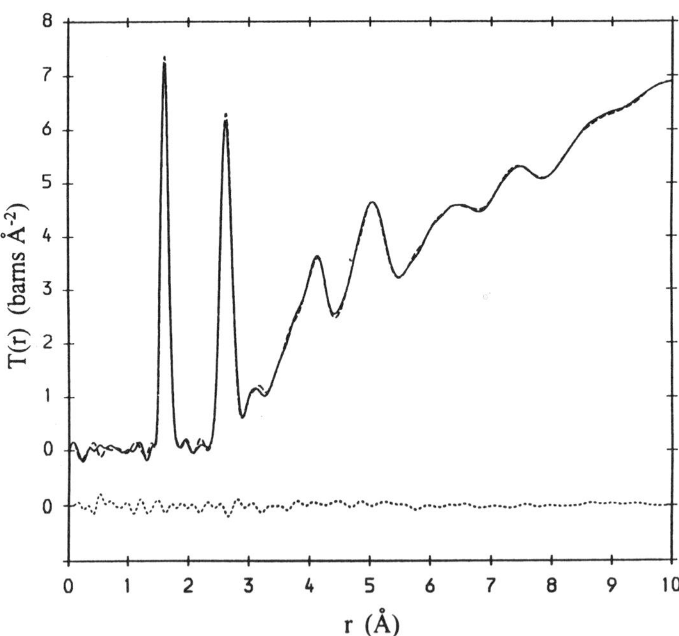

Fig. 10. Two independent determinations of the neutron correlation function for vitreous silica, obtained as described in the text (Q_{max}=45.2Å^{-1}). The difference between the two data sets is indicated by the dotted line.

[13], while the dashed line was obtained using a combination of the D4b diffractometer (ILL) and the LAD spectrometer (ISIS) [23]. The difference between the two correlation functions is denoted by the dotted line, of which the extremities indicate the maximum acceptable deviations for any model that claims to be in agreement with these data (cf. Section 5).

5. Methods of Interpretation

There is an unfortunate tendency in the literature to regard the correlation function, or its individual components, as the final result of a structural study and to merely describe its general form and list a few peak positions. However, the correlation function is closely related to the crystallographer's Patterson function and no crystallographer would ever end a paper with just a Patterson function. The experimental interference and correlation functions should thus be considered the starting point for the real science, once the data reduction is complete, the object being to extract the maximum information concerning the structure of the material in question. The methods for doing this vary for the different ranges of order. In range I, the parameters for the structural unit are best extracted using peak fitting techniques whereas, in ranges II and III, the non-unique, one-dimensional nature of diffraction data leads to an extensive use of modelling techniques. Before discussing each range in detail, however, it is first appropriate to make a few general comments concerning the form of the interference and correlation functions for an amorphous solid and the comparison of peak fits and structural models with experimental data.

Diffraction data are frequently held to be insensitive to the exact nature of the structure of an amorphous solid but, if performed correctly, modern diffraction experiments are able to yield accurate data with high real space resolution, which in practice provide a stringent test for the various structural models proposed in the literature. The main problem with the analysis of diffraction data from amorphous solids is one of uniqueness in that, even if a perfect fit is obtained with experiment, this is no guarantee that there are not other models which would fit equally well. Agreement with diffraction data is thus *a necessary but not sufficient criterion* for any valid structural model. What can be stated with absolute certainty, however, both for diffraction and for the results of any other experimental technique, it that *any model which is not consistent with the data is wrong and must be rejected*. Indeed many models may be rejected on the basis of quite simple measurements, the average number density, ρ°, being an obvious and most important example (cf. Section 6). Too often diffraction data are published, along with the author's favoured model, totally ignoring other published models and/or the results of complementary techniques which would invalidate the model in question.

The amount of structure in the experimental interference and correlation functions for a given material, and hence their information content, varies considerably but tends to be greater for amorphous network solids than for their random close packing based counterparts, as may be seen from Fig. 11 which compares the correlation functions for

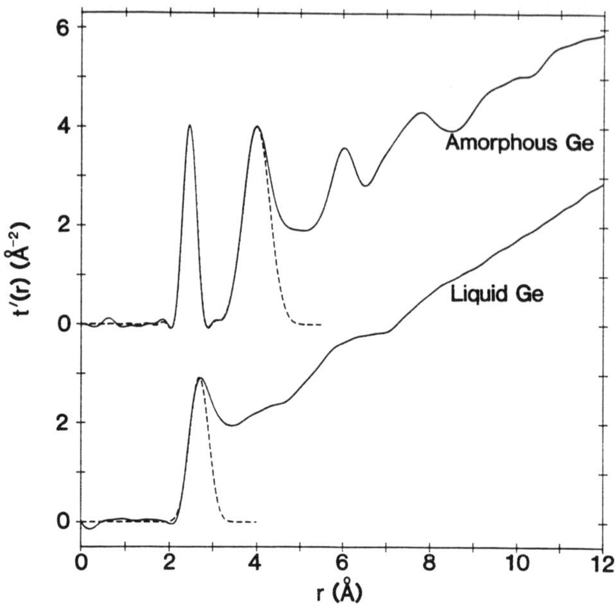

Fig. 11. The correlation function for liquid Ge at 1000°C [24], compared to that for amorphous Ge at ambient temperature [25]. The dashed lines denote peak fits, as discussed in Section 5.1.

amorphous and liquid Ge. The correlation function for liquid Ge is obtained from the data of Salmon [24] and the amorphous solid data are those of [25], Fourier transformed with the same value of Q_{max} (16.1Å^{-1}) so that the curves are directly comparable.

The neutron correlation function for vitreous silica at ambient temperature is shown in Fig. 12, marked with the approximate extent of ranges I, II and III. The high real space resolution (Q_{max} = 45.2 Å^{-1}) means that the characteristics of the three ranges of order can be clearly seen. The two peaks from the intrastructural unit order (range I) are extremely sharp and, even at this value of Q_{max}, there is still a significant contribution to their width from the peak function $P^N_{ij}(r)$ (cf. Section 3, Fig. 5). Ranges II and III overlap but the main features between 3.0 and 5.5Å are due to distances between atoms in adjacent tetrahedra (range II). Their increased breadth compared to the peaks in range I arises from the wide distribution of intertetrahedral Si–O–Si and bond torsion angles. Finally range III is characterised by the broadest features in the correlation function which extend out to ~10-20Å, the limit of the intermediate range order.

The general form of the diffraction pattern, or interference function, for an amorphous solid is also worthy of note since many structural models, particularly those derived from crystalline structures, fail to get this correct. As may be seen from Fig. 12, the absence of long range order leads to a correlation function in which the peaks tend to increase in

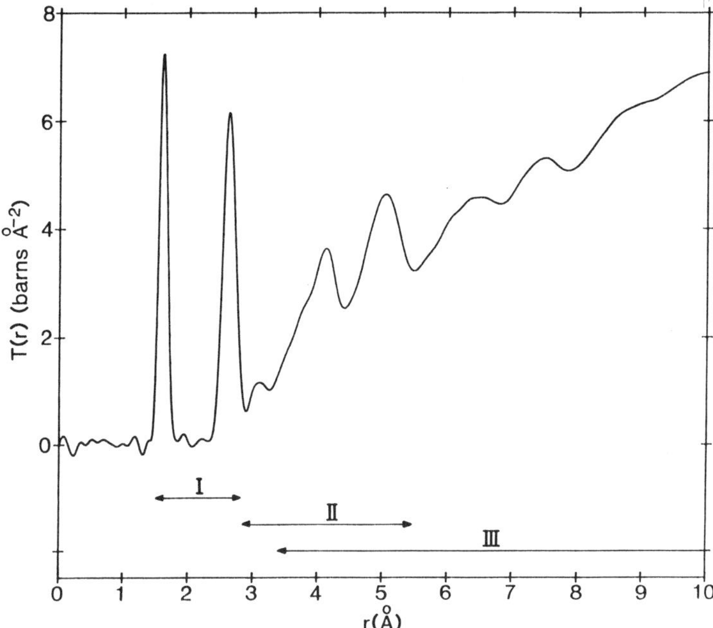

Fig. 12. The correlation function for vitreous silica. The roman numerals indicate the extent of the ranges of order defined in Table 2.

width and decrease in amplitude with increasing r. This means that, in reciprocal space, the highest frequency Fourier components decay the most rapidly and hence the interference function for an amorphous solid is in general characterised by a succession of peaks which get broader with increasing Q until only the Fourier component arising from the sharpest real space peak, usually due to the first interatomic distance, remains (cf. Fig. 7). Thus it is the first few peaks in reciprocal space which are particularly sensitive to the range III order and hence a disagreement between a structural model and experiment in this region ($Q<5Å^{-1}$) is predominantly indicative of incorrect intermediate range order. Similarly the rapid decay of the highest frequency Fourier components means that the first peak in the diffraction pattern from an amorphous material is almost invariably the sharpest (cf. Section 9).

There is considerable misunderstanding in the literature as to what constitutes (good) agreement of a structural model with experiment and the comparison of models with experiment is frequently grossly inadequate. Indeed, for most of the models so far reported, a rigorous comparison reveals serious discrepancies between the model and diffraction data. Many authors term good agreement with experiment as getting the peaks in either Qi(Q) or T(r) in the right place. Only very poor models fail to achieve this. What is necessary is also to get peak shapes and areas correct, which involves including the effects of thermal vibration for static models (e.g. computer relaxed random networks)

104

and in real space folding the model component correlation functions with the correct component peak functions $P_{ij}'(r)$ for the experiment in question. Otherwise, the comparison between model and experiment is meaningless. Note that $P_{ij}'(r)$ has satellite features on either side of the central maximum which depend on the exact form of $M(Q)$ (cf. Fig. 5) and themselves lead to features in $T(r)$. Thus a Gaussian approximation to $P_{ij}'(r)$ is simply not adequate, especially when the width is arbitrarily chosen to give the best agreement with experiment. Resolution effects may also be important in reciprocal space and, if so, the model interference function must be broadened by the experimental reciprocal space resolution function. The functions $Qi(Q)$ and $T(r)$ emphasise different aspects of a given structure and hence it is vital, when comparing models with experiment, to make such comparisons in both real and reciprocal space. It should also be stressed that any valid structural model must not only be consistent with diffraction data but also with all the other available data for the material in question. For example, a reverse Monte Carlo simulation (cf. Section 7.2) may yield a structure which is consistent with diffraction data but has to be rejected due to the presence of too many over or under co-ordinated atoms as determined by magnetic resonance measurements.

One reason why so many inaccurate and irrelevant structural models and computer simulations are accepted for publication in the literature is that authors are not required to include any quantitative measure of the (dis)agreement between their model and experiment. This would never be allowed in a crystallograpic communication, where authors are expected to quote the reliability (R) factor for their structure and compare it with the theoretical minimum R factor for the experimental data. Since the functions used in amorphography have both positive and negative regions, it is convenient to use a slightly different reliability factor [26], which in real space takes the form

$$R_\chi = \left(\sum_i [T_{exp}(r_i) - T_{fit}(r_i)]^2 \Big/ \sum_i T^2_{exp}(r_i) \right)^{\frac{1}{2}}. \tag{16}$$

The discrepancy between the two data sets in Fig. 10, over the range $1 \leq r \leq 8$Å, is characterised by an R_χ factor of 1.5%, which is a measure of the accuracy of the experimental correlation function and indicates the value which should be approached by any structural model which claims to be in agreement with these data.

5.1. RANGE I ORDER

The modern method of extracting range I parameters from diffraction data is via peak fitting techniques. Such fits may be performed either in reciprocal or real space, but the latter has the advantage that the Fourier transformation has the effect of at least partially decoupling the parameters for different peaks. Assuming a Gaussian peak in $t_{ij}(r)$ about a mean interatomic distance r_{ij}, with an r.m.s. deviation of $\overline{u_{ij}^2}$, the contribution to the interference function is given by

$$Qi_{ij}(Q) = n_{i(j)} \overline{a_i(Q)a_j(Q)} \sin(r_{ij}Q) / r_{ij} \exp(-\overline{u_{ij}^2}Q^2/2), \tag{17}$$

in which $n_{i(j)}$ is the co-ordination number of j type atoms about the ith atom in the composition unit. The real space fitting procedure employed by the present author involves folding the Gaussian peak with the peak function $P_{ij}'(r)$ and then optimising the parameters r_{ij}, $\overline{u_{ij}^2}$ and $n_{i(j)}$ using a χ^2 minimisation routine.

Peak fits to the X-ray and neutron correlation functions for vitreous SiO_2, with the Lorch modification function [15], are illustrated in Fig 13, together with the appropriately weighted unbroadened distributions, and the peak parameters are summarised in Table 3. In both fits, parameters for the first Si–Si distance are taken from the work of Konnert and Karle [27] and this distribution is included since its low r tail extends under the O–O peak. For X-rays, the unbroadened peaks are weighted according to the atomic numbers of Si and O and it is interesting to note that the relative heights of the O–O and Si–Si peaks in the experimental correlation function are not what would be expected from the atomic numbers of Si and O due to the satellite features on either side of the central maximum of the component peak functions (Fig. 14 of [2]). The areas under the O–O and Si–Si peaks are $768e^2$ and $784e^2$, respectively.

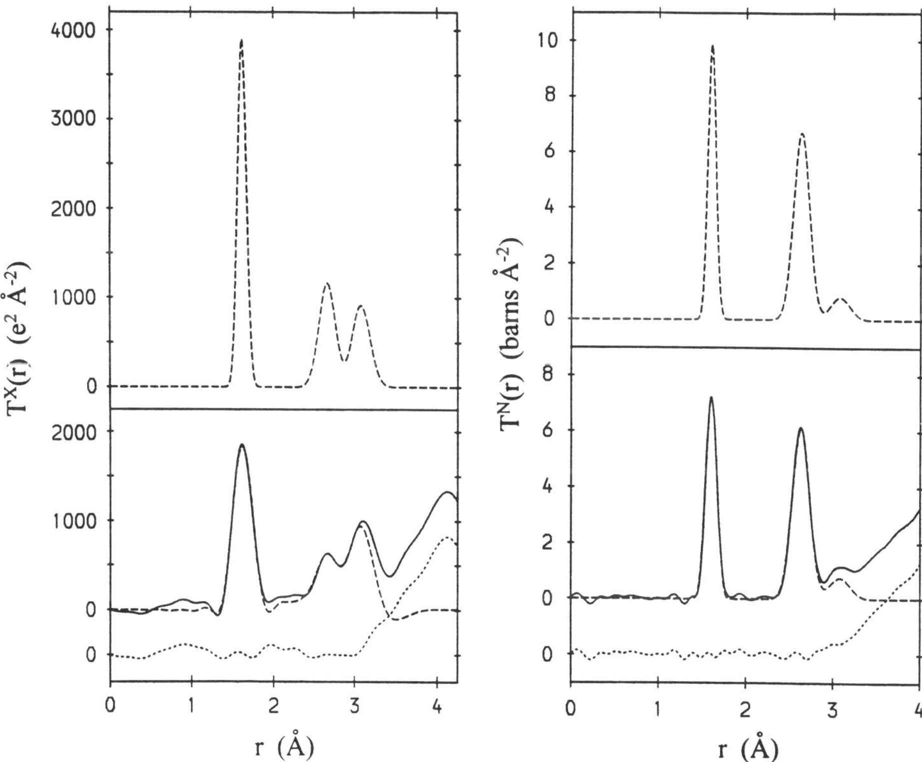

Fig. 13. Peak fits (lower curves) to the X-ray (left) and neutron (right) correlation functions for vitreous silica, obtained using the Lorch [15] modification function, together with the corresponding unbroadened peaks (upper curve). The peak parameters are given in Table 3. ———, experiment; – – – –, fit and ············, residual.

TABLE 3. X-ray and neutron peak parameters for vitreous silica.

Radiation	R_χ	i-j	$n_{i(j)}$	r_{ij} (Å)	$(\overline{u^2_{ij}})^{1/2}$(Å)
X-rays	7.3	Si-O	3.77±0.12	1.626±0.004	0.053±0.009
		O-O	6.15±0.38	2.657±0.012	0.102±0.016
		Si-Si*	4.0	3.077	0.111
Neutrons	3.8	Si-O	3.85±0.16	1.608±0.004	0.047±0.004
		O-O	5.94±0.23	2.626±0.006	0.091±0.005
		Si-Si*	4.0	3.077	0.111

* From [27].

The increase in the width of the second peak in the correlation function over that for the Si–O bond length reflects the variation in the O–Si–O intratetrahedral bond angle and hence is a measure of the tetrahedral distortion. Assuming that the bond length and bond angle variations are uncorrelated, it is possible to use a Monte Carlo technique to extract the mean bond angle and the r.m.s. bond angle variation from the widths of the first two peaks [28]. Using this procedure with the neutron peak parameters yields values of 109.7±0.4° and 4.5±0.4°, respectively, the latter including both thermal and static disorder [28]. The agreement between the former and the expected tetrahedral angle of 109.5° is well within the experimental uncertainty. A fit to the first two peaks for amorphous Ge is shown in Fig. 11 and indicates that the distribution of intra-tetrahedral angles (r.m.s. variation 10.5±0.7° [28]) is much broader than that for vitreous silica, reflecting the very much higher level of network strain associated with its more extreme method of preparation.

The structure of borate glasses is of considerable interest, both due to the change in $n_{B(O)}$ on the addition of network modifiers (cf. Section 1.3) and the possible presence of superstructural units (cf. Section 1.4). On adding PbO to vitreous B_2O_3, the first (B–O) peak in T(r) increases in both width and area, due to the presence of tetrahedrally co-ordinated boron atoms. To investigate the fraction of 4-fold co-ordinated boron atoms, x_4, for a series of PbO-B_2O_3 glasses [29], a fit has been performed to this peak using both a single B–O distance and also independent B–O distances for the BO_3 and BO_4 structural units, as illustrated for vitreous PbO·3B_2O_3 in Fig. 14. The R_χ factor is significantly reduced for the two-distance fit (1.5% vs. 5.9%) and the total peak area from the two-distance fit yields a value for x_4 of 0.25±0.06. The two-distance fit indicates B–O distances of 1.36±0.01 and 1.48±0.01Å for the BO_3 triangles and BO_4 tetrahedra, respectively, in good agreement with the corresponding values for crystalline borates.

The characteristic signature of a superstructural unit is one or more sharp peaks at values of r in excess of the largest distance within the basic structural unit (cf. Section 2.1), such as that at 3.6Å in the correlation function for vitreous B_2O_3 (Fig. 15), which arises from the distance r_4 in Fig. 3. The role played by boroxol groups in the structure of vitreous B_2O_3 is still the subject of considerable controversy in the literature and, in

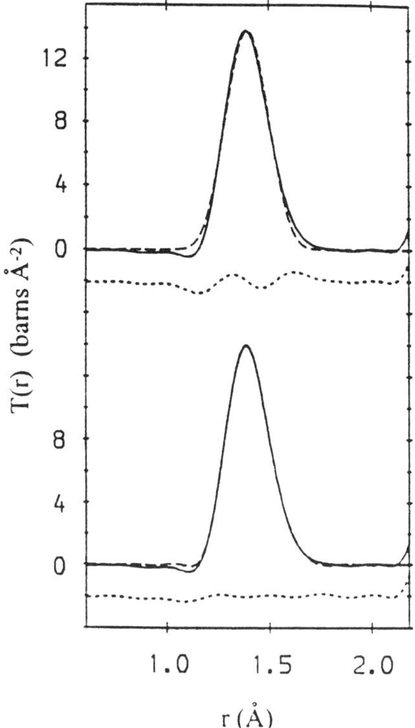

Fig. 14. A fit (dashed line) to the first B–O peak (solid line) for vitreous $PbO\cdot3B_2O_3$ [29], using one (upper curves) and two (lower curves) distances. The R_χ factors are 5.9 and 1.5%, respectively, and the dotted line is the residual.

particular, the fraction, f, of the boron atoms involved in boroxol groups [30]. The value of f has been redetermined from new neutron diffraction data, by performing a fit to the interference function, $Qi^N(Q)$, in the range $5\leq Q\leq23.24\text{Å}^{-1}$. A plot of the R_χ factor vs f indicates a value of 0.80±0.05, in excellent agreement with earlier NMR and neutron diffraction results [30]. The corelation function corresponding to the optimum fit (f=0.80) is also shown in Fig. 15.

5.2. RANGE II ORDER

In addition to interatomic distances and bond angles, the quantification of the order for an amorphous network solid in range II involves the distribution of bond torsion angles $A(\alpha)$ which cannot be obtained directly from diffraction data without assuming some model. For materials such as SiO_2, the difference between the structures of the crystal and glass are that, in the latter, there is a broad distribution of Si–O–Si and bond torsion angles whereas in general the former is characterised by one, or at most a few, discrete values. One simple model for the interconnection of structural units in a random network

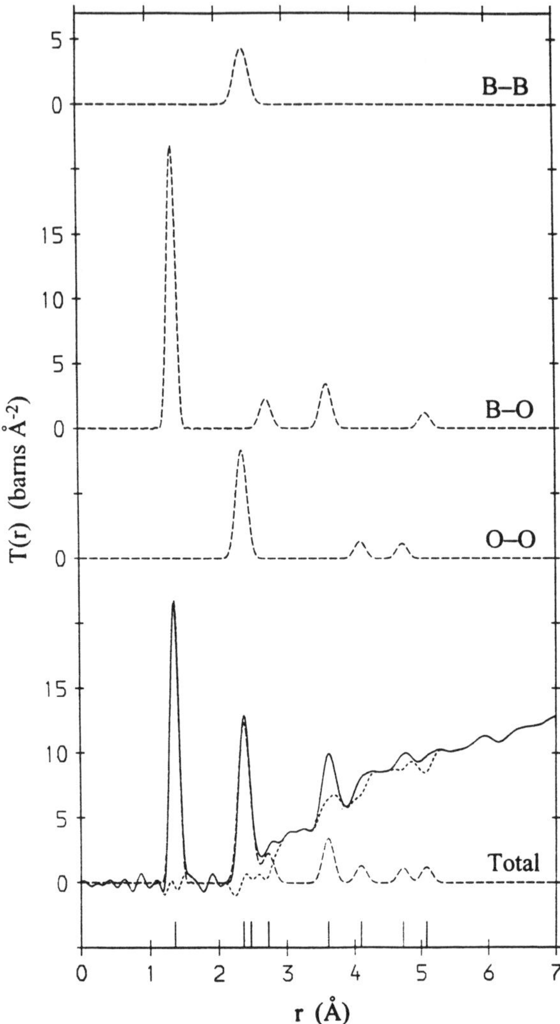

Fig. 15. The real space correlation function for vitreous B_2O_3 [30]. ———, experiment; – – – –, optimum boroxol ring fit (f = 0.80) and ···········, residual. The vertical lines indicate the distances within the boroxol group.

assumes a uniform distribution of torsion angles, except where excluded by steric effects, and has been used by Mozzi and Warren [17] to obtain the Si–O–Si bond angle distribution in Fig. 16. The structural models discussed in the next section on range III order must also, by definition, contain some inherent bond and torsion angle distribution and indeed often fail in this respect, making a comparison of the range III order with experiment pointless.

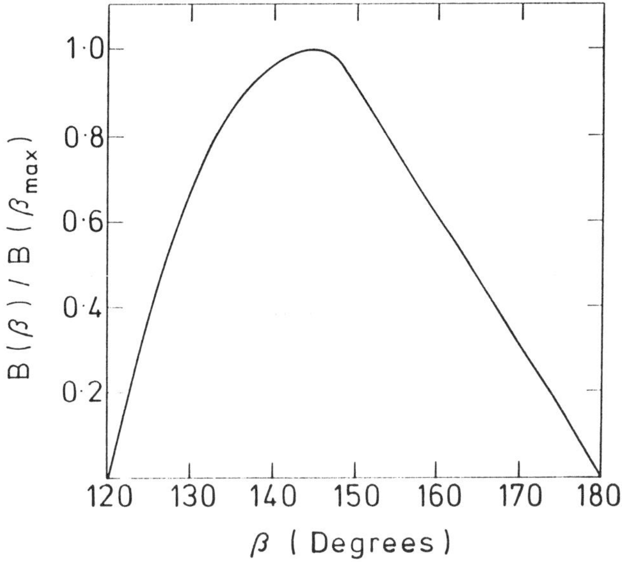

Fig. 16. The Si–O–Si bond angle distribution for vitreous silica as determined by Mozzi and Warren [17].

5.3. RANGE III ORDER

The usual method of investigating the order in range III is via modelling studies, which fall into two groups, those of an analytical nature and those leading to actual atomic co-ordinates, where an attempt is normally made to generate a "typical region" of the structure that can be used both to compare to diffraction data and in the calculation of other properties. The problem with the first group is that there is no guarantee that it is possible to build a real structure corresponding to the analytical formalism and hence the second group is of very much greater use in understanding the structure of amorphous solids. A wide range of structural models of amorphous insulators and semiconductors has been presented in the literature and the most important of these are summarised in Table 4. Within the space available, it is only possible to give a few examples of the use of structural models in interpreting diffraction data for amorphous network solids and these are chosen to illustrate specific points. Further examples can be found in [2,23].

6. Structural Modelling

The importance of modelling techniques in the study of amorphous solid structures cannot be overstated. The simple act of building a model by hand gives a "feel" for the structure which it is difficult to obtain in any other way and may lead to the conclusion that a postulated structure is sterically impossible, even if no quantitative information is extracted from the finished model. The simplest models are of the hand-built variety,

110

TABLE 4. Structural models of amorphous network solids.

Model	Example(s)
Random Network	
(a) Hand built	As, As_2O_3, B_2O_3, Ge, SiO_2
(b) Computer generated	Ge, SiO_2
(c) Geometric transformation	B_2O_3, H_2O, P, P_2O_5
Polymer Models	
(a) Random coil	S, Se
(b) Bundled coil	
Crystal Based Models	
(a) Limited range of order	$GeSe_2$, SiO_2
(b) Strained crystal	SiO_2
Molecular Model	CCl_4, P_4Se_3, $As_{1-x}S_x$
Layer Model	$As_{1-x}S_x$, $Ge_{1-x}Se_x$
Amorphous Cluster	Ge, SiO_2
Monte Carlo	
(a) Energy minimisation	BeF_2, Ge
(b) χ^2 minimisation	$As_{1-x}Se_x$, SiO_2, $ZnCl_2$
Molecular Dynamics	BeF_2, SiO_2, Na_2O-SiO_2

which are constructed using suitable units to reflect the experimental parameters such as the structural unit regularity and the bond angle distribution. The disadvantage of hand building is that the construction and quantitative measurement of large models is extremely tedious and it is difficult to vary the model parameters in a systematic way. The use of computer relaxation techniques (molecular mechanics), however, reduces the accuracy with which it is necessary to measure the atomic co-ordinates and also removes any anisotropy which might result from the effects of gravity on a large model. Alternatively, models may be completely computer generated which removes both the tedium and any unintentional bias associated with hand-built models.

An important aspect of co-ordinate models, as opposed to the analytical variety, is the problem of finite model size, which limits the correlation function to interatomic distances less than or equal to some maximum value, r_{max}. This problem may be approached in two different ways, cluster models and those which employ periodic boundary conditions, and it is vital to understand the restrictions which these place on the model in both real and reciprocal space. It is important to note that most structural modelling is performed in real space, whereas diffraction data are collected in reciprocal space, and hence, to compare model and experiment, it is necessary to perform a Fourier transformation, either explicitly or implicitly. However, while the effect of the finite range of Q on diffraction data is reasonably well understood (cf. Section 3), it is not generally appreciated, particularly by those who build structural models but do not themselves perform diffraction experiments, that exactly the same considerations apply when transforming a structural model to reciprocal space.

The average number density, $\rho°$, is a very simple structural parameter, but nevertheless an extremely important one, since any model must be capable of predicting the correct density. In the words of Finney [31], *"Models are wrong whose density is wrong."* A difficulty in this respect is that the equilibrium bond length for a given material is frequently not known to much better than 1%, which is equivalent to a 3% uncertainty in the density. Uniformity of density may also be a problem for cluster models, particularly for those which are computer generated, in that the value of $\rho°$ is not maintained as extra units are added to an initial nucleus. A useful check [32] is to determine $\rho°$ as a function of model size from the number of atoms inside a sphere of increasing radius R, about the model centre. As R increases, $\rho°$ should approach a constant value and then fall off at the edge of the model. The point just before the density begins to fall off determines the largest possible spherical model for use in calculating the model component correlation functions (cf. Section 6.1).

6.1. CLUSTER MODELS

The simplest method of calculating the component correlation functions, $t_{ij}(r)$, for a cluster model is only to use those atoms as centre which are at distances in excess of r_{max} from the model surface. In this case, there is no effect of the finite model size on $t_{ij}(r)$ for $r \leq r_{max}$ and, on transformation to reciprocal space, the corresponding interference functions, $Qi_{ij}(Q)$, are simply broadened by the real space peak function, $P(Q)$, appropriate to the modification function used in the Fourier transformation; e.g., for a step function cut-off, $P(Q)$ is of the same form as Fig. 5C. The problem with this approach is that only a small fraction of the total atoms are used as centre and hence the statistical accuracy of $t_{ij}(r)$ is very much poorer than if all the whole model had been used. Consequently, it is normal to use all of the atoms in a spherical cluster, of radius R ($r_{max}=2R$), taken from the centre of the model (cf. Section 6), for which the component correlation functions are reduced relative to those for an infinite model by the factor [32]

$$F(r) = (r/R - 2)^2 (r/R + 4)/16 \qquad r \leq 2R$$
$$F(r) = 0 \qquad r \geq 2R. \tag{18}$$

The corresponding reciprocal space peak function is the Fourier cosine transform of $F(r)$ and takes the form

$$P(Q) = [3/(4\pi RQ^2)][1 - 2\sin(2RQ)/(2RQ) + \sin^2(RQ)/(RQ)^2], \tag{19}$$

which is identical to the particle size broadening function for a polycrystalline material with spherical crystallites [33]. The finite model size also leads to small Q scattering appropriate to a uniform sphere of the same size and average scattering density as the original model,

$$I^\circ(Q) = [12\pi\rho^\circ/(R^3 Q^6)] \left[\sum_i \overline{a_i(Q)}\right]^2 [\sin(RQ) - RQ\cos(RQ)]^2. \tag{20}$$

This is equivalent to the average density term, $I^\circ(Q)$, for a real sample (cf. Section 3). Thus, for a cluster model, the Fourier transformation to reciprocal space leads to a broadening of the interference function and the introduction of satellite features, in exactly the same way as for an experimental correlation function in real space. Hence, in order to compare model and experiment in reciprocal space, it is necessary to fold the *experimental* interference function with the reciprocal space peak function, $P(Q)$, appropriate to the model calculation. Similarly the model interference function must be broadened by the reciprocal space resolution function for the instrument used in the experiment, if this is significant.

6.2. PERIODIC BOUNDARY CONDITIONS

The use of periodic boundary conditions in the modelling of amorphous materials is extremely controversial, in that a periodic boundary model is in reality a crystal. The lack of internal symmetry within the unit cell means that the structure belongs to the triclinic P1 space group, although the unit cell itself is frequently pseudo-cubic. Any calculation which uses periodic boundary conditions, therefore, is strictly appropriate to a P1 crystal and not an amorphous solid. For a periodic boundary model, r_{max} is usually taken as half the smallest unit cell dimension and the periodic images are used in calculating $t_{ij}(r)$ to avoid any effects of the finite cell for $r \leq r_{max}$.

In reciprocal space, the application of periodic boundary conditions means that, with the exception of (inelastic) thermal diffuse scattering, the diffraction pattern for such a structure, after averaging over all orientations of the unit cell with respect to the scattering vector $\mathbf{Q}$, is only sampled at discrete values of Q (Q_{hkl}, the δ-function Bragg peaks) and comprises a Debye-Scherrer polycrystalline powder diffraction pattern, not the continuous distribution typical of an amorphous solid. Assuming a pseudo-cubic unit cell, with lattice parameter a_0, the positions of the Bragg peaks are given by

$$Q_{hkl} = (2\pi/a_0)(h^2 + k^2 + l^2)^{\frac{1}{2}} = (2\pi/a_0)m^{\frac{1}{2}}, \tag{21}$$

where m is an integer, and there is no elastic scattering below the Bragg cut-off at $Q = 2\pi/a_0$. The correct method of calculating the interference function for a periodic boundary model is to use the formalism appropriate to crystals such that

$$Qi(Q) = (2\pi^2\rho^\circ/n_u^2)\sum_h\sum_k\sum_l |F_{hkl}|^2 \delta(Q - Q_{hkl})/Q_{hkl}$$
$$- Q\sum_i \overline{a_i^2(Q)}\exp(-u_i^2 Q^2/3), \tag{22}$$

in which $|F_{hkl}|^2$ is the isotropically averaged square of the structure factor

$$|F_{hkl}|^2 = \sum_i \sum_j \overline{a_i(Q)a_j(Q)} \cos 2\pi[h(x_i - x_j) + k(y_i - y_j) + l(z_i - z_j)]$$
$$\times \exp[-(\overline{u_i^2} + \overline{u_j^2})Q_{hkl}^2/6]. \tag{23}$$

The i and j summations are taken over the atoms in the unit cell, with fractional co-ordinates x_i, y_i and z_i, n_u is the number of composition units in the unit cell and h, k and l are the conventional Miller indices. Equations (22) and (23) assume that the atomic rms thermal displacements are spherically symmetric.

In Fig. 17, the spacing of the Bragg peaks at low Q for a model of amorphous Ge by Wooten and Weaire [34] (cf. Section 7.2) is shown, relative to the experimental neutron diffraction pattern for amorphous Ge. Each Bragg peak is represented by a vertical line in the appropriate position, the height of which is proportional to the peak intensity for the static model. The prominent Bragg peak at 1.87Å^{-1} is close to the 111 Bragg reflection for crystalline Ge, suggesting that the model may contain a small amount of residual crystallinity. Conventionally, such a diffraction pattern is artificially broadened, e.g. by summing the δ-function Bragg peaks into relatively broad histogram columns, in which case it is imperative that, before intercomparison, the experimental data are treated identically.

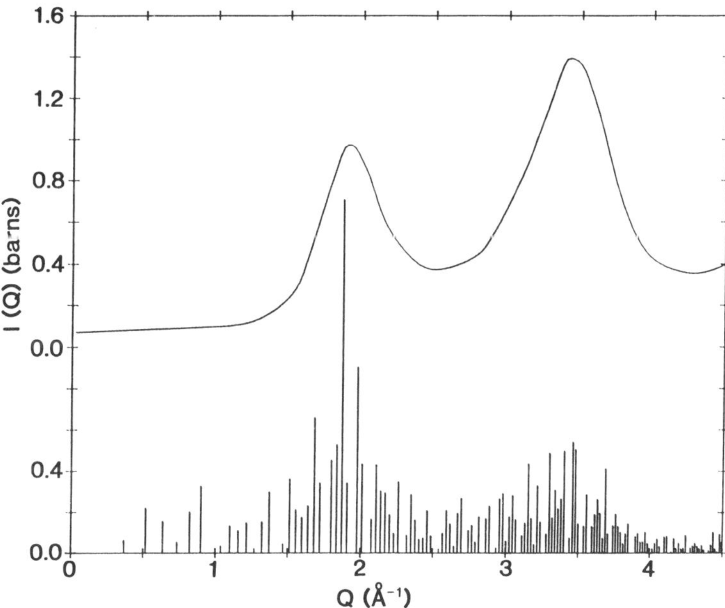

Fig. 17. The positions and intensities of the Bragg peaks (vertical bars) for a model of amorphous Ge, generated by Wooten and Weaire [34] (Model C in Fig. 20), relative to the experimental neutron diffraction pattern for amorphous Ge [25] (solid line).

114

Many workers, using periodic boundary models, incorrectly calculate reciprocal space functions for their model by Fourier transforming the component real space differential correlation functions, $d_{ij}(r)$, over the range $0 \leq r \leq r_{max}$ (cf. Section 6.1), but this results in a modified model, for which the range of order is limited by an *extra* multiplicative factor, $F(r)$, and does *NOT* yield the true interference function appropriate to their model (i.e. the factor of $F(r)$ introduces extra disorder, which is not present in the original model). Similarly, any calculation which truncates the structure (correlation function) at some distance less than a_0 is simply incapable of distinguishing between the crystalline and amorphous states and the limitation of the range of order is a property of the calculation, not the underlying structural model.

6.3. MINIMUM MODEL SIZE

The discussion of the preceeding two sections raises the question as to the minimum size for a model to be truly representative of the structure of the material in question and to accurately predict its properties. There are two aspects to this question. Clearly, the model must be of a sufficient size to yield statistically accurate distributions of all the relevant structural parameters, such as those in Table 2. However, there is also a minimum size to ensure the correct diffraction properties, which will now be discussed. In this connection, it is interesting to consider the structure of triclinic α-tridymite which, at first sight, would seem to satisfy many of the criteria for a periodic boundary model of vitreous silica, in that it has a large unit cell, containing 960 atoms, with a wide distribution of Si–O–Si bond angles ($139.7 \leq \beta_0 \leq 173.2$) [35]. However, the diffraction pattern of α-tridymite is very definitely that of a typical polycrystalline powder.

For a periodic boundary model, one possible criterion for minimum model size is that the calculated diffraction pattern should be continuous over the first peak, within the experimental resolution, so that the only necessary broadening of the model interference function would be by the experimental reciprocal space resolution function $P(Q)$. Differentiating Eq. (21) and taking finite elements, the minimum lattice parameter a_{min} becomes

$$a_{min} = (\pi/Q_1)[2\Delta m/(\Delta Q/Q)]^{\frac{1}{2}}. \tag{24}$$

Since m is an integer, the minimum Δm is unity and the best experimental resolution in the region of the first diffraction peak at Q_1 is ~1% ($\Delta Q/Q = 0.01$). Hence

$$a_{min} \simeq 45/Q_1. \tag{25}$$

This gives values for a_{min} of ~25Å and ~30Å, respectively, for amorphous Ge ($Q_1 = 1.91$Å^{-1}) and vitreous silica ($Q_1 = 1.53$Å^{-1}), although it should be noted that this only yields a continuous diffraction pattern, for which the statistical noise will almost certainly still be far too high. It is interesting to compare this method of calculating a_{min} with the criteriton $r_{max} \leq a_0/2$ (cf. Section 6.2). For vitreous silica, Konnert and Karle [27] report measurable structure out to ~20Å and thus, for a periodic boundary model to be capable of predicting this structure, it should have a minimum lattice parameter of ~40Å.

Similar minimum model size criteria can also be formulated, in both real and reciprocal space, for cluster models. In reciprocal space, the finite model size leads to a broadening of the interference function. The full width at half maximum height of the the the peak function, $P(Q)$, for a spherical model {Eq. (19)}, is 3.477/R. A frequently used criterion for experimental broadening to be neglected is that it should not exceed one tenth of the true width of the peak in question. Applying this criterion, to the finite model size broadening of the first diffraction peak, leads to a minimum radius

$$R_{min} \simeq 35/\Delta Q_1. \qquad (26)$$

The width of the first diffraction peak, ΔQ_1, for vitreous silica is 0.63Å^{-1} yielding a minimum model radius of ~55Å.

In real space, the finite size of a cluster model limits the statistical accuracy with which $T(r)$ can be determined at high r, since the multiplicative correction factor {the reciprocal of $F(r)$ in Eq. (18)} increases rapidly for $r > R$ {$1/F(R) = 3.2$} and hence the model radius should be greater than the correlation length for the real material. For vitreous silica this yields $R_{min} = 20$Å which is much smaller than the value obtained from the reciprocal space criteria. Hence, for models intermediate in size between the real and reciprocal space criteria, the correct broadening of the *experimental* reciprocal space data will be required, before comparison with the model. Note that, for good amorphous diffractometers, the experimental broadening of the first diffraction peak can frequently be neglected and hence it may not be necessary to correct model interference functions for experimental reciprocal space resolution.

For materials which can be prepared as extremely small particles ($R \lesssim 30$Å), the exciting possibility exists of generating cluster models of the same size as the actual particles. In this case, the finite model size is truly representative of the actual sample and the model can be used to investigate the structure of the particles' surface layers.

7. Examples of Structural Models for Amorphous Network Solids

7.1. RANDOM NETWORK MODELS

Two important hand-built structural models, which have been much used in the literature, are the Bell and Dean model for vitreous silica [36] and the Polk and Boudreaux model for amorphous Si and Ge [37]. The correlation function for the Bell and Dean model, after energy minimisation [38] and thermal broadening, is compared to neutron and X-ray data in Fig. 18. The peaks at 4.1Å (mainly Si–O interactions) and 5.0Å (O–O) arise mainly from the range II order and it can be seen that the shoulder on the low r side of the 4.1Å

116

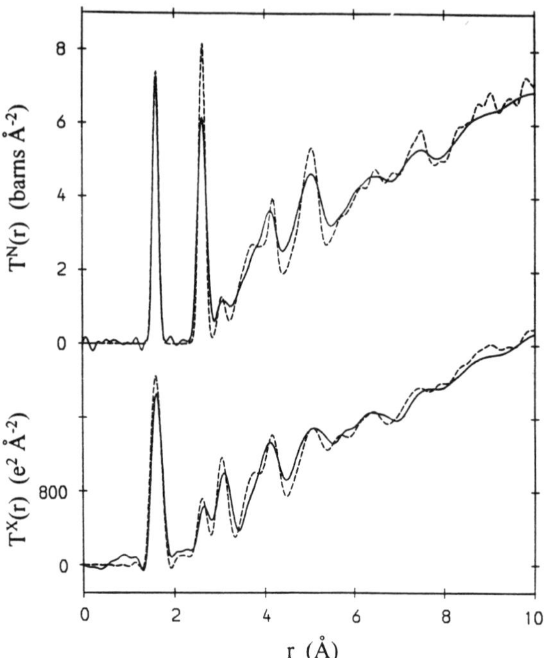

Fig. 18. Neutron (upper curves) and X-ray (lower curves) correlation functions for the Bell and Dean model of vitreous silica [36,38] (dashed lines) compared to experiment (solid lines).

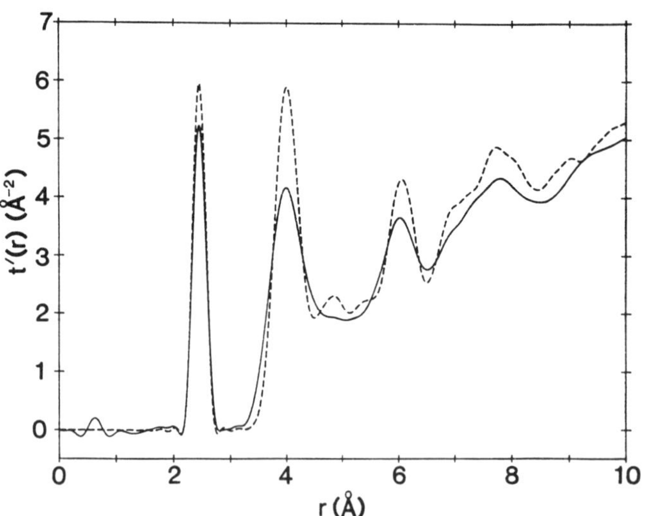

Fig. 19. The correlation function for the Polk and Boudreaux [37] model of amorphous Ge compared neutron data [25]. ———, experiment and – – – –, model.

peak is much too pronounced compared to experiment, indicating that the range II order in the model is incorrect. This is a common failing of models of vitreous silica and a similar discrepancy between model and experiment occurs for almost all the other random network models of this material discussed in the literature [23]. The Polk and Boudreaux model [37] is compared to neutron data for amorphous Ge [25] in Fig. 19 and is interesting because it was constructed from rigid plastic tetrahedra joined together with rigid aluminium sleeves (cf. Fig. 39 of [2]), which had the effect of constraining the Ge–Ge–Ge bond angle distribution to be much too narrow, as may be seen from the second peak in the correlation function. Hence it is a classic example of using inappropriate units in a modelling study, which lead to an incorrect structure.

7.2. MONTE CARLO TECHNIQUES

The time-consuming nature of hand building and the difficulty of controlling the distribution of the various structural parameters during construction means that the future must reside with computer generated models, using either an algorithm which mimics the hand-building process or some form of Monte Carlo technique. The latter are of two main types, according to the quantity which is minimised. In conventional Monte Carlo simulations, the energy is minimised while, in the so-called reverse Monte Carlo technique, the algorithm optimises the agreement with the experimental data. Although usually termed reverse Monte Carlo simulation, this technique does not in practice involve a simulation, since it does not try to generate the structure and properties of the material in question starting from some suitable interatomic potential. Instead, it moves the atoms in a periodic unit cell or cluster to yield the best agreement with experiment, as indicated by a reliability factor such as that given in Eq. (16), and is thus the amorphous solid equivalent of the Rietveld technique used in powder crystallography.

A novel and highly successful Monte Carlo technique, for generating random network models of amorphous Si and Ge ("Sillium" models), has been evolved by Wooten and Weaire [34]. This starts with the ambient crystalline network and then introduces disorder via a series of topological transformations which are tested against a Maxwell-Boltzmann factor as in the case of conventional Monte Carlo simulations. Correlation functions for three models due to Wooten and Weaire are compared to neutron diffraction data in Fig. 20. Models A and B are topologically equivalent, but have been relaxed with different interatomic potentials, while model C contains 4-membered rings which were specifically excluded from A and B. It can be seen that all three models have a bond angle distribution very much closer to that of the real material than the Polk and Boudreaux model [37] (Fig. 19) and in general are in very good agreement with experiment, except that the average model number densities are too high. The additional structure in t'(r), at higher r, is almost certainly statistical in nature, due to the small model size (216 atoms).

The reverse Monte Carlo technique was originally devised by Rechtin et al. [39], but the most recent advances have been made by McGreevy et al. [40], who have introduced an exponential probability for accepting unfavourable moves and used very much larger systems. Clearly, given adequate computer time, reverse Monte Carlo modelling is

118

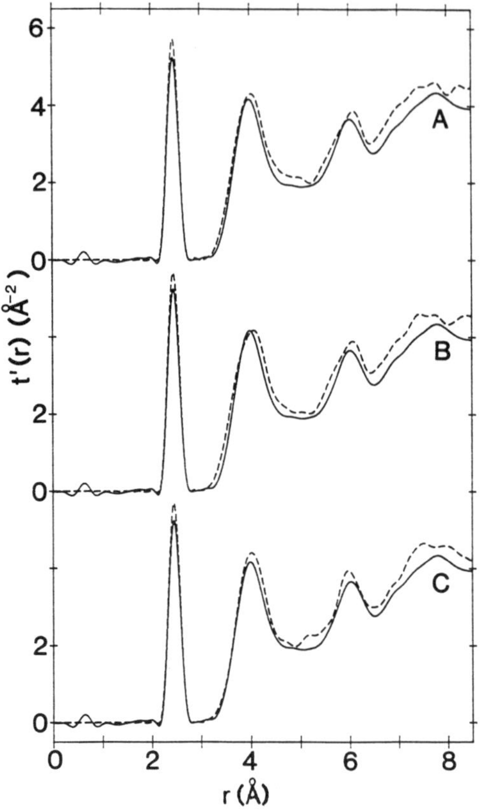

Fig. 20. A series of periodic boundary models of amorphous Ge generated by Wooten and Weaire [34] (dashed lines) compared to experiment (solid lines). (A), model relaxed with Keating potential; (B), as model A but relaxed with Weber potential and (C), model containing 4-membered rings.

capable of yielding excellent agreement with experiment, but this is to be expected since a model with N atoms is equivalent to fitting the experimental data with 3N adjustable parameters (typically several thousand). In comparison, a cubic spline fit to the same data requires only a few hundred parameters and hence the number of excess parameters in the reverse Monte Carlo case is extremely large. For this reason it is imperative, when using reverse Monte Carlo techniques, to examine the resulting configuration to ensure that it is chemically reasonable and that it is in accord with all other available data for the material in question. Thus, for example, the vitreous silica model of Keen et al. [41] must be rejected in that it contains far too many silicon atoms with co-ordination numbers other than 4. Further constraints, such as the maximum percentage of incorrectly co-ordinated atoms, may of course be included in the modelling procedure, but at the expense of considerable extra computational time. In addition to checking that the model is chemically reasonable and does not contain too many defects, it is also very important that it be stable, when relaxed with a reasonable interatomic potential, since an unstable

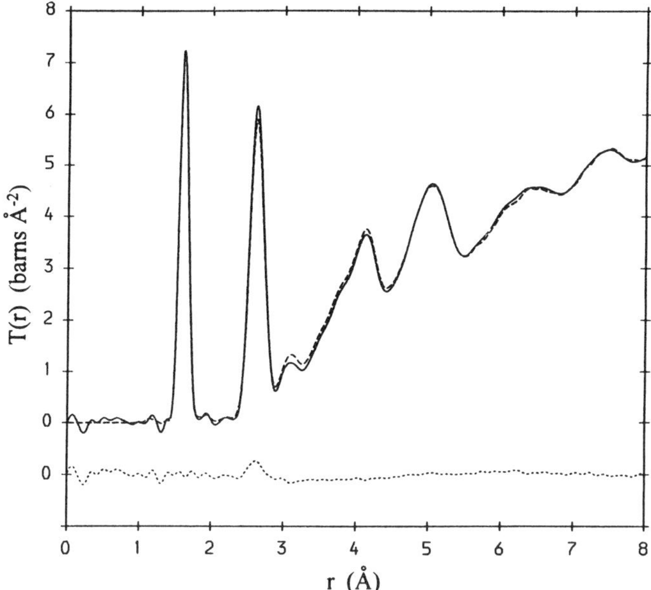

Fig. 21. The neutron correlation function for a vitreous silica model of Gladden [42], refined using the reverse Monte Carlo technique. ————, experiment; – – – –, model and ············, residual.

structure is similarly unacceptable as a model on which to base further calculations. The structure must be stable to small atomic (e.g. thermal) displacements (i.e. not critically metastable) and kinetically accessible.

Instead of starting from random co-ordinates, the reverse Monte Carlo technique may also be employed to refine models generated in other ways. For example, an extremely promising algorithm has been developed by Gladden which mimics the hand building process and, in a recent study of vitreous silica, Gladden [42] has further refined such a model using reverse Monte Carlo techniques. The final model is compared to neutron data in Fig. 21. The agreement between the model and experiment is excellent, yielding an R_χ factor of 2.0%, and the number of "defects" in the model is very much less than that for the model of Keen et al. [41], which started from random atomic co-ordinates.

7.3. MOLECULAR DYNAMICS SIMULATIONS

An alternative approach is to use molecular dynamics simulations, which are discussed in very much greater detail elsewhere in this volume [43]. However, it is worthwhile including a few comments here on the comparison of such simulations with diffraction data. Early molecular dynamics simulations of amorphous network solids employed 2-body (ionic) potentials, with no bond angle restoring forces, but led to structural units that are far too distorted and, as a result, more recent simulations have used 3-body forces. Such a simulation of vitreous silica by Vessal et al. [44] is shown in Fig. 22, together with

120

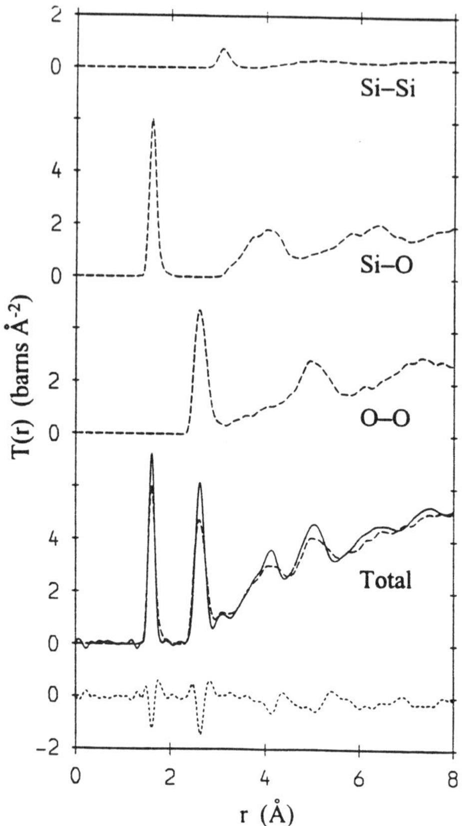

Fig. 22. A comparison of the molecular dynamics simulation of vitreous silica by Vessal et al. [44] with the neutron data from Fig. 12. ————, experimental data; – – – – , simulation and ············, difference curve.

the difference curve (dotted line). The R_χ factor of 9.1% is somewhat lower than that for the relaxed Bell and Dean model (cf. Section 7.1).

Even though the simulation in Fig. 22 employs 3-body potentials, it is obvious from the difference curves that the SiO_4 tetrahedra are still too distorted. The intratetrahedral O–O peak is narrower than for 2-body potentials, although still too broad, but the reduction in the spread of O–Si–O bond angles has been obtained at the expense of an increased spread of bond lengths, such that the Si–O peak is also too broad. The structure in the correlation function at higher r is similarly reduced, relative to that for the real glass. A major problem with molecular dynamics simulations of glasses is that the quench rates from the liquid state are many orders of magnitude faster than that for the real glass and hence the simulated structure is appropriate to a much higher fictive temperature. Thus it is to be expected that the latter will be more disordered, as indeed is the case. The way in which the thermal energy is removed during quenching is also different in that, for the simulation, this is usually done homogeneously by reducing the energy of each atom by

Fig. 23. A molecular graphics representation of the structure of vitreous $Rb_2O \cdot 2SiO_2$, generated by molecular dynamics simulation [45].

the same fraction, whereas the quenching of a laboratory glass relies on conduction to the outer surfaces. An interesting possibility would be to try to refine the structure from a molecular dynamics simulation using reverse Monte Carlo techniques, as Gladden [42] has done for her computer generated model.

One important application of molecular dynamics simulations in the analysis of diffraction experiments, even those using 2-body potentials, is in the interpretation of data for complex multicomponent systems, particularly when coupled with a versatile molecular graphics package. For example, a number of simulations have been performed of alkali silicate systems and a frame from a simulation of a $Rb_2O\text{-}SiO_2$ glass [45] is illustrated in Fig. 23. In the case of vitreous $Na_2O\text{-}SiO_2$, the Na–O distance distribution is asymmetric (cf. Section 8.3, Fig. 27) and this is predicted by the Na–O contribution from a recent molecular dynamics simulation [45], shown as the dashed upper curve in Fig. 24, together with the separate Na–O distributions for non-bridging and bridging oxygen atoms. The disorder inherent to the vitreous state means that not all of the oxygen atoms in the first neighbour shell around the sodium ions will be non-bridging and, similarly, that the cavities in which some of the Na^+ ions reside will not be sufficiently regular for them to be at the ideal distance from every oxygen atom. In this situation, the positively-charged Na^+ ions will tend to adopt the optimum configuration with respect to the negatively-charged non-bridging oxygen atoms, rather than the bridging oxygen atoms

122

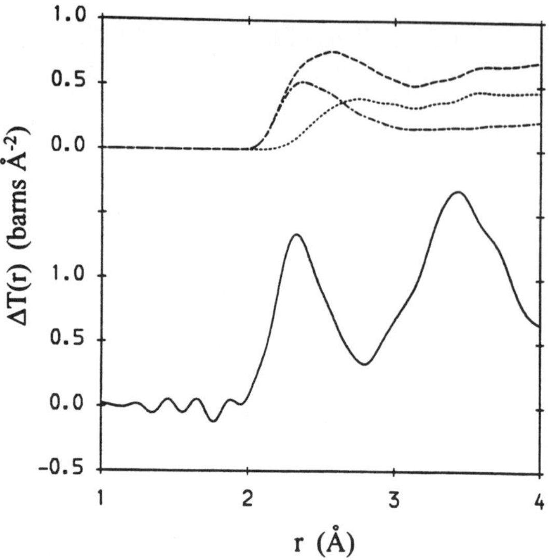

Fig. 24. An analysis of the Na–O distribution for vitreous Na_2O-SiO_2, obtained using the modified difference technique [46] (cf. Section 8.3). Bottom, modified difference correlation function (———); top, molecular dynamics simulation showing the total Na–O distribution (– – – –) and the separate contributions from non-bridging (– · – · –) and bridging (··········) oxygen atoms.

which only carry the much smaller effective charge associated with heteropolar covalent bonding. Thus, it might be expected that the Na–O distribution will comprise an approximately symmetric peak at the non-bridging oxygen atom distance plus a higher r contribution arising from the bridging oxygen atoms, which is consistent both with the molecular dynamics simulation and with the Na(O) co-ordination number of 3.0 ± 0.5 [47] extracted from a symmetric peak fit to the modified difference correlation function, $\Delta T(r)$ (cf. Section 8.3), since in an oversized cavity the Na^+ ion will normally only be "in contact" with three oxygen neighbours.

7.4. ANALYTICAL MODELS

Instead of generating atomic co-ordinates, some modelling techniques generate real or reciprocal space functions analytically, examples being the molecular model and many crystal based models. As indicated in Section 5.3, the problem with this approach is that there is no guarantee that it is possible to generate a real structure (i.e. atomic co-ordinates) consistent with the analytical formalism. For example, many crystal based models calculate $Qi(Q)$ and/or $D(r)$ for crystal-like regions, but completely ignore the necessary interconnecting material which in any real structure may well comprise the major fraction of the total volume. Similar arguments have been put forward by Ziman, in his book "Models of Disorder," who concludes [48]: *"Until a protagonist of these models*

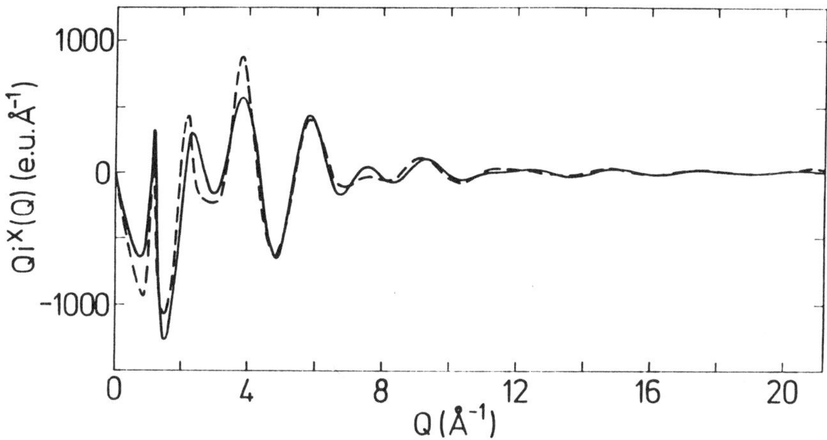

Fig. 25. The X-ray interference function for an As_4S_4 molecular model (solid line) compared with experimental data for thin film $AsS_{1.38}$ (dashed line) [50].

produces an actual three-dimensional structure that conforms to the assumptions he makes we must be doubtful whether it can be done at all, nohow."

Examples of vapour deposited amorphous solids have already been given in Table 1, including arsenic sulphide [49], where the molecules have undergone at least partial polymerisation. Fits to the first peaks in T(r) for vitreous ($As_{0.38}S_{0.62}$) and vapour-deposited ($As_{0.43}S_{0.57}$) arsenic sulphide indicate that, while the bulk glass is chemically ordered, the vapour deposited film contains far more As–As bonds than the minimum required by stoichiometry as revealed by a high r shoulder [49]. Moreover, the mean As–As distance (2.57Å) exceeds the normal covalent bond length (2.49Å) and is much nearer that (2.59Å) found in the As_4S_4 realgar molecule, which is known from mass spectrometric studies to be present in the vapour phase. Hence it can be concluded that vapour deposited arsenic sulphide contains a significant fraction of As_4S_4 molecules and/or molecular fragments. A random orientation molecular model [49,50], based on the realgar molecule, is compared to X-ray data in Fig. 25. At high Q, Qi(Q) is dominated by intra-molecular contributions and in this region the model is in best agreement with experiment. The differences between model and experiment at low Q are the result of polymerisation and orientational effects arising from the fact that the As_4S_4 molecule is in reality not spherical.

8. Separation of Individual Component Correlation Functions

Various methods exist for the separation of individual component correlation functions, or linear combinations of subsets of these components, which involve the variation of the scattering length of one or more elements. The most important of these are isotopic substitution (neutron diffraction) and the anomalous dispersion technique (X-ray and neutron diffraction), but it is also possible to utilise neutron magnetic diffraction and approximate methods such as isomorphous substitution. A first order difference correlation function, involving a variation in the scattering length of element A via either isotopic substitution or anomalous dispersion, is the equivalent of an EXAFS experiment at the absorption edge of A [51].

8.1. ISOTOPIC SUBSTITUTION

The scattering length of an element with more than one isotope may be varied by altering their relative isotopic abundances. Samples may thus be prepared with different values of $\overline{b_A}$. For a few elements (H, Li, Ti, Cr, Ni, Sm, Dy and W), isotopes exist with both positive and negative scattering lengths, the latter corresponding to scattering without the normal phase change of π. This gives rise to a special form of isotopic substitution known as the null technique in which a combination of isotopes is used such that element A has a zero scattering length and hence does not contribute to the measured interference or correlation functions. A measurement of the scattering from such a sample thus yields the X–X component directly, where X indicates any element present other than A.

Figure 26 shows the Dy–Dy + Dy–X (X=Na, Be or F) difference correlation function for vitreous NaF-DyF_3-BeF_2, together with the results of a molecular dynamics simulation using 2-body potentials [52]. Due to slightly too large an ionic radius for Dy^{3+}, the first neighbour Dy–F peak for the simulation is displaced to higher r but its width is in agreement with experiment. The structure for the simulation between 3.3 and 5.0Å is in qualitative agreement with experiment but is more smeared out. The simulation does, however, identify the experimental peak at 3.7Å as being due to the first Dy–Be coordination shell, while the feature between 4 and 5Å arises from higher order Dy–F interactions with a small contribution from Dy–Dy correlations.

8.2. ANOMALOUS DISPERSION

The X-ray anomalous dispersion technique [53] is based on the fact that, near an absorption edge, the X-ray form factor has both real and imaginary parts which are wavelength dependent

$$f(Q) = f^{\circ}(Q) + \Delta f'(\lambda) + i\Delta f''(\lambda) \tag{27}$$

The utilisation of anomalous dispersion to separate component correlation functions has, however, only really become feasible with the advent of suitable diffractometers on synchrotron X-ray sources, which allow the wavelength to be continuously varied to obtain the optimum values of $\Delta f'(\lambda)$ and $\Delta f''(\lambda)$. Anomalous dispersion experiments are also possible with neutrons [54] for which a variation in scattering length occurs close to an absorption resonance and is much larger for the isotope in question than is the case for X-ray scattering.

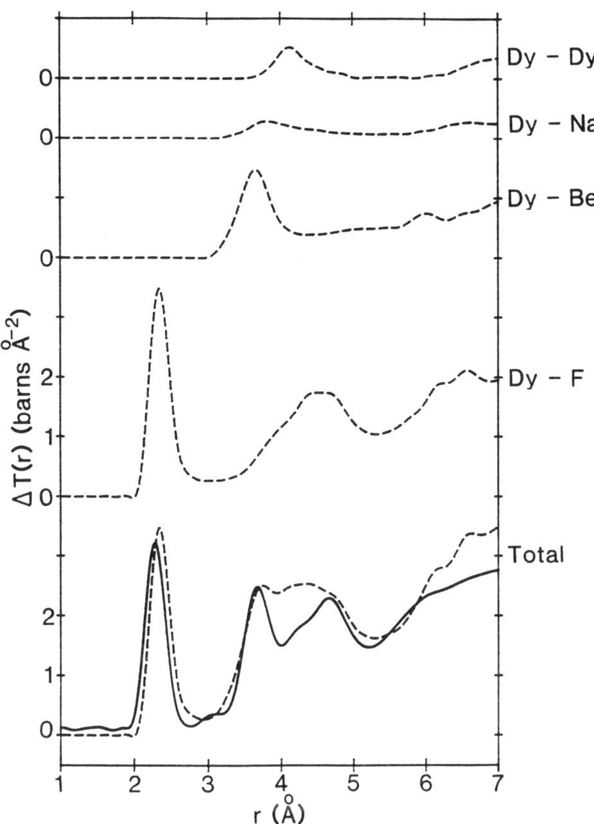

Fig. 26. The Dy–F + Dy–X (X=Na, Be, or F) isotopic difference correlation function for vitreoous NaF-DyF$_3$-BeF$_2$ (solid line) compared to the predictions of a molecular dynamics simulation (dashed lines) [52].

8.3. APPROXIMATE METHODS

Unfortunately, for many amorphous insulators and semiconductors, none of these techniques is feasible, but even in such cases valuable extra information can be obtained from a combination of neutron and X-ray diffraction, for which the component weighting factors are different ($b_i b_j$ and ~$Z_i Z_j$ respectively). For multicomponent systems, the information obtainable from diffraction experiments is greatly increased if a systematic study is carried out as a function of composition. A possible approach, in such cases, is to use the difference technique in which the correlation function for the base amorphous solid is subtracted from that of a more complex material, as illustrated for a sodium silicate glass in Fig. 27. The interpretation of the simple difference curve (dashed line), obtained by subtracting the correlation function for pure vitreous SiO$_2$ from that for 0.328Na$_2$O·SiO$_2$, is complicated by the structure arising from the increase in the average

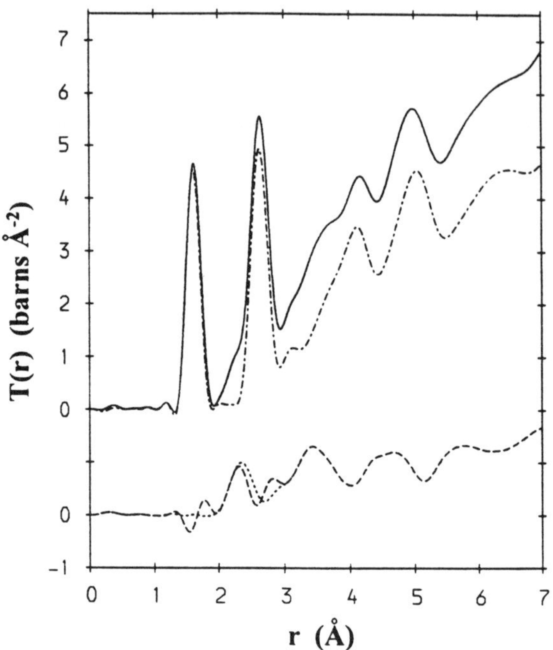

Fig. 27. The application of the difference technique to data for vitreous $0.328Na_2O \cdot SiO_2$ [46]. ———, $0.328Na_2O \cdot SiO_2$; – · – · –, SiO_2 ; – – – –, difference correlation function and ············, modified difference correlation function (see text).

intratetrahedral Si–O and O–O distances, which occurs on adding Na_2O to SiO_2. It is therefore necessary to work with the modified difference curve (dotted line), generated from the peak fit residuals, as discussed in detail in [46]. The latter suggests that the first Na–O distance distribution is asymmetric and this is supported by the molecular dynamics simulation of Fig. 24 (Section 7.3).

9. First Diffraction Peak

For the reasons outlined in Section 5, the rapid decay of the highest frequency Fourier components means that the first peak in the diffraction pattern of an amorphous solid is almost invariably the sharpest and as such it has attracted a great deal of attention in the literature. Before discussing the structural origins of this often named "first sharp diffraction peak" (FSDP), however, it is perhaps worth making a few general comments to dispel some of the mysticism and misunderstanding with which this feature has been associated. First, and this seems to be completely lost on many authors who claim special importance for the first diffraction peak, any diffraction pattern with peaks must by definition have a first peak and the fact that this is usually the sharpest also means that it

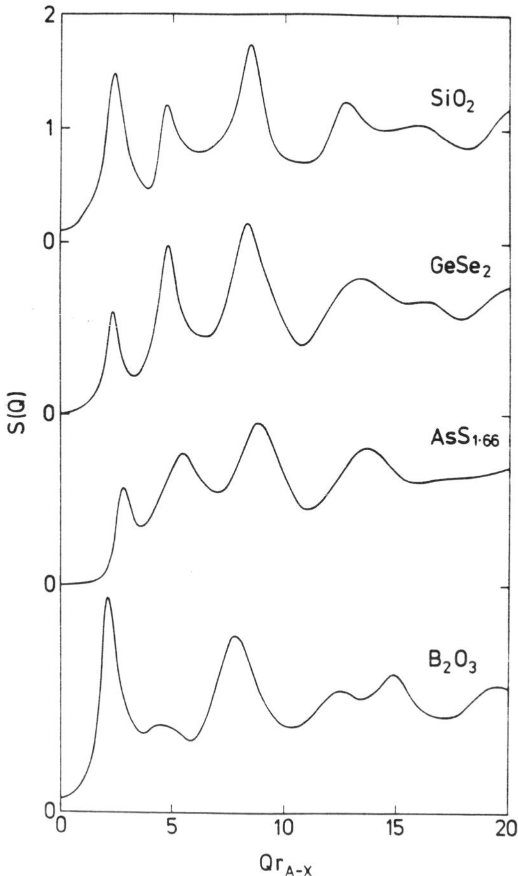

Fig. 28. Reduced diffraction patterns for vitreous SiO_2, GeS_2, $AsS_{1.66}$ and B_2O_3 [50].

is the most complicated since it has contributions from the reciprocal space Fourier components arising from all of the peaks in T(r). Second, the first diffraction peak does not arise from a ***distance*** in real space but from the longest period, most slowly decaying (assuming that the first peak is indeed the sharpest) real space ***Fourier component***, the period being given by $2\pi/Q_1$ where Q_1 is the peak position in reciprocal space. As for crystalline Bragg peaks, therefore, the first diffraction peak for a glass is evidence of a ***periodicity*** in real space, the much greater width relative to crystalline materials indicating that the associated Fourier component decays after relatively few periods. There is of course absolutely no reason why the origin of the longest period Fourier component should be the same for all amorphous solids, which indicates the futility of searching for some universal theory for the first diffraction peak, with which many workers seem so obsessed. A review of the theories of the first diffraction peak for amorphous materials is given by Moss and Price [55].

Many workers, in interpreting diffraction data from amorphous chalcogenides, have placed great emphasis on the first diffraction peak which it is claimed is anomalous in that it occurs at very low Q and is very sharp. In the case of vitreous arsenic sulphide (Fig. 28), this peak is close to the 020 reflection for the mineral orpiment, which has a layer structure, the 020 reflection corresponding to the interlayer spacing. A similar situation exists for other chalcogenides (e.g. As_2Se_3, GeS_2 and $GeSe_2$) and has consequently led to the suggestion that the glasses also have layer structures. The first neutron diffraction peak for vitreous $GeSe_2$ is indeed at a lower value of Q than that for vitreous SiO_2, but this would be expected, since the Ge–Se bond is much longer than the Si–O bond (2.370Å as opposed to 1.608Å). If the vitreous SiO_2 structure were expanded to give the Ge-Se bond length, the Q scale would contract by the same factor. Thus, in order to compare the neutron diffraction patterns for the two materials, it is necessary to plot the structure factor S(Q) against $r_{A-X}Q$ where r_{A-X} is the A–X bond length (A = Si or Ge; X = O or Se). This is done in Fig. 28 [50] and it can be seen that the first three neutron diffraction peaks for $GeSe_2$ are almost exactly coincident with those of vitreous SiO_2, which certainly does not have a layer structure. The situation for vitreous $AsS_{1.66}$ is also ambiguous in that its first diffraction peak can equally well be predicted by a molecular model as demonstrated in Section 7.4. For amorphous network solids, the first diffraction peak has been associated with the periodicity arising from the boundaries between a succession of the cages which comprise the structure of a 3-dimensional covalent network (cf. Fig. 1). In the case of 3-fold co-ordinated structural units, the cages tend to become flattened and a few such adjacent cages, flattened in the same direction, will approximate a set of layers and give rise to the necessary Fourier component to generate the first diffraction peak, without the need to resort to what is traditionally understood by a layer structure.

The first peak in the neutron diffraction pattern for vitreous silica is very closely Lorentzian [28] and a Lorentzian fit yields $Q_1=1.53\pm0.02Å^{-1}$, which is equivalent to a real space periodicity of 4.1Å, and a full width at half maximum height, ΔQ_{fwhm}, of $0.63\pm0.02Å^{-1}$. The first point to note is that the Fourier component arising from the first diffraction peak is not at all obvious in the real space correlation function of Fig. 12 and that it will only become obvious at much higher r when the Fourier components from the broader, higher Q diffraction peaks have effectively decayed to zero. The predominant periodicity at higher values of r in Fig. 12 (~1.2Å) is in fact associated with the strong diffraction peak at ~5.3Å^{-1} (cf. Fig. 7).

Given a Lorentzian peak in reciprocal space, the envelope of the corresponding real space Fourier component is

$$F(r) = \exp(-\Delta Q_{fwhm}|r|/2) \tag{28}$$

This will have decayed to 10% of its value at zero r at a real space correlation length, $L_{0.1}$, of 7.3Å, which corresponds to an average of less than 2 cages. If the boundaries between successive cages are approximated to planes of atoms then it is possible to further analyse the first diffraction peak using the theory appropriate to strained crystals [2,23]. Starting from a fixed origin plane, if it is assumed that subsequent planes exhibit a Gaussian distribution about their average position, the Lorentzian shape of the first diffraction peak then indicates that the width of this Gaussian distribution increases as the square root of the distance from the origin plane {i.e. the width is proportional to $\sqrt{n}$ if the planes are numbered n = 0 (origin), 1, 2, ...} and it is interesting to note that this type of disorder is similar to that inherent in the paracrystalline model of Hosemann [56].

10. Conclusions

The examples discussed in this chapter and in [2-4,23] demonstrate the role of X-ray and neutron diffraction techniques in investigations of the structure of amorphous insulators and semiconductors. The best modern diffraction experiments are capable of providing accurate data with high real-space resolution, which if used correctly are an extremely fine filter for the various models proposed in the literature. The greatest barrier to progress in understanding the structure of these materials lies not so much with the diffraction technique itself but in the development of modelling procedures, in which model parameters can be varied in a systematic way until agreement with experiment is obtained, in both real and reciprocal space, within the known experimental uncertainties. This situation is unique to amorphous materials and has no parallel in corresponding studies of the crystalline state.

11. References

1. Zachariasen, W.H. (1932) The Atomic Arrangement in Glass, *J. Amer. Chem. Soc.* **54**, 3841-3851.
2. Wright, A.C. (1993) Neutron and X-Ray Amorphography, in: *Experimental Techniques of Glass Science*, eds Simmons, C.J. and El-Bayoumi, O.H., Amer. Ceram. Soc., Westerville, Chap. 8.
3. Wright, A.C. (1974) The Structure of Amorphous Solids by X-ray and Neutron Diffraction, *Adv. Struct. Res. Diffr. Meth.* **5**, 1-84.
4. Wright, A.C. and Leadbetter, A.J. (1976) Diffraction Studies of Glass Structure, *Phys. Chem. Glasses* **17**, 122-145.
5. Simmons, J.H., Simmons, C.J., Ochoa, R. and Wright, A.C. (1991) Fluoride Glass Structure, in: *Fluoride Glass Fiber Optics*, Eds. Aggarwal, I.D. and Lu, G., Academic Press, San Diego, 37-84.
6. Greaves, G.N., Fontaine, A., Lagarde, P., Raoux, D. and Gurman, S.J. (1981) Local Structure of Silicate Glasses, *Nature* **293**, 611-616.
7. Vedishcheva, N.M., Shakhmatkin, B.A., Shultz, M.M., Vessal, B., Wright, A.C., Bachra, B., Clare, A.G., Hannon, A.C. and Sinclair, R.N. (1995) A Thermodynamic, Molecular Dynamics and Neutron Diffraction Investigation of the Distribution of Tetrahedral $\{Si^{(n)}\}$ Species and the Network Modifying Cation Environment in Alkali Silicate Glasses, *J. Non-Cryst. Solids* **192&193**, 292-297.
8. Wright, A.C., Vedishcheva, N.M. and Shakhmatkin, B.A. (1995) Vitreous Borate Networks Containing Superstructural Units: A Challenge to the Random Network Theory?, *J. Non-Cryst. Solids* **192&193**, 92-97.
9. Bray, P.J. (1985) Structural Models for Borate Glasses, *J. Non-Cryst. Solids* **75**, 29.
10. Sinclair, R.N. and Wright, A.C. (1983) Neutron Scattering from Vitreous Silica I. The Total Cross-Section, *J. Non-Cryst. Solids* **57**, 447-464.
11. Johnson, P.A.V., Wright, A.C. and Sinclair, R.N. (1983) Neutron Scattering from Vitreous Silica II. Twin-Axis Diffraction Experiments, *J. Non-Cryst. Solids* **58**, 109-130.

12. Wright, A.C. and Sinclair, R.N. (1985) Neutron Scattering from Vitreous Silica III. Elastic Diffraction, *J. Non-Cryst. Solids* **76**, 351-368.

13. Grimley, D.I., Wright, A.C. and Sinclair, R.N. (1990) Neutron Scattering from Vitreous Silica IV. Time-of-Flight Diffraction, *J. Non-Cryst. Solids* **119**, 49-64.

14. Finbak, C. (1949) The Structure of Liquids I, *Acta Chem. Scand.* **3**, 1279-1292.

15. Lorch, E.A. (1969) Neutron Diffraction by Germania Silica and Radiation-Damaged Silica Glasses, *J. Phys. C* **2**, 229-237.

16. Warren, B.E. and Mavel, G. (1965) Elimination of the Compton Component in Amorphous Scattering, *Rev. Sci. Instrum.* **36**, 196-197.

17. Mozzi , R.L. and Warren, B.E. (1969) The Structure of Vitreous Silica, *J. Appl. Crystallogr.* **2**, 164-172.

18. Bushnell-Wye, G., Finney, J.L., Turner, J., Huxley, D.W. and Dore, J.C. (1992) The Use of Synchrotron X-rays to Exploit the Warren-Mavel Fluorescence Detection Technique in Studies of Disordered Systems, *Rev. Sci. Instrum.* **63**, 1153-1155.

19. Krogh-Moe, J. (1956) A Method for Converting Experimental X-ray Intensities to an Absolute Scale, *Acta Crystallogr.* **9**, 951-953.

20. Norman, N. (1957) The Fourier Transform Method for Normalizing Intensities, *Acta Crystallogr.* **10**, 370-373.

21. Filon, L.N.G. (1929) On a Quadrature Formula for Trigonometric Integrals, *Proc. Roy. Soc. (Edinburgh)* **49**, 38-47.

22. Dixon, M., Wright, A.C. and Hutchinson, P. (1977) The Smoothing and Fast Fourier Transformation of Experimental X-Ray and Neutron Diffraction Data from Amorphous Materials, *Nucl. Instrum. Meth.* **143**, 379-383.

23. Wright, A.C. (1994) Neutron Scattering from Vitreous Silica V. The Structure of Vitreous Silica: What Have We Learned from 60 Years of Diffraction Studies?, *J. Non-Cryst. Solids* **179**, 84-115.

24. Salmon, P.S. (1988) A Neutron Diffraction Study on the Structure of Liquid Germanium, *J. Phys. F* **18**, 2345-2352.

25. Etherington, G., Wright, A.C., Wenzel, J.T., Dore, J.C., Clarke, J.H. and Sinclair, R.N. (1982) A Neutron Diffraction Study of the Structure of Evaporated Amorphous Germanium, *J. Non-Cryst. Solids* **48**, 265-289.

26. Wright, A.C. (1993) The Comparison of Molecular Dynamics Simulations with Diffraction Experiments, *J. Non-Cryst. Solids* **159**, 264-268.

27. Konnert, J.H. and Karle, J. (1973) The Computation of Radial Distribution Functions for Glassy Materials, *Acta Crystallogr.* **A29**, 702-710.

28. Wright, A.C., Hulme, R.A., Grimley, D.I., Sinclair, R.N., Martin, S.W., Price, D.L. and Galeener, F.L. (1991) The Structure of Some Simple Amorphous Network Solids Revisited, *J. Non-Cryst. Solids* **129**, 213-232.

29. Vedishcheva, N.M., Shakhmatkin, B.A., Wright, A.C., Grimley, D.I., Etherington, G. and Sinclair, R.N. (1993) Structural Order in Lead Borate Glasses: A Test of Recent Thermodynamic Predictions, in: *Fundamentals of Glass Science and Technology*, ESG, Venice, 459-462.

30. Hannon, A.C., Grimley, D.I., Hulme, R.A., Wright, A.C. and Sinclair, R.N. (1994) Boroxol Groups in Vitreous Boron Oxide: New Evidence from Neutron Diffraction and Inelastic Scattering Studies, *J. Non-Cryst. Solids* **177**, 299-316.

31. Finney, J.L. (1981) Amorphous Polymorphism: Structural Variability and Characterisation in Amorphous Systems, in: *Diffraction Studies of Non-Crystalline Substances*, Eds. Hargittai, I and Orville-Thomas, W.J., Elsevier, Amsterdam, 440-490.

32. Mason, G. (1968) Radial Distribution Functions from Small Packings of Spheres, *Nature* **217**, 733-735.

33. Warren, B.E. (1978) Calculation of Powder-Pattern Intensity Distributions, *J. Appl. Crystallogr.* **11**, 695-698.

34. Wooten, F. and Weaire, D. (1987) Modeling Tetrahedrally Bonded Random Networks by Computer, *Solid State Phys.* **40**, 1-42.

35. Konnert, J.H. and Appleman, D.E. (1978) The Crystal Structure of Low Tridymite, *Acta Crystallogr.* **B34**, 391-403.

36. Bell, R.J. and Dean, P. The Structure of Vitreous Silica: Validity of the Random Network Theory, *Philos. Mag.* **25**, 1381-1398.

37. Polk, D.E. and Boudreaux, D.S. (1973) Tetrahedrally Coordinated Random-Network Structure, *Phys. Rev. Lett.* **31**, 92-95.

38. Gaskell, P.H. and Tarrant, I.D. (1980) Refinement of a Random Network Model for Vitreous Silicon Dioxide, *Philos. Mag.* **B42**, 265-286.

39. Rechtin, M.D., Renninger, A.L. and Averbach, B.L. (1974) Monte Carlo Models of Amorphous Materials, *J. Non-Cryst. Solids* **15**, 74-82.

40. McGreevy, R.L. and Pusztai, L. (1988) Reverse Monte Carlo Simulation: A New Technique for the Determination of Disordered Structures, *Mol. Simulation* **1**, 359-367.
41. Keen, D.A. and McGreevy, R.L. (1990) Structural Modelling of Glasses Using Reverse Monte Carlo Simulation, *Nature* **344**, 423-425.
42. Gladden, L.F. (1992) Structure and Dynamics of 4-2 Coordinated Glasses, in: The Physics of Non-Crystalline Solids, Eds. Pye, L.D., LaCourse, W.C. and Stevens, H.J., Taylor and Francis, London, 91-95.
43. Vashishta, P. (1996) Molecular Dynamics Method and Large Scale Simulation of Amorphous Materials, this volume.
44. Vessal, B., Amini, M. and Catlow, C.R.A. (1993) Computer Simulation of the Structure of Silica Glass, *J. Non-Cryst. Solids* **159**, 184-186.
45. Wright, A.C., Vessal, B., Bachra, B., Hulme, R.A., Sinclair, R.N., Clare, A.G. and Grimley, D.I. (1995) Neutron Scattering Studies of Network Glasses, in: *"Neutron Scattering in Materials Science II"*, Eds Neumann, D.A., Russell, T.P. and Wuensch, B.J., *Mater. Res. Soc. Proc.* **376**, 635-659.
46. Wright, A.C., Clare, A.G., Bachra, B., Sinclair, R.N., Hannon, A.C. and Vessal, B. (1991) Neutron Diffraction Studies of Silicate Glasses, *Trans. A.C.A.* **27**, 239-253.
47. Wright, A.C., Bachra, B., Vessal, B., Clare, A.G., Sinclair, R.N. and Hannon, A.C. (1993) A Combined Neutron Diffraction and Molecular Dynamics Study of the Structure of Alkali Silicate Glasses and of the Mixed Alkali Effect, in: *Fundamentals of Glass Science and Technology*, ESG, Venice, 211-216.
48. Ziman, J.M. (1979) *Models of Disorder*, Cambridge University Press, Cambridge, 72.
49. Daniel, M.F., Leadbetter, A.J., Wright, A.C. and Sinclair, R.N. (1979) The Structure of Vapour-Deposited Arsenic Sulphides, *J. Non-Cryst. Solids* **32**, 271-293.
50. Wright, A.C., Sinclair, R.N. and Leadbetter, A.J. (1985) Effect of Preparation Method on the Structure of Amorphous Solids in the System As-S, *J. Non-Cryst. Solids* **71**, 295-302.
51. Hayes, T.M. and Wright, A.C. (1983) Diffraction and EXAFS Spectroscopy as Structural Probes of Amorphous Solids, in: *The Structure of Non-Crystalline Materials 1982*, Eds. Gaskell, P.H., Parker, J.M. and Davis, E.A., Taylor and Francis, London, 108-119.
52. Clare, A.G., Etherington, G., Wright, A.C., Weber, M.J., Brawer, S.A., Kingman, D.D. and Sinclair, R.N. (1989) A Neutron Diffraction and Molecular Dynamics Investigation of the Environment of Dy^{3+} Ions in a Fluoroberyllate Glass, *J. Chem. Phys.* **91**, 6380-6392.
53. Krogh-Moe, J. (1966) A Method for the Resolution of Composite Radial Pair Distribution Functions, *Acta. Chem. Scand.* **20**, 2890-2891.
54. Wright, A.C., Etherington, G., Desa, J.A.E. and Sinclair, R.N. (1982) Neutron Diffraction Studies of Rare Earth Ions in Glasses, *J. Phys. Coll.* **C9**, 31-34.
55. Moss, S.C. and Price, D.L. (1985) Random Packing of Structural Units and the First Sharp Diffraction Peak in Glasses, in: *Physics of Disordered Materials*, Eds. Adler, D., Fritzsche, H. and Ovshinsky, S.R., Plenum, New York, 77-95.
56. Hosemann, R., Hentschel, M., Lange, A., Uther, B. and Brückner, R. (1984) Dreidimensionale Analyse der Parakristallinen Struktur von Kieselglas, *Z. Kristallogr.* **169**, 13-33.

COMPUTER MODELLING OF GLASSES AND GLASSY ALLOYS

NORMAND MOUSSEAU[1]

Département de physique and
Groupe de recherche en physique et
technologie des couches minces (GCM)
Université de Montréal
C.P. 6128, succ. Centre-ville
Montréal, (Québec)
Canada H3C 3J7

July 5, 1996

1. Introduction

The construction of glassy networks as well as the study of their dynamics pose multiple challenges for the computational physicist. After more than two decades of simulations, there still exists no fully satisfactory technique for even obtaining the static properties of disordered materials, let alone their electronic and dynamical ones. A great deal of progress toward this goal has been achieved, however, and this will be the subject of this chapter.

The basic difficulty in modelling glasses is the lack of knowledge about the exact position of the atoms on the lattice. Since these are disordered networks, one cannot use symmetries to extract the information and so we have to rely heavily on comparison between simulation and experiment to improve our understanding about these materials. We know, however, that glasses and amorphous systems do possess more order than liquid, on short- and medium–range at least. This statement is particularly true of covalent materials which keep a local environment very similar to the corresponding crystalline state.

Before getting to the dynamical and electronic properties of a glass, it becomes therefore necessary to construct an appropriate network. One can perform this task using either static, where only the final result is impor-

133

M. F. Thorpe and M. I. Mitkova (eds.), Amorphous Insulators and Semiconductors, 133–150.
© 1997 *Kluwer Academic Publishers. Printed in the Netherlands.*

134

tant, or dynamical methods, which gives a representation for the evolution toward these phases.

In this chapter, I cover the methods for computer modelling of the networks themselves. In the next section, I discuss the static and dynamical approaches. In the third one, I present different interaction potentials used to simulate these materials. Finally, I comment on some specific problems which underline the work which still needs to be done and mention some avenues for future studies of these systems. Most of the discussion will be related to amorphous semiconductors with some mention of metallic glasses.

2. Static procedures

For most static procedures the end justifies the means and, therefore, the intermediate steps necessary to reach a satisfactory structure are considered irrelevant: the steps for statically obtaining a good amorphous or glassy network do not have to mimic any physical process. This freedom has allowed the development of methods generally specific to certain types of interactions or networks, some of which have produced excellent agreement with experiment.

One distinguishes generally two families of random networks representing real glasses which are simulated differently: The *Dense Random Packing* (DRP), which is appropriate for metallic glasses and the like, and the *Continuous Random Network* (CRN), which applies to covalent glasses and amorphous semiconductors (see Fig 2(a) and (b), respectively).

2.1. DENSE RANDOM PACKING

The first models of DRP were constructed by Bernal [2, 3] to mimic the highly coordinated metallic liquids where the atomic structure is only constrained by space-filling restrictions. These models were obtained by pouring thousands of steel balls into a flexible container itself placed on an irregular surface in order to prevent crystallisation. After some kneading to maximise the density, wax or paint was poured over the balls to fix the configuration. The analysis then proceeded by hand, with the careful measurement of the position of every ball. With such a frozen configuration, it was a simple step to propose that a model of liquid could also be applied to amorphous metallic alloys. A hypothesis which was justified in 1970 by comparison with the experimental radial distribution functions of amorphous $Ni_{76}P_{24}$ [4].

This success pushed physicists to try to develop faster methods, using newly available computers. The first algorithms simply added hard-spheres one by one, radially [5], on an initial random cluster. However, these methods proved unable to produce the high density random packings seen in

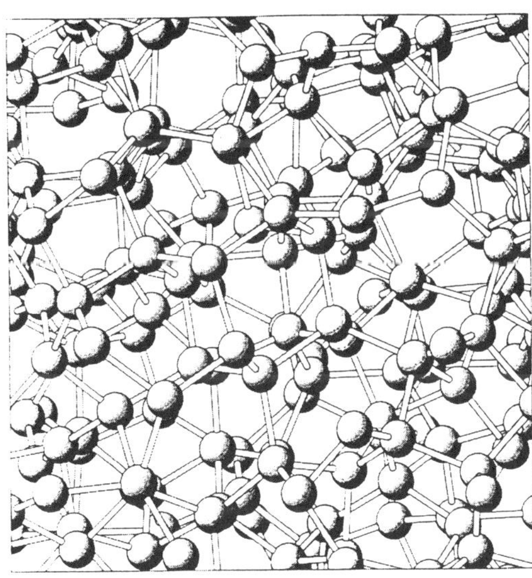

Figure 1. Two models of glasses. Top: Bernal's Dense Random Packing model, here a hand-build representation made of steel balls (reprinted from Ref. 6). Bottom: Zachariasen's Continuous Random Network as obtained following the Wooten, Winer and Weaire algorithm (reprinted from Ref.7).

hand-build models. No static approach, with or without local relaxation, was able to give both a high density and a cluster that could fill the whole space. It is only with molecular dynamics, coupled with softer interactions, that these two requirements were filled simultaneously [6].

2.2. CONTINUOUS RANDOM NETWORK

While in the DRP model, atomic structure is restricted by local packing, the continuous random network is defined by perfect coordination in the first neighbour shell. More important short-range order in the CRN is a consequence of the strongly covalent nature of semiconductors. Proposed in the early thirties by Zachariasen [8] to described SiO_2, the first realisation of the CRN was achieved in a hand–built model of Bell and Dean [9]. Soon after, the first hand-build model of elemental amorphous semiconductor was assembled by Polk [10]. Starting with a randomised core of a few atoms, Polk proceeded to add further elements so that all bonds at the surface were satisfied. Following this simple algorithm, Polk assembled 440 units and measured, by hand, the position of each atom. The radial distribution function of the model turned out to be in reasonable agreement with experimental data.

Immediately following this work other groups, like Shevchik [11], and Henderson and Herman [12], began to replicate Polk's procedure with computers, increasing the rapidity of the construction while saving the trouble to have to measure the position of every atom by hand. Moreover, with the coordinates on computer, it was also possible to slightly displace atoms in order to minimise the width of length and bond angle distributions, which were the main criteria for judging of the quality of these models.

A better approach for optimising the structure was taken by Steinhart *et al.* [13] who used an empirical potential especially developed for tetravalent semiconductors [14] to relax the atom positions of one of Polk's hand-build model. Keating's potential includes both a radial and an angular energy terms which are needed to give rigidity to fourfold coordinated lattice:

$$V = \frac{3}{16}\frac{\alpha}{d^2}\sum_{\langle ij \rangle}(\mathbf{r}_{ij} \cdot \mathbf{r}_{ij} - d^2)^2 + \frac{3}{8}\frac{\beta}{d^2}\sum_{\langle jik \rangle}(\mathbf{r}_{ij} \cdot \mathbf{r}_{ik} + \frac{1}{3}d^2)^2, \qquad (1)$$

where ratio α and β are force constants which depend on the material modelled, d is the ideal nearest-neighbour distance, and $-1/3$ is the cosine of the tetrahedral angle. This two-part energy form has been retained in most empirical potentials for semiconductors.

With a fixed list of neighbours, the relaxation technique used by Steinhart *et al.* and also, later, by Wooten, Winer and Weaire [15, 16], was simple and efficient: Each atom in turn was moved to its equilibrium position as given by

$$dx_i = -\frac{\mathbf{F}_i}{\nabla \cdot \mathbf{F}_i}. \qquad (2)$$

Although very efficient for harmonic potentials, this technique cannot be used with more realistic potentials which contain an inflection point in the energy curve, since it leads to a divergence of the right-hand side of the equation above. Other techniques of convergence like Monte-Carlo moves [17] or the conjugate-gradient method must be used.

These computer models, however, were only mimicking the hand-building approach, i.e. atoms were added one by one on the outside of the amorphous cluster, leaving up to half the atoms at the surface and raising two problems: (1) Only a small portion of the atoms could be considered "in the bulk", affecting the quality of the statistics. (2) It was still not clear whether a CRN could span the whole space or had a maximum size before exploding under strain. It is therefore worth mentioning Henderson's *tour de force* who succeeded in building, by hand, an amorphous network of 61 atoms with *periodic boundary conditions* (PBC) [18]! Besides having PBC, this model what considerably less strained then a similar model obtained via computer by Henderson and Herman a couple of years before, showing that CRN could fill the whole space with only finite strain.

Guttman was the first to introduce a computer approach which broke with hand-building procedures. Starting from a crystalline cell, with PBC, the structure was amorphised through a series of bond exchanges and relaxation [19]. This way, Guttman succeeded in constructing a series of PBC configurations with a tolerable level of strain. However, Guttman's bond switching algorithm was not well defined and it was almost impossible to amorphise cells containing more than about 60 atoms.

It was not before 1985 that a clear and straightforward algorithm was proposed for building amorphous networks of any size. The Wooten, Winer and Weaire scheme [15] is based on ideas similar to Guttman's but with a very systematic approach which prevents a structure from getting stuck into overstrained configurations. The whole algorithm draws on a list of nearest-neighbours which is first defined from a crystalline structure and then modified only through allowed bond switches. The relaxation following each bond exchange minimises the energy of a network with the topology constrained by the list of neighbours. It can be described in four steps:

1. Choose at random from the neighbour's list a 4-atom chain. For example, atoms 1-2-5-6 in Fig 2.2.
2. To minimise distorsion, the chosen bonds must be as parallel as possible. (Bonds 1-2 and 5-6 fulfill this requirement but not 1-2 and 5-7.)
3. To reduce the strain it is important that the two bonds to be exchanged are not part of the same 5, 6 or 7-member ring.

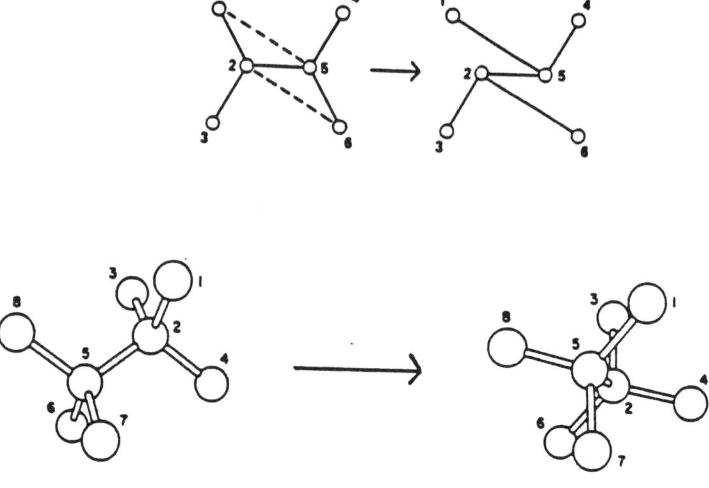

Figure 2.   Moves allowed by the Wooten, Winer and Weaire algorithm. At the top, a two-dimensional representation, without relaxation; at the bottom, three dimensional with relaxation after bond permutation as discussed in the text (reprinted from Ref. 16).

4. In the same spirit, bond exchanges between atoms which are more than 1.7 times the ideal bond length apart before relaxation are rejected.

Starting from a crystal, these steps are applied iteratively to the network. Even with the restrictions, accepting all the legal moves would rapidly yield a completely entangled network. Keeping only the moves which decrease the energy would, on the other hand, prevent the system from leaving local minima. A solution to this dilemma is to use a simulated annealing technique which allows the system to move out of local minima by allowing some moves which increase the temperature. Simulated annealing, proposed by Kirkpatrick *et al.* [20], tries to reproduce the temperature effects of real annealing by proposing to use a Boltzmann factor as an accept/reject criterion of a move: After every move, a random number $0 < p < 1$. If.

$$p < \exp\left[\frac{-\Delta E}{kT}\right],\tag{3}$$

where ΔE is the energy lost or won after a move, then the move is accepted; otherwise it is rejected. A careful choice of temperature will help the system get out of local minima and reach very low energy configurations. There exists a large literature about how to optimise the temperature but, in the end, it often depends on the specificity of the problem studied.

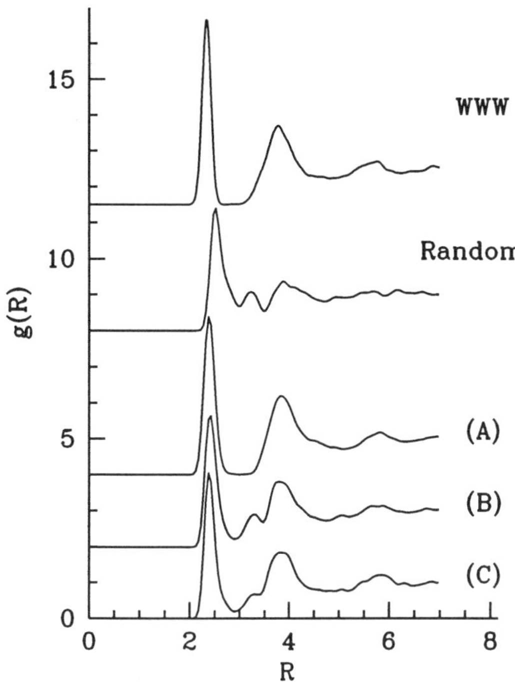

Figure 3. Radial distribution function for simulated *a*-Si. The top curve shows the RDF as obtained using the Wooten-Winer-Weaire algorithm (data from Ref. 7). The other curves are for a 1000-atom *a*-Si cell as obtained via activation-relaxation technique (ART, discussed later in the text) with a Stillinger-Weber potential. The curve labelled "Random" shows the initial random packed configuration. Curve (A) is for a network relaxed the the usual SW parameter set while curve (B) is for a modified SW potential as discussed in the text.

A regular WWW run would proceed in three or four steps: a randomnisation at 1 eV, some search in the phase space at 0.4 eV and "cooling" at 0.25 and 0 eV. Each step requires normally a few moves per atom on average. It must be understood here that the "temperature" is fictitious since the bond exchange is not based on physical moves. The RDF for a lattice obtained this way is displayed in Fig. 2.2(top curve). It is in very good agreement with experiment. Other topological properties of this network are equally excellent: ring distribution, width of bond angle distribution, etc. The successes of the WWW algorithm are such that it has been used to this day for a range of problems from a Si to a Si$_{1-x}$Ge$_x$ [21], including a-C and a-SiO$_2$ [7].

There are problems, however, with the WWW algorithm. First, the unphysical moves require a fully attractive potential of the Keating type, i.e.

140

without a cut-off. Because of the harmonic potential, the resulting lattice is generally overstrained by comparison with more realistic empirical potentials like Stillinger-Weber's or Tersoff's. Moreover, the specific algorithm retained, besides forcing the use of a nonphysical potential, can only be applied to the diamond network. This is not a portable solution. Finally, the unphysicalness of the moves prevents the use of this technique to study the dynamics of the network.

3. Molecular dynamics

Although molecular dynamics was very successful with liquids and gases, it was not applied to the study glasses before the mid-seventies. One of the reasons for this is that glasses and amorphous require long simulation time because of their slow dynamics; the other being that interest in disordered systems is fairly recent.

Molecular dynamics consists simply in the integration of the newtonian equations of motion for a many body system:

$$F_i = ma_i = - \sum_{\{j\}} \nabla_i V(\{j\}) \tag{4}$$

where V is the interaction potential and $\{j\}$ denotes the ensemble of all coordinates. So the quality of the results of MD depends for a large part on the interaction potential and, for disordered system, the time scale of the simulation.

The conceptual simplicity of molecular dynamics is what makes its strength. One can understand clearly what is the meaning of time, fluctuations, etc. when solving equation 4. This approach reproduces most closely the way Nature proceeds and all the information produced by MD can be used to get a better understanding of experimental phenomena. This same simplicity is also what limits MD, the single most important inconvenience being the time scale of a run. Since we are solving the newtonian equation of motion at the atomic level, the timescale we have to work at, when integrating Eq. 4, is determined by the phonon frequency. This means that simulations must be done with a timestep typically between one and ten femtosecond. The maximum time that can be reached is therefore almost completely determined by the brute force we can to apply to the problem. In general, most simulations are run in the range of a couple tens of picoseconds to the nanosecond. A recent state–of–the–art simulation of NiP glass [22], for example, reached a real time equivalent to about 3×10^{-8}s. There is no hope at the moment to run MD simulations on much longer timescales.

This is a major obstacle to studying many of the features of glassy systems since for them, the dynamics happens on a very slow time scale.

Other techniques, involving thermodynamics, for example, cannot easily be used because of the inherent disorder in the glassy materials. For a more detailed discussion of the impact of time limitations, see the recent review paper by Angell [23].

4. Interaction potentials

Time is not the only obstacle for computer modelling of glasses and amorphous materials. Finding the appropriate interaction potential can also pose problems. Especially for alloys where one has to deal with a considerable number of different interactions. In this section, I will present some of the recent development in this field: empirical potentials, *ab initio* and tight-binding interactions. As the last two method has been almost exclusively applied to semiconductors, all the examples will be for these materials.

4.1. EMPIRICAL POTENTIALS

The availability of interaction potentials has determined in large part the approach followed for studying materials and disordered systems more particularly. For example, it was not before 1985 that the first empirical potential valid in more than one phase were introduced for tetravalent semiconductors. The very first potential, and the one which remains favourite for silicon today, was proposed by Stillinger and Weber [24] and included a two-body plus a three-body part with an exponential decay:

$$E = \sum_{<ij>} \epsilon f_2(r_{ij}/\sigma) + \sum_{<jik>} \epsilon f_3(r_{ij}/\sigma, r_{ik}/\sigma, \theta_{jik}) \tag{5}$$

where

$$f_2(r) = A\left(Br^{-p} - 1\right)\exp[\frac{1}{r-a}] \tag{6}$$

and

$$f_3(r_{ij}, r_{ik}, \theta_{jik}) = \lambda\exp[\frac{1}{r_{ij}-a}]\exp[\frac{1}{r_{ik}-a}]\left(\cos\theta_{jik} + \frac{1}{3}\right)^2. \tag{7}$$

These two functions have a cut–off radius at $r = a$.

This was followed immediately by other propositions from Biswas and Hamann [25] and Tersoff [26]. With the help of these potentials MD simulations of a-Si and other semiconductors could at last take place [27, 28, 29, 30, 31]. The usual approach for obtaining an amorphous network with MD is to first melt the crystalline solid and then cool the liquid down, rapidly. The resulting network is generally highly strained and much closer to the liquid phase than to a crystal. In order to release the strain and decrease

142

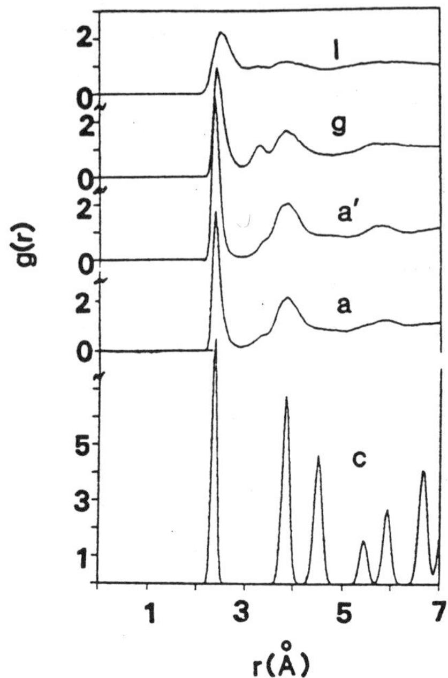

Figure 4. Radial distribution function for several phases of silicon as obtained via molecular dynamics with the Stillinger-Weber potential: crystalline (c), amorphous (a') and (a), glass (g) and liquid (l). The amorphous configuration (a) and (a') were obtained using different quenching procedures. Note the shoulder on the left of the second neighbour peak, indicating some remnants of the liquid state (reprinted from Ref. 33).

the coordination (which is higher in liquid then in solid for tetravalent semiconductors), the cell is often annealed at a temperature slightly below the melting point and then cooled back again for structural studies, following a process also used experimentally to release the strain in samples (see Fig. 4.1). As all simulations for amorphous semiconductors led to essentially the same qualitative results, I will discuss them as a whole.

Although these empirical potentials give respectable results for the amorphous phase, none of them produces excellent comparison with the experimental radial distribution function, which is the simplest quantity with which one can judge of the quality of the structure of amorphous and glasses. The general shape of the RDF is in agreement with experimental measurements but the first neighbour peak is not as well defined, indicating a large density of defects, much higher than in well annealed experimental samples. More worrisome, most of these defects consist of overcoordinated atoms, with so-called floating bonds, for which there is not experimental trace. This result surprises since these potentials tend to produce undercoordinated liquids [32]. The question of whether the "wrong" defects in amorphous cells are due to the slow dynamics and rapid quenching being used at the present or a defective potential remains to be answered. It was generally thought that if one could wait long enough, the Stillinger-Weber potential would lead to the correct amorphous structure, unattainable with the limited time scale of MD [33]. Recent results using the activation-relaxation technique described in the last section of this chapter, however, indicate that this is probably not a problem of large phase space but rather of bad potential [34].

Besides elemental semiconductors, many attempts have been made over the last decade to develop similar set of parameters for semiconductor compounds like SiC, GeSi [26] as well as impurities like H in Si [17, 35] with limited success. For most of these interactions, detailed knowledge of their nature is missing and one has to proceed by analogy or simple guess.

In spite of their limitations, empirical potentials have two advantages over more accurate interactions: simplicity and low computing requirements. In metallic glasses following the dense packing model, it is still generally believed that long simulations are preferable to an exact but complicated potential. Many of the simulations, therefore have been performed using simple potentials like the Weber-Stillinger potential [36] (not to be confused with the Stillinger-Weber one!) or even Lennard Jones interactions [22]. In the case of tetravalent amorphous networks, however, it is clear that the present empirical potentials are not sufficient to produce via MD the "ideal" CRN from which one could start to understand the numerous and important opto-electronic properties of amorphous semiconductors. In systems with slow dynamics, both amorphous and glasses, speed always needs to be weighted carefully against the precision and accuracy of semi-empirical tight-binding and *ab-initio* techniques and no unique answer exists about which approach to choose.

4.2. *AB-INITIO* APPROACHES

Contemporarily with the development of empirical potentials for the tetravalent semiconductor, the first *ab initio* molecular dynamics technique was proposed by Car and Parrinello [37] who rapidly used it to study *a*-Si [38, 39] and *a*-Si:H [40]. This technique is based on the fast relaxation time of electrons compared to that of ions. It minimises the electronic energy at each time step, using density-functional theory, and calculates the corresponding force to evolve the ions. In recent years, many other groups repeated *ab-initio* MD calculations for silicon [41] and *a*-Ge [42]. Most of these simulations followed the usual MD procedure: first production of a liquid phase, then rapid quench, of the order of $10^{14} K/s$ and relaxation at

300K. Instead of going into the liquid phase, Drabold *et al.* [41] preferred quenching the lattice while it was still trying to equilibrate at 8000K, starting with a more ordered lattice than other groups.

All these simulations, performed on cells of about 64 atoms, produced very similar results. The first- and seconds–neighbour shell is better defined that with empirical potential. In particular, the small r shoulder present on the second–neighbour peak of an amorphous relaxed with the Stillinger-Weber potential does not appear. Moreover, the first- and second–neighbour peaks are better distanciated that for the SW potential. In general, the coordination number of these models varies between 3.97 and 4.06 with, in all cases except one realisation of a-Ge, more 5-fold than 3-fold defects. The width of the bond angle distribution was found to be between 16 and 17% in all samples, somewhat larger than the 10% found experimentally. This can be due to extremely short the time scale on which these studies happen: runs typically last less than 25 ps. In addition, unit cells are also extremely small, at the limit of the feasible. In spite of these problems, the fact that results are in better agreement with experiment than those using empirical potentials is very positive. The next step is therefore to provide for larger systems and longer runs to make sure that the network has time to relax fully and minimise finite-size effects.

4.3. SEMI-EMPIRICAL INTERACTIONS

One avenue which could provide a good balance between empirical potentials and *ab initio* approach is tight-binding molecular dynamics where the interactions are easier to compute than the full *ab initio* scheme while still providing for some quantum mechanical effects by introducing a certain number of orbitals for each atoms. With little effort, it is generally possible to study samples of 64 atoms on time scales 10 times longer than with *ab-initio* MD.

The basic tight-binding expression for the binding energy of N atoms is given by

$$E_{binding} = \int_{-\infty}^{\epsilon_F} \epsilon \left[n^{out}(\epsilon) - n^{at}(\epsilon) \right] d\epsilon + \sum_{\{ij\}} \phi(r_{ij}) + CN \qquad (8)$$

where n^{out} is the density of states (DOS) for the interacting atoms, n^{at}, the DOS for free atoms, $\phi(r_{ij})$ is a repulsive potential and C is a constant. The first term, representing the quantum-mechanical bond energy is expressed by a tight-binding Hamiltonian comprising 4 or 5 orbitals for semiconductors. However, the tight-binding interaction does not include a dependence on distances. This does not cause problem for crystals, where neighbour shells are well defined. It does, however, for liquids and amorphous phases.

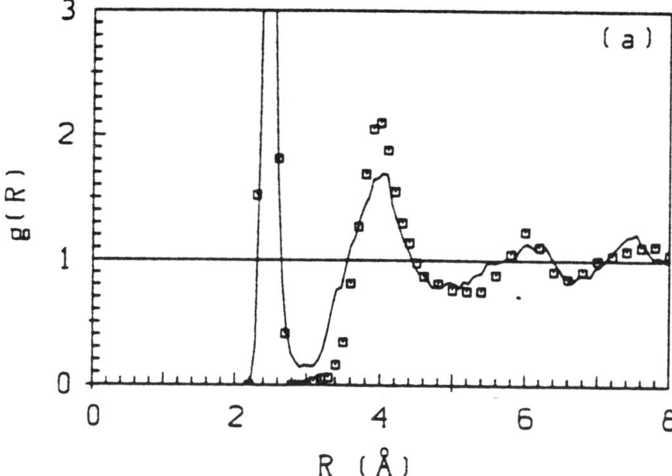

Figure 5. Radial distribution function for a 64-atom quenched *a*-Ge cell obtained via molecular dynamics with tight-binding parameters. The square symbols are from experiment (reprinted from Ref. 42.)

One has therefore to add an empirical decaying function of r_{ij} which includes a cut-off term somewhere between the first and second-neighbour shells (which is not necessarily well defined in disordered systems). Moreover, the repulsive part, which also has an empirical form is defined as a two-body interactions and merely keeps the lattice from collapsing; all the angular contributions are taken care by the bonding energy. The two main forms which are used today for semiconductors are those proposed by Goodwin *et al.* [43] and by Molteni *et al.* [44].

In spite of their considerable empirical side, these interactions have led to better results than empirical potentials both in liquid and amorphous silicon [45, 46, 50](Fig. 4.3). If the procedure followed by Kim and Lee [50] led to very large number of coordination defects, Servalli and Colombo [46] obtained a 64-atom cell comparable in quality to what is seen using *ab-initio* MD. They also showed that it was possible to construct larger unit cells of *a*-Si and presented some results for a 216-atom cell.

Tight-binding can also be used to simulate binary compounds as was demonstrated recently by a work on *a*–GaAs for which no satisfactory empirical potential exists [48, 49]. This approach should work best for very covalent alloys since charge transfer is not taken into account by traditional tight-binding equations and must therefore be added empirically.

With the development of order N tight-binding and *ab-initio* techniques, i.e. approaches for which computing demands scale, for large enough cells,

TABLE 1. Approximate number of rings per atom in a crystalline and amorphous silicon network.

Ring size	4	5	6	7	8
Crystalline	0	0	2	0	3
Amorphous	0	0.45	0.85	1.05	2.0

linearly with the number of atoms [51, 52, 53, 54, 55, 56, 57], one should be able to go beyond the present size and time limitations which affect both semi-empirical and *ab-initio* techniques. In the near future we should see semi-empirical MD runs of cells containing many hundreds of atoms over a few hundreds of picoseconds.

5. A few specific problems

As mentioned previously, alloys and impurities are particularly difficult to deal with both via empirical and tight-binding potentials. In this section, I describe two outstanding fundamental problems: a-GaAs and a–Si:H. The solution to the first one is within reach but will require considerable effort. The second one remains elusive but is of fundamental technological importance.

5.1. GAAS

If the CRN model seems appropriate for elemental amorphous semiconductors, there are still questions raised about its adequacy to represent binary semiconductors like GaAs. While simulations of CRN using either the WWW algorithm or molecular dynamics indicate a large and universal presence of odd-membered rings (see Table 1), the chemical ordering seen in a-GaAs would imply that they are almost absent in this material.

One cannot measure experimentally the ring distribution in elemental amorphous. It is possible, however, to extract some bounds in binary compounds like GaAs where one can test the nature of the bonds. Experimental results from EXAFS have indicated repeatedly that contrary to binary amorphous semiconductors like InP or GaP, GaAs retains a strong chemical ordering even in the amorphous phase while presenting a RDF qualitatively identical to a-Si (see [58, 59], for example). This would mean that the a-GaAs network contains almost no odd-membered ring.

Although Connell and Temkin [60] showed some time ago that it was possible to construct by hand a CRN with no odd-membered ring, no computer simulation to date has been able to replicate these results. Only recently was this problem studied via MD, due to the absence of good potential for this compound. However, TB simulations of Molteni *et al.* [48] and Seong and Lewis [49] as well as *ab-initio* MD [61] display a concentration of wrong bonds between 10 and 13 %, corresponding to the number of wrong bonds found in a CRN with the ring statistics given in Table 1. This proportion is somewhat larger that experimental bounds.

We still ignore what is the exact topology of the *a*-GaAs network and whether or not one needs a second model besides the CRN. Solving this question will require long runs, allowing a better topological and chemical relaxation. This should be possible today, however, although with some good deal of computer time.

5.2. A-SI:H

The problems with hydrogenated amorphous silicon are somewhat different. The main interest here is to understand where exactly is the hydrogen going in the *a*-Si network as well as what is its role in the many electro-optic effects measures in *a*-Si:H. In order to simulate this, one therefore needs to know first what the defects are in *a*-Si. Even with this knowledge, the problem of designing satisfactory interaction potentials between H-H, and H-Si remains complete.

Static methods for introducing hydrogen in the network, which have the advantage of simplifying the type of interaction needed, have been able to give some indication about its effects on reducing the strain in the network [17, 31] as well as defect formation and excitation [35, 62]. However, to really understand the role of hydrogen, given the problem with designing interaction potential, one needs to go to either semi-empirical or *ab initio* technique. There has been some work using the Car-Parrinello technique [40] but the limited size of the lattice (64 Si plus 8 H) makes it difficult to extract any significant results as regards to the type of defects and the role hydrogen plays in this materials. Significant efforts are therefore needed before one can address properly the role of hydrogen in amorphous networks. With order N methods, one could hope that *ab initio* MD will be able to simulate cells with a few hundred atoms in a near future and then study defect and impurities properly.

6. Future directions

The simulation approaches to glasses have not changed very much over the last few years. The usual methods have been refined and better interactions proposed but this does not provide for the simulation time needed to address many of the most fundamental questions about glasses. Order N methods for tight-binding and *ab initio* molecular dynamics will not be able to go much further than a few hundred ps in simulation time. Even state of the art MD with simpler empirical potentials [63, 22] are limited to the nanosecond range. This is not long enough for the slow relaxation which happens in these glassy systems. Other procedures are therefore needed so as to extend the simulation timescales. To reach this goal, we have to concentrate the effort on events which are important to the dynamics and neglect the others, like local vibrations which limit so much MD.

One of these methods under development is the activation-relaxation technique (ART) of Barkema and Mousseau [34]. The approach is based on a series of Monte-Carlo moves controlled by simulated annealing. Contrary to the static methods used until now, the moves are realistic in the sense that there is a search for local saddle-points over which the configuration is pushed; there is no artificial list of neighbours or bond-switching. The basic algorithm can be described in few steps:

1. Relax the configuration with the potential of your choice;
2. Move this configuration to a saddle point, moving any number of atoms needed on your way;
3. Push the configuration over this saddle point and go back to (1).

The generality of this method is such that it can be used with any type of interaction, from Lennard Jones to *ab-initio* interactions. Moreover, by restricting the types of moves to activation-relaxation, we concentrate on the significant moves, contrary to more traditional MC. And doing this, we can have single moves that displace any number of atoms depending on the specific event.

Fig. 2.2 shows the radial distribution function of a sample of silicon as it is relaxed with this technique and goes from a randomly packed configuration to amorphous in a few attempts per atom.

Since the moves in ART are physical and use saddle-points, it is possible to think of using a rare-event dynamics (RED) approach [34] together with this technique in order to follow the dynamics of the system on long time scale.

7. Conclusion

Due to their disordered nature glasses and amorphous materials need, even more than other phases of matter, computer simulations in order to be understood properly. In many instances, computer models play a fundamental role as a bridge between experimental results and theory. Over the last twenty years or so, this field has benefited enormously from numerical work, especially in the short time dynamics range.

To modelise successfully the time dynamics in glasses and amorphous materials, computational physicists still need to find new approaches which will allow them to go beyond what is feasible at the moment. For some questions, the speed increase of the hardware will suffice, for others, we need new concepts. This is where methods like ART and others which are being developed at this moment will play a significant part in the study of disorder materials.

8. Acknowledgements

I would like to thank Laurent Lewis for providing a number of references for this review. I want also to acknowledge financial support from the Natural Sciences and Engineering Research Council of Canada as well as from the "Fonds pour la formation des chercheurs et l'aide à la recherche" of the Province of Québec.

References

1. Internet address: Mousseau@PhysCN.UMontreal.CA
2. Bernal, J. D., Nature **188**, 910 (1960).
3. Bernal, J. D., Proc. R. Soc. **A284**, 299 (1964).
4. Cargill, G. S. (III), J. Appl. Phys. **31**, 12 (1970).
5. Bennett, C. H., J. Appl. Phys. **43**, 2727 (1972).
6. For a review of these models, see Finney, J. L., *in* Amorphous Metallic Alloys, Ed. F. E. Luborsky, Butterworths, London (1983).
7. Djordejvić, B. R. and Thorpe, M. F., Phys. Rev. B **52**, 5685 (1995).
8. Zachariasen, W. H., J. Am. Chem. Soc. **54**, 3841 (1932).
9. Bell, R. J. and Dean, P., Nature (London) **212**, 1354 (1966).
10. Polk, D. E., J. Non-Cryst. Sol. **5**, 365 (1971).
11. Shevchik, N. J., Phys. Stat. Sol (b) **58**, 111 (1973).
12. Henderson, D. and Herman, F., J. Non-Cryst. Sol. **8-10**, 359 (1972).
13. Steinhardt, P. J., Alben, R., Duffy, M. G., and Polk, D. E., Phys. Rev. B **8**, 6021 (1973).
14. Keating, P. N., Phys. Rev. **145**, 637 (1966).
15. Wooten, F., Winer, K., and Weaire, D., Phys. Rev. Lett. **54**, 1392 (1985).
16. Wooten, F. and Weaire, D., Sol. Stat. Phys. **40**, 1 (1986).
17. Mousseau, N. and Lewis, L. J., Phys. Rev. B **41**, 3702 (1990).
18. Henderson, D., J. Non-Cryst. Sol. **16**, 317 (1974).
19. Guttman, L., Ching, W. Y., and Rath, J., Phys. Rev. Lett. **44**, 1513 (1980).
20. Kirkpartrick, S., Gelatt, D. C. and Vecchi, M. P., Science **220**, 671 (1983).
21. Mousseau, N. and Thorpe, M. F., Phys. Rev. B **48**, 5172 (1993).
22. Kob, W. and Andersen, H. C., Phys. Rev. B **51**, 4626 (1994).
23. Angell, C. A., Comput. Mat. Sci. **4**, 285 (1995).
24. Stillinger, F. H. and Weber, T. A., Phys. Rev. B **31**, 5262 (1985).
25. Biswas, R. and Hamann, D. R., Phys. Rev. Lett. **55**, 2001 (1985).
26. Tersoff, J., Phys. Rev. B **37**, 6991 (1988).
27. Ding, K. and Andersen, H. C., Phys. Rev. B **34**, 6987 (1986).
28. Biswas, R., Grest, G. S., and Soukoulis, C. M., Phys. Rev. B **36**, 7437 (1987).
29. Kluge, M. D., Ray, J. R. and Rahman, A., Phys. Rev. B **36**, 4234 (1987).
30. Luetdke, W. D. and Landman, U., Phys. Rev. B **37**, 4656 (1988).

31. Mousseau, N. and Lewis, L. J., Phys. Rev. B **43**, 9810 (1991).
32. Cook, S. J. and Clancy, P., Phys. Rev. B **47**, 7686 (1993).
33. Luetdke, W. D. and Landman, U., Phys. Rev. B **40**, 1164 (1989).
34. Barkema, G. T. and Mousseau, N., to be published.
35. Kwon, I, Biswas, R., and Soukoulis, C. M., Phys. Rev. B **45**, 3332 (1992).
36. Lewis, L. J., Phys. Rev. B **44**, 4245 (1990).
37. Car, R. and Parrinello, M., Phys. Rev. Lett. **55**, 2471 (1985).
38. Car, R. and Parrinello, M, Phys. Rev. Lett. **60**, 204 (1988).
39. Štich, I., Car, R. and Parrinello, M., Phys. Rev. B **44**, 11092 (1991).
40. Buda, F., Chiarotti, G. L., Car, R. and Parrinello M., Phys. Rev. B **44**, 5908 (1991).
41. Drabold, D. A., Fedders, P. A., Sankey, O. F.
42. Kresse, G. and Hafner, J., Phys. Rev. B **49**, 14251 (1994).
43. Goodwin, L., Skinner, A. J. and Pettifor, D. G., Europhys. Lett. **9**, 701 (1989).
44. Molteni, C., Colombo, L. and Miglio, L., J. Phys. C: Cond. Matt. **6**, 5243 (1994).
45. Mercer, J. L. Jr. and Chou, M. Y., Phys. Rev. B **43**, 6768 (1991).
46. Servalli, G. and L. Colombo, Europhys. Lett. **22**, 107 (1993).
47. Kim, E. and Lee, Y. H., Phys. Rev. B **49**, 1743, (1994).
48. Molteni, C., Colombo, L, and Miglio, L., Phys. Rev. B **50**, 4371 (1994).
49. Seong, H. and Lewis, L. J., Phys. Rev. B **53**, 4408 (1996).
50. Kim, E. and Lee, Y. H., Phys. Rev. B **49**, 1743, (1994).
51. Yang, W. Phys. Rev. Lett. **66**, 1438 (1991).
52. Alavi, A. and Frenkel, D., J. Chem.Phys. **97**, 9249 (1992).
53. Aoki, M., Phys. Rev. Lett. **71**, 1438 (1993).
54. Li, X.-P., Nunes, W. and Vanderbilt, D., Phys. Rev. B **47**, 10981 (1993).
55. M. S. Daw, Phys. Rev. B **47**, 10895 (1993).
56. Goedecker, S. and Colombo, L., Phys. Rev. Lett. **73**, 122 (1994).
57. Kim, J., Mauri, F., and Galli, G., Phys. Rev. B **52**, 1640 (1995).
58. Udron, D., Flank, A.-M., Lagarde, P., Raoux, D., and Thèye, M. L., J. Non-Cryst. Sol. **150**, 361 (1992).
59. gheorghiu, A., Driss-Khodja, K, Fisson, S., Thèye, M. L. and Dixmier, J., J. Phys. (Paris) **46**, C8 545 (1985).
60. Connell, G. A. N. and Temkin, R. J., Phys. Rev. B **9**, 5323 (1974).
61. Fois, E., Selloni, A., Pastore, G., Zhang, Q. M., and Car, R., Phys. Rev. B **45**, 13378 (1992).
62. Lutz, R. and Lewis, L. J., Phys. Rev. B **47**, 9896 (1993).
63. Lewis, L. J. and Wahnström, G., Phys. Rev. B **50**, 3865 (1994).

MOLECULAR DYNAMICS METHODS AND LARGE-SCALE SIMULATIONS OF AMORPHOUS MATERIALS

PRIYA VASHISHTA, RAJIV K. KALIA, AIICHIRO NAKANO
WEI LI AND INGVAR EBBSJÖ[(*)]

Concurrent Computing Laboratory for Materials Simulations
Department of Physics & Astronomy and Department of Computer Science
Louisiana State University, Baton Rouge, Louisiana 70803, U.S.A.

[(*)]*Studsvik Neutron Research Laboratory*
University of Uppsala, S-611 82 Nyköping, Sweden

1. Introduction

Computer simulation is a very powerful tool for studying physical and chemical phenomena in amorphous materials. In these lectures we shall review the molecular dynamics method and its implementation on parallel computer architectures. Using the molecular dynamics method we will study a number of materials in different ranges of density, temperature, and uniaxial strain. These include structural correlations in silica glass under pressure, internal fracture in porous silica, nature of low frequency floppy modes in high temperature ceramic silicon nitride, crack propagation in silicon nitride films, sintering of silicon nitride nanoclusters, amorphization and fracture in silicon diselenide nanowires, amorphous carbon, and dynamic fracture in graphite.

Recent advances in computing technology - hardware, especially parallel computer architectures, software and development of robust O(N) algorithms - have revolutionized the field of computer simulation. Some of the reasons for doing computer simulation studies are:

- Computer simulations bridge the gap between theory and experiment.

- Nonlinearities, lack of symmetry, and a large number of degrees of freedom can be handled much more easily by computer experiments than by analytical methods.

- The hypothetical universe available to computer experimentation is not limited to the processes occurring in nature.

Some scientific disciplines where computer simulations have been used widely are

M. F. Thorpe and M. I. Mitkova (eds.), Amorphous Insulators and Semiconductors, 151–213.

152

- Astrophysics

- Biology

- Chemistry

- Condensed matter physics - especially disordered materials which lack symmetry

- Lattice gauge theory

- Statistical physics [Classical and quantum liquids, plasmas, superionic conductors, glasses, high T_c superconductors, *etc.*].

In these lectures we will deal with the simulation of classical systems, *i.e.*, systems in which the dynamics of particles is governed by classical mechanics. For simulations of time-independent and time-dependent quantum systems we suggest the reader to look into other review articles and research papers.

In many materials, such as semiconductor binary oxides (SiO_2, GeO_2), chalcogenides ($GeSe_2$, $SiSe_2$), III-V semiconductors (GaAs, InI, InSb), fast-ion conductors (AgI, CuBr, Ag_2S, Ag_2Se), and high temperature ceramic materials (Si_3N_4, SiC, Al_2O_3, AlN, GaN) one of the dominant interaction is the Coulomb potential arising as a result of charge-transfer effects. As the Coulomb potential is a long range interaction, each particle interacts with all the other particles in the system. Therefore, the computing time involved in the evaluation of interaction increases as N^2, where N is the number of particles in the system. This makes the large-scale simulation of Coulombic systems difficult.

Special techniques have been developed to reduce the computational complexity from $O(N^2)$. The Ewald summation is one of the most widely used approaches for the Coulomb interaction in molecular-dynamics and Monte Carlo simulations for bulk systems. With this method, the total potential energy can be transformed into rapidly convergent sums of short-ranged terms in real space and a local sum in the Fourier space. By a suitable choice of the separation constant controlling the relative overheads of the two contributions, the total computing time grows as $O(N^{3/2})$ [1]. Kalia et al. designed parallel algorithms to implement the Ewald summation approach on distributed-memory MIMD machines [2]. The execution times for these algorithms scale linearly with N. Recently, several divide-and-conquer schemes have been implemented in the evaluation of long range Coulomb interaction. The basic idea of these schemes is to separate the Coulomb potential into near-field and far-field interactions. The near-field terms are evaluated directly, whereas the far-field interaction is calculated by means of a hierarchical grouping of particles. Since the number of nearby particles is much less than that of faraway particles, the portion of operations for direct calculation will be reasonably small. Also as the force is inversely proportional to the square of distance, direct calculation in near-field interaction assures a certain precision. Many different approaches have been designed for the far-field interactions. Appel [3] and Barnes and Hut [4] used a hierarchical tree-based algorithm which reduces the computational complexity to O(NlogN). Fox and coworkers [5] used this algorithm to perform galaxy merger simulation for 180,000 bodies on a distributed-memory parallel machine. Using this algorithm, Pfalzner and Gibbon studied the dynamics of dense plasmas in three dimensions [6]. Recently this algorithm was implemented on Caltech's Touchstone Delta machine. Simulations involving over 60 million particles were performed [7]. Greengard and Rokhlin have proposed the Fast Multipole Method (FMM) [8-11]. The far-field contribution to the Coulomb interaction is calculated with the multipole expansion for the Coulomb potential. On sequential machines, the complexity of the FMM is proportional to N if the number of particles equals the number of boxes. Schmidt and Lee evaluated the performance of this algorithm on a Cray 2 [12]. They also extended it to incorporate Ewald summation for bulk Coulombic systems. Zhao and Johnsson have implemented the FMM for three dimensional systems on a Connection machine [13].

The goals of this article are: (1) to discuss the molecular dynamics (MD) method and to design and implement a parallel MD algorithm including calculation of long range Coulomb interaction for large scale systems; (2) to investigate the structural and dynamical correlations in a variety of systems (glasses, ceramic materials, nanowires, amorphous carbon, and graphite) under extreme conditions of density, temperature, and external strain using MD simulations.

This article contains seven Sections including the References. A list of topics contained in Sections 2 to 6 is given below:

2. **Molecular Dynamics (MD) Methods**

2.1. MOLECULAR DYNAMICS FORMULATION

2.2. NUMERICAL SOLUTIONS FOR THE EQUATIONS OF MOTION

2.3. EVALUATION OF PHYSICAL PROPERTIES

3. **Implementation of the Molecular Dynamics (MD) Method on Parallel Computer Architectures**

3.1. THE DOMAIN DECOMPOSITION SCHEME

3.2. MULTIRESOLUTION MOLECULAR DYNAMICS (MRMD) ALGORITHM

4. **Interaction Potentials for Silica, Silicon Nitride, and Silicon Diselenide**

4.1. CRYSTALLINE STRUCTURES OF SiO_2, Si_3N_4 AND $SiSe_2$

4.2 INTERATOMIC POTENTIAL

5. **Molecular Dynamics Simulations of SiO_2 Glass, Si_3N_4, $SiSe_2$, and Graphite**

5.1. STRUCTURAL CORRELATIONS IN SILICA GLASS UNDER PRESSURE AND INTERNAL FRACTURE IN POROUS

5.2. SILICON NITRIDE: LOW-FREQUENCY FLOPPY MODES, FRACTURE, AND SINTERING OF NANOCLUSTERS

5.3. AMORPHIZATION AND FRACTURE IN SiSe$_2$ NANOWIRES

5.4. AMORPHOUS CARBON AND FRACTURE IN GRAPHITE

6 . Research in Progress and Future Plans

6.1. MEMS SUBMICRON GEAR TRAINS -- WEAR, FRICTION AND SELF-
ASSEMBLED MONOLAYERS AS LUBRICANTS

2. Molecular Dynamics (MD) Methods

Molecular-dynamics simulations for complex physical systems such as nanosized (diameter scale of a few 10^{-9} m) materials and complex phenomena like crack propagation and fracture often require large system sizes (with number of particles ranging from N=10^5 to 10^7) and long simulation time. In phenomena involving atoms and molecules, to prepare a well-thermalized MD configuration and calculate the mechanical and thermal properties, it is often necessary to investigate time evolution over hundreds of picoseconds (10^{-12} sec). Since the time increment needed to solve Newton's equations is usually on the order of a few femtoseconds (10^{-15} sec), simulations involving 10^4-10^6 time intervals are necessary [14]. Clearly, it is not feasible to simulate large systems over long intervals of time on sequential machines, unless the interparticle interactions are fairly simple. Large-scale MD simulations for real systems can only be performed on emerging parallel computer architectures [15]. In this section we shall discuss general aspects of molecular dynamics simulation. Implementation of MD on parallel architectures is discussed in Section 3.

2.1. MOLECULAR DYNAMICS FORMULATION

From the statistical mechanics point of view, MD simulation is a method to compute the phase-space trajectory [16-18]. For a system of N particles, the phase-space is a 6N-dimensional hyperspace consisting of positions and momenta of the N particles. Based on different statistical ensembles, MD simulation can derive reliable macroscopic physical properties by sampling on different phase-space surfaces. Each ensemble has its own conserved thermodynamic quantity such as energy, temperature, and pressure. In conventional MD simulation, averages of physical quantities are measured in the microcanonical ensemble (NVE) where the particle systems are considered to be isolated. In simulations based on canonical ensembles (NVT), auxiliary variables are introduced to allow thermal coupling between particle systems and heat reservoir. In simulations involving external stress/pressure, isoenthalpic-isotension ensemble (HtN) and isothermal-isobaric ensemble (NPT) are used to allow mechanical coupling between particle systems and external sources. In the following sections, we will review the basic formalisms of MD simulation in these ensembles.

2.1.1. Fundamental Equations for the Microcanonical Ensemble (NVE)

The basic formalism for molecular dynamics is based on the conventional microcanonical ensemble. For a system of N particles interacting with a potential $\mathcal{V}$ within a fixed volume V, the Lagrangian equation of motion is

$$\frac{d}{dt}\left(\frac{\partial \mathcal{L}}{\partial \dot{q}_k}\right) - \left(\frac{\partial \mathcal{L}}{\partial q_k}\right) = 0 \ , \tag{2.1}$$

where $\mathcal{L}(\vec{q}, \dot{\vec{q}})$ is defined in terms of kinetic energy $\mathcal{K}$ and potential energy $\mathcal{V}$. In Cartesian coordinates, we have

$$\mathcal{L} = \mathcal{K} - \mathcal{V} \ , \tag{2.2}$$

where

$$\mathcal{K} = \sum_{i=1}^{N} \frac{m_i}{2} \dot{\vec{r}}_i^2 \ , \tag{2.3}$$

$$\mathcal{V} = \sum_{i=1}^{N} v_1(\vec{r}_i) + \sum_{j>i}^{N} v_2(\vec{r}_i, \vec{r}_j) + \sum_{k>j>i}^{N} v_3(\vec{r}_i, \vec{r}_j, \vec{r}_k) + \dots \ , \tag{2.4}$$

In Eq.(2.4), v_1, v_2 and v_3 represent one-body (or say, external field), two-body and three-body interactions, respectively. When we substitute (2.2)-(2.4) into (2.1), we get the Newton equations of motion

$$m_i \ddot{\vec{r}}_i = -\nabla_{\vec{r}_i} \mathcal{V} \ , \qquad\qquad i=1, \ \dots, \ N. \tag{2.5}$$

When forces among atoms are conservative, the Hamiltonian is a constant of the motion. In the case of microcanonical ensemble, the total energy is conserved, i.e.,

$$\mathcal{H} = \mathcal{K} + \mathcal{V} = E \ . \tag{2.6}$$

Note here that from the general physics point of view, Eq.(2.5) holds for systems in all ensembles. Next we will see that in ensembles other than the microcanonical ensemble, the MD equations of motion are modified by introducing additional dynamical variables that act as monastat/thermostat to couple with the particle system.

2.1.2. Extended Equations for Canonical Ensemble (NVT)

To treat the dynamics of a system in contact with a heat reservoir, an additional degree of freedom, f, is introduced [19]. f can be interpreted as an exchange of heat between particles and the heat reservoir. The total energy of the particle system is allowed to flow dynamically back and forth from the heat reservoir. In this case, the Lagrangian becomes

$$\mathcal{L} = \sum_{i=1}^{N} \frac{m_i}{2} f^2 \dot{\vec{r}}_i^2 - \mathcal{V} + \frac{Q}{2}\dot{f}^2 - (3N+1)k_B T_{eq} \ln f \; , \tag{2.7}$$

where Q corresponds to the 'mass' of the heat reservoir (in unit of energy·(time)^2), T_{eq} is desired equilibrium temperature, and k_B is the Boltzmann constant. The corresponding equations of motion for particles and heat reservoir are

$$m_i \ddot{\vec{r}}_i = -\frac{1}{f^2} \nabla_{\vec{r}_i} \mathcal{V} - 2m_i \dot{\vec{r}}_i \frac{\dot{f}}{f} \; , \qquad\qquad i=1, \dots, N, \tag{2.8}$$

$$Q\ddot{f} = \sum_{i=1}^{N} m_i f \dot{\vec{r}}_i^2 - \frac{(3N+1)k_B T_{eq}}{f} \quad . \tag{2.9}$$

These equations generate a canonical ensemble. By specifying a temperature, T_{eq}, the thermal coupling between particle system and heat reservoir will bring the total system to thermoequilibrium. The constant of motion for systems in the canonical ensemble is

$$\mathcal{H} = \sum_{i=1}^{N} \frac{m_i}{2} f^2 \dot{\vec{r}}_i^2 + \mathcal{V} + \frac{Q}{2}\dot{f}^2 + (3N+1)k_B T_{eq} \ln f \; . \tag{2.10}$$

When the additional freedom is set to be $f = 1.0$ and $\dot{f} = 0.0$, Eq.(2.7) becomes the Lagrangian for the microcanonical ensemble as in Eq.(2.2).

2.1.3. Extended Equations for Isoenthalpic-Isotension Ensemble (HtN)

In molecular-dynamics simulations, a system of N particles is usually considered to be contained in a simulation box. In the study of solids, however, changing of structural phases usually involves changes in the unit cell dimensions and angles. Recently, new and powerful techniques have been developed that allow variations in the shape of the MD cell under conditions of constant stress and/or hydrostatic pressure. Here we review the Parrinello-Rahman approach (first developed by Parrinello and Rahman and later modified by Ray and Rahman) [18, 20-27]. This method allows the mechanical coupling between system and external pressure/stress. A 3×3 matrix, **h**, which defines the shape of the simulation box, is included in the Lagrangian. **h** is formed by the vectors ($\vec{a}$, $\vec{b}$, $\vec{c}$) that span the edges of the simulation box, see Figure 2.1. The 3-D Cartesian coordinate, $\vec{r}_i$, of the i th particle is related to its dimensionless coordinate, $\vec{s}_i$, through the relation

$$\vec{r}_i = \mathbf{h}\, \vec{s}_i \; , \qquad\qquad i = 1, 2, \dots, N. \tag{2.11}$$

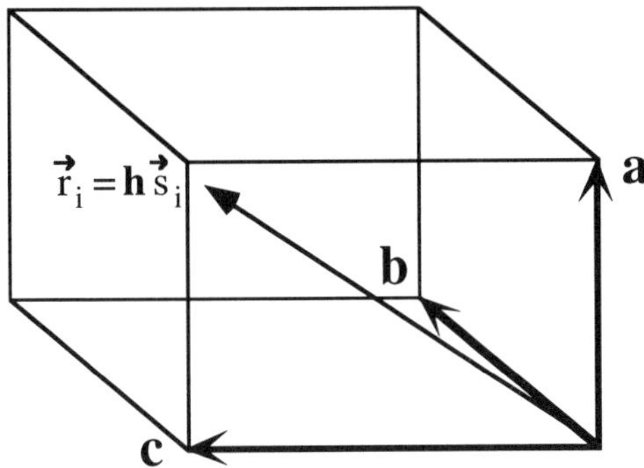

Figure 2.1   A MD box is formed by three vectors, $\vec{a}, \vec{b}$ and $\vec{c}$.
The coordinates of particles are scaled by $\mathbf{h} = (\vec{a}, \vec{b}, \vec{c})$.

With this expression, the Lagrangian becomes

$$\mathcal{L} = \sum_{i=1}^{N} \frac{m_i}{2} f^2 \dot{\vec{s}}_i{'} \mathbf{G} \, \dot{\vec{s}}_i - \mathcal{V} + \frac{Q}{2}\dot{f}^2 - (3N+1)k_B T_{eq}\ln f$$
$$+ \frac{W}{2}\mathrm{Tr}(\dot{\mathbf{h}}{'}\dot{\mathbf{h}}) - V_0 \mathrm{Tr}(\mathbf{t}\boldsymbol{\varepsilon}), \tag{2.12}$$

where

$$\mathbf{G} = \mathbf{h}{'}\mathbf{h}, \tag{2.13}$$

$$\boldsymbol{\varepsilon} = \frac{1}{2}(\mathbf{h}_0^{'-1}\mathbf{G}\mathbf{h}_0^{-1} - \mathbf{I}). \tag{2.14}$$

In the above equations, $\boldsymbol{\varepsilon}$ is the strain tensor, $\mathbf{t}$ is the external stress tensor, W is the 'mass' of simulation box (in unit of mass), and V_0 is the reference volume of the simulation box determined from the reference box variable $\mathbf{h}_0$

$$V_0 = \det (\mathbf{h}_0). \tag{2.15}$$

In these equations, vectors and matrices with prime are their transpose. The equations of motion for particle system, heat reservoir, and simulation box are

$$m_i\ddot{\vec{s}}_i = -\frac{1}{f^2}\mathbf{h}{'-1}\nabla_{\vec{r}_i}\mathcal{V} - 2m_i\dot{\vec{s}}_i\frac{\dot{f}}{f} - m_i\mathbf{G}^{-1}\dot{\mathbf{G}}\dot{\vec{s}}_i \quad , \qquad i=1,...,N, \tag{2.16}$$

158

$$Q\ddot{f} = \sum_{i=1}^{N} m_i f \dot{\vec{s}}_i' \mathbf{G}\, \dot{\vec{s}}_i - \frac{(3N+1)k_B T_{eq}}{f} \quad ,$$
(2.17)

$$W\ddot{\mathbf{h}} = \mathcal{P}\mathbf{A} - \mathbf{h}\Gamma,$$
(2.18)

where

$$\mathcal{P} = \frac{1}{V} \sum_{i=1}^{N} [m_i f^2 \mathbf{h}\dot{\vec{s}}_i \dot{\vec{s}}_i' \mathbf{h}' - (\nabla_{\vec{r}_i} \mathcal{V})\vec{s}_i' \mathbf{h}'] \quad ,$$
(2.19)

$$\mathbf{A} = V\mathbf{h}'^{-1},$$
(2.20)

$$\Gamma = V_0 \mathbf{h}_0^{-1} \mathbf{t} \mathbf{h}_0'^{-1}.$$
(2.21)

In these equations, $\mathcal{P}$ is the instantaneous internal stress tensor, $\mathbf{A}$ is the area tensor, Γ is a constant matrix related to the external stress applied to the system, and V is the volume of the system $V = \det(\mathbf{h})$.

The constant of motion for systems in the constant-stress ensemble is

$$\mathcal{H} = \sum_{i=1}^{N} \frac{m_i}{2} f^2 \dot{\vec{s}}_i' \mathbf{G}\, \dot{\vec{s}}_i + \mathcal{V} + \frac{Q}{2}\dot{f}^2 + (3N+1)k_B T_{eq}\ln f$$

$$+ \frac{W}{2} \mathrm{Tr}(\dot{\mathbf{h}}' \dot{\mathbf{h}}) + V_0 \mathrm{Tr}(\mathbf{t}\boldsymbol{\varepsilon}).$$
(2.22)

If we consider a hydrostatic pressure, P, applied to the system in addition to the presence of a stress $\mathbf{t}$, the Lagrangian equation (2.12) then involves an additional term PV. The nine equations of motion for the simulation box is then modified as

$$W\ddot{\mathbf{h}} = (\mathcal{P}-P\mathbf{I})\mathbf{A} - \mathbf{h}\Gamma,$$
(2.23)

where $\mathbf{I}$ is the 3×3 unit matrix. The equations of motion for particles and heat reservoir remain the same as in Eqs.(2.16) and (2.17). If we set $f = 1.0$ and $\dot{f} = 0.0$, then the constant of motion will be

$$\mathcal{H} = \sum_{i=1}^{N} \frac{m_i}{2} \dot{\vec{s}}_i' \mathbf{G}\, \dot{\vec{s}}_i + \mathcal{V} + \frac{W}{2}\mathrm{Tr}(\dot{\mathbf{h}}' \dot{\mathbf{h}}) + V_0\mathrm{Tr}(\mathbf{t}\boldsymbol{\varepsilon}) + PV \quad ,$$
(2.24)

or

$$\mathcal{H} \approx E + V_0 Tr(\mathbf{t}\boldsymbol{\epsilon}) + PV, \tag{2.25}$$

which is the enthalpy of the system. Here we have ignored the kinetic energy term of the simulation box which is relatively small due to the slow motion of simulation box. Thus, Eqs.(2.18), (2.19), and (2.23) generate a thermodynamically consistent ensemble (HtN).

2.2.   NUMERICAL SOLUTIONS FOR THE EQUATIONS OF MOTION

To obtain the trajectories of each particle we need to introduce a numerical method to integrate the equations of motions. As we described above, equations of motion are second-order differential equations. To solve them numerically, at first we need to specify the initial conditions as well as the boundary conditions. The basic idea of a numerical solution is: given positions and velocities at time t, by following some specific algorithm, equations of motion are integrated over the time interval, Δt, to obtain new positions and velocities at time $t + \Delta t$. The central part of any MD program is the algorithm which carries out this integration. In general, the actual time one spends on integrating the equations of motion is negligible compared with the time it takes to compute the interatomic forces. In the following subsections, we will first address the usual way to setup the initial and boundary conditions in MD simulations. Then we introduce the 'Velocity Verlet' algorithm to integrate the equations of motion. Finally, we discuss the issue of saving computing time on force calculation.

2.2.1. Initial Conditions and Boundary Conditions

Before starting MD simulations, one needs to specify the initial positions and velocities of particles, heat reservoir and simulation box. For the simulation of solids, it is very natural to initially place the particles at their equilibrium lattice positions, set the simulation box with length of multiple of lattice-parameters, and set the heat reservoir variable as f=1.0. For the simulation of liquids and glasses, the initial positions of particles and the shape of simulation box can be set in the same way. The system is then gradually heated to the desired temperature (usually above the melting point of the material) by increasing the temperature of the heat reservoir, and the box variables are scaled to obtain the correct density of the liquid or glass.

Usually the initial velocities are chosen randomly according to a Maxwell-Boltzmann distribution. The magnitude of the initial velocities of the particles and the box variables are scaled to the desired temperature according to the equipartition theorem. The net linear and angular momenta are set to be zero.

A useful simulation should reflect all of the relevant features of a real physical system of interest. In reality, the macroscopic size of sample that can be studied is limited. In order to minimize surface effects to simulate the behavior of an infinite system more reasonably, periodic boundary conditions (pbc) are usually used. Under pbc, the simulation box of length L is taken as a basic unit. The whole space is filled by periodically repeated images of the basic unit. Each image cell contains N particles with the same relative positions and corresponding velocities as in the basic unit. When a particle leaves from one face of the basic unit, an image of that particle enters through the opposite face to balance the move, see Figure 2.2.

In principle, each particle interacts with all other particles in the same cell and with all the other images including its own. If the range of the forces is smaller than L/2, we only need to consider interactions with the closest images. In the case of long-range forces, we need to take into account interactions with all the images. In the Parrinello-Rahman approach for constant stress and/or pressure ensemble, the shape of the unit cell is allowed to change, but the basic idea of periodic boundary conditions is still the same.

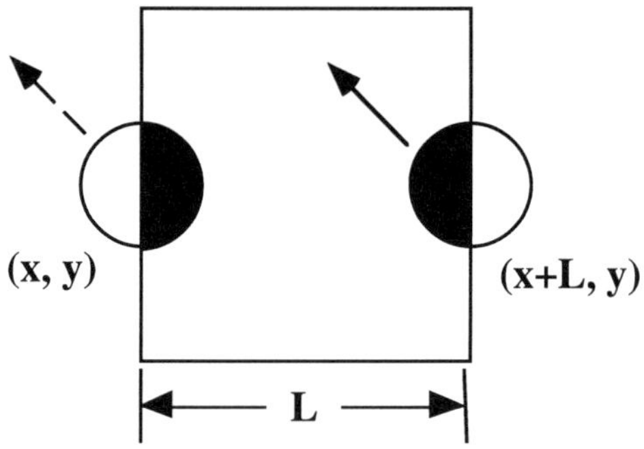

Figure 2.2 Periodic boundary conditions.

2.2.2. *Integration algorithms*

There are many integration algorithms for Newton's equations of motion. All of them convert the differential equations into finite-difference equations. The commonly used algorithms are Gear's predictor-corrector method; Verlet's algorithms; Beeman algorithms, etc..

Recently, Tuckerman and co-workers have generated reversible MD integrators from the Trotter factorization of the Liouville propagator [28]. Here we briefly describe their derivation of the Velocity-Verlet algorithm.

In Cartesian coordinates, the Liouville operator L for a system of N particles is

$$iL = \sum_{j=1}^{3N} \left[\dot{p}_j \frac{\partial}{\partial p_j} + \dot{x}_j \frac{\partial}{\partial x_j} \right] = iL_1 + iL_2 \; . \tag{2.26}$$

The corresponding classical propagators are $U(t) = e^{iLt}$, $U_1(t) = e^{iL_1 t}$, and $U_2(t) = e^{iL_2 t}$. These operators are reversible in time in the sense, $U(-t)=U^{-1}(t)$. The states of the system at time t are given by $\Gamma(t,\Gamma(0)) = U(t)\Gamma(0)$, $\Gamma_1(t,\Gamma(0)) = U_1(t)\Gamma(0)$, and $\Gamma_2(t,\Gamma(0)) = U_2(t)\Gamma(0)$. For a small time interval Δt, the Trotter expansion yields

$$e^{i(L_1 + L_2)\Delta t} = e^{iL_1\Delta t/2}e^{iL_1\Delta t/2}e^{iL_2\Delta t}$$

$$= e^{iL_1\Delta t/2}e^{iL_2\Delta t}e^{iL_1\Delta t/2} + O(\Delta t^3). \tag{2.27}$$

Therefore

$$U(\Delta t) = U_1(\Delta t/2)U_2(\Delta t)U_1(\Delta t/2) + O(\Delta t^3). \tag{2.28}$$

Starting from the state $\Gamma(0)$, the above operator generates a new state at time Δt:

$$\Gamma(\Delta t,\Gamma(0)) = U(\Delta t)\Gamma(0)$$

$$= U_1(\Delta t/2)U_2(\Delta t)U_1(\Delta t/2)\Gamma(0) \tag{2.29a}$$

$$= U_1(\Delta t/2)U_2(\Delta t)\Gamma_1(\Delta t/2,\Gamma(0)) \tag{2.29b}$$

$$= U_1(\Delta t/2)\Gamma_2[\Delta t,\Gamma_1(\Delta t/2,\Gamma(0))] \tag{2.29c}$$

$$= \Gamma_1\{\Delta t/2,\Gamma_2[\Delta t,\Gamma_1(\Delta t/2,\Gamma(0))]\}. \tag{2.29d}$$

This general expression is simplified when we substitute L_1 and L_2 into the above equations. For an initial state described by positions and momenta, $\{x_j(0),p_j(0); j=1,2,..,3N\}$, the sequence of operations (2.29a)-(2-29d) is equivalent to

$$x_j(\Delta t) = x_j(0) + \Delta t\frac{p_j(0)}{m} + \frac{(\Delta t)^2}{2}\frac{\dot{p}_j(0)}{m}, \tag{2.30}$$

$$p_j(\Delta t) = p_j(0) + \frac{\Delta t}{2}\dot{p}_j(0) + \frac{\Delta t}{2}\dot{p}_j(\Delta t). \tag{2.31}$$

There are three steps involved in this algorithm. At first the new positions at time Δt are calculated using Eq.(2.30). Then the velocities (momenta) at mid-step are computed using

$$p_j(\frac{\Delta t}{2}) = p_j(0) + \frac{\Delta t}{2}\dot{p}_j(0). \tag{2.32}$$

The forces $\dot{p}_j(\Delta t)$ at time Δt are then evaluated and finally the velocities (momenta) at Δt are completed from

$$p_j(\Delta t) = p_j(\tfrac{\Delta t}{2}) + \tfrac{\Delta t}{2}\dot{p}_j(\Delta t). \tag{2.33}$$

Note that $e^{iL_1\Delta t/2}e^{iL_2\Delta t}e^{iL_1\Delta t/2}$ ensures that the algorithm is time reversible.

2.2.3. Special Numerical Treatment in Computing Forces

Since the force calculation is the most time-consuming part in an MD simulation, an efficient way in handling force calculation is very crucial. There are many different strategies in dealing with different interatomic potentials. For a system of N particles, the conventional straightforward evaluation of two-body interactions takes $O(N^2)$ operations, and $O(N^3)$ operations are needed for three-body interactions. With the use of a special technique one can convert the three-body forces into two-body forces, therefore reduce the computational complexity from $O(N^3)$ to $O(N^2)$. Furthermore, short range and long range interactions are treated differently. For short range interactions, the link-cell-list method is used to reduce the computational complexity from $O(N^2)$ to $O(N)$. In addition, a multiple-time-step technique is implemented to further reduce the cost on computing time. For long range interactions, an optimized Ewald summation method reduces the number of operations for two-body interactions to $O(N^{3/2})$. Several divide and conquer algorithms are designed for long range interactions to obtain computational complexity as $O(N)$. In this section, we describe these methodologies in detail.

Two-body vs. Three-body potentials. Monatomic systems with close-packed structures can be described reasonably well by two-body interaction potentials. Usually two-body potentials depend only on the interparticle distances, r_{ij}. In many physical systems, these are short-range interactions. the Lennard-Jones potential is a typical pairwise potential that is frequently used for computer simulations [29]. It takes a form

$$\mathcal{V}(r) = 4\,\varepsilon\,[(\tfrac{\sigma}{r})^{12} - (\tfrac{\sigma}{r})^6], \tag{2.34}$$

where ε represents the depth of the potential well and σ is the characteristic length in the potential. This potential has a cutoff at r_c depending on different material properties. The first term, i.e., the repulsive part, describes the scale of atomic sizes. The second term, i.e., the attractive part, represents the van der Waal's attractive interaction. This potential has been used to successfully provide reasonable agreement with the experimental properties of liquid argon (Ar). The calculation of forces is very straight forward,

$$\vec{F}_i = -\nabla_i\,\mathcal{V} = -(\frac{\partial \mathcal{V}}{\partial x_i}\hat{i} + \frac{\partial \mathcal{V}}{\partial y_i}\hat{j} + \frac{\partial \mathcal{V}}{\partial z_i}\hat{k}), \qquad i=1, 2, ..., N. \tag{2.35}$$

Here $\hat{i}$, $\hat{j}$ and $\hat{k}$ are unit vectors in Cartesian coordinate system. In conventional computer programs, this calculation takes $O(N^2)$ operations.

For molecular systems involving covalent bonds, one needs to consider the interactions among triplets of particles. Usually the three-body interaction contains two terms, namely the bond stretching term and the bond bending term. As an example, the Stillinger-Weber potential [30] has been used widely in the simulations of covalent materials. The three-body potential is given by

$$\mathcal{V}_3(r_{ij}, r_{ik}, \theta_{jik}) = \lambda \exp\left[\gamma(r_{ij} - a)^{-1} + \gamma(r_{ik} - a)^{-1}\right]$$
$$\cdot (cos\theta_{jik} - cos\bar{\theta}_{jik})^2 , \tag{2.36}$$

where

$$\cos\theta_{jik} = \frac{\vec{r}_{ij} \cdot \vec{r}_{ik}}{r_{ij} r_{ik}} . \tag{2.37}$$

In conventional computer programs, the three-body force calculation takes $O(N^3)$ operations, which is the most time consuming part in the MD simulations. In the case of AX_2-type systems (SiO_2, $SiSe_2$, and so on), there are six different combinations of three-body forces, namely, A-A-A, A-A-X, A-X-A, A-X-X, X-A-X, and X-X-X. But the two most important terms are A-X-A and X-A-X because A-X is the shortest bond with the strongest attractive energy. Ignoring other terms, the three-body form can be rewritten by decomposing the inner product, $\vec{r}_{ij} \cdot \vec{r}_{ik}$,

$$\vec{r}_{ij} \cdot \vec{r}_{ik} = x_{ij} x_{ik} + y_{ij} y_{ik} + z_{ij} z_{ik}. \tag{2.38}$$

This makes it possible for this three-body potential to be separated into a two-body form. Therefore the computational complexity can be reduced from $O(N^3)$ to $O(N^2)$ [31, 32].

Short-Range vs. Long-Range Potentials. For a short-range pair-additive potential, $\mathcal{V}$, with a cutoff radius r_c, the use of r_c always produces truncation errors in the calculation of forces. This in turn causes extra fluctuation and/or drift in the total energy of the system. Usually the potential is modified in such a way that both potential and force vanish at r_c:

$$\mathcal{V}(r) \Rightarrow \mathcal{V}(r) - \mathcal{V}(r)\Big|_{r=r_c} - (r - r_c)\frac{d\mathcal{V}}{dr}\Big|_{r=r_c} . \tag{2.39}$$

Advances in software technology have resulted in many fast algorithms which employ multiple length scales. The link-list-cell method and the multiple-time-step scheme are the two widely used strategies. They exploit multiple scales in either space or time.

Link-cell-list method. In a typical MD simulation, as much as 95% of computing time is spent in computing pair interactions, which includes examining the complete set of $O(N^2)$ pairs, identifying a subset of those pairs separated by less than r_c, and computing the

164

forces for this subset. The rest of the pairs separated by more than r_c does not contribute to the forces, therefore has no influence on the dynamics of the system. To speed-up the calculation of forces, we can use link-cell-list method [33], as shown in Figure 2.3.

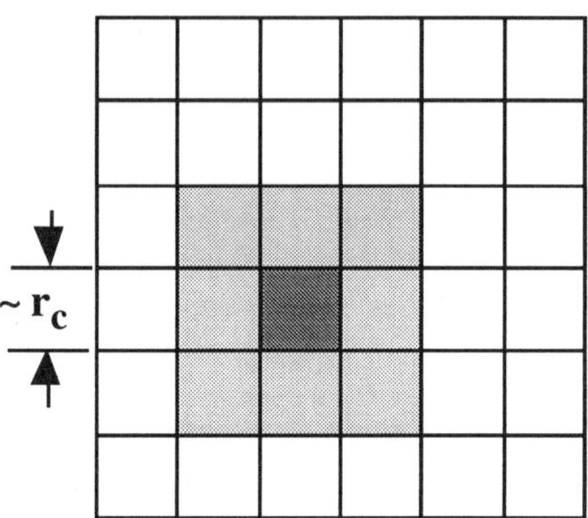

Figure 2.3 Link-cell-list method. The lightly shaded boxes
are the nearest neighbor boxes of the darkened box.

The total volume (L^3) of an N-particle system is divided into n^3 cells of equal volume with edge L/n which is slightly larger than r_c. Particles residing in each cell are identified to construct a linked-cell-list using two integer arrays. The first integer array identifies the particles at the top of the list in each cell, and the second integer array links particles belonging to the same cell. Since particles in one cell will interact only with the particles within their nearest neighbor cells, a neighbor-list is then constructed for each particle according to the linked-cell-list. At each MD step, forces are examined only within the neighbor-lists. The linked-cell-list can be updated every few tens of time steps. Suppose the maximum number of neighbors for a given system is N_b, which is independent of N, then the number of operations will be proportional to $N_b N$. Therefore the time complex is reduced from $O(N^2)$ to $O(N)$.

Multiple-time-step method. Multiple-time-step (MTS) approach is an efficient way to further speedup the calculation of short-range (with a cutoff radius at r_c) pairwise forces. The MTS method is based on the fact that the farther the distance between particles the slower is the change in forces [33]. Therefore, for different interparticle separations we use different time steps to compute forces. In the MTS method, the force $\vec{F}_i$ acting on the i th particle is divided into two groups. The primary force $\vec{F}_i^p$ arises from the interaction with particles lying in a sphere of radius r_a around the i th particle. The secondary force $\vec{F}_i^s$ is due to pairwise interactions in the range of [r_a, r_c] around the i th particle, see Figure 2.4.

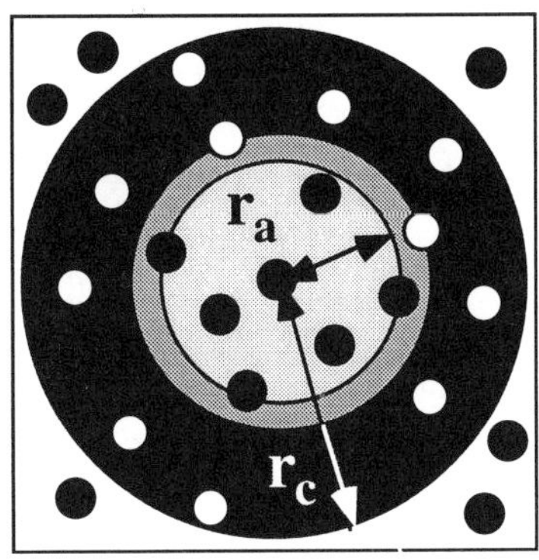

Figure 2.4 Multiple-time-step method. Forces acting on the centered particle are separated into two parts. Forces due to particles inside the sphere of radius r_a are updated every timestep. Forces due to particles in the shell $[r_a, r_c]$ are calculated every n steps.

Since the secondary force varies much slower than the primary force, we calculate $\vec{F}_i^p$ every MD time step while we evaluate $\vec{F}_i^s$ every n steps. At the intermediate steps, $\vec{F}_i^s$ is extrapolated according to the Taylor series expansion,

$$\vec{F}_i^s(t + m\Delta t) = \sum_{l=0}^{l_{max}} \frac{(m\Delta t)^l}{l!} \frac{d^l \vec{F}_i^s(t)}{dt^l}, \qquad m=1, 2, ..., n\text{-}1, \tag{2.40}$$

where l_{max} is the cutoff of the Taylor series. For a fixed time step, the efficiency and accuracy of this method are very sensitive to the choice of r_a, n, and l_{max}. In many cases, r_a is slightly larger than $r_c/2$, n varies from 5-15, and l_{max} can be chosen from 2-4, depending on the requirement of energy conservation. With the MTS algorithm, one typically can achieve a 2- to 5-fold speedup.

Long-range forces, such as Columbic interaction, are a serious problem for the computer simulations since their ranges are greater than half the box length. Materials, such as plasma, electrolyte solutions, biological systems, gravitational many-body systems, and certain solids, have long range interactions. Using periodic boundary conditions, the summation over infinitely repeated image charges must be carried out. Several algorithms have been designed to efficiently calculate long-range forces. The Ewald method is a technique to transform the sum into rapidly convergent sums in real and Fourier spaces. The Fast Multipole Method is based on the rapid decay of the multipole expansion of the Coulomb interaction. These two methods enable one to calculate the long-range interaction efficiently with a specific level of precision. In

Section 3, we will discuss the implementation of the Fast Multipole Method on parallel architectures.

2.3.   EVALUATION OF PHYSICAL PROPERTIES

The physical properties of a system can be calculated with the knowledge of the positions and velocities of particles. In the following sections, we will introduce structural properties such as the pair-distribution function g(r) and static structure factor S(q); time correlation functions such as the velocity auto-correlation function and mean square-displacement; thermodynamical properties such as temperature, pressure, specific heat; and mechanical properties.

2.3.1. Structural Correlations Functions

For a system with N particles in two species α and β contained in volume V, two-body structural correlation is analyzed by using partial pair-distribution functions, $g_{\alpha\beta}(r)$. It is calculated from [34]

$$\langle n_{\alpha\beta}(r) \rangle = 4\pi r^2 \, \Delta r \, \rho \, c_\beta \, g_{\alpha\beta}(r), \tag{2.41}$$

where $n_{\alpha\beta}(r)$ denotes the number of particles of species β in the shell between r and r+Δr around a particle of species α. The angular brackets $\langle \, \rangle$ represent the ensemble average and an average over all the particles of species α. ρ is the total number density (= N/V, $N = N_\alpha + N_\beta$) and c_β is the concentration of species β (= N_β/N). The total pair-distribution function g(r) is defined as

$$g(r) = \sum_{\alpha,\beta} c_\alpha \, c_\beta \, g_{\alpha\beta}(r). \tag{2.42}$$

For materials in the crystalline phase the structure can be well characterized by well-defined peaks in g(r). The Fourier transform of the pair correlation functions can be measured by x-ray and neutron diffraction experiments. A brief description of how this is done can be found in Chandler's text on Statistical Mechanics [35]. During MD simulations, we usually calculate the appropriately weighted structure factors from partial static structure factors which are the Fourier transformations of corresponding partial pair-distribution functions:

$$S_{\alpha\beta}(q) = \delta_{\alpha\beta} + 4\pi\rho(c_\alpha \, c_\beta)^{1/2}$$
$$\cdot \int_0^R r^2 [\, g_{\alpha\beta}(r) - 1] \, \frac{\sin(qr)}{qr} \, \frac{\sin(\pi r / R)}{\pi r / R} \, dr \, , \tag{2.43}$$

where the window function, $\dfrac{\sin{(\pi r/R)}}{\pi r/R}$, has been introduced to include the finite-size effect. The cutoff length R is chosen to be half the length of the cubic simulation box (R = L/2). The total static structure factor is given by

$$S(q) = \sum_{\alpha,\beta} (c_\alpha c_\beta)^{1/2} S_{\alpha\beta}(q).$$ (2.44)

The neutron-scattering static structure factor $S_N(q)$ can be obtained from the partial static structure factors by weighting them with coherent neutron-scattering length:

$$S_N(q) = \frac{\sum_{\alpha,\beta} b_\alpha b_\beta (c_\alpha c_\beta)^{1/2} [\, S_{\alpha\beta}(q) - \delta_{\alpha\beta} + (c_\alpha c_\beta)^{1/2}]}{\left[\sum_\alpha b_\alpha c_\alpha\right]^2},$$ (2.45)

where b_α denotes the coherent neutron-scattering length of species α. When neutron scattering experiment data are available for materials of interest, comparing the data with results calculated from Eq.(2.45) is usually a way to calibrate the interatomic potential model in MD simulations.

2.3.2. Time Correlation Functions

The ensemble average of a quantity A is given by the time average over the configurations generated as

$$<A> = \overline{A} = \lim_{t \to \infty} \frac{1}{(t - t_0)} \int_{t_0}^{t} dt' A(t').$$ (2.46)

Ergodicity ensures that the time average over a sufficiently long trajectory is equivalent to the ensemble average. The fluctuation of A is readily computed as:

$$\langle \delta A^2 \rangle = \langle A^2 \rangle - \langle A \rangle^2.$$ (2.47)

Correlations between two quantities A and B are measured via a correlation coefficient c_{AB},

$$c_{AB} = \frac{\langle A - \langle A \rangle \rangle \langle B - \langle B \rangle \rangle}{\sqrt{\langle \delta A^2 \rangle \langle \delta B^2 \rangle}}.$$ (2.48)

If we consider A and B to be measured at two different times, t_A and t_B, then c_{AB} becomes a time correlation function of the time difference $t = |t_B - t_A|$, $c_{AB}(t)$. The value of c_{AB} lies between 0 and 1, with values close to 1 indicating a higher degree of time correlation. The time correlation functions are of great interest in computer simulations because they often relate directly to macroscopic transport coefficients. For instance, the

168

velocity auto correlation function $Z_\alpha(t)$ is related to the constants of self-diffusion D_α as in

$$D_\alpha = \frac{k_B T}{m_\alpha} \int_0^\infty Z_\alpha(t)\, dt, \tag{2.49}$$

where

$$Z_\alpha(t) = \frac{\langle \vec{v}_i(0) \cdot \vec{v}_i(t) \rangle_\alpha}{\langle v_i^2(0) \rangle_\alpha}, \tag{2.50}$$

where N_α and m_α are the number of particles and mass of species α. In MD simulations, the constants of self-diffusion can also be calculated from mean-square displacements

$$D_\alpha = \lim_{t \to \infty} (\langle r^2 \rangle_\alpha / 6t), \tag{2.51}$$

where

$$\langle r^2 \rangle_\alpha = \left\langle \frac{1}{N_\alpha} \sum_{j(\alpha)} [\vec{r}_j(t+s) - \vec{r}_j(s)]^2 \right\rangle. \tag{2.52}$$

2.3.3. *Thermodynamic Properties*

Temperature. In MD simulations based on the microcanonical ensemble, the temperature of the system is given by the mean kinetic energy $< \mathcal{K} >$ via the equipartition theorem

$$< \mathcal{K} > = \frac{3N}{2} k_B T. \tag{2.53}$$

Therefore, by scaling the velocities of particles with a factor of $(T / \mathcal{T})^{1/2}$, where T is the desired thermodynamic temperature and $\mathcal{T}$ is the current kinetic temperature of the system, one is able to investigate behavior of the system at different temperatures.

Mean Pressure. The mean pressure, P, can be calculated from the virial theorem:

$$PV = N k_B T + \frac{1}{3} \left\langle \sum_{i=1}^N \vec{r}_i \cdot \vec{F}_i \right\rangle, \tag{2.54a}$$

where N, V, T are number of particles, volume, and mean temperature of the system, respectively. $\vec{r}_i$ and $\vec{F}_i$ are the position of the i th particle and total force acted on it due to all the other particles. If the N-body interaction is pairwise, it is more convenient to express the above equation in a form which is explicitly independent of the origin of coordinates:

$$PV = N k_B T + \frac{1}{3} \left\langle \sum_{j>i}^N \vec{r}_{ij} \cdot \vec{F}_{ij} \right\rangle, \tag{2.54b}$$

where $\vec{r}_{ij} = \vec{r}_i - \vec{r}_j$, and $\vec{F}_{ij}$ is force between ith and jth particle. It is essential to use (2.54b) in a simulation that employs periodic boundary conditions.

Specific Heat. In the microcanonical ensemble, the mean-square fluctuation of the kinetic energy K at fixed energy E is related with c_v [36] by

$$\frac{\langle \delta K^2 \rangle}{\langle K \rangle} = k_B T \, [1 - \frac{3k_B}{2c_v}], \qquad (2.55)$$

where T is the temperature corresponding to the energy E.

In the canonical ensemble, the specific heat can be derived from the fluctuation of total energy

$$c_v = \frac{\langle \delta E^2 \rangle}{N k_B T^2} = \frac{3}{2} k_B + \frac{\langle \delta \mathcal{V}^2 \rangle}{N k_B T^2}, \qquad (2.56)$$

where T is the temperature of the heat reservoir.

Young's Modulus. For a solid bar of length L subject to a stretching force $\vec{F}$ along the bar direction, we calculate tensile stress and strain as Stress $= \frac{F}{A}$ and Strain $= \frac{\Delta L}{L}$, where A, L, and ΔL represent cross section area, original length, and change of length in the object, respectively. For a typical solid, up to a critical strain, the stress vs. strain is linearly related. Young's modulus is defined as the ratio of stress over strain in the linear region $Y = \frac{\text{Stress}}{\text{Strain}}$. In MD simulations, either stress or strain is specified. In the case of strain being specified, the stretching force is evaluated from the internal energy at equilibrium configurations, $F = \frac{dE}{dL}$. And the Young's modulus is obtained from $Y = \frac{L}{A} k$, where $k = \frac{d^2E}{dL^2}$ is the force constant in the linear region of force vs. length.

3. Implementation of the Molecular Dynamics (MD) Method on Parallel Computer Architectures

Despite significant recent developments in materials-simulation techniques [14, 15, 33, 37, 38], the goal of reliably predicting the properties of new materials in advance of fabrication and measurement has not yet been achieved. The primary reason for this lack of success is the inability of sequential machines to handle large-scale simulations. For example, molecular dynamics (MD) simulations for long-range interactions scale as N^2 where N is the number of particles in the system. In many physical systems, the desired system sizes are in the range of 10^6 particles. These are beyond the compute power of most sequential machines. However, the MD technique has considerable inherent

170

parallelism. By exploiting this parallelism on emerging parallel architectures, it is possible to perform large-scale simulations for complex materials [2, 4, 8, 14, 15, 31, 39-41].

We have developed a highly efficient MD algorithm based on multiresolutions in both space and time [2, 14, 15, 31, 39-41]. The long-range Coulomb potential for periodic systems are computed with the fast multipole method (FMM) [8] and reduced cell multipole method (RCMM) [42], while the medium-range non-Coulombic potentials are computed by the multiple time step (MTS) method [31, 43]. The three-body interactions are calculated with the separable tensor decomposition [31, 44]. The performance of the multiresolution molecular dynamics (MRMD) algorithm is tested on the 512-node Intel Touchstone Delta machine at Caltech and the 128-node IBM SP1 system at Argonne National Laboratory [2, 14, 15, 31, 39-41].

3.1.   THE DOMAIN DECOMPOSITION SCHEME

For a system of N particles, a domain-decomposition algorithm is used to balance the computation load among p processors [15]. In this algorithm, the total system is divided into p subdomains of equal volume. These subdomains are geometrically mapped onto the p processors, see Figure 3.1.

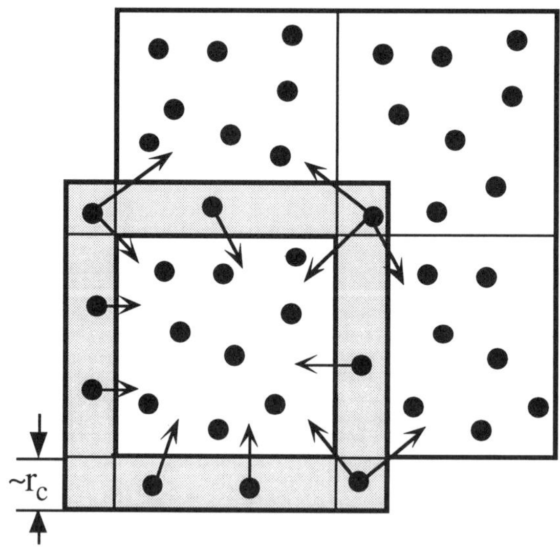

Figure 3.1 Domain decomposition on 4 processors. The arrows
indicate the message-passing direction.

Coordinates of particles and other relevant data are assigned to a processor if the coordinates fall into the spatial domain of that processor. Relevant data for particles that move out of a subdomain boundary into a neighboring subdomain are transferred using message-passing technique (for example, by subroutines **csend** and **crecv** on iPSC/860 systems).

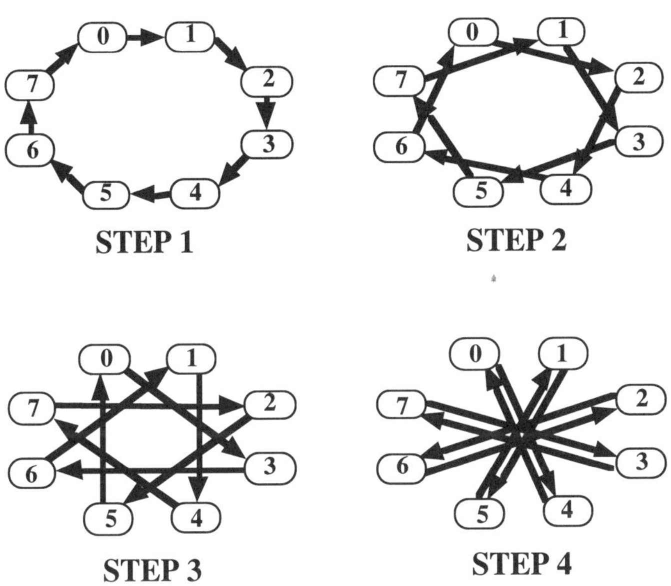

Figure 3.2 Internode communication strategy within eight processors. Arrows indicate the direction of message passing.

The potential energy and forces are computed with the link-cell-list scheme and the multiple-time-step method. Interactions involve those among particles on the same processor (intranode) and on different processors (internode). For short range interaction with a cutoff of r_c, to calculate the forces on particles in a subdomain, the coordinates of the boundary particles in the neighbor subdomains must be copied from the corresponding nodes. Here we distinguish the primary- and secondary-boundary particles. In a subdomain, primary boundary particles to the kth neighbor (k=1, 2, ..., 26) subdomain are located within a distance $r_c+\Delta$ from the boundary with the kth neighbor, where Δ is a small positive value. Similarly, secondary boundary particles are those which are within a distance $r_c+\Delta'$ from the boundary but are not the primary particles, where Δ' is another small distance. Note that a particle can be a boundary particle to several neighbor subdomains. Using the multiple-time-step method, the primary boundary particles are copied every step, and the secondary boundary particles are copied every n-steps. Domain decomposition can be updated every few steps. For long range interactions, the coordinates of particles from all other subdomains must be copied.

It is important to design an efficient strategy for internode communication. For short-ranged interaction, communications are within the 26 neighboring processors. Newton's third law is used to reduce the number of message-passing operation by half. Each processor only communicates with its east (or west), left (or right) and upper (or lower) neighboring processors. Coordinates and other relevant data are sent forward to destination processor. Calculated forces and potential are sent backward to the source processor. For long-ranged interaction, communications are carried out among all of the p processors. We arrange the processors in a ring topology [45], see Figure 3.2.

First the data from node 0 are sent to node 1, data from node 1 to node 2,..., and data from node p-1 to node 0, synchronously. Then, using the link-cell-list method, the primary and secondary contributions to forces, potential energy, and other information are calculated. Node 0 sends back the calculated contributions to node p-1 while receiving the contributions calculated at node 1. Similar message-passing of calculated contributions takes place synchronously at other nodes as well. Next node 0 receives data from node p-2, node 1 from node p-1, ..., and node p-1 from node p-3. The contributions to forces, potential energy and others are calculated synchronously and the results are sent back to the nodes from which the data had been received. This procedure is continued until all the necessary interactions have been computed. The entire communication process takes p/2 steps.

Parallel algorithms based on domain decomposition scheme have been developed to implement MD simulations on the distributed-memory 8-node Intel iPSC/860 machine. Using the link-cell-list method and the multiple-time-step scheme, the combined approach reduces computation time significantly. The total execution time scales linearly with the number of particles and is inversely proportional to the number of processors [2, 15, 45].

3.2.　MULTIRESOLUTION MOLECULAR DYNAMICS (MRMD) ALGORITHM

3.2.1. Efficiency of Parallel algorithms

The speed-up of a parallel algorithm is defined as the ratio between the time to execute the algorithm on one processor to the execution time on p processors. The parallel efficiency, η, is the speed-up divided by p. Smaller grain size leads to less optimized utilization of the computing resource. It is more relevant to use a parallel efficiency with constant grain size to assess the performance of an algorithm on many processors. To do so, we first define the speed as a product of the total number of particles and the time steps executed in a second. The constant-grain speed-up is given by the ratio between the speed of p processors and that of one processor. The constant-grain efficiency η' is the constant-grain speed up divided by p.

3.2.2. *Parallel algorithms for Coulombic Systems Using Ewald Method*

Recently, we have also implemented MD simulations involving the Ewald summation for Coulomb interaction on distributed-memory MIMD machines [2]. The Ewald summation [1] is the most widely used approach for the long-range Coulomb interaction in bulk systems. In this approach the interaction is written as a sum of a constant term, a sum in the Fourier space, and a sum of "short range" terms in real space. The parallel algorithm we have designed for the Ewald summation reduces the computational complexity from $O(N^{3/2})$ to $O(N)$. This is achieved by ensuring that both the real-space and Fourier-space contributions scale linearly with the size of the system. In real space, the potential energy and force calculations are truncated at $r_c = 5r_0$, where r_0 is the ion-sphere radius, $r_0 = (3/4\pi\rho)^{1/3}$, and ρ is the number density. This cutoff maintains the desired level of precision $\sim 0.01\%$ for all system sizes. The real-space contributions are then calculated with the domain decomposition and the linked-list methods which scale as $O(N)$. The computation of the Fourier-space contributions reveals that only a certain number of wave vectors need to be included. We performed simulations as a function of the number of Fourier components and system sizes. It is found that an increase in the number of wave vectors from 309 to 2,192 produces a change of 0.01% in the total potential energy. The computation time for the k-space calculation increases linearly as the number of wave vectors increases. Thus, the total execution time for the Ewald sum scales linearly with the number of particles (see Figure 3.3) and is inversely proportional to the number of processors.

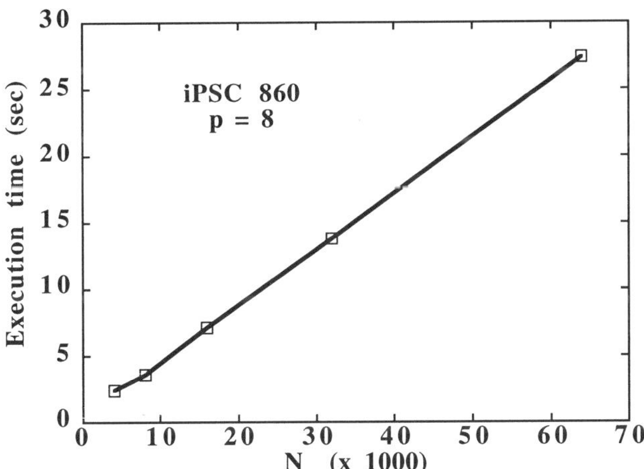

Figure 3.3 Execution time on the 8-node Intel iPSC/860 hypercube as a function of the number of particles in a bulk Coulombic system.

3.2.3. *Parallel algorithms for Coulombic Systems Using Fast Multipole Method*

Recently, several divide-and-conquer schemes have been implemented to reduce the computational complexity for the long-range Coulomb interaction [4, 8]. The Fast Multipole Method (FMM) is based on the multipole expansion of the Coulomb interaction [8]. The rapid decay of the multipole expansion enables one to calculate the Coulomb interaction efficiently for a specific level of precision. On a sequential machine, it reduces the computational complexity from $O(N^2)$ to $O(N)$ for systems with free-boundary conditions [8]. It involves: i) Decomposition of the MD box into a hierarchy of cells; ii) calculation of multipole moments; and iii) Taylor series expansion of the Coulomb potential. The next three sections describe these steps.

Cell Decomposition. In the FMM, the MD box is decomposed into a hierarchy of cells according to a tree structure [4, 8]. The root of the tree is at level 0, and it corresponds to the MD box shown in Fig. 3.4.

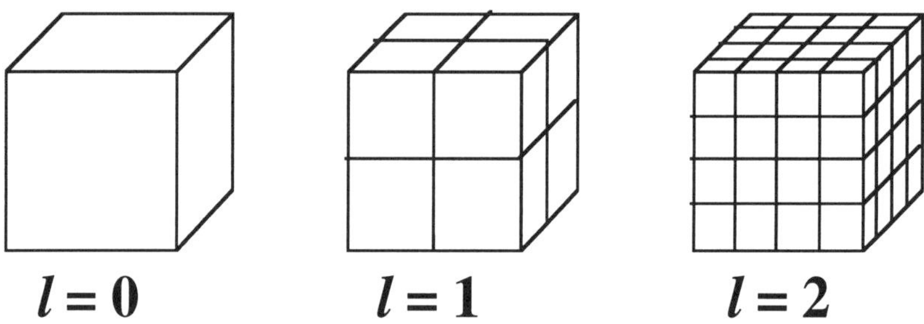

Figure 3.4 Hierarchy of cell decomposition starting from the MD box at level $l = 0$

Figure 3.5 Two-dimensional schematic representation of the original cell (dark gray), near-neighbor 8 cells (light gray), and well-separated cells. In three dimensions the number of near-neighbor cells is 26

In the presence of periodic boundary conditions, the images of the MD box are repeated infinitely. The interactions of particles in the MD box with those in the image boxes is delt with using the reduced cell multipole method (RCMM) [40-42] as described in the Sec. 3.2.4.

The MD box is divided into 2x2x2 cells at level 1, see Fig. 3.4. At level 2, each of these cells is further decomposed into 8 cells of equal volume, see Fig. 3.4. Repeating this procedure, one obtains 8^l cells at level l. The recursive decomposition is stopped at the **leaf level**, $l = L$, where further decomposition would cause the sides of the cells to be less than the cutoff for the non-Coulombic interaction (i. e., charge-dipole, steric, and three-body), see Fig. 3.5. Note that particles experience only the Coulomb potential when they reside in non **nearest-neighbor** or **well-separated cells**. (In three dimensions, for a given cell, the nearest-neighbor cells are defined as the cell itself and the surrounding 26 cells with which it shares a corner. Other cells at the same level are said to be well-separated, see Fig. 3.5.)

Near and Far Fields. The total potential energy can be written as,

$$V = \frac{1}{2} \sum_{i=1}^{N} \sum_{j \in \{nn(c(i))\}} V^{(2)}(\vec{r}_{ij}) + \sum_{i,j<k} V^{(3)}(\vec{r}_{ij}, \vec{r}_{ik}) + \frac{1}{2} \sum_{i=1}^{N} Z_i \Psi_{L,c(i)} \tag{3.1}$$

where c(i) denotes a cell at the leaf level L, Z_i is the charge on the ith atom in cell c, and nn(c) are the nearest neighbors of the cth cell. The first and second terms constitute the **near-field** contribution due to particles in nearest-neighbor cells. The near-field contribution completely includes the three-body interaction and the two body charge-dipole and steric terms. It also includes the Coulomb contribution from particles in nearest-neighbor cells. The near field is calculated directly at the leaf level, and it requires O(N) operations.

The third term in Eqn. (3.1) is called the **far-field** contribution to the potential energy. It only contains the contribution to the Coulomb potential, $\psi_{L,c}(\boldsymbol{r})$, at point $\boldsymbol{r}$ in the cth cell at the leaf level from particles in well-separated cells:

$$\Psi_{L,c}(\vec{r}) = \sum_{c' \notin \{nn(c)\}} \sum_{j \in c'} \frac{Z_j}{\left| \vec{r} - \vec{r}_j - \vec{R}_{c'-c} \right|} \tag{3.2}$$

where $\vec{R}_{c'-c}$ is a vector joining the centers of the cth and c'^{th} cells.

In the next three sub-sections, we describe how the FMM enables the calculation of the far-field contribution in O(N) operations.

Multipole Expansion. Equation (3.2) is evaluated by expanding the Coulomb field in terms of multipoles:

$$\Psi_{lc}(\vec{r}) = \frac{Z}{R} + \frac{\mu_\alpha R_\alpha}{R^3} + \frac{Q_{\alpha\beta} R_\alpha R_\beta}{R^5} + \ldots \tag{3.3}$$

where $R = |r - R_{c'c}|$ and $\Phi \equiv \{Z, \mu_\alpha, Q_{\alpha\beta}, \ldots\}$ are the charge, dipole, and quadrupole, etc. In Eqn. (3.3), α and β are the Cartesian components and there is an implicit summation over repeated indices.

The multipoles Φ are calculated hierarchically for every cell at every level. We first determine multipoles, $\Phi_{L, c}$, for all the cells at the leaf level. Next the multipoles at coarser levels are calculated by combining the multipoles at deeper levels (see Fig. 3.6).

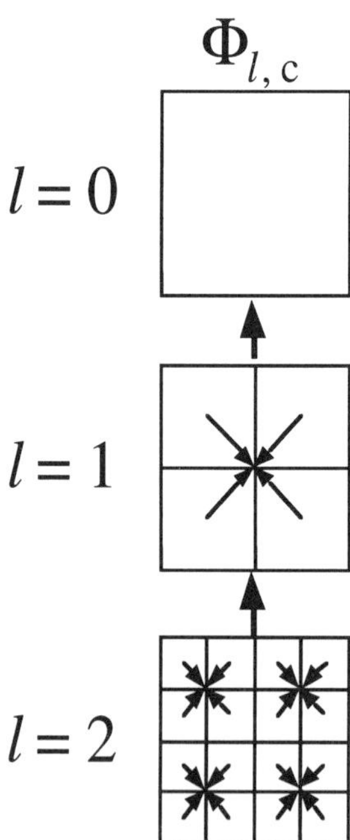

Figure 3.6 Schematic of the far-field computation in a two dimensional. system. Starting from the leaf level, the tree of cells is traversed upward to compute the multipoles for all the cells at all the levels.

At level l, $\Phi_{l,\,c}$ is calculated by shifting and adding up the multipoles of its 8 children cells, $\Phi_{l+1,\,c'}$. By traversing the tree upward, $l = L,\ L\text{-}1,...,1,\ 0$, one computes the multipoles, $\Phi_{l,\,c}$, for all the cells at all levels.

Local Expansion. The separation between particles in a given cell, c, is small compared to the separation between particles in well-separated cells (Fig. 3.7a).

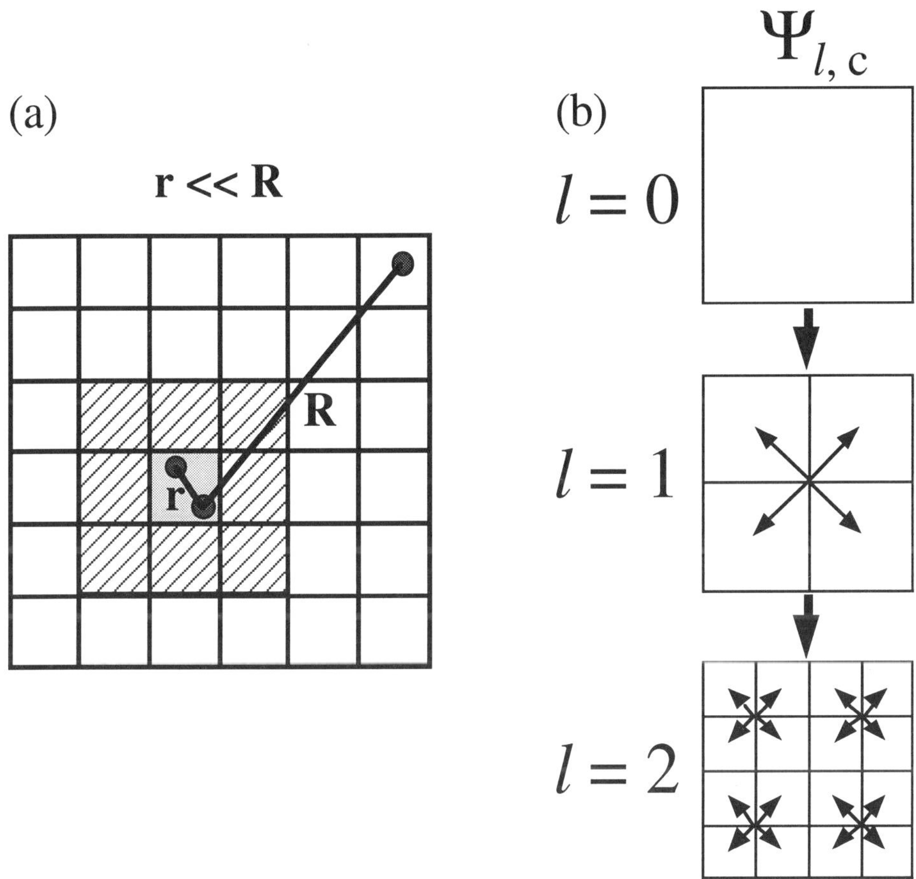

Figure 3.7 (a) Original cell , near-neighbor cells, and well-separated cells are shown. Using the fact that r<<R, the Taylor expansion can be used to any desired accuracy; (b) The tree is traversed downward to compute the local expansions for all the cells at all the levels.

Therefore, instead of repeating the sum over all well-separated cells for each atom in the cth cell, the field $\psi_{l,c}(\mathbf{r})$ is expanded in Taylor series about the center of the cell c.

$$\Psi_{l,c}(\vec{r}) = \Psi_{l,c}^{(0)} + \Psi_{l,c,\alpha}^{(1)} r_\alpha + \Psi_{l,c,\alpha\beta}^{(2)} r_\alpha r_\beta + \ldots \tag{3.4}$$

where $\mathbf{r}$ is the position of a particle in the cth cell relative to the center of the cell. Starting at the root level ($l = 0$) and traversing the tree downward, we compute the Taylor coefficients, $\Psi_{l,c}$, from $\Phi_{l,c}$ for all the cells at all levels (Fig. 3.7b). The Taylor coefficients are stored, and are used to calculate the far field from Eqn. (3.4) at the positions of the particles in cell c.

Computation of Far Field. Suppose we wish to compute the far-field contribution to particles in a cell c_0 at level 4. There are 8^4 or 4,096 cells at level 4, and Eqn. (3.2) implies a summation over 4,069 (= 4,096 - 27) well-separated cells. Direct summation of the far-field contribution from these 4,069 cells is a highly inefficient approach.

To compute far-field efficiently, one starts at level 2 where there are 64 cells (Fig. 3.8). One of these cells is a grandparent of c_0, and 26 of these cells are nearest neighbors of the grandparent. The remaining 37 cells at level 2 are well-separated from the grandparent of c_0. We first compute the far-field contribution of these well-separated cells at level 2 using multipole and local expansions.

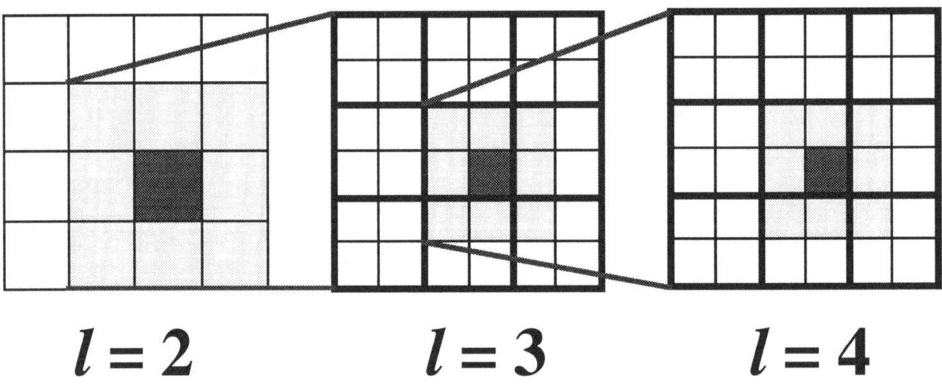

Figure 3.8 Two-dimensional representation of the hierarchy of cells for an efficient computation of the far-field.

Now we are left with 27 cells at level 2. We compute their far-field contribution at levels 3 and 4. At level 3, these 27 cells give rise to 216 cells. One of these 216 cells is a parent of c_0, 26 of them are nearest neighbors of the parent, and the remaining 189 cells are well-separated from the parent of c_0. The far-field contribution to c_0 from these 189 well-separated cells at level 3 is computed with multipole and local expansions.

Finally we are left with 27 cells at level 3. Their far-field contribution to c_0 is calculated at level 4 where they have 216 children cells. Twenty-six of these children cells are nearest neighbors of c_0, and particles in c_0 and in those 26 cells contribute to the near-field. The far-field contribution from 189 well-separated cells of c_0 is again computed with multipole and local expansions.

The above procedure accounts for the whole far-field contribution to the Coulomb field. At each level, the contribution arises from a fixed number of far cells -- at most 37 at level 2 and 189 at each subsequent level. This is far more economical than summing over all well-separated cells at the leaf level (e. g., 4,069 cells at level 4; 32,741 at level 5; or 262,117 at level 6).

3.2.4. Reduced Cell Multipole Method (RCMM)

In simulations of bulk systems, periodic boundary conditions are imposed by repeating the MD box infinitely as shown in Fig. 3.9.

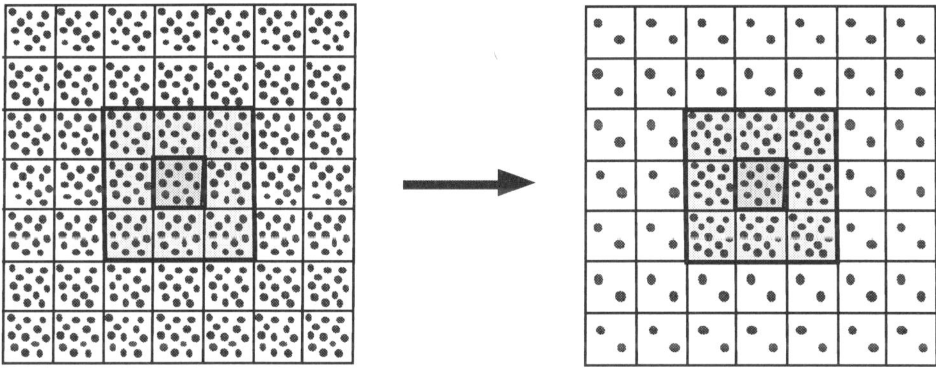

Figure 3.9 Computation of the Coulomb potential from the periodically repeating MD box by the reduced cell multipole method (RCMM). At left, central cell (full MD box), near-neighbor images of the MD box, and well-separated images. At right, FMM is used for the computation of the Coulomb potential in the original cell (MD box), and near-neighbor images. The well-separated images are replaced by reduced cells containing a small number of charged particles to match the lowest K multipoles.

In this case, the Coulomb interaction involves the well-known Ewald summation which expresses the interaction as a local sum in the Fourier space and a sum of "short-range" terms in real space [1].

The Ewald summation is calculated with the so-called reduced cell multipole method (RCMM) [40, 42]. The main idea of the RCMM is that even though the MD box contains multimillion particles, the contribution from ∞ - 27 well-separated image cells to the far field is described accurately by the first few terms of the multipole expansion. Therefore it is reasonable to replace each well-separated image cell by a reduced cell (see Fig. 3.9) containing a small number of randomly positioned fictitious particles whose first K multipole moments are the same as those of the original N particle system. To reproduce the multipole moments up to order K, only $P = (K+1)(K+2)(K+3)/6$ fictitious particles are needed. (For $K = 4$, only 35 fictitious particles are needed.) The charges, $\{q_i\}$, of these P fictitious particles are obtained by equating their first K multipole moments to the corresponding multipole moments of N image particles.

In the RCMM, the Coulomb potential is a sum of two terms: a) the potential generated by charges in the central MD box and image particles in the 26 nearest-neighbor boxes; and b) the potential due to P fictitious charges in the remaining ∞ - 27 cells (see Fig. 3.9). The first term (a) is computed with the fast multipole method. The second term (b) involves Taylor series of the Ewald summation over P fictitious particles.

3.2.5. *The Speed of MRMD Algorithm on Touchstone Delta and IBM SP machines*

The multiresolution molecular dynamics algorithms cause a significant increase in the efficiency of MD simulations. The reduced cell multipole method decreases the execution time by a factor of 6 over the usual implementation [40, 42] of the Ewald sum, the multiple time-step MD is 3 to 5 times faster than the conventional approach [31], and the separable three-body interaction scheme speeds up the calculation by a factor of 2 [31]. Figure 3.10 shows the overall performance of this parallel MRMD implementation on the Touchstone Delta machine at California Institute of Technology and IBM SP1 at Argonne National Laboratory. For a 4.2 million particle silica glass (Coulomb interactions and three-body covalent forces are included) on 512 nodes Touchstone Delta, the execution time for a single MD time step is only 4.84 seconds! On the IBM SP1 which has faster nodes (for our MRMD algorithm, 4.7 times faster than the i860 nodes at the Delta) the execution time on 512 SP1 nodes will be 1.1 seconds per step. The execution time scales linearly with the size of the system and the computation dominates the communication time [2, 14, 15, 31, 39, 40].

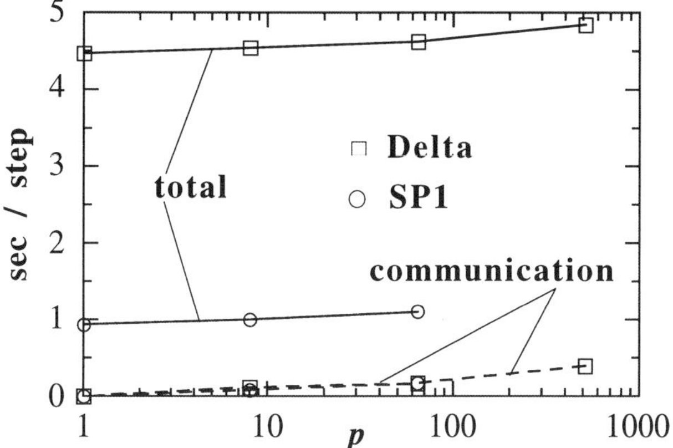

Figure 3.10 Execution time and communication time per MD time step in SiO_2 (Coulomb interactions and three-body covalent forces are included) on the Touchstone Delta machine (squares) at California Institute of Technology, and IBM SP1 (circles) at Argonne National Laboratory. Here p is the number of nodes on the Touchstone Delta and IBM SP1. The size of the system, N, increases as $8232p$. Because of excellent scalability of multiresolution algorithms, the execution time and communication time remain almost constant with increasing p and N.

4. Interaction Potentials for Silica, Silicon Nitride, and Silicon Diselenide

In this section, we focus on the effective interatomic potentials for three important materials - SiO_2, Si_3N_4, and $SiSe_2$. The structural unit in all these three materials is tetrahedron. There are two possible connectivities for the tetrahedra, namely edge-sharing and corner-sharing connection [46], see Figure 4.1. Among all of the binary chalcogenide materials, SiO_2, $GeSe_2$, and $SiSe_2$ are usually considered as typical models with unique structures. The crystalline forms of SiO_2 and Si_3N_4 are three-dimensional networks of only corner-sharing tetrahedra, while crystalline $GeSe_2$ consists of two-dimensional planes that are made of both corner-sharing and edge-sharing tetrahedra. In contrast, crystalline $SiSe_2$ is built exclusively of edge-sharing tetrahedra, which leads to a structure consisting of non-intersecting one-dimensional chains.

182

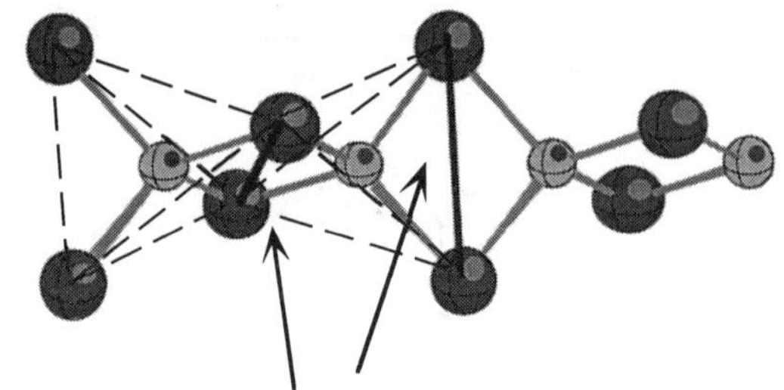

edge-sharing tetrahedra

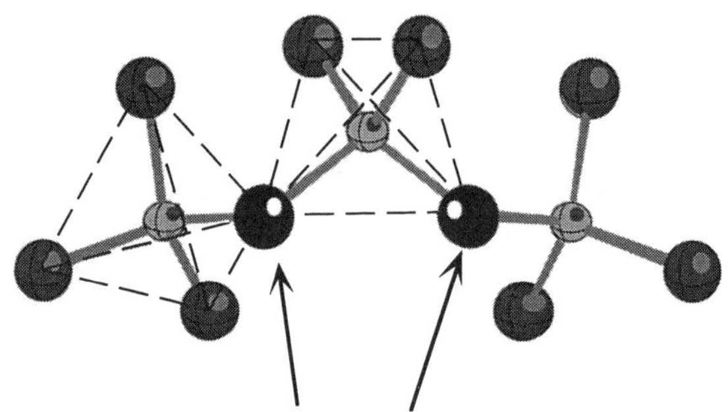

corner-sharing tetrahedra

Figure 4.1 Two possible connectivities of tetrahedra: edge-sharing and corner-sharing configurations.

4.1. CRYSTALLINE STRUCTURES OF SiO_2, Si_3N_4 AND $SiSe_2$

4.1.1. Structures of SiO_2

Silicon dioxide is one of the most extensively studied materials in condensed-matter physics, chemistry, materials science, and engineering [47, 48]. Although crystalline SiO_2 is known to have as many as 40 different structures [49, 50], only cristobalite [51], quartz [50, 52], coesite [53], and stishovite have temperature-density field of thermodynamic stability for chemically pure SiO_2 (no other element added for structural stability). Structures of α- and β-cristobalite and α- and β-quartz at atmospheric pressure and coesite at high pressure, involves different arrangements of nearly ideal corner-sharing $Si(O_{1/2})_4$ tetrahedra [49, 54]. Since we are going to discuss only the interaction potential for SiO_2 glass (2.2 g/cm^3), it is useful to mention the crystal structure of β-cristobalite which happens to have its density very close to that of SiO_2 glass. β-cristobalite at normal density of 2.2 g/cm^3 (number density = 6.618 x $10^{22} cm^{-3}$) has cubic (Fd3m) structure with 8 molecules per unit cell. In this structure Si-O bond length is 1.61 Å and the internal tetrahedral bond angle, O-Si-O, has values 107.8° and 112.8° whereas Si-O-Si bond angle is 146.7°.

4.1.2. Structures of Si_3N_4

Silicon nitride has been at the forefront of research for high-temperature, high-strength materials owing to its outstanding properties [55, 56]. The combination of low thermal expansion and high strength makes silicon nitride one of the most thermal-shock-resistant materials currently available. The strong covalent bonding between the atoms results in a superb resistance of mechanical deformation and chemical corrosion. Additionally, silicon nitride based films have found a number of applications in microelectronics technology [55, 57]. Crystalline silicon nitride exists in two polymorphs: the α- and -β phases [58-60]. The crystal structure of both phases consists of slightly distorted, corner-sharing $Si(N_{1/3})_4$ tetrahedra. In the β-phase Si-N layers parallel to the hexagonal basal plane are stacked in an alternate sequence ABABAB... In the α-phase the sequence becomes ABCDABCD... where the CD layers are related to the AB layers by a c-glide plane. As a result, the c-axis of the α-phase (unit cell containing $Si_{12}N_{16}$) is about double that of the β-phase (unit cell containing Si_6N_8). The enthalpy difference between the two phases was estimated to be only 30 kJ/mol [60]. Therefore, the coexistence of both phases in bulk Si_3N_4 often occurs, and it is hard to avoid small amounts of impurities (usually oxygen).

4.1.3. Structures of $SiSe_2$

Crystalline silicon diselenide is orthorhombic with four $Si(Se_{1/2})_4$ tetrahedra per unit cell [61]. At room temperature, T=293K, the lattice constants are known as a=9.669Å, b=5.998Å, and c=5.851Å. Crystalline density is 3.64 g/cm³. The interatomic distances in c-SiSe₂ are Si-Se= 2.275Å, and Si-Si nearest neighbor distance along a chain and between chains are 2.926Å and 5.689Å, respectively. Se-Se distances are in the range of 3.484 to 4.530Å. The coordination number for Si atoms is 4. For an ideal tetrahedron, the Se-Si-Se angle is 109°. In c-SiSe₂, the Se-Si-Se bond angles (100.0°, 112.2°, 116.7°) indicate the distortion of tetrahedra in c-SiSe₂. Among these angles, 100° is an intra-tetrahedral angle. This angle and the Si-Se-Si bond angle (80°) form a planar two-fold ring. In c-SiSe₂, each Si atom has only two two-fold rings, which implies the non-intersecting feature of chains.

4.2   INTERATOMIC POTENTIAL

Our MD simulations are based on an effective interatomic potential which includes both 2-body and 3-body interactions. The 2-body potential function consists of steric repulsion due to atomic sizes, screened Coulomb interaction caused by the charge transfer effect, and charge-dipole interaction resulting from the large electronic polarizability of O^{2-}, N^{3-}, and Se^{2-} [62, 63]. The 3-body covalent contribution takes into account the bond bending and bond stretching effects, shown schematically in Figure 4.2.

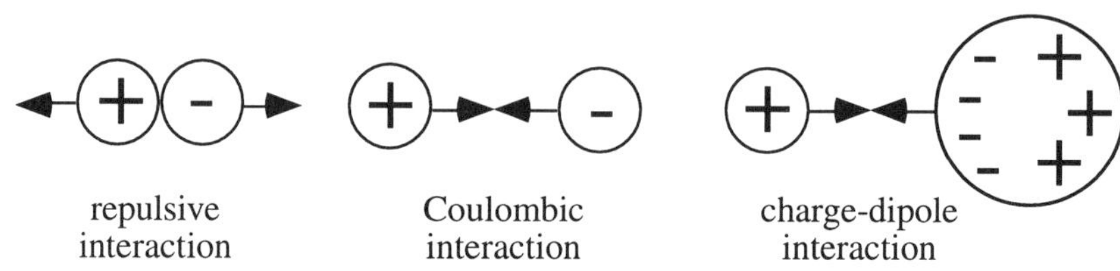

(a) two-body interaction

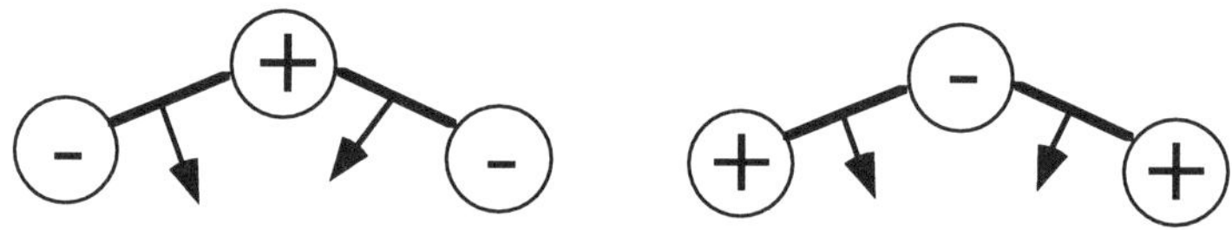

(b) three-body interaction

Figure 4.2 Schematic diagram of interatomic interactions.

For a system with N particles, the 2-body interaction takes the form of

$$V_2 = \sum_{i<j}^{N} V_{ij}^{(2)} \tag{4.1}$$

where

$$V_{ij}^{(2)} = \frac{H_{ij}}{r_{ij}^{\eta_{ij}}} + \frac{Z_i Z_j}{r_{ij}} e^{-r_{ij}/r_{1s}} - \frac{P_{ij}}{r_{ij}^4} e^{-r_{ij}/r_{4s}}, \tag{4.2}$$

$$H_{ij} = A_{ij}(\sigma_i + \sigma_j)^{\eta_{ij}}, \tag{4.3}$$

$$P_{ij} = (\alpha_i Z_j^2 + \alpha_j Z_i^2)/2, \qquad\qquad i, j = 1, 2, ..., N. \tag{4.4}$$

This 2-body term is a short-range pair-additive potential with a cutoff radius r_c. (See the modification according to Eq.(2.39)).

The 3-body interaction takes the form of

$$\mathcal{V}_3 = \sum_{i<j<k}^{N} \mathcal{V}^{(3)}_{jik} , \qquad (4.5)$$

where

$$\mathcal{V}^{(3)}_{jik} = B_{jik}\, f(r_{ij}, r_{ik})\, (\cos\theta_{jik} - \cos\bar{\theta}_{jik})^2 , \qquad (4.6)$$

and

$$f(r_{ij}, r_{ik}) = \exp\left(\frac{l}{r_{ij} - r_{c3}} + \frac{l}{r_{ik} - r_{c3}}\right), \qquad r_{ij}, r_{ik} < r_{c3}. \qquad (4.7)$$

$$\cos\theta_{jik} = \frac{\vec{r}_{ik}\cdot\vec{r}_{ij}}{r_{ik}r_{ij}} . \qquad r_{ij}, r_{ik} < r_{c3}. \qquad (4.8)$$

This potential has been used in the molecular-dynamics simulations to study structural correlations and dynamical properties of many covalent materials under various conditions of densities and temperatures. The MD simulation results for structural correlations, and static structure factors for SiO_2, $GeSe_2$, Si_3N_4, and $SiSe_2$ are in good agreements with corresponding experimental results [34, 64-68].

4.2.1. Interatomic Potential for SiO_2 Glass

The parameters defined in the potential, Eqns. (4.1)-(4.8), such as ionic radii σ_i, repulsive exponents η_{ij}, repulsive strength A_{ij}, effective charges Z_i, polarizabilities α_i, strength of 3-body interaction B_{jik}, average bond angles $\bar{\theta}_{jik}$, and the cut-off distances for 2-body and 3-body interactions are tabulated in Table 4.1.

TABLE 4.1 Parameters in the interaction potential for SiO_2 glass.

$A_{ij}(erg)$	$r_{s1}(\text{Å})$	$r_{s4}(\text{Å})$	$r_c(\text{Å})$	$l(\text{Å})$	$r_{c3}(\text{Å})$
1.242×10^{-12}	4.43	2.5	5.5	1.0	2.6

	$\sigma_i(\text{Å})$	$Z_i(e)$	$\alpha_i(\text{Å}^3)$
Si	0.47	1.20	0.00
O	1.20	-0.60	2.40

	η_{ij}		B_{jik} (erg)	$\bar{\theta}_{jik}$
Si-Si	11	**Si-O -Si**	3.2×10^{-11}	141.00
Si-O	9	**O -Si-O**	0.8×10^{-11}	109.47
O -O	7			

4.2.2. Interatomic Potential for Si₃N₄

The parameters defined in the potential, Eqns. (4.1)-(4.8), are tabulated in Table 4.2.

TABLE 4.2 Parameters in the interaction potential for Si3N4.

$A_{ij}(erg)$	$r_{s1}(\text{Å})$	$r_{s4}(\text{Å})$	$r_c(\text{Å})$	$l(\text{Å})$	$r_{c3}(\text{Å})$
2.00×10^{-12}	2.5	2.5	5.5	1.0	2.6

	$\sigma_i(\text{Å})$	$Z_i(e)$	$\alpha_i(\text{Å}^3)$
Si	0.47	1.472	0.00
N	1.30	-1.104	3.00

	η_{ij}		B_{jik} (erg)	$\bar{\theta}_{jik}$
Si-Si	11	**Si-N-Si**	2.0×10^{-11}	120.00
Si-N	9	**N-Si-N**	1.0×10^{-11}	109.47
N-N	7			

4.2.3. Interatomic Potential for SiSe₂

In the case of GeSe$_2$ and SiSe$_2$ (in general with all materials with edge-sharing tetrahedra) it is found that the Stillinger-Weber type bond stretching term in Eqn. (4.7) makes the three body part of the interaction potential produce an unexpected extra peak in the pair distribution function of $g_{Ge-Se}(r)$ and $g_{Si-Se}(r)$ for amorphous GeSe$_2$ and SiSe$_2$. Using the 3-body interaction as given below in Eqns.(4.9)-(5.0), however, we have reproduced correct structural correlations of SiSe$_2$ at the experimental glass density of 3.25g/cm^3. According to Eqns.(2.42)-(2.45), the neutron static structure factor, $S_n(q)$, is calculated for a bulk system. This bulk SiSe$_2$ system contains 13,824 particles in 3-D with periodic boundary conditions (pbc). Simulations were done with the variable-shape MD method (constant stress/pressure ensemble). The results from MD simulations are in good agreement with the diffraction data for amorphous SiSe$_2$ [69]. We also calculated the melting temperature for bulk SiSe$_2$ at normal density of 3.25g/cm^3. Finite diffusivities for Si and Se at temperatures higher than 1250K±50K indicate that the system is in the molten states. On the other hand, the small values of diffusion constants at temperatures lower than 1250K±50K imply that the system is in solid state. These results suggest a melting temperature, T_m=1250K±50K, which agrees very well with the experimental result T_m=1233K±5K [70].

The potential function used for SiSe$_2$ is different from the potential proposed in the in Eqns. (4.6)-(4.8) in the form of 3-body interaction. Here the 3-body interaction contains only the bond bending term (the bond stretching term: has been removed).

$$V^{(3)}_{jik} = B_{jik} \left(\cos \theta_{jik} - \cos \bar{\theta}_{jik} \right)^2 , \tag{4.9}$$

$$\cos \theta_{jik} = \frac{\vec{r}_{ik} \cdot \vec{r}_{ij}}{r_{ik} r_{ij}} , \qquad\qquad r_{ij}, r_{ik} < r_{c3} . \tag{5.0}$$

The parameters defined in the potential, Eqns. (4.1)-(4.4) and (4.9)-(5.0), are tabulated in Table 4.3.

TABLE 4.3 Parameters in the interaction potential for SiSe₂.

$A_{ij}(erg)$	$r_{s1}(\text{Å})$	$r_{s4}(\text{Å})$	$r_c(\text{Å})$	$r_{c3}(\text{Å})$
1.610x10^{-12}	4.43	4.43	9.0	3.0

	$\sigma_i(\text{Å})$	$Z_i(e)$	$\alpha_i(\text{Å}^3)$
Si	0.675	1.3456	0.00
Se	2.045	-0.6728	7.00

	η_{ij}		$B_{jik}(\text{erg})$	$\bar{\theta}_{jik}$
Si-Si	13	**Si-Se-Si**	1.973×10^{-11}	80.00
Si-Se	11	**Se-Si-Se**	0.986×10^{-11}	109.47
Se-Se	9			

5. MOLECULAR DYNAMICS SIMULATIONS OF SiO_2 GLASS, Si_3N_4, $SiSe_2$, AND GRAPHITE

5.1. STRUCTURAL CORRELATIONS IN SILICA GLASS UNDER PRESSURE AND INTERNAL FRACTURE IN POROUS SILICA

Structure of binary chalcogenide glasses (e.g. SiO_2, $GeSe_2$, $SiSe_2$) has been widely studied by neutron and X-ray scattering measurements [71]. These measurements reveal many similarities in the static structure factor, $S(q)$, when expressed in terms of the dimensionless quantity $Q_0 = qr_0$, where r_0 is the bond length. The most celebrated structural feature in these glasses is the first sharp diffraction peak (FSDP) observed between 1.0 and 1.5 Å^{-1}, which is believed to be a signature of intermediate range correlations. Although the chemical constituents and topologies of these glasses are widely different, the dimensionless quantity Q_0 for the FSDP is nearly the same. So the key issues are: (i) What is the nature of structural correlations that give rise to the FSDP? (ii) How are these correlations related to the connectivity of glasses when described in terms of n-fold rings.

Pore interface growth and the roughness of internally fractured surfaces in silica glasses are also investigated. The internal surface area, pore surface-to-volume ratio, pore-size distribution, and other structural correlations are determined as a function of the density.

Molecular dynamics simulations in conjuction with the interatomic potentials described in Sec. 4 is used to address these and related issues. Reactive empirical bond-order potential (REBOP) model for hydrocarbons [72] and large scale MD simulations of carbon systems are used to study amorphous carbon and dynamical fracture in graphite [73].

5.1.1. Structural Correlations and First Sharp Diffraction Peak in SiO_2 Glass at Normal Density

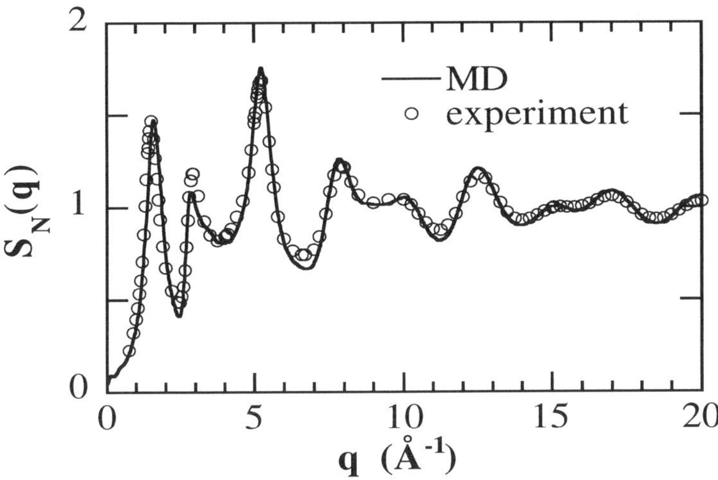

Figure 5.1 Neutron-scattering static structure factor, $S_N(q)$, of SiO_2 Glass. Solid curve, the MD result for a 41,472 particle system at 300 K; open circles, neutron diffraction experiment at 10 K [P. A. V. Johnson, A. C. Wright, and R. N. Sinclair, J. Non-Cryst. Solids 58 (1983) 109].

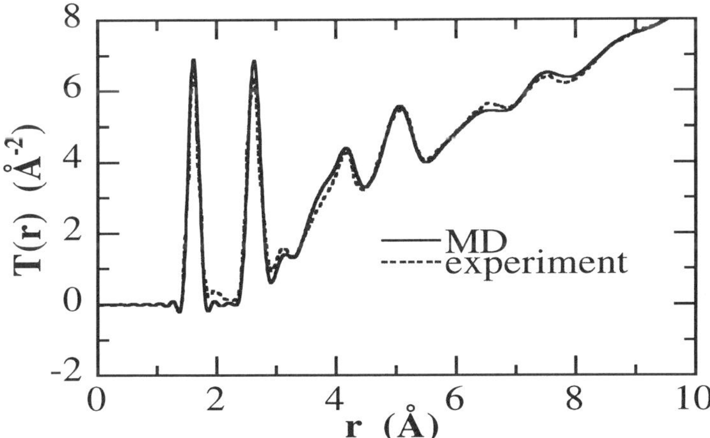

Figure 5.2 Neutron-scattering pair correlation function, $T(r)$, of SiO_2 Glass. Solid curve, the MD result for a 41,472 particle system at 300 K; dotted curve, neutron diffraction experiment [S. Susman, K. J. Volin, D. G. Montague, and D. L. Price, Phys. Rev. B 43 (1991) 11076] at 298 K.

190

The structure factor from MD simulations at the normal density (2.2 g/cm^3) is in excellent agreement with neutron diffraction experiments, including the height of the FSDP (Fig. 5.1). The dominant contribution to the FSDP at q~1.5 Å^{-1} is related in *r* -space to density-density correlations in the region of 4-8 Å [34, 64, 66, 74, 75]. The FSDP provides information about the intermediate range order in glasses, although the nature and how the structural entities give rise to the FSDP is still controversial. The FWHM is small and remains nearly constant with the density. From FWHM we can infer the correlation length of the intermediate range order, ~2π/(FWHM/2), which is in the range 15-25 Å [75, 76]. The agreement between the MD and experimental results can be quantified in terms of the R_χ factor introduced by Wright [77]. The R_χ factor for the comparison of Fig. 5.2 between 1 < r < 10 Å is 4.4 %.

5.1.2. Distribution of n-Fold Rings and the Height of the FSDP in SiO$_2$ Glass Under Pressure

The structure and network topology of a glass can be further characterized through the distribution of *n* -fold rings.

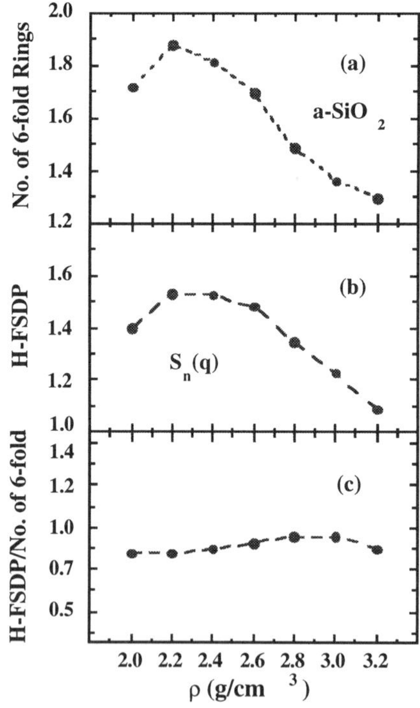

Figure 5.3 (a) Number of 6-fold rings, (b) height of the FSDP, and (c) ratio of height of the FSDP and the number of 6-fold rings as a function of the density.

An n-fold ring is defined as the shortest closed path of alternating Si-O bonds. Therefore, an n-fold ring consists of $2n$ alternating Si-O bonds. For a system with N 4-fold coordinated Si atoms there are 6N rings in the glass. For example, α- and β-cristobalite have only six 6-fold rings, whereas α- and β-quartz have four 6-fold and two 8-fold rings [54, 78] Using the MD method, we have studied the effect of pressure on the distribution of rings and their relationship to intermediate range correlations manifested as the FSDP for SiO_2 glass. A systematic analysis of the modifications observed in the FSDP for densities ranging from 2.0 to 3.2 g/cm^3 and temperatures from 0 to 1500 K has been carried out [75].

We have calculated the distribution of rings from $n = 2$ to 12. The distribution of distances and angles for each n-fold ring are also calculated [75]. We have explored the relationship between the height of the FSDP and the number of 6-fold rings per Si atom in SiO_2 glass. In Figs. 5.3 (a) and (b) we plot the density dependence of the number of 6-fold rings and the height of the FSDP. The ratio of the height of the FSDP and the number of 6-fold rings is shown in Fig. 5.3 (c). It is quite remarkable that even though the number of six-fold rings and the height of the FSDP vary substantially with density, their ratio is almost constant for the entire range of density, from 2.0 g/cm^3 to 3.2 g/cm^3.

Correlations in SiO_2 glass span two domains: (i) the short range order which extends to 4 Å, and (ii) intermediate range order between 4 Å and 25 Å, where dominant contribution to the FSDP comes from 4-8 Å region which is related to the diameter of six-fold rings. It has been shown by MD simulations that beyond this length (L/2 > 25 Å, L being the length of the MD box) there is no contribution to the FSDP [75, 76].

5.1.3. SiO_2 Glass Under Very High Pressure

Recent in-situ high-pressure X-ray measurements at the National Synchrotron Light Source at Brookhaven National Laboratory revealed dramatic changes in the structure of amorphous silica. Molecular-dynamics simulations were performed to gain insight into the microscopics of structural changes at high pressures [79]. Under normal conditions, amorphous silica is a disordered network of corner-sharing SiO_4 tetrahedra [67, 75, 80, 81] . However, at high pressures exceeding 40 GPa, the simulations reveal a new disordered network consisting of corner-sharing and edge-bonded SiO_6 octahedra [80], see Fig. 5.4.

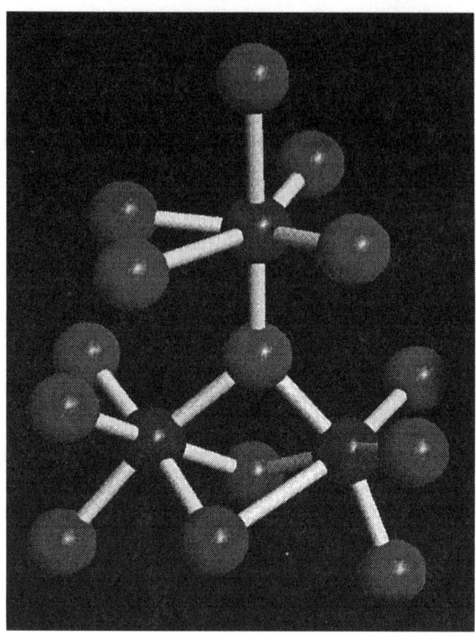

Figure 5.4 Corner-sharing and edge-bonded SiO_6 octahedra in the new high-pressure phase of silica glass at a density of 4.28 g/cm^3.

5.1.4. Internal Fracture in Porous Silica

Aerogel silica, a porous form of SiO_2 prepared by hypercritical drying of an alcoholic silica gel, is an environmentally safe material with a large thermal resistance which makes it a suitable alternative to chlorofluorocarbon (CFC)-foamed plastic in thermal insulation of commercial and household refrigerators [82, 83]. Scientists at Lawrence Livermore National Laboratory and Lawrence Berkeley Laboratory are investigating various other commercial uses of aerogel silica. These include passive solar energy collection devices, catalysis and chemical separation, and optical switches where porous materials are used as embedding frameworks for quantum-confined semiconducting microclusters.

We have performed MD calculations on 1.12-million particle silica glass systems, to investigate the growth of pores with a decrease in the density of the system [84]. (The low-density MD glasses are akin to aerogel silica.) As the normal-density glass is uniformly expanded, the pores begin to form when the density of the system is reduced to 1.8 g/cm^3. Further decrease in the density of the system causes an increase in the number of pores and also the pores coalesce to form larger entities. There is a dramatic increase in the size of pores when the mass density is reduced to the critical value, $r_c = 1.4$ g/cm^3. At the critical density, some pores percolate through the entire system causing fracture.

In Fig. 5.5 we show one of the surfaces of the percolating pore in the $1.12{\times}10^6$-particle system.

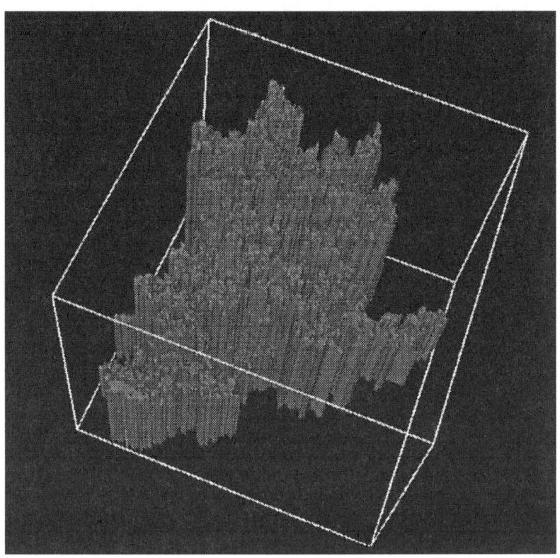

Figure 5.5 Fracture surface of a percolating pore in the 1.12-million atom
MD simulation of silica glass.

The roughness of a fractured surface is calculated from the height-height correlation function, $G(\sigma) = \left\langle [h(y+y_0, z+z_0) - h(y_0, z_0)]^2 \right\rangle^{1/2}$, where $\sigma = (y^2 + z^2)^{1/2}$ and $h(y, z)$ is the highest vertical coordinate at the point (y, z). The MD results for $G(\sigma)$ are well-described by the relation, $G(\sigma) \sim \sigma^\alpha$ with $\alpha = 0.87 \pm 0.02$ for $\sigma < 100$ Å. Experimental measurements on bakelite, concrete, steel, and aluminum alloys indicate that the roughness exponent α has a universal value 0.8. The MD results for the roughness exponent agree with experimental measurements, thus lending further support to experimental claims that the roughness exponent of fracture surfaces is a material-independent quantity. Moreover, the MD results indicate that the universality of the roughness exponent may prevail even at length scales ≤ 10 nm.

5.2. SILICON NITRIDE: LOW-FREQUENCY FLOPPY MODES, FRACTURE, AND SINTERING OF NANOCLUSTERS

5.2.1. Low-Frequency Floppy Modes in Si_3N_4

Silicon nitride is a very promising material for high-temperature applications. It has a strong resistance against thermal shock, high strength at high temperatures, corrosion resistance against fused metals, chemical inertness, and oxidation resistance at high temperatures. Because of these excellent characteristics, there are numerous applications of silicon nitride as engineering components. Amorphous silicon nitride, prepared by chemical vapor deposition (CVD), is used as a dielectric layer in microelectronics technology. Recently laser-induced CVD has been used to synthesize Si_3N_4 nano-powders with particle size as small as 7 nm. Promising application areas of nanocrystalline Si_3N_4 include pyroceramics as well as pressure sensors.

We have performed MD simulations for crystalline and amorphous silicon nitride [68, 85]. Figure 5.6 (a) shows the MD results for the static structure factor in amorphous Si_3N_4. These results are in good agreement with neutron scattering measurements [86]. The specific heat of the α-crystal at normal density has been calculated, and the results along with experimental data are shown in Fig. 5.6 (b). The MD results for the specific heat agree very well with the experimental results [87]. Low-energy phonon modes play an important role in determining the physical properties of ceramics [87].

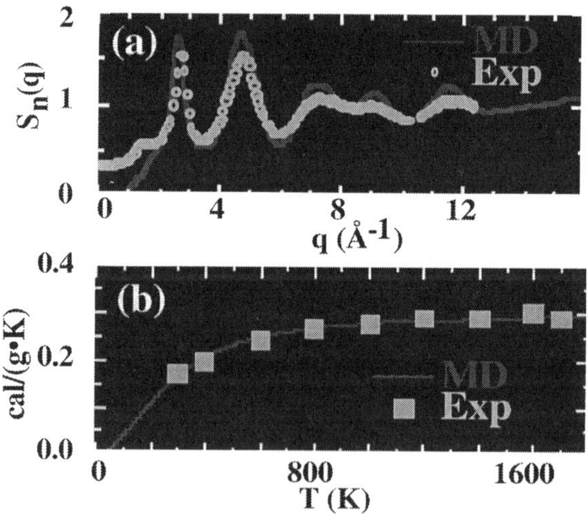

Figure 5.6 (a) MD and experimental results for the static structure factor of a-Si_3N_4 at a density of 2.8 g/cm^3. (b) MD and experimental results for the specific heat of α-Si_3N_4.

In our MD simulations of Si$_3$N$_4$, low-energy modes emerge in the amorphous and crystalline systems under uniform dilation. In the amorphous state, the number of low-energy floppy modes increases continuously with a decrease in the density and the average coordination of the system. Figure 5.7 shows the phonon density-of-states (DOS) in the frequency region below 15 meV: As the connectivity is reduced, low-frequency floppy modes appear and their DOS varies linearly with energy.

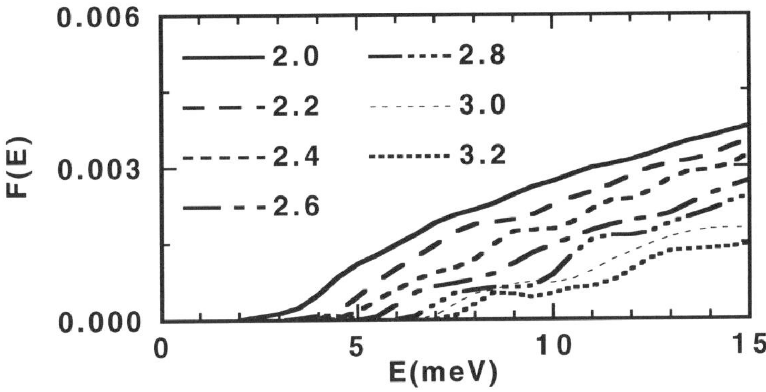

Figure 5.7 Linear behavior of the low-energy part of the phonon DOS, F(E), for amorphous Si$_3$N$_4$ at densities between 3.2 and 2.0 g/cm^3.

Results for the low-energy floppy modes in the α- crystal are shown in Figs. 5.8 (a) - (d). The DOS for the normal crystal is in good agreement with Raman and IR experiments [88]. Figure 5.8 (b) shows that although the system at 2.8 g/cm^3 is stretched, there are still sharp peaks in the DOS indicating the crystalline nature of the system. However, at 2.8 g/cm^3 the 10 meV gap in the DOS of the normal crystal disappears, the point group symmetry of the original crystal is destroyed, and the maximum frequency in the DOS is reduced from 140 to 110 meV. Note that the Si-N bond is robust and the nearest neighbor N-N and Si-Si distances change to accommodate expansion, thereby distorting the tetrahedra. (This is different from the early stages of uniform expansion of a monatomic crystal where all bonds stretch uniformly and the point group symmetry of the crystal is maintained.) As the density is further reduced to 2.4 g/cm^3, the system undergoes an abrupt decohesion. This represents the most profound rearrangement of the system - voids/pores grow, and bonds are broken and reassembled within the clusters and in the intercluster regions. The crystalline symmetry is completely lost and a linear variation of the DOS appears below 20 meV. At the same time the maximum frequency in the DOS increases from 110 meV to 150 meV. As the connectivity of the system is further reduced at 2.0 g/cm^3 (Fig. 5.8 (d)), the magnitude of the linear part of the DOS increases, the spectrum from 110 to 150 meV acquires more weight, and the maximum frequency increases to 165 meV. Floppy modes have been found to cause a substantial enhancement in the specific heat over the temperature range of 50-400K.

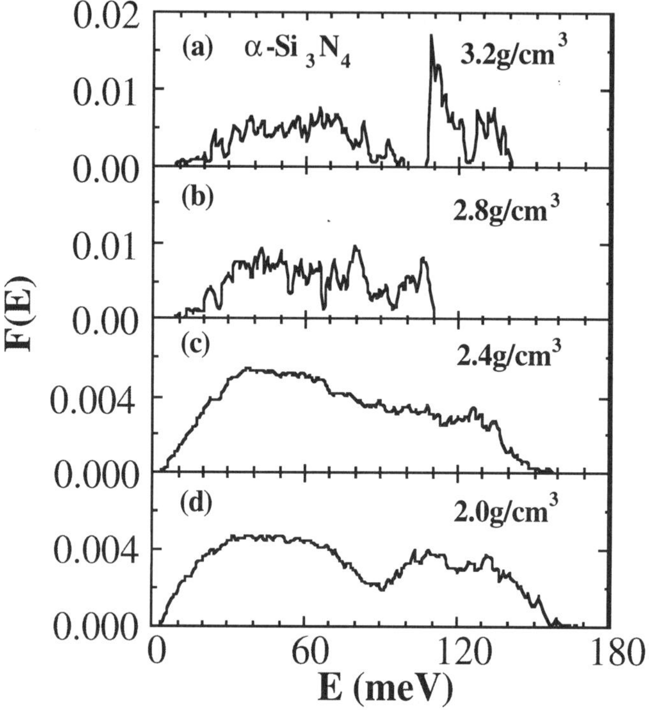

Figure 5.8 Phonon DOS at different stages of decohesion of the α-crystal due to volume expansion at 300K. (a) DOS for the crystal at normal density, 3.2 g/cm³. (b) Expanded crystal at 2.8 g/cm³. (c) At 2.4 g/cm³ there is an abrupt decohesion of the crystal and the emergence of a linear region in the DOS due to low-energy floppy modes. (d) DOS for the system at 2.0 g/cm³.

5.2.2. *Fracture in Si₃N₄ Films*

The morphology of fracture surfaces has drawn a great deal of attention in recent years. It is now well-established that a fracture surface, $z(x,y)$, is a self-affine object in that it remains invariant under the transformation, $(x, y, z) \rightarrow (ax, ay, a^\alpha z)$, where α is the roughness exponent. Bouchaud et al. carried out measurements for aluminum alloys with different heat treatments and in each case they obtained $\alpha = 0.8$ [89, 90]. Måløy et al. measured six different brittle materials and found α to be 0.87 ± 0.07 [91, 92].

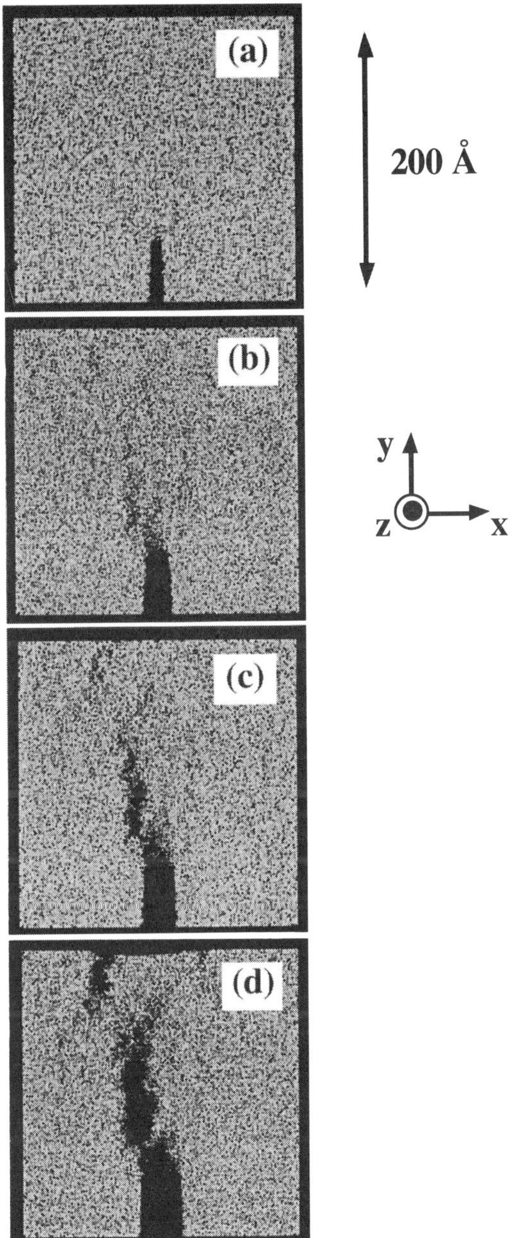

Figure 5.9 Snapshots of crack propagation in an amorphous Si_3N_4 film at time
(a) 4.5 ps; (b) 9.7 ps; (c) 12.3 ps; and (d) 16.4 ps.

198

Based on these experiments it was conjectured that fracture surfaces had a "universal" roughness exponent, independent of material characteristics and the mode of fracture. Milman et al. questioned the validity of the "universality" of α, especially at microscopic length scales, by pointing out that their scanning tunneling microscopy data for MgO, Si, and Cu revealed the roughness exponent to be around 0.6 [93, 94].

We have performed MD simulations of fracture [95] in amorphous Si_3N_4 films involving 100,352 atoms (typical dimensions of a film were 220Å×220Å×20Å). We first prepared well-thermalized bulk system by quenching the molten state and then periodic boundary conditions were removed and the systems were relaxed with MD and conjugate-gradient methods. These well-thermalized films were subjected to uniaxial tensile loads by displacing atoms uniformly in the leftmost and rightmost layers (thickness ~ 5.5Å each) along the x direction. The strain was applied at a constant rate while maintaining the temperature at 300 K.

To investigate crack propagation, we insert a crack in an uniaxially stretched film (strain ~ 4 %) by removing particles within a region whose projection onto the xy plane is 4Å×50Å, and we use a strain rate of 0.01 ps^{-1}. The crack plane is parallel to the yz plane, and the crack propagates along the y direction. Figure 5.9 shows snapshots of the amorphous Si_3N_4 film projected onto the xy plane. For the first 4.5 ps the crack propagates straight (Fig. 5.9a). At 9.7 ps, we observe the formation of voids in front of the crack tip, see Fig. 5.9b. These voids grow and form a secondary crack, and at 12.3 ps the secondary crack and the primary crack coalesce as shown in Fig. 5.9c. The resulting crack surface is very rough, see Fig. 5.9d at t = 16.4 ps.

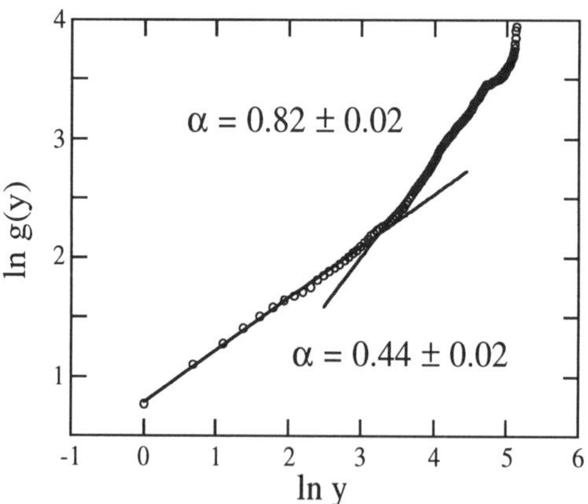

Figure 5.10 Log-log plot of the height-height correlation function, g(y) (open circles). The solid curves represent the best fit, g(y) ~ y^{α}, with α = 0.44 ± 0.02 for y < 25Å, and α = 0.82 ± 0.02 for y > 25Å.

We have calculated the height-height correlation function g(y) of the crack surface. For a self-affine surface, we expect the scaling relation, $g(y) \sim y^{\alpha}$ [95]. Figure 5.10 shows a log-log plot of the height-height correlation function for the crack surface in the amorphous film. There are two well-delineated regimes in this figure. For smaller length scales, we observe a rather smooth crack surface with a smaller roughness exponent. By linear fitting, we obtain a roughness exponent of 0.44 ± 0.02 for $y < 25\text{Å}$. Bcyond 25Å, the surface is rougher with a larger exponent, $\alpha = 0.82 \pm 0.02$. Detailed analysis reveals that the smaller exponent corresponds to "slow" crack propagation inside microcracks, while the larger exponent is due to coalescence of microcracks which causes rapid propagation. Similar crossover from slow to rapid fracture was observed in recent experiments by Bouchaud and Navéos [90].

5.2.3. Early Stages of Sintering of Si3N4 Nanoclusters

Sintering [96] of fine grains is of great importance in the formation of high-density ceramics. In recent years, a great deal of effort has been made to synthesize less brittle ceramics by consolidation of nanometer size clusters [97]. Many different experimental probes [98] have been used to determine the properties of so-called *nanophase materials*. The sintering behavior of nanosize particles is found to be different from that of macroscopic particles.

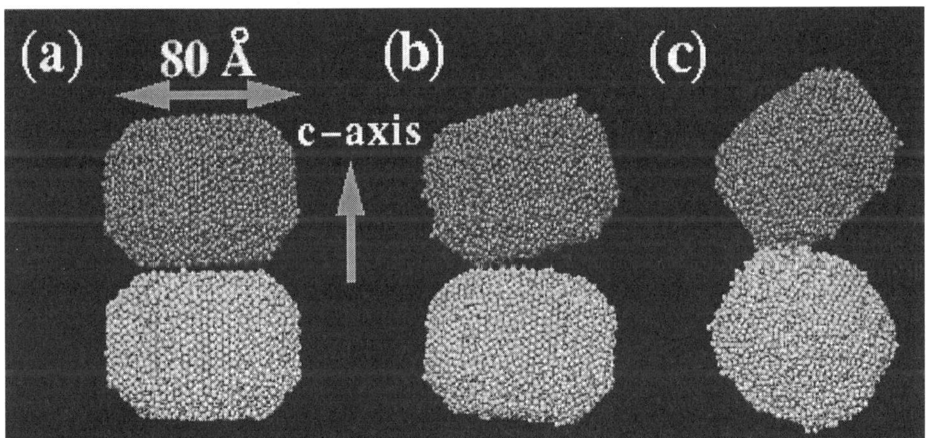

Figure 5.11 Snapshots of Si3N4 nanocrystals at 2,000K: (a) at time t = 0; (b) after 40 ps; (c) after 100 ps. Small spheres represent Si atoms, and large spheres denote N atoms.

We investigate early stages of sintering of silicon nitride nanoclusters. Within 100 pico seconds, an asymmetric neck is formed between nanocrystals at 2,000K, see Fig. 5.11. In the neck region, there are more four-fold than three-fold coordinated Si atoms [99]. In contrast, amorphous nanoclusters develop a symmetric neck, which has nearly the same number of three-fold and four-fold coordinated Si atoms. In the case of sintering among three nanoclusters, a chain-like structure forms in 200 pico seconds. The present study shows that sintering is driven by rapid diffusion of surface atoms and cluster rearrangement.

5.2.4. Nanophase Si_3N_4

In recent years, a great deal of effort has been made to synthesize less brittle ceramics by consolidation of nanometer size clusters [100]. A variety of these so-called nanophase ceramics have been synthesized. It is found that ceramics with ultrafine microstructures are more ductile and have lower sintering temperatures than their coarse-grained counterparts.

Figure 5.12 Million-atom MD simulation of fracture in nanophase Si_3N_4.

Using million-atom MD simulations Tsuruta and collaborators are investigating the structure and dynamics of atoms in the interface regions, the distribution and morphology of pores, and the effects of porosity and the size of nanoclusters on mechanical properties of nanophase Si_3N_4 (see Fig. 5.12). The MD simulations reveal: i) the interface regions in nanophase Si_3N_4 are amorphous, and have nearly 50% three-fold coordinated Si atoms; ii) the consolidation of nanophase Si_3N_4 is caused by atomic self-diffusion, which is estimated to be 10^{-6} cm^2/s in the interface regions; iii) the average pore radius R scales with the pore volume V as, $R \sim V^{1/d}$, where the fractal dimension d $= 2.0 \pm 0.2$ at all densities; and iv) the elastic modulii scale with the density as r^t, where the elastic exponent t varies considerably with the size of nanoclusters.

5.3. AMORPHIZATION AND FRACTURE IN $SiSe_2$ NANOWIRES

We have carried out MD simulations on amorphization and fracture in silicon diselenide nanowires [101, 102]. Our MD simulations reveal that all nanowires fracture at the same critical strain (ε_{zz}=15%), independent of the number of chains. We keep the strain constant after the critical strain is reached. A series of snapshots of a 64-chain nanowire projected onto the a-b plane were constructed. Here the elapsed time is measured from the moment when critical strain is reached. The number of particles in this nanowire is 230,400. At t=6.5 ps, we observe disordered structure in one of the chains at the outermost layer. As time evolves, the disorder spreads over an increasing number of chains. At t=10.9 ps, all of the chains across the nanowire become disordered. We also constructed the same series of snapshots projected onto the b-c plane. In these snapshots, we zoom into the disordered structure (about one fifth of the whole nanowire ~ 4,000 Å). We observe that disorder is localized along the c-axis and it expands longitudinally with time. A sandwich-like structure is formed with the disordered region in the middle. At 29.0 ps, the nanowire fractures at one of the two boundaries of this region, see Fig. 5.13.

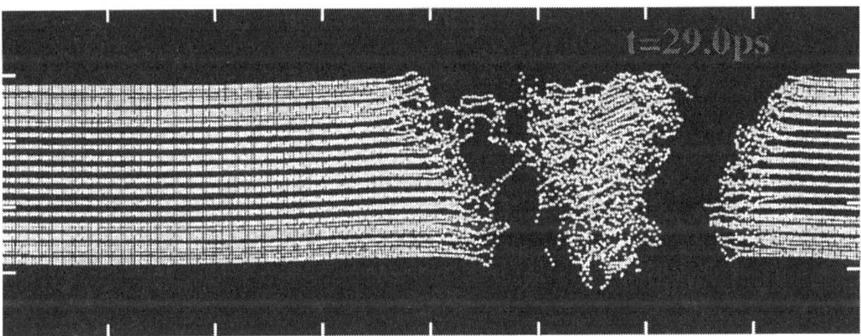

Figure 5.13 Amorphization and fracture in a 64-chain nanowire. Only a segment (~ 800 Å) of the nanowire is presented.

In Figure 5.14 we show the time evolution of local "temperature" and "density". Here the nanowire is sliced into 300 blocks along the c-axis, and the "temperature" and "density" are averaged over each block. The x- and y-axis correspond to the time and the c-coordinate along the 64-chain nanowire, respectively. The z-axis represents the average local "temperature" and "density". (Here the "temperature" is denoted by shading; lighter shading corresponds to higher temperature. In the case of "density" darker shading represents lower density) We have found that the high-temperature region is localized at the fracture site. As time evolves, this region expands along the nanowire while the temperature increases. From the local bond-angle distribution, we find that up to about 10ps the nanowire in the high-temperature region is in amorphous state. The rest of the nanowire remains crystalline.

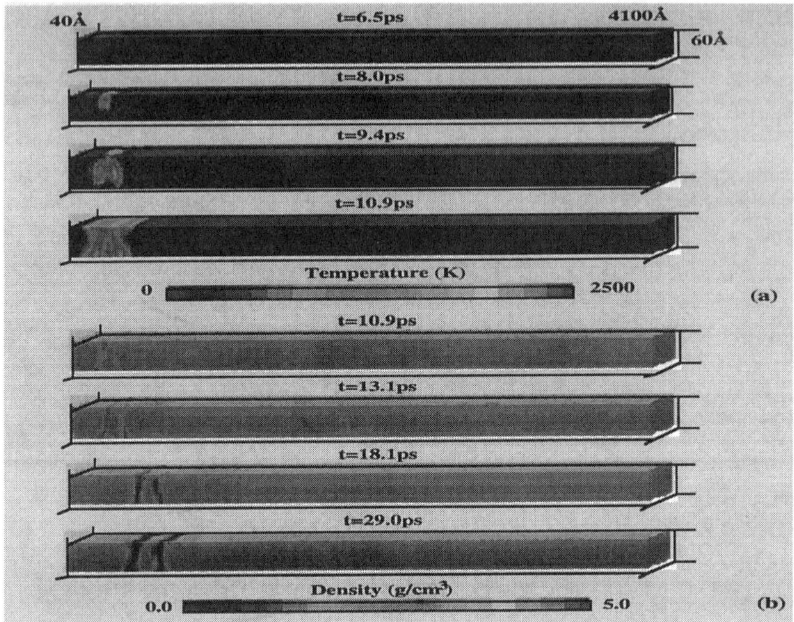

Figure 5.14 Time evolution of temperature profiles (a), and of density profiles (b), for the 64-chain nanowire.

5.4.   AMORPHOUS CARBON AND FRACTURE IN GRAPHITE

Carbon is unique among all the elements in that it can form strong covalent bonds with various coordination numbers. Carbon can not only form many different kinds of chemical compounds, it also has a rich structure and complex phase diagrams. New structures of carbon, such as fullerenes and graphitic tubules, and various new compounds of carbon are continuously being discovered.

Using a reactive empirical bond-order potential (REBOP) model for hydrocarbons [72], large scale MD simulations of carbon systems are carried out on parallel machines. Structural and dynamical correlations of amorphous carbon at various densities are studied [73]. The calculated structure properties agrees well with neutron scattering experiments and the results of tight-binding molecular dynamics simulations [103].

The MD simulations of amorphous carbon were performed on 32,768-particle systems at various densities ranging from 2.0 to 3.2 (g/cm³). The structure factor $S(q)$ is shown in Fig. 5.15. The simulation result agrees well with the neutron scattering data measured by Li and Lannin [104].

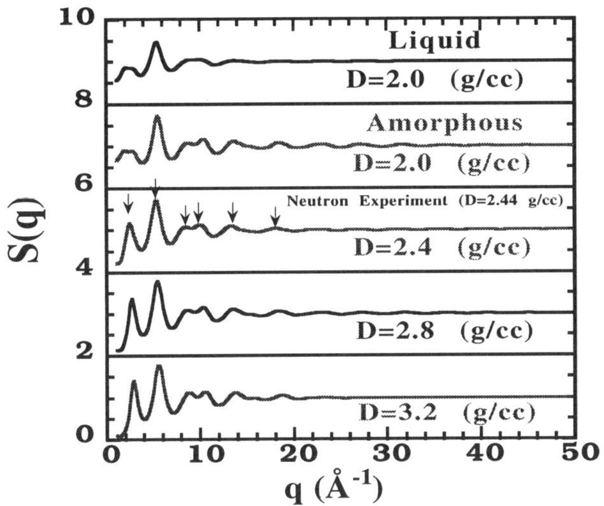

Figure 5.15 The structure factor of liquid carbon (2.0g/cm^3) and amorphous carbon at various densities.   The arrows indicate the peak positions of the neutron scattering experimental results measured by Li and Lannin (density of the experiment sample was estimated to be between 2.0 and 2.44 g/cm^3).

The dynamic behavior of crack propagation through graphite sheet is also investigated with the MD method. Effects of external stress and initial notch shape on crack propagation in graphite are studied. It is found that graphite sheet fractures in a cleavage-like or branching manners depending on the orientations of the graphite sheet with respect to the external stress The roughness of crack surfaces is analyzed. Two roughness exponents are observed in two different regions [73].

Molecular dynamics simulations were performed on 100,800-particle systems. The size of the sheet was approximately 500 Å x 600 Å. The graphite sheet was initially laid out on the XY-plane, but the atoms were allowed to move in the Z-direction as well. No periodic boundary conditions were imposed and the boundaries were terminated with hydrogen atoms. The external strain was applied in the X-direction. We considered two different orientations of the graphite sheet in the XY plane. In one case, which we call G(1,1), there are bonds which are parallel to the X-axis. The other case we considered is G(1,0), where the bonds make an angle of 30° with respect to the X-axis, and there are bonds parallel to the Y-axis. Recently million atom simulations of dynamic fracture of graphite sheet have also been completed by Omeltchenko and collaborators.

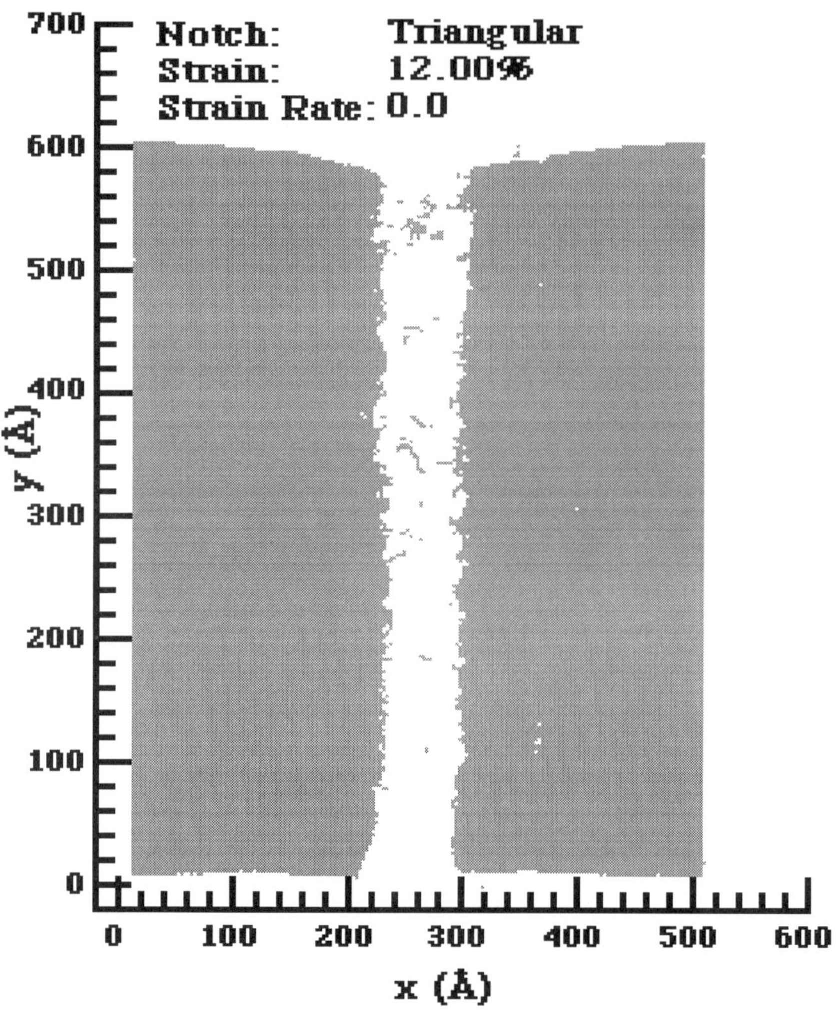

Figure 5.16 Snapshot of the crack propagation in graphite: G(1,1) under constant strain.

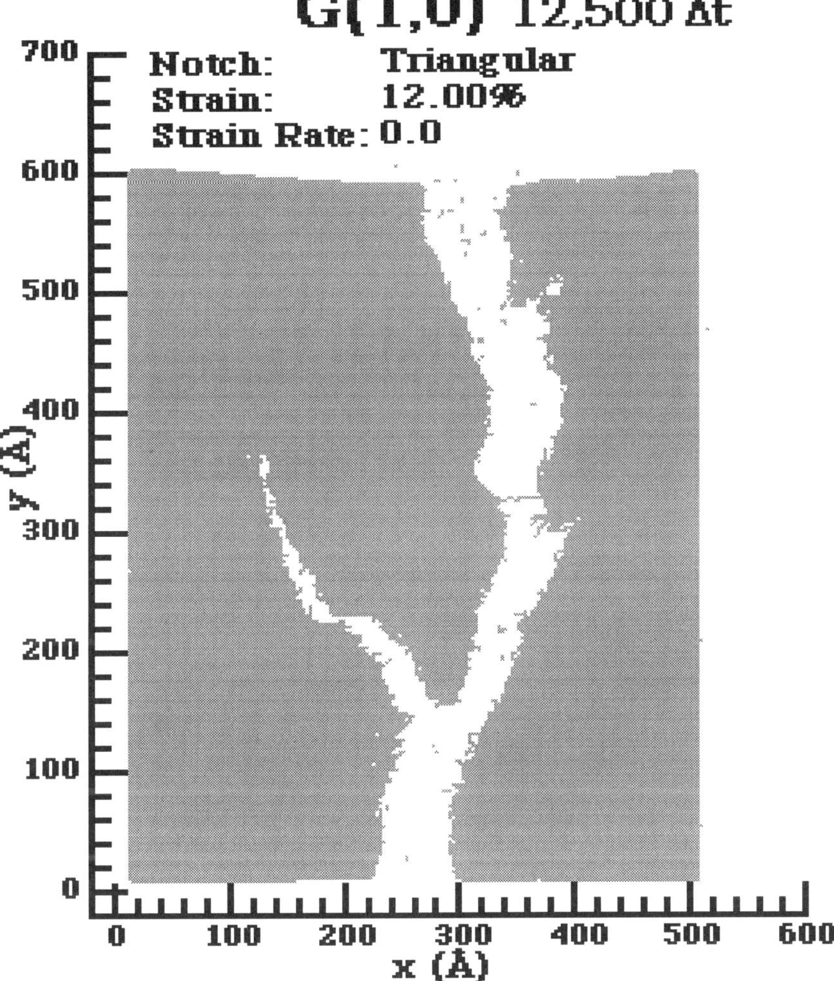

Figure 5.17 Snapshot of the crack propagation in graphite G(1,0) under constant strain.

In Figure 5.16 we show the snapshot of the graphite sheet with the orientation G(1,1) after 10,000 steps at constant strain of 12% and with the triangular notch. In this case, the crack propagates straight and the fracture is cleavage-like. Similar behavior was observed for a square notch under the same conditions.

The situation is quite different for the G(1,0) orientation. Figure 5.17 shows the crack in the graphite sheet with the orientation G(1,0) and triangular notch after 12,500 steps at 12% strain. The crack branches and changes direction. Even more interesting is the behavior for the square notch. It turns out that for the G(1,0) orientation and square notch the material does not break at 12% strain. In order to induce the fracture, we applied a strain rate of 1% per 5,000 MD steps. The crack started to propagate only after the strain exceeded 17%. For fracture at such a high strain and finite strain rate, the branching is dramatic. Along with the obvious branches, we also see a number of periodically spaced attempted branches.

To investigate the roughness of fracture surfaces we have calculated the roughness exponents from the height-height correlation function of the crack surface. For a self-affine surface, the height-height correlation is expected to obey the scaling relation $g(y) \sim y^{\alpha}$. The correlation function can be fitted with two straight lines with different slops (roughness exponents) characterizing two well-delineated regimes: For shorter length scales (< 30 Å), we find a smaller roughness exponent of 0.51 ± 0.03, whereas for larger length scales (> 30 Å), we obtain a larger roughness exponent of 0.72 ± 0.03.

6.   Research in Progress and Future Plans

Some of the most interesting phenomena in materials science require the confluence of ideas and expertise from diverse points of view, ranging from continuum macroscopic approaches to atomistic descriptions. Fracture of solids, friction, brittle versus ductile behavior under external loading are characteristic examples of phenomena that demonstrate the interplay of microscopic processes and macroscopic behavior. Particularly challenging problems of this type arise in relation to interfaces between materials. The different properties of the two components of an interface allow for a rich variety of phenomena, possibly quite distinct from that observed in either component alone. Of special scientific and technological importance is the behavior of interfaces between materials that are used to protect, insulate or otherwise change the properties of the entire system.

Our future research will focus on three interrelated materials simulation problems:

- Delamination in metal-ceramic and semiconductor/ceramic coatings;

- Dynamic fracture in cluster-assembled nanophase ceramic composites;

- Effects of wear, lubrication, and friction on Micro-Electro-Mechanical Systems (MEMS);

These problems are bound by commonalties such as the underlying surface and interface phenomena. We will study these problems with both atomistic (first-principles electronic structure, tight binding, and classical MD) and macroscopic (finite element) approaches using multiscale algorithms on heterogeneous parallel machines.

6.1. MEMS SUBMICRON GEAR TRAINS -- WEAR, FRICTION AND SELF-ASSEMBLED MONOLAYERS AS LUBRICANTS

There is tremendous growth and excitement in the area of Micro-Electro-Mechanical-Systems (MEMS), see Fig. 6.1. The idea is to reduce the size of coupled electro/magnetic systems (accelerometers, pumps, actuators, sensors etc.) down to micron-sized dimensions. Much of the expertise here has originated out of the microfabrication technology of the semiconductor revolution where the ability to control 3D shapes for the purposes of chip design has reached a very high degree of control and sophistication. On the same piece of "real estate", the MEMS device (the machine) can be controlled by a semiconductor device (the brain).

Figure 6.1 Microgear fabricated in Guckel's group at Wisconsin.

We are presently applying atomistic methods to study the behavior of quartz crystal oscillators (QCO) via direct simulation on parallel machines [105-107]. The rough "rule of thumb" is that system sizes should be in the 1-10 million atom regime and total elapsed time scales should be 10^{-9} to 10^{-6} second. QCOs are examples of accelerometers and are being manufactured on micron length scales.

Recent developments in multiprocessor nodes for proposed parallel computer architectures and advances in efficient space-time multiresolution MD algorithms lead us to believe that in the next few years we can expect that 100 million to 1 billion atoms may be studied and time scales of 10^{-6} sec will be possible. This will allow realistic MD simulations of submicron size gear trains including the effects of lubricants, friction, and wear.

ACKNOWLEDGMENT

This work was supported by DOE (Grant No. DE-FG05-92ER45477), NSF (Grant No. DMR-9412965), AFOSR (Grant No. F 49620-94-1-0444), and USC-LSU Multidisciplinary University Research Initiative (Grant No. F 49620-95-1-0452). One of the authors (Wei Li) would like to acknowledge support from NSF Graduate Research Traineeship (Grant No. GER93550007). Simulations were performed on the 128-node IBM SP computer at Argonne National Laboratory and the parallel machines in the Concurrent Computing Laboratory for Materials Simulations (CCLMS) at Louisiana State University. The facilities in the CCLMS were acquired with equipment enhancement grants awarded by the Louisiana Board of Regents through Louisiana Education Quality Support Fund. Help in preparation of this manuscript from Dr. Kenji Tsuruta and Andrey Omeltchenko is gratefully acknowledged.

7.   References

1.   S.W. de Leeuw, J.W. Perram and E.R. Smith (1980) Simulation of Electrostatic Systems in Periodic Boundary Conditions. II. Equivalence of Boundary Conditions, *Proc. R. Soc. London*, **A373**, 27.

2.   R.K. Kalia, S.W. de Leeuw, A. Nakano and P. Vashishta (1993) Molecular-dynamics Simulations of Coulombic Systems on Distributed-Memory MIMD Machines, *Computer Physics Communications*, **74**, 316.

3.   A.W. Appel (1985) *SIAM J. Sci.Stat. Comput*, **6**, 85.

4.   J. Barnes and P. Hut (1986) A hierarchical O(N log N) force-calculation algorithm, *Nature*, **324**, 446.

5.   G.C. Fox, P. Hipes and J. Salmon (1989), in *Practical Parallel Supercomputing: Examples from Chemistry and Physics*, I.C.S.a.A. SIGARCHs, Eds., Supercomputing ' 89, ACM Press.

6.   S. Pfalzner and P. Gibbon (1992) A 3D Hierarchical Tree Code for Dense Plasma Simulation, *Computer Physics Communications*, July.

7.   M. Warren and J. Salmon (1992) Astrophysical N-body Simulations on the Delta, *CSCC Update*, **13**, 1.

8.   L. Greengard and V. Rokhlin (1987) A Fast Algorithm for Particle Simulations, *J. Computational Physics*, **73**, 523.

9.   L. Greengard (1987), *The Rapid Evaluation of Potential Fields in Particle Systems*, The MIT Press, Cambridge, Massachusetts, London, England.

10.   L. Greengard and W.D. Gropp (1989), in *A Parallel Version of the Fast Multipole Method*, G. Rodrigues, Eds., Parallel Processing for Scientific Computing, SIAM.

11.   L. Greengard (1990) The Numerical Solution of the N-body Problem, *Computers in Physics*, **4**, 142.

12.   K.E. Schmidt and M.A. Lee (1991) Implementing the Fast Multipole Method in Three Dimensions, *J. Statistical Physics*, **63**, 1223.

13.   F. Zhao and S.L. Johnsson (1991) The Parallel Multipole Method on the Connection Machine, *SIAM J. Sci. Stat. Comput*, **12**, 1420.

14.   P. Vashishta, R.K. Kalia, S.W. de Leeuw, D.L. Greenwell, A. Nakano, W. Jin, J. Yu, L. Bi and W. Li (1994) Computer Simulation of Materials Using Parallel Architectures, *Computational Materials Science*, **2**, 180.

15.   R.K. Kalia, A. Nakano, D.L. Greenwell and P. Vashishta (1993) Parallel Algorithms f or Molecular Dynamics Simulations on Distributed-Memory MIMD Machines, *Supercomputer 54*, **X**, 11.

16.   A. Rahman (1977), in *Correlation Functions and Quasiparticle Interactions in Condensed Matter*, J.W. Halley, Eds., Plenum, New York, 417.

17.   A. Rahman and P. Vashishta (1983), in *The Physics of Superionic Conductors and Electrode Materials*, J.W. Perram, Eds., Plenum, New York, 93.

18.   M. Parrinello and A. Rahman (1981) Polymorphic Transitions in Single Crystals: A New Molecular Dynamics Method, *J. Appl. Phys*, **52**, 7182.

19.   S. Nosé (1984) A Molecular Dynamics Method for Simulations in the Canonical Ensemble, *Molecular Physics*, **52**, 255.

20.   M. Parrinello and A. Rahman (1982) Strain Fluctuations and Elastic Constants, *J. Chem. Phys.*, **76**, 2662.

21.   J.R. Ray and H.W. Graben (1981) Direct Calculation of Fluctuation Formulae in the Microcanonical Ensemble, *Molecular Physics*, **43**, 1293.

22.   J.R. Ray (1982) Fluctuations and Thermodynamics Properties of Anisotropic Solids, *J. Appl. Phys.*, **53**, 6441.

23.   J.R. Ray and A. Rahman (1984) Statistical Ensembles and Molecular Dynamics Studies of Anisotropic Solids, *J. Chem. Phys.*, **80**, 4423.

24.   J.R. Ray and A. Rahman (1985) Statistical Ensembles and Molecular Dynamics Studies of Anisotropic Solids. II, *J. Chem. Phys.*, **82**, 4243.

25.   J.R. Ray, M.C. Moody and A. Rahman (1985) Molecular Dynamics Calculation of Elastic Constants for a Crystalline System in Equilibrium, *Phys. Rev.*, **B32**, 733.

26.   J.R. Ray, M.C. Moody and A. Rahman (1986) Calculation of Elastic Constants Using Isothermal Molecular Dynamics, *Phys. Rev.*, **B33**, 895.

27.   J.R. Ray (1988) Elastic Constants and Statistical Ensembles in Molecular Dynamics, *Computer Physics Reports*, **8**, 109.

28.   M. Tuckerman and B.J. Berne (1992) Reversible Multiple Time Scale Molecular Dynamics, *J. Chem. Phys.*, **97**, 1990.

29. L. Verlet (1967) Computer "Experiments" on Classical Fluids. I. Thermodynamical Properties of Lennard-Jones Molecules, *Phys. Rev.*, **159**, 98.

30. F.H. Stillinger and T.A. Weber (1983) Dynamics of Structural Transitions in Liquids, *Phys. Rev.*, **A28**, 2408.

31. A. Nakano, P. Vashishta and R.K. Kalia (1993) Parallel Multi-time-Step Molecular Dynamics with Three-body Interaction, *Computer Physics Communications*, **77**, 303.

32. A. Nakano, L. Bi, R.K. Kalia and P. Vashishta (1994) Molecular-dynamics Study of the Structural Correlation of Porous Silica with use of a Parallel Computer, *Phys. Rev.*, **B49**, 9441.

33. M.P. Allen and D.J. Tildesley (1987), *Computer Simulation of Liquids*, Clarendon Press, Oxford.

34. P. Vashishta, R.K. Kalia, J.P. Rino and I. Ebbsjö (1990) Interaction Potential for SiO_2: A Molecular-Dynamics Study of Structural Correlations, *Phys. Rev.*, **B41**, 12197.

35. D. Chandler (1987), *Introduction to Modern Statistical Mechanics*, Oxford University, New York, Oxford.

36. J.L. Lebowitz, J.K. Percus and L. Verlet (1967) Ensemble Dependence of Fluctuations with Application to Machine Computations, *Phys. Rev.*, **153**, 250.

37. R. Car and M. Parrinello (1985) Unified Approach for Molecular Dynamics and Density-Functional Theory, *Phys. Rev. Lett.*, **55**, 2471.

38. D.C. Rapaport (1991) Multi-Million Particle Molecular Dynamics II. Design Considerations for Distributed Processing, *Comput. Phys. Commun.*, **62**, 217.

39. W. Li, R.K. Kalia, S.W. de Leeuw, A. Nakano, D. Greenwell and P. Vashishta (1992), in *Parallel Algorithms for Molecular-Dynamics Simulations of Coulombic Systems*, P.D. Bristowe, J. Broughton and J.M. Newsams, Eds., Mat. Res. Soc. Symp. Proc.

40. A. Nakano, P. Vashishta and R.K. Kalia (1994) Massively Parallel Algorithms for Computational Nanoelectronics Based on Quantum Molecular Dynamics, *Computer Physics Communication*, **83**, 181.

41. A. Nakano, R.K. Kalia and P. Vashishta (1994) Muitiresolution Molecular Dynamics Algorithm for Realistic Materials Modeling on Parallel Computers, *Computer Physics Communication*, **83**, 197.

42. H.-Q. Ding, N. Karasawa and W.A.G. III (1992) The Reduced Cell Multipole Method for Coulomb Interactions in Periodic Systems with Million-atom Unit Cell, *Chemical Physics Letters*, **196**, 6.

43. W.B. Streett, D.J. Tildesley and G. Saville (1978) Multiple Time-Step Methods in Molecular Dynamics, *Mol. Phys.*, **35**, 639.

44. D. Frenkel (1989), in *Simple Molecular Systems at Very High Density*, A. Polian and P. Loubeyre, Eds., Plenum, New York, 411.

45. R.K. Kalia, W. Jin, S.W. de Leeuw, A. Nakano and P. Vashishta (1993) Atomistic Simulations on Parallel Architectures, *International Journal of Quantum Chemistry: Quantum Chemistry Symposium*, **27**, 781.

46. P. Vashishta, R.K. Kalia, G.A. Antonio, J.P. Rino, H. Iyetomi and I. Ebbsjö (1990) Molecular Dynamics Study of the Structure and Dynamics of Network Glasses, *Solid State Ionics*, **40/41**, 175.

47. D. Adler, H. Fritzsche and S.R. Ovshinsky (1985), *Physics of Disordered Materials*, Plenum, New York.

48. R.A.B. Devine (1988), *The Physics and Technology of Amorphous SiO_2*, Plenum, New York.

49. F. Liebau (1988), in *The Physics and Technology of Amorphous SiO$_2$*, R.A.B. Devine, Eds., Plenum, New York.

50. R.W.G. Wyckoff (1965), *Crystal Structures,* Wiley, New York.

51. J.J. Pluth, J.V. Smith and J. J. Faber (1985) Crystal Structure of Low Cristobalite at 10, 293, and 473 K: Variation of Framework Geometry with Temperature, *J. Appl. Phys.*, **57**, 1045.

52. L. Levien, C.T. Previtt and D.J. Weidner (1980) Structure and Elastic Properties of Quartz at Pressure, *American Mineralogist*, **65**, 920.

53. L. Levien and C.T. Previtt (1981) High-Pressure Structure and Compressibility of Coesite, *American Mineralogist*, **66**, 324.

54. J.A.E. Desa, A.C. Wright, J. Wong and R.N. Sinclair (1982) A Neutron Diffraction Investigation of the Structure of Vitreous Zinc Chloride, *J. Non-Crystalline Solids*, **51**, 57.

55. I. Chen, P.F. Becher, M. Mitimo, G. Petzow and T. Yen (1993), in *Silicon Nitride Ceramics, Scientific and Technological Advances*, Mat. Res. Soc. Symp. Proc.,

56. M.J. Hoffmann, P.F. Becher and G. Petzow (1994), *Silicon Nitride 93,* Trans. Tech. Publications, Switzerland.

57. M.J. Hoffmann and G. Petzow (1994), *Tailoring of Mechanical Prpperties of Si$_3$N$_4$ Ceramics,* Kluwer, The Netherlands.

58. F.H.P.M. Habraken (1991), *LPCVD Silicon Nitride and Oxynitride Films,* Springer-Verlag, Berlin.

59. J.T. Milek (1971), *Silicon Nitride for Microelectronic Applications,* IFI/Plenum, New York.

60. R. Grün (1979) The Crystal Structure of ß-Si$_3$N$_4$; Structural and Stability Considerations Between α- and ß-Si$_3$N$_4$, *Acta Cryst.*, **B35**, 800.

61. J. Peters and B. Krebs (1982) Silicon disulphide and Silicon diselenide: A Reinvestigation, *Acta Cryst.*, **B38**, 1270.

62. P. Vashishta and A. Rahman (1978) Ionic Motion in α-AgI, *Phys. Rev. Lett.*, **40**, 1337.

63. P. Vashishta and A. Rahman (1979), in *Fast Ion Transport in Solids*, P. Vashishta, J.N. Mundy and G.K. Shenoy, Eds., Elsevier North Holland, 527.

64. G.A. Antonio, R.K. Kalia, A. Nakano and P. Vashishta (1992) Crystalline fragments in glasses, *Phys. Rev.*, **B45**, 7455.

65. A. Nakano, L. Bi, R.K. Kalia and P. Vashishta (1993) Structural Correlations in Porous Silica: Molecular Dynamics Simulation on a Parallel Computer, *Phys. Rev. Lett.*, **71**, 85.

66. W. Jin, R.K. Kalia, P. Vashishta and J.P. Rino (1993) Structural Transformation, Intermediate-Range Order, and Dynamical Behavior of SiO$_2$ Glass at High Pressures, *Phys. Rev. Lett.*, **71**, 3146.

67. W. Jin, P. Vashishta, R.K. Kalia and J.P. Rino (1993) Dynamic Structure Factor and Vibrational Properties of SiO$_2$ Glass, *Phys. Rev.*, **B48**, 9359.

68. C.-K. Loong, P. Vashishta, R.K. Kalia and I. Ebbsjö (1995) Crystal Structure and Phonon Density of States of High-Temperature Ceramic Silicon Nitride, *Europhys. Lett.*, **31**, 201.

69. R.W. Johnson, D.L. Price, S. Susman, M. Arai, T.I. Morrison and G.K. Shenoy (1986) The Structure of Silicon-Selenium Glasses I: Short-range order, *J. Non-Crystalline Solids*, **83**, 251.

70. R.W. Johnson, S. Susman, J. McMillan and K.J. Volin (1986) Preparation and Characterization of Si_xSe_{1-x} Glasses and Determination of the Equilibrium Phase Diagram, *Mat. Res. Bull.*, **21**, 41.

71. S.C. Moss and D.L. Price (1985), in *Physics of Disordered Materials*, D. Adler, H. Fritzsche and S.R. Ovshinsky, Eds., Plenum, New York, 77.

72. D.W. Brenner (1990) Empirical potentials for hydrocarbons for use in simulating the chemical vapor deposition of diamond films, *Phys. Rev. B*, **42**, 9458.

73. J. Yu, A. Omeltchenko, R.K. Kalia, P. Vashishta and D.W. Brenner (1996) Large Scale Molecular Dynamics Study of Amorphous Carbon and Graphite on Parallel Machines, *Mat. Res. Soc. Symp. Proc.*, **408**, 113.

74. P. Vashishta, R.K. Kalia, G.A. Antonio and I. Ebbsjö (1989) Atomic Correlations and Intermediate Range Order in Molten and Amorphous $GeSe_2$, *Phys. Rev. Lett.*, **62**, 1651.

75. J.P. Rino, G. Gutiérrez, I. Ebbsjö, R.K. Kalia and P. Vashishta (1996) Distribution of Rings and Intermediate Range Correlations in Silica Glass under Pressure - A Molecular Dynamics Study, *Mat. Res. Soc. Symp. Proc.*, **408**, 333.

76. A. Nakano, R.K. Kalia and P. Vashishta (1994) First Sharp Diffraction Peak and Intermediate-range Order in Amorphous Silica: Finite-size Effects in Molecular Dynamics Simulations, *J. Non-Crystalline Solids*, **171**, 157.

77. A.C. Wright (1993) The Comparison of Molecular Dynamics Simulations with Diffraction Experiments, *Journal of Non-Crystalline Solids*, **159**, 264.

78. J.P. Rino, I. Ebbsjö, R.K. Kalia, A. Nakano and P. Vashishta (1993) Structure of Rings in Vitreous SiO_2, *Phys. Rev.*, **B47**, 3053.

79. C. Meade, R.J. Hemley and H.K. Mao (1992) High-Pressure X-Ray Diffraction of SiO_2 Glass, *Phys. Rev. Lett.*, **69**, 1387.

80. W. Jin, R.K. Kalia, P. Vashishta and J.P. Rino (1993) Structural Transformation, Intermediate-Range Order, and Dynamical Behavior, *Phys. Rev. Lett.*, **71**, 3146.

81. W. Jin, R.K. Kalia, P. Vashishta and J.P. Rino (1994) Structural Transformation in Densified Silica Glass: A Molecular-Dynamics Study, *Phys. Rev.*, **B50**, 118.

82. J. Fricke (1990) SiO_2-Aerogels: Modifications and Applications, *Journal of Non-Crystalline Solids*, **121**, 188.

83. J.M. Drake and J. Klafter (1990) Dynamics of Confined Molecular Systems, *Physics Today*, **43**, 46.

84. A. Nakano, R.K. Kalia and P. Vashishta (1994) Growth of Pore Interfaces and Roughness of Fracture Surfaces in Porous Silica: Million Particle Molecular-Dynamics Simulations, *Phys. Rev. Lett.*, **73**, 2336.

85. P. Vashishta, R.K. Kalia and I. Ebbsjö (1995) Low-Energy Floppy Modes in High-Temperature Ceramics, *Phys. Rev. Lett.*, **75**, 858.

86. M. Misawa, T. Fukunaga, K. Niihara, T. Hirai and K. Suzuki (1979) Structural Characterization of CVD Amorphous Si_3N_4 by Pulsed Neutron Total Scattering, *J. Non-Cryst. Sol.*, **34**, 314.

87. P.T.B. Shaffer and A. Goel (1993) Silicon Nitride, Advanced Refractory Technologies Inc., March/93.

88. N. Wada, S.A. Solin, J. Wong and S. Prochazka (1981) Raman and IR Absorption Spectroscopic Studies on α, β, and Amorphous Si_3N_4, *Journal of Non-Crystalline Solids*, **43**, 7.

89. E. Bouchaud and J.-P. Bouchaud (1994) Fracture Surfaces: Apparent Roughness, Relevant Length Scales, and Fracture Toughness, *Phys. Rev.*, **B50**, 17752.

90. E. Bouchaud and S. Navéos (1995) From Quasi-Static to Rapid Fracture, *J. Phys. I France*, **5**, 547.

91. K.J. Måløy, A. Hansen, E.L. Hinrichsen and S. Roux (1992) Experimental Measurements of the Roughness of Brittle Cracks, *Phys. Rev. Lett.*, **68**, 213.

92. K.J. Måløy, A. Hansen, E.L. Henrichsen and S. Roux (1992) Experimental Study of The Geometrical Effects in The Localization of Deformation, *Phys. Rev. Lett.*, **68**, 213.

93. V.Y. Milman, N.A. Stelmashenko and R. Blumenfeld (1994) Fracture Surfaces: A Critical Review of Fractal Studies and a Novel Morphological Analysis of Scanning Tunneling Microscopy Measurements, *Progress in Materials Science*, **38**, 425.

94. V.Y. Milman, R. Blumenfeld, N.A. Stelmashenko and R.C. Ball (1993) Comment on "Experimental Measurements of the Roughness of Brittle Cracks", *Phys. Rev. Lett.*, **71**, 204.

95. A. Nakano, R.K. Kalia and P. Vashishta (1995) Dynamics and Morphology of Brittle Cracks: A Molecular-Dynamics Study of Silicon Nitride, *Phys. Rev. Lett.*, **75**, 3138.

96. S. Somiya and Y. Moriyoshi, Eds., (in)*Sintering Key Papers* (Elsevier Applied Science, London, 1990).

97. R.W. Siegel (1994), in *Physics of New Materials*, F.E. Fujita, Eds., Springer-Verlag, Heidelberg, 65.

98. J. Karch, R. Birringer and H. Gleiter (1987) Ceramics Ductile at Low Temperature, *Nature*, **330**, 556.

99. K. Tsuruta, A. Omeltchenko, R.K. Kalia and P. Vashishta (1996) Early Stages of Sintering of Silicon Nitride Nanoclusters: A Molecular-Dynamics Study on Parallel Machines, *Europhysics Letters*, **33**, 441.

100. R.W. Siegel (1992), in *Materials Interfaces: Atomic-Level Structure and Properties*, D. Wolf and S. Yip, Eds., Chapman and Hall, London, 431.

101. W. Li, R.K. Kalia and P. Vashishta (1996) Dynamical Fracture in $SiSe_2$ Nanowires - A Molecular Dynamics Study, *Europhys. Lett.* (in press).

102. W. Li, R.K. Kalia and P. Vashishta (1996) Amorphization and Fracture in Silicon Diselenide Nanowires - A Molecular Dynamics Study, *Phys. Rev. Lett.* (in press).

103. C.Z. Wang and K.M. Ho (1993) Structure, Dynamics, and Electronic Properties of Diamondlike Amorphous Carbon, *Phys. Rev. Lett.*, **71**, 1184.

104. F. Li and J.S. Lannin (1990) Radial Distribution Function of Amorphous Carbon, *Phys. Rev. Lett.*, **65**, 1905.

105. J.Q. Broughton, P. Vashishta and R.K. Kalia (1995), in *High Performance Computing. Grand Challenges in Computing Simulations*, A. Tentner, Eds., SCS, San Diego, 25.

106. J.Q. Broughton, P. Vashishta and R.K. Kalia (1995), in *Condensed Matter Theory*, E. Ludena, Eds., Plenum, New York, **11**, 218.

107. J.Q. Broughton, C.A. Meli, P. Vashishta and R.K. Kalia (1996) Direct Atomistic Simulation of Quartz Crystal Oscillators. I. Bulk Properties, *Phys. Rev.*, **B** (in press).

REVERSE MONTE CARLO SIMULATIONS FOR DETERMINING DISORDERED STRUCTURES
Basics, and application for amorphous semiconductors

L. PUSZTAI
Studsvik Neutron Research Laboratory (NFL)
University of Uppsala
S-611 82 Nyköping, Sweden
(On leave from: Laboratory of Theoretical Chemistry, L. Eötvös
University, Budapest 112., P.O.B. 32, H-1518, Hungary)

Abstract
The technique of Reverse Monte Carlo (RMC) modelling of structural disorder is introduced. Emphasis is placed on the principles, particularly on the uniqueness of structural models obtained. Some potentialities of RMC, apart from structural modelling, are also mentioned. The structure of amorphous silicon (and, to some extent, a-Ge and tetrahedral a-C) is discussed in some detail.

1. Introduction

The atomic (microscopic) structure of a material is one of its most basic properties. Understanding the structure of a material often provides the key for understanding other, perhaps for most people, more interesting, properties. Trivial examples of the latter can be transport properties, or (chemical) reactivity.

Diffraction methods [1] (neutron, X-ray or electron) are widely used for studying the atomic structure of materials. They have been rather successful for crystalline systems, where the time average of all the atomic positions can be determined, in most cases with high precision. However, for non-crystalline materials in condensed phases, such as the subjects of the present Proceedings, diffraction patterns cannot be directly related to atomic co-ordinates. Although in the case of disordered systems, especially that of liquids, the exact positions of particles may not be relevant, the knowledge of the local environment of particles is of importance. Unfortunately, diffraction studies of non-crystalline materials in themselves can only provide information on the average local structure, via the structure factor, $F(Q)$, and its sine Fourier-transform, the pair correlation function, $g(r)$ [1,2]. In order to obtain deeper insights of the structure, one needs to turn to structural modelling [2].

The most usual means of generating atomistic models of materials is via computer simulation methods [3]. Conventional Monte Carlo (MC) and molecular dynamics (MD) methods need a physical model of the interactions between particles, which are usually in the form of effective interatomic potentials (most commonly, pairwise additive ones), and which are not a trivial task to build in general. In addition, the agreement between simulated and experimental structure factors (or pair correlation functions) is usually only qualitative, apart from a relatively small, although increasing, proportion of simulation studies.

215

M. F. Thorpe and M. I. Mitkova (eds.), Amorphous Insulators and Semiconductors, 215–224.

It is not intended to overemphasise here the importance of quantitative agreement between model and experiment while interpreting diffraction results. Nevertheless, it should be noted that a model that does agree with the experiment within errors is more likely to be close to the real structure than the ones which do not. Using the Reverse Monte Carlo technique seems to guarantee at least this basic (but, naturally, not sufficient) requirement; this is why it is suggested that it has a major role to play in the determination of the structure of amorphous insulators and semiconductors.

2. RMC method

The aim of RMC modelling is to generate three dimensional ensembles of particles that possess structure factors that agree with the experimental structure factor within errors. These errors are supposed to be purely statistical, and to have a normal distribution at each experimental point. A possible algorithm for achieving this goal can be formulated as follows [4]:

1. Start with an initial collections of points, that would represent particles, in a simulation box. The number density in the box should correspond to the real density of scattering centres (as opposed to the macroscopic density of the material). Calculate the pair correlation function, $g^C(r)$, for this configuration, and Fourier transform it to the structure factor, $F^C(Q)$. Calculate the difference, χ^2, between the experimental structure factor, $F^E(Q)$, and $F^C(Q)$. (The superscripts `C' and `E' refer to `calculated' and `experimental', respectively.)
2. Move one atom at random, and calculate $g^C(r)$, then $F^C(Q)$ for the new particle configuration. Take the difference, $\chi^2\,(new)$, of $F^E(Q)$ and the new $F^C(Q)$.
3. If $\chi^2\,(new) < \chi^2$, then accept move immediately, otherwise accept move with a probability that is proportional to a normal distribution.
4. Repeat from 2., until χ^2 is converged.

Note that the above algorithm uses a proper Markov chain, just as conventional Monte Carlo simulations do. Therefore in principle it is possible that during an RMC calculation all the structures (points in the configuration space) would be visited that are consistent with a given experimental structure factor, even if the time necessary for that would be unfeasibly long.

The above scheme considers the case of a one component system with one single measurement; extension for more component systems with the necessary number of independent (sticking to the above formalism: isotopic substitution neutron diffraction) experiments is straightforward. Moreover, it is possible to combine different kinds of experiments: the only requirement is that the measured signal be calculated unequivocally from the positions of particles. At the moment, any number of neutron and x-ray diffraction, and x-ray absorption fine structure (EXAFS) experiments can be modelled simultaneously, as it has been done for AgI-AgPO$_3$ superionic conductor glasses [5].

It should also be noted that RMC, since it calculates (partial) pair correlation function directly from the atomic co-ordinates, is capable of providing a set of partial pair correlation functions even if the number of independent measurement is less than it would be required for the direct separation of the partials. Naturally, not all of these partials would be necessarily meaningful, but in fortunate cases, at least the ones with high relative weights (when composing the total) would be reliable. (Some considerations on the reliability of the partial pair correlation functions obtained this way can be found in Ref. [6]).

Once particle configurations are produced, the next step is to analyse them, so that structural features that are beyond pair correlations could be revealed. This is a very important issue, since in the absence of proper interpretation even the best structural model would be useless. Some general methods are described in Ref. [7], but in many cases, individual tools are necessary (see e.g. Ref. [5]).

There have been many papers published on the details of RMC (see e.g. Refs. [4,7-9]); for a more practical description the reader can also refer to the work of Zotov [10] in these Proceedings, so that in this study no more technicalities will be mentioned. There is one point, however, that concerns the practical usefulness of RMC very much, and that needs some further discussion here.

3. Constraints

In order to find particle arrangements that are consistent with a set of diffraction data, one has to constrain the space of available configurations. If the single constraint applied to the (RMC) simulated system is the input experimental data, that is that the structure factor calculated for the box should be consistent with the measurement within errors, then the range of possible structures is just too large. Using the constraint of the *density of scattering centres* (number density) makes the range considerably smaller. (Note that the data in itself would not define the density unambiguously in most cases, see Ref. [11].)

An additional powerful constraint is that in the majority of calculations the excluded volume effect can be taken into account directly, via imposing *distances of closest approach* between particles. By this constraint the volume of the configuration space available is reduced considerably, in a rather straightforward manner. In the presence of sufficient number (e.g. 3 for a two component system) of independent data sets of good quality, the application of closest approaches simply speeds the calculation up (see, for example, some of the data modelled in Ref. [7]). On the other hand, where the necessary number of independent experiments cannot be carried out (for the lack of suitable isotopes, or, more and more frequently, for the large number of components) then distances of closest approaches have to play the important role of differentiating between different kinds of atoms. This problem is discussed in more detail in Ref. [6].

If some associations of particles (molecules, or well defined covalent networks, for instance) are present in a material, then diffraction data in itself, which contains only a kind of two dimensional projection of the higher order correlation functions, would not be sufficient to uniquely define them. Consequently, RMC, with the above constraints only, would fail to form these units to the extent they are present in reality. Also, RMC tends to give the structure that is nearly the most disordered (or, in other words, that possesses nearly the maximum of configurational entropy) among the ones consistent with an experiment. On the other hand, in many cases the existence of the associations is unquestionable, such as in molecular liquids (or solids), or in network glasses, so that a structural model would not be complete without them, not even if the structure factor of the model would match the experimental one perfectly. After a while it has become evident that these units had to be incorporated in RMC models, and this is why the method of *co-ordination constraints* was developed (see e.g. Ref. [12]). Practical aspects are given in more detail in Ref. [10]. Briefly, co-ordination constraints are implemented in a way that central particles have to have a given number of neighbours within specified distances, and to the specified extent (types of centres and neighbours are also to be specified for multicomponent systems). During an RMC calculation only moves that are consistent with (or leading towards the realisation of) these constraints would be accepted. Co-ordination constraints proved very useful, indeed, for studying materials with covalent networks (see e.g. Ref. [5]), so that without them RMC studies on amorphous semiconductors and insulators would not have been possible.

Applying constraints (including the experimental data) greatly reduces the number of possible structural models, but still not to just one single model. This is why the discussion of the uniqueness of RMC solutions is unavoidable.

4. Uniqueness

The question that has been raised the most frequently about RMC concerns the uniqueness of the particle configurations (structural models) obtained, and can be formulated as follows: how many, and how much, different structures are consistent with a given experimental structure factor? Before going into details, it has to be stated clearly that RMC solutions are *not* unique, but simply structural models that are consistent with a (set of) diffraction experiment(s) *and* additional constraints applied.

Shortly after the appearance of the RMC method the question of uniqueness was discussed, from a theoretical point of view, by Evans [13]. He showed that if pairwise additive interparticle interactions determine the structure of a system entirely, as in the case of conventional MC and MD simulations with pair potentials, then the pair correlation function itself will determine all the higher order correlation functions. That is, the Reverse Monte Carlo technique, that works on the basis of information closely related to the pair correlation function, should *in principle* be unique for systems where there are only pairwise additive interactions present.

Some test RMC calculations, based on resulting pair correlation functions and structure factors of conventional Monte Carlo simulations, have been carried out for hard sphere and Lennard-Jones liquids, and for some molten salts with Coulomb interactions between ions [14,15]. After the RMC calculations, bond angle distribution functions, and other characteristics of higher order correlations, of original (MC simulated) and model (RMC) particle configurations were compared, to see how much of the original features could be retrieved by RMC, based only on the pair correlation function/structure factor. As is evident from Refs. [14,15], in the studied cases the original and model structures were almost indistinguishable at the level of three particle correlations. On the one hand, this means that the RMC method works, since it is consistent with basic statistical mechanical considerations. On the other hand, it was shown that even `unconstrained' RMC, therefore, is able to provide a great deal of information on higher order correlation functions of materials in which pair interactions have a dominant role. Examples of such systems are condensed phases of noble gases [4], or the large family of molten salts [7].

During the studies of Ref. [15], however, it became clear that even the simplest molecules, hard dumbbells, cannot be formed during an RMC calculation on the sole basis of the structure factor of hard dumbbells, unless the molecules have been defined in some ways. More precisely, it was possible to achieve an excellent fit to the model structure factor, and the first co-ordination number has come out as exactly 1, but the system contained not only pairs of atoms, but single atoms and chains, as well. Soon it was evident that in any material where higher order correlations are dominant, and they are important wherever covalent bonding is present, RMC *without constraints* would yield models that would *not* meet even the simplest requirements about apparent regular associations of particles. This is why constraints are indispensable when condensed phases of molecules or covalent networks are dealt with.

Some thoughts about the role of the starting configuration, as far as the problem of uniqueness is concerned: in general, the starting configuration does not have any effect on the final structure of the RMC model (for some tests, see e.g. Ref. [14]). There are, however, some exceptional cases where the unusually high packing fraction, or perhaps the complicated shape of molecules, might effectively prevent the system from exploring configuration space sufficiently. An example for the former case can be the $Ni_{81}B_{19}$ amorphous metallic alloy (or `metallic glass'), where starting from the crystalline structure of $Ni_3B$ retained the layer structure of boron atoms, whereas starting from a random (`hard sphere') configuration did not produce the same kind of layered structure [16]. Note, however, that at the level of local three particle correlations, the models were indistinguishable.

The non-uniqueness of RMC-generated structural models is often mentioned as the greatest disadvantage of the method – sometimes as so great that it would render the RMC method even useless. It should be noted, however, that it is not the method, but in fact the *diffraction data themselves* that (in some cases) allow such diversity of models (see Section 6., where the example of amorphous silicon is considered). RMC has just happened to be the method in connection with this problem, or rather, this feature, of the data could be exposed. It also has to be seen clearly that there are *no techniques* available that would be able to provide the one and only answer to a structural problem – not even the highly sophisticated *ab initio* MD methods [17] are such. All the techniques of structural modelling are only capable of providing models that are consistent with the constraints, like physical models (interatomic potentials, in however elaborated form), applied, and – in fortunate cases – with the diffraction data, as well. RMC is just a method that always ensures this latter – which, naturally, does not automatically mean that a given RMC model is necessarily physically meaningful. For that, all physically meaningful constraints have to be incorporated into the model – and RMC is readily applicable for this, as it has the ability of incorporating interatomic potentials, as well.

In this way, the fact that RMC usually is able to generate different models for a given data, by using different sets of constraints, can be turned to be an advantage: this is the very feature of the method that allows us the exploration of the *range of structural models* that are consistent with a particular diffraction data set. If a given constraint cannot be made consistent with the diffraction data then the model, represented by that constraint, has to be abandoned as unsuitable. Therefore RMC, instead of providing a unique structural model of a material, can be used to select out the structural models that are certainly *not* sensible models of that material.

5. Possible uses of RMC during diffraction data analysis

5.1. A SIMPLE CONSISTENCY CHECK OF MEASURED STRUCTURE FACTORS

An infinitely long Monte Carlo simulation of a system containing an infinite number of perfect gas particles (particles with no volume, and no interaction between particles) would visit every single conceivable particle configuration (every point in the configuration space), including all the configurations that would correspond to the structure factor resulting from any diffraction measurement. Using a particular data

220

set, as a constraint, for the above simulation, would therefore select all the configurations that are consistent with that data. Physically possible configurations could then be obtained by incorporating excluded volume effects in conjunction with the appropriate number density of the system. Consequently, an infinitely long Monte Carlo simulation, applying a structure factor as constraint, with hard sphere particles at the correct density, in an infinitely large box, would produce all the physically meaningful particle arrangements that correspond to the structure factor in question.

The Reverse Monte Carlo method can be viewed as a hard sphere Monte Carlo algorithm, with the experimental data as constraint on the acceptable particle configurations. According to the above, an infinitely long, and infinitely large RMC calculation then should, in principle, produce all the particle configurations that are consistent with a given diffraction experiment. On the other hand, if an experimental structure factor cannot be fitted by such an RMC calculation then that structure factor cannot correspond to a real system [18].

Naturally, infinitely long and infinitely large RMC calculations cannot be carried out in practice. However, for non-crystalline materials, several thousand particles can easily be applied in simulations lasting for several million accepted moves. It is quite possible that if such a practically feasible RMC calculation, particularly if using possible smaller than realistic particles (allowing for an enhanced mobility), cannot fit a particular experimental result within errors, then, according to the above, the data in question will not correspond to any real system. Needless to say, successful fit would not guarantee that the data was perfect – but still, this would be a feasible method for deciding on the self-consistency of measured structure factors.

5.2. PAIR CORRELATION FUNCTIONS FROM STRUCTURE FACTORS MEASURED OVER VERY LIMITED Q-RANGES

If a structure factor is available over only very limited Q-range then the pair correlation function might still be fairly accurately obtainable if, instead of the conventional direct Fourier transform, inverse methods are applied. The most reliable inverse method for this purpose known at present is the RMC technique, as was shown in a number of model calculations, involving also an amorphous silicon like model [19].

The tests that have been carried out followed a rather simple fashion: experimental structure factors (or more precisely, very close RMC fits to experimental structure factors, in order to ensure that the structure factors correspond to physically possible particle arrangements) were taken initially. Then, shorter and shorter versions of them were generated by chopping off roughly 2 Å^{-1} parts from the high Q end of the structure factors. All of the shortened versions were handled as independent `experimental' results, that is, all of them were modelled by RMC, starting off with identical initial configurations (for a given model system), and lasting for identical numbers of accepted moves. Finally, pair correlation functions obtained by using the shortened structure factors were compared to the original g(r)'s. The range of systems tested so far consists of simple liquids (liquid Ga), amorphous covalent networks (a-Si), and two component metallic glasses ($Ni_{62}Nb_{38}$); note that in the last case the separation of the partial pair correlation functions also had to be managed.

The general picture emerging from these tests was, at first sight, quite surprising: in all the studied cases the shortening could go on quite far down in the upper limit of the scattering vector, without the resulting pair correlation functions changing very much. In the case of a-Si, for instance, a g(r) almost identical to the original one could be obtained on the basis of a structure factor extending from 0.5 Å^{-1} up to only 6 Å^{-1}, and comparable ranges were obtained for other test systems, as well [19]. On the other hand, if direct Fourier transformation was used for obtaining the pair correlation function from the same structure factors then these g(r)'s deviated considerably from the original ones if the above Q range was applied [18,19].

It seems that, at least for all the investigated systems, short parts of the full structure factor may be sufficient for reproducing the correct r-space information, provided that appropriate methods are applied for deriving the pair correlation function. This finding reflects the well known fact that each point of the structure factor contains information on the full pair correlation function.

It has to be mentioned that other inverse methods, that avoid the direct Fourier transform from Q space to r space, might also possess similar properties. For instance, the MCGR method [20], which also uses a Monte Carlo method for obtaining the pair correlation function, has also been shown to have this feature, even if to a somewhat lesser extent. On the other hand, MCGR is faster by some orders of magnitude, which, especially for quick tests, might make it then more attractive than RMC.

6. The structure of some tetrahedral amorphous semiconductors, as seen by RMC

There have been a number of RMC studies carried out on a-Si, a-Ge and 'tetrahedral' a-C [21-28]. (For a recent review on the structural modelling of these systems, see Ref. [29].) These materials share very similar structural features, based on a diamond-like tetrahedral network. There are, however, still unresolved problems about the details, concerning mainly the first co-ordination number of atoms and the extent of tetrahedral local ordering. It was these two characteristics that have been investigated in detail by means of the RMC technique. But before going into any details, it should be mentioned that since the preparation of these materials is far from obvious, and also, they are not in thermodynamic equilibrium, the details of their structure may depend on the actual way of their preparation – and some of the confusion about their structure might be explained by this.

The exact value of the first co-ordination number greatly depends on the value of the microscopic density applied during modelling. As it is usually not a straightforward matter to measure the number density, first a consistency check, applying the RMC method, was carried out [11,23]. It was found that the microscopic density of scattering centres for a-Si and for tetrahedral a-C were in agreement with the values that had been found previously, but for a-Ge, a value about 10 % higher appeared to be more satisfactory.

For each material, a series of RMC calculations have been carried out, with varying starting configurations and co-ordination constraints. It was evident that different models could be made equally consistent with the experimental structure factor in each case. However, the main alterations could easily be located: First, the distribution of the number of nearest neighbours was wide, showing the existence of not only three- and four-, but a whole range from zero- to eight-fold co-ordinated atoms, whenever the appropriate co-ordination constraint was absent. Secondly, the fraction of 'hard sphere like' bond angles (angles of about 60°) was always unrealistically high whenever the appropriate constraints *and* a corresponding starting configuration (diamond lattice, or a structure related to the Wooten-Winer-Weaire (WWW) model [30], for a-Si and a-Ge) were not applied simultaneously.

As a test on the flexibility of the data, some unrealistic co-ordination constraints were also applied in a few calculations. The weirdest of all was the one where no fourfold co-ordinated atoms were allowed to occur at all. Such models were created for all of the three materials, with structure factors that were in excellent agreement with the experimental data. This experience should give us a warning against overinterpreting results of diffraction experiments, whichever method is chosen for modelling the structure [23].

There was one feature, nevertheless, that was not possible to exclude from the models, and that was the always (even in the above case, with no fourfold co-ordinated atoms present) high (at least 40 %, but could easily be made 100 %) fraction of tetrahedral bond angles. This, on the one hand, indicates that previous ideas about the structure of these materials had grossly been correct, and on the other hand, that it is, indeed, possible to infer basic structural features from diffraction data, even by unconstrained RMC. The exact fraction of tetrahedral angles, however, cannot be determined on the sole basis of diffraction data. Similar conclusion can be drawn concerning the first co-ordination number, with the addition that diffraction data would allow for an even greater flexibility, the fraction of exactly fourfold co-ordinated being allowed to fall between 0 and 100 % [23].

An interesting result of the above studies was that they revealed some apparent differences between the structures a-Si and a-Ge, and a-C. Whereas the structures of the former two are nearly identical, apart from a scaling factor originated to the size difference between Si and Ge atoms, the structure factor of tetrahedral a-C could not be made consistent either with 100 % fourfold co-ordination or with a fraction of 100 % of tetrahedral bond angles. It seemed to be necessary to actually break the diamond network for fitting the data on tetrahedral a-C. This could well be the consequence of the material being not purely 'diamond-like', but containing also regions of 'graphite-like' particle arrangements.

Several attempts have been made for further refining the details of the structure of amorphous Si [24,26], mostly by calculating quantities related to the full three particle correlation function. Although the information obtained via these routes proved valuable, widespread applications would probably be barred by the high computational costs. On the basis of the three body correlation function of different models, it was suggested [26] that possibly the most reliable model of amorphous silicon is the result of the RMC calculation that started from the WWW-model [30], and was conducted by requiring a 100 % proportion of fourfold co-ordinated particles.

7. Summary – why use RMC?

There are several advantages of using the Reverse Monte Carlo method for modelling structurally disordered materials:

- the full range of experimental (mostly diffraction) data is made use of, in a quantitative manner;
- the application of interatomic potentials is not necessary for producing structural models (although they can be used, if required);
- structural models provided by RMC are self-consistent, and they always correspond to physically possible structures; also, they possess the correct density of scattering centres;

- different types of data (neutron-, x-ray diffraction, and EXAFS, at present) can be combined simultaneously;
- via appropriately chosen geometrical constraints, many other types of data (e.g. NMR) can also be incorporated in the models;
- RMC can be useful while deriving the pair correlation function(s): for instance, data available only over limited Q-ranges can be used for obtaining reliable g(r); or multicomponent systems can be modelled, and, in fortunate cases, partial pair correlation functions of them can be determined, even if the number of independent experiments is less than it would be required for conventional data analysis.

However, it should be noted that apart from its conceptional simplicity, RMC cannot be used as a simple 'black box' software, since many technical details (even simple ones, like number of particles vs. r-spacing, or the choice of closest approaches, for instance) may have unexpectedly deep consequences when interpreting the structure.

It should also be stressed that wherever it is possible to describe the experimental structure factor quantitatively via standard computer simulation techniques, using interatomic potentials, these models are, naturally, superior to the ones generated by RMC, since RMC does not allow for modelling time dependent properties directly.

Finally, it has to be emphasised that RMC is a practical tool for structural modelling, or sometimes, for helping the analysis of diffraction data prior to the stage of structural modelling, but in no circumstances it should be considered as a theory. (Even if there are obvious analogies to, for instance, the standard Monte Carlo method.) As is shown by the large number of successful applications, it can be extremely helpful when other methods are not applicable, but still, it is entirely empirical. This is the reason why applications to any new type of problem, just as described in Section 5.2., for example, must be experimented beforehand on a wide range of model systems .

References

1. see e.g. Atkins, P.W. (1994) *Physical Chemistry*, Longman, London.
2. see e.g. Elliott, S.W. (1993) *The Physics of Amorphous Materials*, Longman, London.
3. Allen, M.P. and Tildesley, D.J. (1987) *Computer Simulation of Liquids*, Clarendon Press, Oxford.
4. McGreevy, R.L. and Pusztai, L. (1988) Reverse Monte Carlo simulation: A new technique for the determination of disordered structures, *Molec. Simul.* 1, 359
5. Wicks, J.D., Börjesson, L., Bushnell-Wye, G., Howells, W.S. and McGreevy, R.L. (1995) Structure and ionic conduction in $(AgI)_x(AgPO_3)_{1-x}$ glasses, *Phys. Rev. Lett.* 74, 726
6. Pusztai, L. (1995) Partial pair correlation functions of some Ge-Sb liquid alloys from single diffraction measurements, *ACH – Models in Chemistry* 132, 99
7. McGreevy, R.L. and Pusztai, L. (1990) The structure of molten salts, *Proc. Roy. Soc. Lond. A* 430, 241
8. McGreevy, R.L. and Howe, M.A. (1992) RMC: Modelling disordered structures, *Annu. Rev. Mater. Sci.* 22, 217
9. McGreevy, R.L. (1995) RMC: Progress, problems and prospects, *Nucl. Inst. Meth. A* 354, 1
10. Zotov, N. (1996), these Proceedings
11. Gereben, O. and Pusztai, L. (1996) The determination of the microscopic density in liquids and other disordered materials using Reverse Monte Carlo simulation, *Phys. Chem. Liq.* 31, 159
12. Howe, M.A., McGreevy, R.L., Pusztai, L. and Borzsák, I. (1993) Determination of three body correlations in simple liquids by RMC modelling of diffraction data. II. Elemental liquids, *Phys. Chem. Liq.* 25, 205
13. Evans, R. (1990) Some remarks on the Reverse Monte Carlo simulation, *Molec. Simul.* 4, 409
14. Pusztai, L. and Tóth, G. (1991) On the uniqueness of the Reverse Monte Carlo simulation. I. Simple liquids, partial radial distribution functions, *J. Chem. Phys.* 94, 3042
15. McGreevy, R.L. and Howe, M.A. (1991) Determination of three body correlations in simple liquids by RMC modelling of diffraction data. I. Theoretical tests, *Phys. Chem. Liq.* 24, 1
16. Iparraguirre, E.W., Sietsma, J., Thijsse, B.J. and Pusztai, L. (1993) A Reverse Monte Carlo study of amorphous $Ni_{81}B_{19}$, *Computational Materials Sci.* 1, 110

17. Car, R. and Parrinello, M. (1985) Unified approach for molecular dynamics and density functional theory, *Phys. Rev. Lett.* **55**, 2471; for an up-to-date application, see Benoit, M., Bernasconi, M., Focher, P. and Parrinello, M. (1996) New High Pressure Phase of Ice, *Phys. Rev. Lett.* **76**, 2934

18. Gereben, O., Pusztai, L. and Baranyai, A. (1995) Some remarks on the measured structure factor, *Physica Scripta* **T57**, 69

19. Gereben, O. and Pusztai, L. (1995) Determination of the atomic structure of disordered systems on the basis of limited Q-space information, *Phys. Rev. B* **51**, 5768

20. Howe, M.A., McGreevy, R.L. and Pusztai, L. (1995) MCGR manual (pre-print), Studsvik NFL (Copies of the manual and the computer programme are available from the author on request, at: *laci@studsvik.uu.se*); the basis of MCGR is the MCGOFR method, see Soper, A.K. (1989) Extracting the Pair Correlation Function from Structure Factor Data by Monte Carlo Simulation, *Static and Dynamic Properties of Liquids (eds.: Davidovic, M. and Soper, A.K.), Springer Proceedings in Physics* **40**, 189

21. Kugler, S., Pusztai, L., Rosta, L., Chieux, P. and Bellisent, R. (1993) Structure of evaporated pure amorphous silicon: Neutron diffraction and Reverse Monte Carlo investigations, *Phys. Rev. B* **48**, 7685

22. Pusztai, L. and Kugler, S. (1993) Reverse Monte Carlo simulation: the structure of amorphous silicon, *J. Non-Cryst. Solids* **164-166**, 147

23. Gereben, O. and Pusztai, L. (1994) Structure of amorphous semiconductors: Reverse Monte Carlo studies on a-C, a-Si and a-Ge, *Phys. Rev. B* **50**, 14136

24. Gereben, O., Pusztai, L. and Baranyai, A. (1994) Calculation of the three-particle contribution to the configurational entropy for two different models of amorphous Si, *Phys. Rev. B* **49**, 13251

25. Pető, G., Horváth, Zs. F., Gereben, O., Pusztai, L., Hajdú, F. and Sváb, E. (1994) Implantation-induced structural changes in evaporated amorphous Ge, *Phys. Rev. B* **50**, 539

26. Gereben, O., Pusztai, L. and Baranyai, A. (1994) Possible quantitative measures of order/disorder in models of liquid and amorphous structures, *J. Phys.: Condens. Matter* **6**, 10939

27. Walters, J.K and Newport, R.J. (1996) Reverse Monte Carlo modelling of amorphous germanium, *Phys. Rev. B* **53**, 2405

28. Gilkes, K.W.R., Gaskell, P.H. and Robertson, J. (1995) Comparison of neutron-scattering data for tetrahedral amorphous carbon with structural models, *Phys. Rev. B* **51**, 12303

29. Pusztai, L. (1995) On the atomic structure of amorphous semiconductors, *Solid State Phenomena* **44-46**, 25

30. Wooten, F, Winer, K. and Weaire, D. (1985) Computer generation of structural models of amorphous Si and Ge, *Phys. Rev. Lett.* **54**, 1392

REVERSE MONTE CARLO SIMULATIONS OF GLASSES - PRACTICAL ASPECTS

N. ZOTOV*

Departamento de Fisica de la Materia Condensada

University of Cadiz, 11510 Puerto Real (Cadiz), Spain

1. Introduction

The structure of glasses is usually described in terms of the pair distribution functions $g_{ij}(r)$, which give the probability to find an atom of type j at a distance r from arbitrary chosen atom of type i. The pair distribution functions represent one-dimensional projections of the three dimensional structure of the glass. Therefore, large number of different structural models would be in general consistent with the experimental $g(r)$ functions, measured by neutron and/or X-ray diffraction.

Numerous methods for modelling the structure of amorphous materials have been proposed in the literature [1]. The relatively novel Reverse Monte Carlo (RMC) method [2] has several advantages over other methods. The number of atoms is usually much larger than for hand-build models. No interaction potentials are needed, the choice of which in the molecular dynamics of multi-component covalent glasses is not strightforward. The RMC method generates atomic configurations by fitting the whole data set(s) and not just interpreting particular features like the area under the first peak in the total radial distribution function (RDF) as in the quasi-crystalline methods.

Permanent Address: Central Laboratory of Mineralogy and Crystallography,

Bulgarian Academy of Sciences, Rakovski str. 92, Sofia 1000, Bulgaria

M. F. Thorpe and M. I. Mitkova (eds.), Amorphous Insulators and Semiconductors, 225–234.
© 1997 *Kluwer Academic Publishers. Printed in the Netherlands.*

While the general principles of the RMC method are well described in the literature [2,3], the proper choice of fitting parameters and constraints, especially for covalent glasses are still not well understood, partially because they depend to some extend on the system studied.

The RMC algorithm represents a non-linear fit of the pair distribution function of the model to the experimental data, which inevitably contain statistical errors and even can contain some residual systematic errors. Therefore, the range of the possible structural models, generated by the RMC method, will depend on the information content of the experimental data. Besides that, the presence and the amount of statistical errors not only determine the quality of the final fit but also influence the choice of the fitting parameters.

Therefore, the aim of the present paper is to give some practical considerations for the application of the RMC simulations which are based on the work done recently on ternary chalcogenide glasses at the Department of Condensed Matter Physics of the University of Cadiz, Spain using the program RMCA [4].

2. The RMC algorithm

2.1 PAIR DISTRIBUTION FUNCTIONS

The experimentally measured total pair distribution function g(r) for an n-component glass can be written in the form:

$$g(r) - 1 = \sum_{i=1}^{n} \sum_{j=1}^{n} P_{ij}(r) * (g_{ij}(r) - 1) \tag{1}$$

The partial pair distribution functions $g_{ij}(r)$ represent the Fourier transform of the corresponding partial structure factors $F_{ij}(Q)$:

$$g_{ij}(r) - 1 = \frac{1}{2\pi^2 \rho_o r} \int_0^{\infty} Q \, (F_{ij}(Q) - 1) \, sin(Qr) \, dQ \tag{2}$$

where ρ_o is the total number density of atoms in the system, $Q = 4\pi sin(\Theta)/\lambda$ is the scattering vector, Θ is the diffraction angle and λ is the wavelength used. The total structure factor, F(Q) is a linear sum of all partial structure factors

$$F(Q) = \sum_{i=1}^{n} \sum_{j=1}^{n} \gamma_{ij}(Q) \, (F_{ij}(Q) - 1) \tag{3}$$

where the weighting coefficients $\gamma_{ij}(Q) = c_i c_j f_i f_j / <f>^2$ depend on the atomic concentrations (c_i) and the scattering form-factors (f_i) of the atoms. The resolution functions $P_{ij}(r)$ in Eq. (1) represent the Fourier transform of the coefficients $\gamma_{ij}(Q)$ multiplied by the box function M(Q), defined to be unity for $Q < Q_{max}$ and zero for $Q > Q_{max}$, where Q_{max} is the maximum measured Q vector. In the case of X-ray diffraction the weighting coefficients $\gamma_{ij}(Q)$ depend on the scattering vector and it is highly recommended to perform a final fit using directly the structure factors.

2.2 STARTING CONFIGURATION

The initial atomic configuration with periodic boundary conditions can be a random ensemble of atoms or previous model generated by molecular dynamics, energy minimization, etc. or even a crystalline structure. In general, the RMC algorithm uses a weighted sampling procedure known as a Markovian chain [2], so that the "final" configuration must be independent of the initial configuartion.

However, in practice, we have established that the time necessary to obtain a fit at a given statistical level depends on the size and to some extend on the type of the initial model. The size of the models is defined in terms of the number of atoms N = ΣN_i, (where N_i is the number of atoms of type i in the configuration) and the length of the box edge L. Although they are related through the atomic number density ρ_o, N = $\rho_o L^3$, there are seperate arguments for the choice of N and L. Evidently, for multi-component glasses in which some of the atoms have low concentration, the size of the model must be large enough so that $(N_i)_{min}$ can be statistically significant.

The size of the models depends also on the amount of structural oscillations at high r-values. In the RMCA program [4] it is assumed that $g(r) \equiv 1$ for $r > L/2$ and as a rule of thumb L can be chosen according to the condition $L \geq 2\,L_c$, where the overall correlation length L_c is defined so that $g(r) \approx 1$, for $r > L_c$.

Finally, the total time necessary to obtain a good fit also depends on N. RMC simulations of the structure of the $Cu_{0.20}As_{0.26}Te_{0.54}$ chalcogenide glass using only one g(r) function, measured by neutron diffraction, indicate that the time necessary to reach structural equilibrium increases as N^2 (Fig. 1)

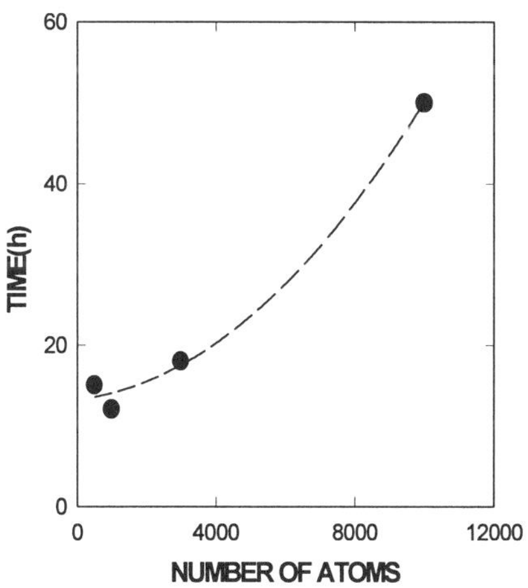

Figure 1

228

Therefore, the final choice of the size of the model is a compromise between the statistical accuracy desired and the computational time required.

2.3 REFINEMENT

The refinement of the initial configuration includes several steps:

Step 1: Using the starting configuration, the initial pair distribution function $g_c(r)$ or structute factor $F_c(Q)$ are calculated;

Step 2: A goodness-of-fit factor χ^2, is calculated then by

$$\chi^2 = \sum_e \frac{1}{\sigma_e^2} \sum_m \left(A_c(m) - A_o^e(m) \right)^2 \tag{4}$$

where A denotes g(r) and/or F(Q), σ_e is the standard deviation of the experimental data set e and the sum is over all measured data points (m) and all data sets (e).

Step 3: An atom is moved at random from it's initial position $\mathbf{r}_i$ to $\mathbf{r}_i + \mathbf{r}_\Delta$ and the new A_c and χ^2 (χ^2_n) are calculated;

Step 4: If $\chi^2_n < \chi^2_o$ the move is accepted, since it improves the fit. If the new value χ^2_n is larger than the old value χ^2_o , the move is accepted with probability $p = \exp(-\frac{1}{2}(\chi^2_o - \chi^2_n))$. Replacing χ^2 by U/kT, where U is the total potential energy of the system, the similarity with the standard Metropolis Monte Carlo method becomes obvious.

Repeating steps 3 and 4 many times the system will finally reach a state in which the goodness-of-fit factor will start to oscillate around a given value, which will depend on σ_e, without further significant improvement of the fit. This state will be referred to as structural equilibrium.

The choice of the standard deviations σ_e and the magnitude of the maximum random moves Δ as well as the use of constraints will be discussed in the next section on the example of the ternary $Cu_{0.20}As_{0.26}Te_{0.54}$ chalcogenide glass prepared by melt quenching in ice water. The X-ray diffraction mesurements were performed in the range $0.77\text{-}16\text{Å}^{-1}$ using Mo K_α radiation on a Siemens D500 automatic powder diffractometer equipped with diffracted beam monochromator. The neutron diffraction measurements were performed at the neutron spallation source ISIS using the LAD time-of-flight diffractometer. Further experimental details are given elsewhere [5].

3. Choice of fitting parameters

3.1 STANDARD DEVIATIONS

In the case of single-wavelength X-ray or neutron diffraction experiment the standard deviation $\sigma(F(Q))$ can be calculated by:

$$\sigma^2(F(Q)) \;=\; \frac{\beta^2 A^2 P^2}{\langle f \rangle^4}\left(\sigma_I^2 \;+\; \sigma_B^2\right) \tag{5}$$

where β is the normalization constant, $A(Q)$ is the absorption and multiple scattering correction, $P(Q)$ is the polarization correction, σ_I and σ_B are the standard deviations of the raw intensity and the background scattering, respectively. Small steps ΔQ (high resolution in reciprocal space) and small standard deviations for all Q values, are necessary when a fit is made in r-space, because the statistical errors in $F(Q)$ are redistributed in $g(r)$:

$$\sigma^2(g(r)) \;=\; \left(\frac{\Delta Q}{2\pi^2 \rho_o r}\right)^2 \sum_Q \sigma^2(F(Q))\, sin^2(Qr) \tag{6}$$

Since a uniform error distribution is used in the RMCA program, the standard deviations, calculated from Eq. (4) or Eq. (5) can be averaged to give an initial value for σ_e. Taking into account that Eq. (4) and Eq. (5) do not include all possible sources of errors, we have found that the best fitting strategy is to start with the calculated σ_e values and then to reduce gradually σ_e until a satisfactory fit is obtained.

3.2 MAXIMUM RANDOM MOVES

In a glass containing n different types of atoms, each of them can be separately moved at random. Therefore, n different maximum random moves must be specified. The main factor, which influences their choice is the number of accepted moves. It was established that the number of accepted moves increases exponentially with decreasing Δ (Fig. 2). In other words, a decrease of Δ increases the rate of sampling the configurational space and thus decreases the total time necessary to reach equilibrium.

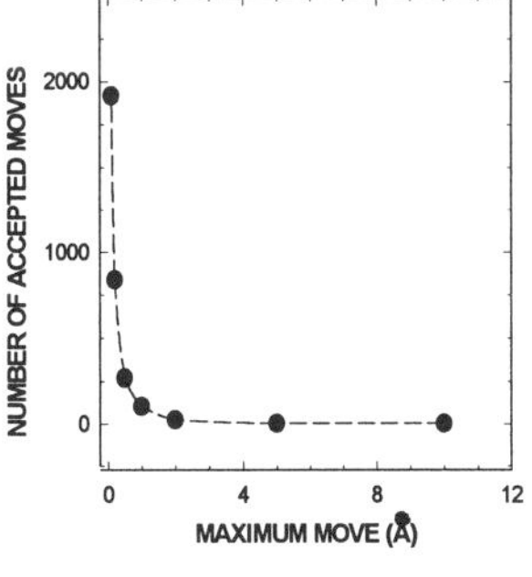

Figure. 2

230

However, there are at least two cases when large Δ ($\Delta \approx L/2$) can be applied for some time:(i) if the system has fallen in a local minimum or (ii) at the initial stages of constrained refinements, especially when starting from a random initial model.

3.3 CONSTRAINTS

In a glass containing n different atoms, there will be $N_c = n (n + 1) /2$ different partial pair distribution functions (PPDF), which evidently cannot be determined from single-wavelength experiment. For binary compositions ($N_c = 3$) the three PPDF can be, in principle, determined from isotopic substitution neutron diffraction measurements and then used for RMC simulations (see Ref. [3] and references therein). However, partial structure factors are unavailable for large number of important multi-component glasses. Nevertheless, the use of m (m $\leq$ n) different diffraction data sets can reduce substantially the range of possible structural models, because the PPDFs of the configuration must satisfy simultaneously m equations of type (1) with different coefficients $\gamma_{ij}(Q)$.

The second type of constraints that can be used in the RMCA [4] program, is related to the area under a given peak in the total radial distrubution function $4\pi\rho_o r^2 g_{ij}(r)$ which is equal to the average number of atoms of type j around atoms of type i. The importance of these coordination constraints arises from the fact that the average coordination numbers can be determined independently by other methods like Extended X-ray Absorption Fine Structure (EXAFS), Nuclear Magnetic Resonance (NMR) or vibrational spectroscopy. A modification of the average coordination number constraints can be used in which we allow only a given fraction of the atoms i to have given coordination. This type of constraint would be very useful, for example in alkali oxide - silicate glasses, where the fraction of the different Q^n-species (Q^n denotes a Si tetrahedron with n bridging oxygen atoms) is accurately known from NMR measurements [6]. The limits R_1 and R_2, which define the range of the selected RDF peak, can be determined, for example, by Gaussian fitting. Since g(r) depends on the resolution function the best would be to fit the RDF peak performing at each cycle a convolution with the resolution function of the diffractometer.

The low-r limit of the first RDF peak, R_{min} , below which there are no structural oscillations, reflects the fact that any two atoms cannot get closer than R_{min}. This idea is exploited in the so-called cut-off constraints, assuming that for all PPDF there exist N_c, in general different, hard-sphere distances r_c. They can be estimated from the sum of the corresponding ionic or covalent radii (r_i) as well as from the shortest nearest-neighbour distances (r_{NN}) in compositionally related crystalline compounds (see Table 1). Comparing these data with the position and the full width at half maximum (FWHM) of the first RDF peak it is also possible to draw conclusions whether models with random bonding or with *chemical* short-range ordering would better describe the structure of the investigated glass.

Table 1

Atom Pair	Cu-Cu	Cu-As	Cu-Te	As-As	As-Te	Te-Te
Σr_i	2.34	2.36	2.52	2.38	2.54	2.70
r_{NN}	2.40	2.38	2.65	2.44	2.78	2.82

In the case of the Cu-As-Te glass, the data in Table 1 suggests that all types of bonds are possible in the range of the first RDF peak between 2.2 and 3.1Å. That is why we have performed combined X-ray and neutron RMC simulations using only cut-off constraints with r_c = 2.2Å for all atoms pairs. The final RMC fits to the corresponding structure factors, using a random starting configuration of 966 atoms (L = 29.9Å) and σ_N = σ_X = 0.005 are shown in Fig. 3. The combined RMC refinements lead to significant changes especially in shape of the Te-Te PPDF in comparison with the separate X-ray or neutron RMC simulations.

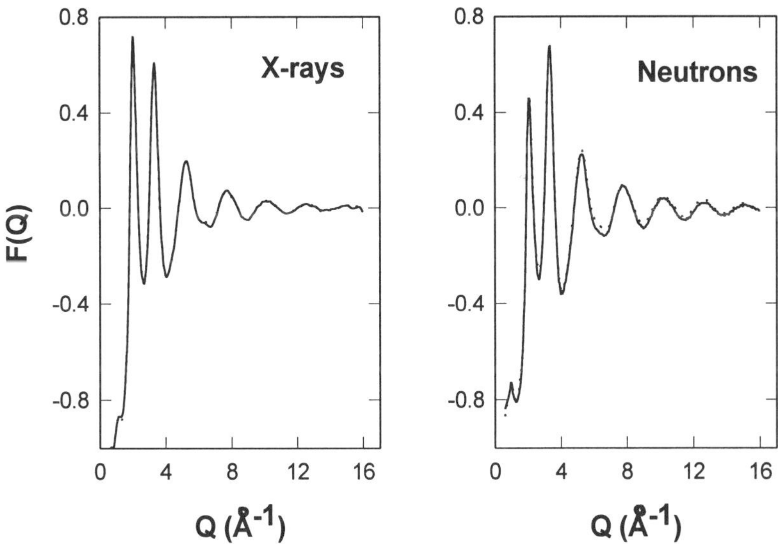

Figure. 3 RMC fits (dotted lines) to the X-ray and neuton diffraction structure factors (full lines) of the ternary chalcogenide glass $Cu_{0.20}As_{0.26}Te_{0.54}$.. Goodness-of fit-factors χ_X^2 = 1.3, χ_N^2 = 8.8, respectively.

4. Analysis of the models

The RMC models contain normally several thousands of atoms and models with up to 30000 atoms are reported in the literature [2]. The geometrical analysis of such large ensembles of atoms is very difficult and is made either qualitatively from a topological point of view or in a purely statistical manner.

4.1 TOPOLOGICAL ANALYSIS OF THE MODELS

Three different r-scales are commonly used to describe the local order in glasses: (i) short-range order (typically 1-3Å scale); (ii) medium-range order (3-10Å scale) and global range order ($\geq$ 10Å).

The character of the SRO can be determined by identifying the typical 'molecular-like' units (tetrahedra, pyramids, etc.) from which the atomic configuration is build. Going to the 3-10Å scale one could try to distinguish some secondary structural units, formed by the coordination polyhedra, like chains, rings, layers etc. Finally, considering the glass network as a whole, it is instructive to analyse the connectivity of the network, the presence of "rigid" and "floppy" regions and the distribution of voids. This is especially important in order to correlate the results obtained with theoretical predictions based on the theory of continuous deformations of glassy networks [7], for example.

4.2 STATISTICAL ANALYSIS OF THE MODELS

The relative abundance of different coordination polyhedra can be quantified by determining the coodination number distributions, from which the average coordination numbers $\langle n_i^j \rangle$ and their standard deviations can be calculated. The estimated standard deviations do not practically depend on the number of atoms N in the model (see Table 2).

Table 2

Number of Atoms	$\langle n_{Cu}^{Cu} \rangle$	$\langle n_{Cu} \rangle$
500	0.7±0.8	3.7±1.4
1000	0.7±0.8	3.8±1.3
3000	0.6±0.7	3.7±1.3
10000	0.7±0.7	3.8±1.3

Additional fractional coordination number constraints must be imposed in order to reduce the FWHM of the coordination number distribution. Analysis of the coordination number distributions should be done in the process of the RMC simulations in order to exclude the formation of under- or overcoordinated atoms, if necessary.

The coodination number distributions characterize the SRO of the network. An important characteristic of the models, containing information about the medium-range order which cannot be determined from the RDFs, are the bond angle distributions. The noise in the bond angle distributions, as in the PPDF, decreases with increasing the size of the model and the short-range ordering. Very often the bond angle distributions exhibit some sharp features, which indicate the presence of specific coordination polyhedra. That is why we have found very instructive to plot the average bond angle distribution separately for the different types of coordination polyhedra (Fig. 4). Such plots give a better idea for the nature and the amount of angular distortions in the network.

In principle, the bond angle distributions can be used also as additional constraints, for example requiring a given bond angle distribution to have a predefined FWHM [8].

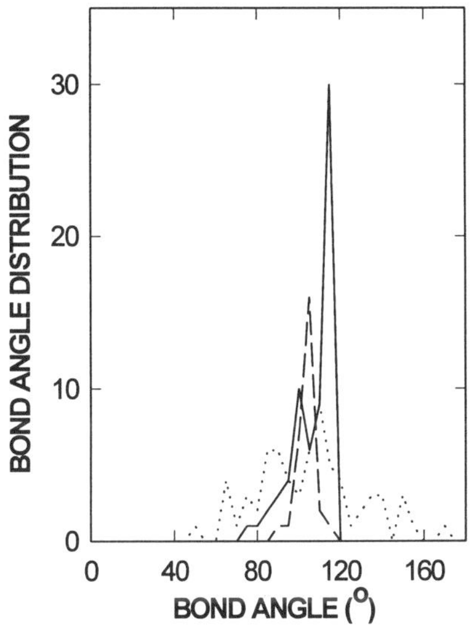

Figure 4. Te-Cu-Te bond angle distribution for 2 (···),

3 (—) and 4 (---) coordinated copper atoms.

5. Concluding remarks

Despite the apparent simplicity of the method, the application of the RMC simulations is not straightforward and requires: (i) The influence of the Q-space and r-space resolution functions as well as the statistical errors in the experimental data to be taken carefully into account; (ii) The standard deviations (σ) and the maximum random moves (Δ) as well as the coordination constraints to be regarded as additional parameters of the model rather than as predefined values and (iii) The topological and statistical analysis of the models, providing various means for internal check of their consistency, to be regarded as an integral part of the RMC simulations.

234

Acknowledgements: This work was done with the financial support of the Spanish Interministerial Commission on Science and Technology for N. Zotov.It is a pleasure to thank Dr. F. Bellido and Prof. Dr. R. Jimenez-Garay for the help in the preparation of the Cu-As-Te sample as well as for the helpful discussions. The assistance of the staff of the neutron spalation source ISIS, UK, and especially Dr. A. Hannon, during the neutron diffraction measurements is highly appreciated. The technical support of the Computer Centre of the University of Cadiz is also acknowledged.

6. References

1. Elliot, S.R. (1990) *Physics of amorphous materials*, Longman, Harlow

2. McGreevy, R.L. and Pusztai, L. (1988) Reverse Monte Carlo simulation: a new technique for the determination of disordered structures, *Mol. Simulations* **1**, 359-367

3. McGreevy, R.L. and Howe, M. (1992) Modelling disordered structures, *Ann. Rev. Mat. Sci.* **22**, 217-242

4. McGreevy, R.L., Howe, M.A., Wicks, J.D. (1993) RMCA-A General Purpose Reverse Monte Carlo Code

5. ISIS Annual Report (1996), Experiment RB 6975, in press

6. Sprenger, D., Bach, H., Meisel, W., Gütlich, P. (1993) Discrete bond model (DBM) of sodium silicate glasses derived from XPS, Raman and NMR measurements, *J. Non-Cryst. Solids*, **159**, 187-203

7. Thorpe, M.F. (1983) Continuous deformations in random networks, *J. Non-Cryst. Solids* **57**, 355-370

8. Pusztai, L. and Kugler, S. (1993) Reverse Monte Carlo simulation: the structure of amorphous silicon, *J. Non-Cryst. Solids* **164-166**, 147-150

A THERMODYNAMIC MODEL FOR THE CALCULATION OF THE PHYSICAL PROPERTIES AND STRUCTURAL CHARACTERISTICS OF GLASSES AND MELTS

Natalia M. VEDISHCHEVA and Boris A. SHAKHMATKIN
*Institute of Silicate Chemistry of the Russian Academy of Sciences,
Odoevskogo Str. 24/2, St. Petersburg, 199155, Russia.*

and

Adrian C. WRIGHT
*J.J. Thomson Physical Laboratory, University of Reading,
Whiteknights, Reading, RG6 6AF, U.K.*

1. Introduction

Since the mid-1970s, systematic studies of the thermodynamic properties of glass-forming systems (melts, glasses and crystals) have been performed using a variety of techniques. As a result, extensive information on thermodynamic potentials has been accumulated for such binary and ternary systems as M_2O-B_2O_3 (M=Li, Na, K, Rb and Cs), MO-B_2O_3 (M=Mg, Ca, Sr and Ba), M_2O-SiO_2 (M=Li and Na), Na_2O-GeO_2, M_2O-B_2O_3-SiO_2 (M=Li and Na) and M_2O-Al_2O_3-SiO_2 (M=Li and Na). The analysis of the experimental results and the available literature data has led to a search for a model that could adequately describe the properties of these systems, using the minimum number of physically meaningful assumptions.

The model of associated solutions [1-4], based on the rigorous approach of De Donder, describes systems formed from components with different chemical natures and is valid for systems with any number of components. According to this model, which has no adjustable parameters, glasses and melts are considered as solutions formed from (unreacted) oxide components and the salt-like products of their interaction, the latter being similar in their stoichiometry to the crystalline compounds which exist in the phase diagram of the system in question. Their structural similarity may also be assumed. When applied to oxide systems, the model has the following advantages: (1) the salt-like products forming the solutions in these systems are strictly defined according to their phase diagrams and (2) reliable information is available on the Gibbs free energies of formation for many crystalline compounds in a range of different systems, which allows the use of adjustable parameters to be avoided.

A system, which is single-phase over wide concentration and temperature regions, can be described using the ideal approximation of the model. This implies that the salt-

235

M. F. Thorpe and M. I. Mitkova (eds.), Amorphous Insulators and Semiconductors, 235–243.
© 1997 *Kluwer Academic Publishers. Printed in the Netherlands.*

like products and the unreacted oxides form an ideal solution. As shown in previous publications [1,2,4], this approximation is valid for a large variety of the systems that form the basis of many industrial glasses (alkali borates, silicates, germanates, phosphates, etc.). If the system in question is phase-separated, the immiscibility of the species forming the solutions presents no special problem and can be taken into account using the approximation of regular associated solutions.

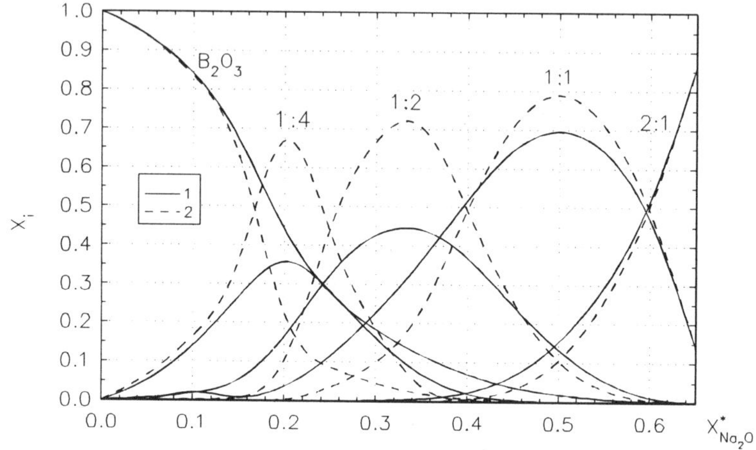

Fig. 1. The chemical structure of sodium borate melts. The solid lines are for a temperature of 1473 K and the dashed lines 1073 K. The sodium borate species are denoted m:n (i.e. $mNa_2O \cdot nB_2O_3$).

2. Basis of the Model

A full derivation of the formalism of the model of associated solutions has been given elsewhere [4] and will not be repeated here. It consists of solving the system of equations for the mass balance of the components and the equations for the law of mass action, written for all the reactions proceeding in the system considered. At constant temperature and pressure, this system of equations has only one solution, which provides information on the chemical structure of glasses and melts of given compositions. The notion of the chemical structure implies the relative content, in a particular glass or melt, of the salt-like products (groupings) and the unreacted oxides. The relative content is represented by the number of moles or mole fraction, X_i, of each species. An example is given in Fig. 1, which shows the chemical structure of sodium borate melts at temperatures of 1073 and 1473 K. The species denoted m:n are the salt-like groupings of the same stoichiometry ($mNa_2O \cdot nB_2O_3$) as the crystalline compounds which occur in the Na_2O-B_2O_3 system. A knowledge of the chemical structure enables a comprehensive description to be made of the various properties of glasses and melts, as will now be demonstrated. In addition, if all the crystal structures are known, this information can also be used to extract structural parameters, such as the distribution of superstructural units (cf. Section 5). Hence both properties and structure are considered on a unified basis.

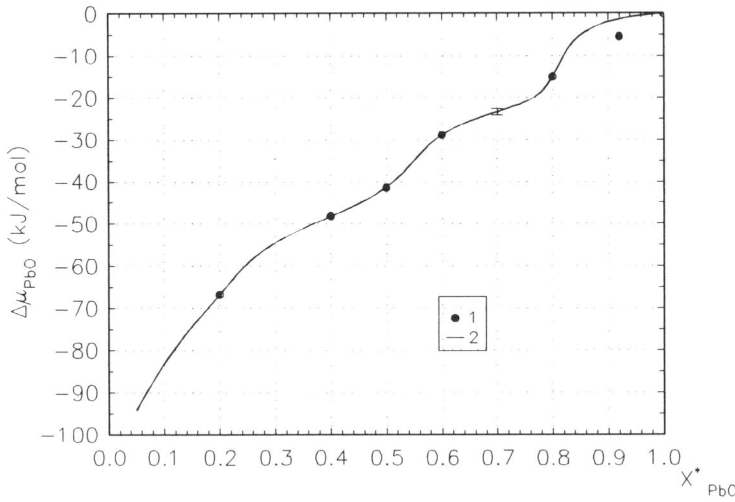

Fig. 2. The chemical potential of PbO in liquid PbO-B_2O_3 at 1373 K. Solid line, model and circles, experiment. (After [7].)

3. Thermodynamic Properties

Using information on the numbers of moles of the unreacted oxides, together with well-known thermodynamic relationships, it is possible to determine the entire set of thermodynamic functions for a glass or melt of a given composition, viz. the partial potentials of the components, $\Delta\mu_i$, $\Delta\overline{H}_i$ and $\Delta\overline{S}_i$, and the integral potentials of mixing, ΔG^{mix}, ΔH^{mix} and ΔS^{mix}.

The thermodynamic potentials have been modelled for a wide range of systems, including alkali borates [2,3,5-7], alkali silicates [1,5], sodium and zinc phosphates, lead borates [7] and lithium borosilicates [8]. All of the modelled values are in good agreement with the experimental data, the uncertainty for the former being in many cases less than that for the latter. Thus the model of associated solutions provides a useful check of the reliability of the experimental data available for various glass-forming systems. This is illustrated by Fig. 2, where the chemical potential of PbO in lead borate melts is given for a temperature of 1373 K. The complex step-like shape of the curve may be explained by changes in the structural motifs existing in the melts as a function of composition, as illustrated in Fig. 1 for the sodium borate system. The

238

deviation of the highest PbO content point from the line is due to uncertainty in the experimental data.

4. Physical Properties

As follows from the model of ideal associated solutions, the expression for the Gibbs free energy of mixing, ΔG^{mix}, written in terms of the chemical structure of the glass or melt, is the equation of state for the system considered, since all of the terms in the expression for ΔG^{mix} depend on the composition, temperature and pressure in a known way. This means that, using this equation together with appropriate thermodynamic relationships, it is possible to determine various physical properties of glasses and melts from the derivatives of the potential ΔG^{mix}, with respect to different variables (e.g. temperature and pressure), such as the heat capacity, C_p, the coefficient of thermal expansion, α, the isothermal compressibility, β, the molar volume, V, and the molar refractivity, R. From these parameters, it is also possible to calculate other properties, such as the density and the refractive index.

The expessions for the above properties, given in terms of the chemical structure, have been derived in an explicit form [3,4]. These equations describe the concentration and temperature dependences of the properties of glasses and melts, incuding the jump observed in the curves for the temperature dependence of such properties as C_p, α and β as well as the break in those for the enthalpy, H, and the molar volume, V, in passing from melts to glasses. In addition, it has been shown [4] that, unlike the properties of the melts, those of the glasses are additive functions of the contributions from all species forming the solutions. For example, the equation for the molar volume, V, takes the form

$$V = \sum_j n_j \, V^0_j, \tag{1}$$

where n_j is the number of moles of the various species (salt-like groupings and unreacted oxides) present in the glass and V^0_j is the molar volume of the corresponding crystalline compound or the molten oxides (this normalisation follows from the formalism of the model).

Using the equations equivalent to (1), various properties have been calculated for the glasses in a number of systems: V, R and C_p for sodium silicates [4], V for lithium borates [6], densities for sodium phosphates and refractive indices for zinc phosphates [7], etc. In all cases, a comparison of the model and experimental values indicates that the discrepancy between them does not exceed the range of the literature values reported by different authors [7] and that the uncertainty in the calculations is mainly determined by the experimental error in the determination of the corresponding properties of the crystalline compounds [4].

As an example, the refractive index of zinc phosphate glasses is illustrated in Fig. 3, where the calculated values are given together with corresponding experimental data for both glasses and crystals. The error bar represents the range of experimental data reported by different authors and crystalline compositions are denoted m:n ($mZnO \cdot nP_2O_5$, etc.). The shape of the calculated curve is somewhat less detailed than the experimental one, although the fluctuations in the experimental curve are mostly well within the error bars. Taking this into account, the agreement between the calculated and experimental data is satisfactory.

It should be noted that, in theory, the range of physical properties of glasses which can be calculated using the model of ideal associated solutions is considerably wider than that given above. However, in practice its use has been limited by the availability both of the necessary data for the crystalline compounds used in the calculations and of the data for the corresponding glasses with which to compare the model predictions. Nevertheless, analytical expressions have been derived for determining mechanical, magnetic, electrical and transport properties and work on the modelling of these properties is currently in progress.

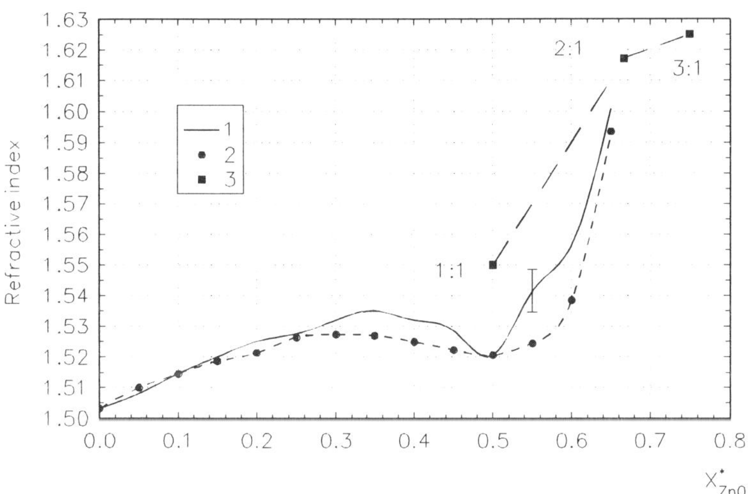

Fig. 3. The refractive index of zinc phosphate glasses. Solid line, experiment (glasses); circles, model (glasses) and squares, experiment (crystals). The dashed lines are a guide to the eye. (After [7].)

5. Structural Information

The concept of chemical structure used in the present approach is very close to the traditional view of the structure of glasses and melts. In the general case, the redistribution of electrons in the course of the interaction between basic and acidic oxides does not depend on the state of aggregation of the substances involved. Hence, it can be concluded that the structure of glasses and melts is closely related to that of the relevant crystals, at least in respect of the distrubution of their fundamental structural units. These units are determined by the first co-odination sphere of the glass-forming atoms and include BO_3 triangles and BO_4 and SiO_4 tetrahedra.

Using information on the chemical structure of the glasses (melts) and the structure of the corresponding crystalline compounds, the following parameters have been determined: the fraction of 4-fold coordinated boron atoms for five alkali borate systems [9] and in lead borate glasses [4], the distribution of 3-fold and 4-fold coordinated boron atoms among the groupings (superstructural units) suggested by Bray for sodium borate glasses [9] and the distribution of the SiO_4 tetrahedra with different numbers of non-bridging oxygen atoms in sodium and potassium silicate glasses [7].

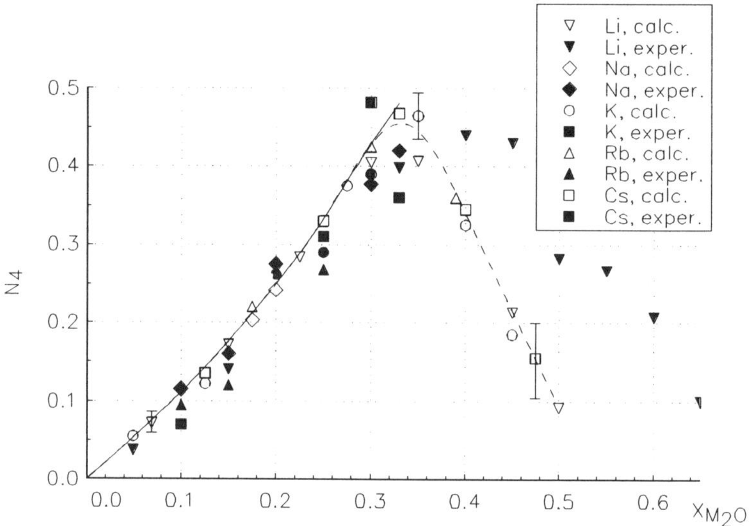

Fig. 4. The fraction of 4-fold co-ordinated boron atoms, N_4, in alkali borate glasses. Filled symbols, NMR data and open symbols, model. (After [9]).

In all cases, the calculated values are in good agreement with the results obtained by NMR spectroscopy and neutron diffraction. This indicates that the model of ideal associated solutions does not contradict the conventional concept of glass structure. Moreover, the model enables the character and concentration of the superstructural units/salt-like groupings (formed from the fundamental structural units) in the glasses to be determined. Thus, the structure of the glasses (melts) is described not only in terms of the short-range order but also of the medium-range order.

Examples from the above are illustrated in Figs 4-6. Figure 4 shows the distribution of 4-fold co-ordinated boron atoms for five alkali borate glass sysems. In Fig. 5, the distribution of $Si^{(n)}$ species in potassium silicate glasses is presented, while Fig. 6 gives the calculated temperature dependence of the $Si^{(n)}$ distribution for sodium silicate glasses. For the region up to ~35 mole % M_2O, the model predictions in Fig. 4 are in excellent quantitative agreement with the relevant NMR data, while at higher M_2O contents the discrepancy between model and experiment is due to the lack of reliable thermodynamic and phase diagram data at the pyroborate (2:1) and orthoborate (3:1) compositions. A similar situation exists in Fig. 5, where the distribution of $Si^{(n)}$ species as a function of composition is again in excellent agreement with experiment, in this case up to ~40 mole % K_2O.

From the distribution of $Si^{(n)}$ species in Fig. 6, it follows that an increase in temperature leads to a wider distribution of $Si^{(n)}$ species, resulting from reactions of the form

$$2Si^{(n)} = Si^{(n+1)} + Si^{(n-1)}. \tag{2}$$

This result is very interesting in that it is not only consistent with trends observed experimentally but also it allows the fictive temperature of molecular dynamics simulations of alkali silicate glasses to be estimated, once the resulting distribution of $Si^{(n)}$ species has been determined, as discussed in [10].

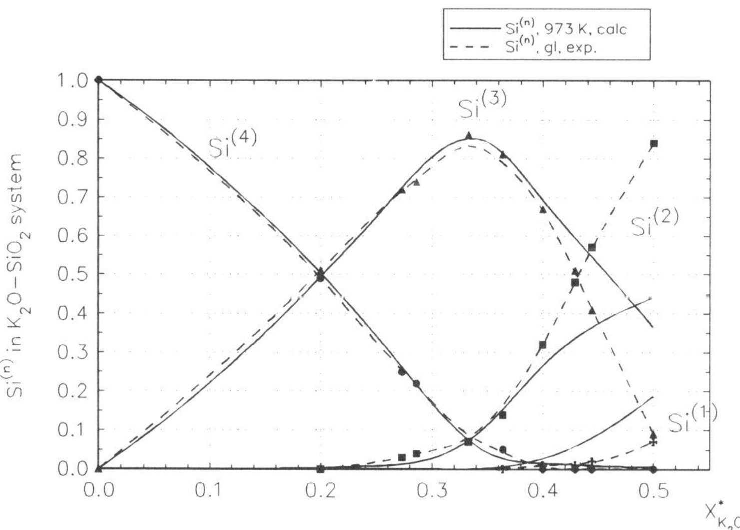

Fig. 5. The distribution of $Si^{(n)}$ species for the potassium silicate system. The solid lines are for the model (liquid) at 973 K and the dashed lines are experimental data for the glass. (After [7].)

6. Conclusions

A rigorous thermodynamic model has been developed, which is based on information for the corresponding properties of related crystalline compounds and enables a wide range of the properties of glasses and melts to be described on the unified basis. These include the entire set of thermodynamic potentials and various physical properties and the structural characteristics generally used in explaining these experimental data. The model is valid for systems with any number of components and does not use adjustable parameters. The calculations of various properties performed for several systems indicate that, in addition to an adequate description of the experimental data, the model is also valid for predicting the properties of systems which have not been adequately studied, if at all.

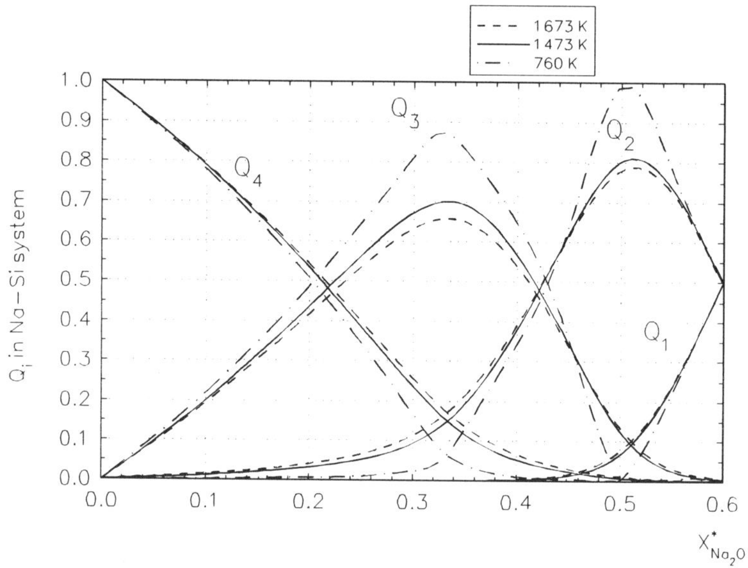

Fig. 6. The temperature dependence of the distribution of $Si^{(n)}$ (here denoted Q_n) species for the sodium silicate system, as predicted by the model of ideal associated solutions.

7. References

1. Shakhmatkin, B.A. and Shultz, M.M. (1980) The thermodynamic properties of glass-forming melts in the system Na_2O-SiO_2 at 800-1200°C, *Fiz. Khim. Stekla* **6**, 129-135.
2. Shakhmatkin, B.A. and Shultz, M.M. (1982) The thermodynamic properties and structure of alkali borate melts, *Fiz. Khim. Stekla* **8**, 270-276.
3. Shakhmatkin, B.A., Vedishcheva, N.M. and Shultz, M.M. (1992) Simulation of thermodynamic properties of slag melts, in: *Proc. 4th Intern. Conf. on Molten Slags and Fluxes*, Iron and Steel Inst. of Japan, Tokyo, pp. 73-78.
4. Shakhmatkin, B.A., Vedishcheva, N.M., Shultz, M.M. and Wright, A.C. (1994) The thermodynamic properties of oxide glasses and oxide glass-forming liquids and their chemical structure, *J. Non-Cryst. Solids* **177**, 249-256.
5. Shakhmatkin, B.A. and Vedishcheva, N.M. (1994) Thermodynamic studies of oxide glass-forming liquids by the electromotive force method, *J. Non-Cryst. Solids* **171**, 1-30.

6. Vedishcheva, N.M., Shakhmatkin, B.A. and Shultz, M.M. (1993) A simulation of thermodynamic properties of oxide melts and glasses, in: A.K. Varshneya, P.P. Bihuniak and D.F. Bickford (eds.), *Advances in the Fusion and Processing of Glass III*, Amer. Ceram. Soc., Columbus, pp. 283-288.

7. Vedishcheva, N.M., Shakhmatkin, B.A., Shultz, M.M. and Wright, A.C. The thermodynamic modelling of glass properties: a practical proposition?, *J. Non-Cryst. Solids*, in press.

8. Kozhina, E.L. and Shakhmatkin, B.A. (1990) The thermodynamic properties and structure of lithium borosilicate melts, *Fiz. Khim. Stekla* **16,** 363-368.

9. Shakhmatkin, B.A., Vedishcheva, N.M., Shultz, M.M. and Wright, A.C. (1994) The chemical structure and physical properties of alkali borate glasses, *Glastechn. Ber.* **67C**, 191-196.

10. Vedishcheva, N.M., Shakhmatkin, B.A., Shultz, M.M., Vessal, B., Wright, A.C., Bachra, B., Clare, A.G., Hannon, A.C. and Sinclair, R.N. (1995) A thermodynamic, molecular dynamics and neutron diffraction investigation of the distribution of tetrahedral $\{Si^{(n)}\}$ species and the network modifying cation environment in alkali silicate glasses, *J. Non-Cryst. Solids* **192&193**, 292-297.

HIGH RESOLUTION AND MULTIDIMENSIONAL NUCLEAR MAGNETIC RESONANCE PROBES OF GLASS STRUCTURE

J. W. ZWANZIGER, K. K. OLSEN, S. L. TAGG AND R. E. YOUNGMAN
Department of Chemistry, Indiana University
Bloomington, IN 47405 USA

1. Introduction

Elements of order in glass structure appear over several length scales, from short-range order at chemical bond distances to intermediate-range order at the 20–50Å scale. The ordering at short range is relatively well-defined, because it is controlled largely by chemical interactions which are themselves short-ranged. Several complementary probes of structure are effective at this length scale, including Raman and IR spectroscopies, neutron scattering, and nuclear magnetic resonance (NMR) spectroscopy.

NMR enjoys several advantages as a probe at this scale, in particular chemical specificity and the possibility to control with great precision the spin interactions probed in a particular experiment. In favorable cases one can determine the various magnetic and electronic interactions each nucleus in the sample experiences, including chemical shift, electric quadrupole, and magnetic dipole interactions. These quantities can be correlated with atomic scale structure through both *ab initio* calculations and comparison with studies of model compounds. In this way many important contributions to the study of glass structure have been made with NMR [1, 2].

The anisotropy of the interactions probed by NMR in solids is always challenging, leading as it does to spectral congestion and lower resolution. The development of pulse sequences like WAHUHA, and sample modulation schemes like Magic Angle Spinning (MAS), showed ways to suppress the anisotropy and hence relieve this congestion [3]. These methods are particularly effective for spin-1/2 nuclei, such as ^{1}H, ^{29}Si and ^{31}P. The past several years have seen an explosion of new methods to control and even take advantage of the anisotropies, particularly for quadrupolar nuclei [4]. These methods accomplish for nuclei like ^{11}B, ^{17}O, and ^{23}Na what MAS did for spin-1/2 nuclei.

245

M. F. Thorpe and M. I. Mitkova (eds.), Amorphous Insulators and Semiconductors, 245–254.
© *1997 Kluwer Academic Publishers. Printed in the Netherlands.*

These new techniques were designed with polycrystalline materials in mind, but we have shown that they can be applied advantageously to the study of glasses as well. Our purpose in this paper is to demonstrate some of these methods, by showing some of the insights into glass structure that they provide. The information these experiments provide is difficult to obtain in other ways, either with more conventional NMR methods or with optical or scattering techniques. Our examples are drawn from high-resolution multidimensional NMR studies of phosphates, tellurites, and borates. First, we show how the phosphate chain configurations in phosphate glasses may be estimated from the ^{31}P chemical shift tensors. We then discuss the bonding of modifying cations in tellurite glasses. In borates, we show how elements of the short and intermediate range structure may be inferred from high-resolution NMR, and propose a new model of chemical modification. We close with a summary.

2. Applications to Glass

2.1. SPIN 1/2 NUCLEI: VARIABLE ANGLE CORRELATION SPECTROSCOPY

The standard NMR tool for study of spin-1/2 nuclei in glasses is MAS, which has become invaluable for nuclei such as ^{29}Si, ^{31}P, and ^{1}H [2]. Here we will focus on an extension of MAS to two spectral dimensions, called Variable Angle Correlation Spectroscopy (VACSY) [5, 6, 7]. The experimental goal is to separate the anisotropic chemical shift interactions from the isotropic terms, so that each resonance in a high-resolution spectrum may be correlated with its own powder pattern. Such a separated spectrum allows the precise determination of the principal chemical shift tensor elements characterizing each site, which are then used to derive detailed information about the local chemical bonding. VACSY accomplishes this separation in the following way. Under conditions of rapid sample rotation the resonance frequency ν of an arbitrary site in the sample is given by $\nu = \nu_{\text{iso}} + P_2(\cos\beta)\nu_{\text{aniso}}$, where ν_{iso} is the isotropic frequency, ν_{aniso} is the anisotropic contribution, and $P_2(\cos\beta)$ is the second Legendre polynomial of $\cos\beta$ where β is the orientation of the sample rotation axis. Note that at the magic angle, $P_2(\cos\beta_m) = 0$. The key in VACSY is to record data with the sample spinning at β_m, but also at other values of β. This set of experiments can be viewed as a modulation of anisotropy, by the experimentally controlled factor $P_2(\cos\beta)$. From such a data set a two-dimensional spectrum is constructed with ν_{iso} along one frequency axis, and ν_{aniso} along the other. The result is the desired separation of the high resolution and powder pattern dimensions.

The value of obtaining the individual powder patterns is that from them

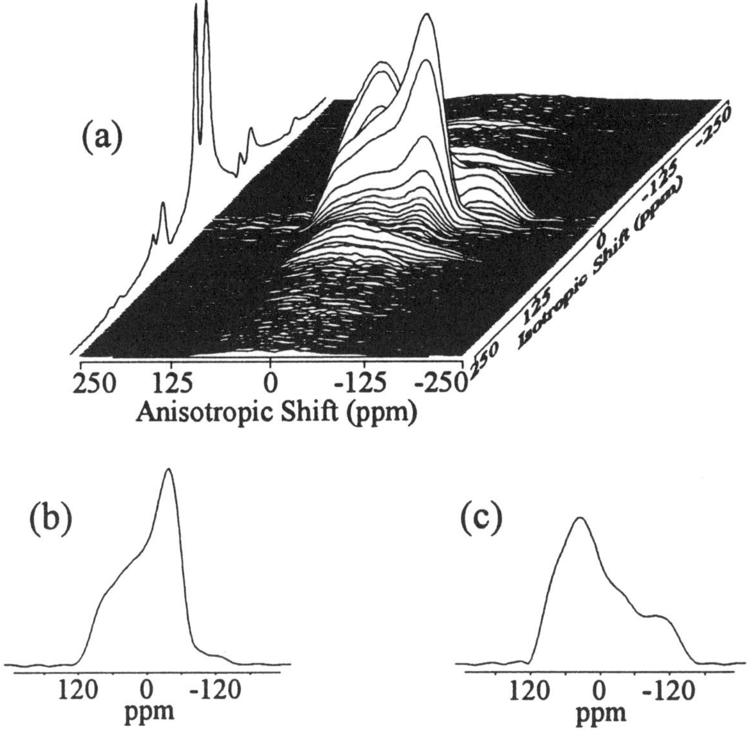

Figure 1. (a) ^{31}P VACSY spectrum of $(Na_2O)_{56}(P_2O_5)_{44}$ glass. The projection along the left edge is the isotropic spectrum, which corresponds to MAS. The powder patterns for each of the two sites are clearly resolved in the stack plot, and also shown as slices below in (b), the $Q^{[1]}$ site, and (c), the $Q^{[2]}$ site.

the principal values of the chemical shift tensor can be determined. These quantities correlate with more detailed elements of the local structure than does the isotropic chemical shift, measurable in conventional MAS. We have used VACSY with success to study silver phosphate-silver iodide glasses [6, 7]. Here we discuss recent applications of this technique to samples in the sodium phosphate system.

Figure 1 shows the VACSY spectrum of a sodium phosphate glass, $(Na_2O)_{56}(P_2O_5)_{44}$ (we will refer to this composition as 56/44). Based on earlier work this composition is expected to contain two distinct phosphorous sites: one type, $Q^{[1]}$, is a chain-terminating unit including three non-bridging oxygen and one bridging oxygen, and the second, $Q^{[2]}$, an intra-chain unit with two bridging oxygen. The VACSY spectrum resolves these sites in its high-resolution dimension (as in MAS). The ratio of integrated intensities of these two sites is $[Q^{[1]}]/[Q^{[2]}] \approx 0.5$, which can be

Composition	Site	δ_{11}	δ_{22}	δ_{33}	δ_{iso}	$\Delta\delta$	η
56/44 glass	$Q^{[1]}$	−37	−37	79	1	116	0
	$Q^{[2]}$	76	18	−133	−17	−180	0.5
50/50 glass	$Q^{[2]}$	75	18	−160	−22	−207	0.4
50/50 crystal	$Q^{[2]}$	73	31	−169	−22	−221	0.3

TABLE 1. Chemical shift tensor components of ^{31}P in $(Na_2O)_x(P_2O_5)_{100-x}$ glasses. The composition is expressed as $x/(100-x)$, and shifts (uncertainty ±5) are given in ppm relative to 85% H_3PO_4. The tensor elements are ordered as $|\delta_{33} - \delta_{iso}| \geq |\delta_{11} - \delta_{iso}| \geq |\delta_{22} - \delta_{iso}|$. The anisotropy is defined as $\Delta\delta = \delta_{33} - \frac{1}{2}(\delta_{11} + \delta_{22})$, and the asymmetry as $\eta = (\delta_{22} - \delta_{11})/(\delta_{33} - \delta_{iso})$.

interpreted simply as arising from phosphate chains of length 6 (e.g., 2 end groups capping 4 links).

This interpretation of the structure of the 56/44 glass assumes monodisperse chains. Of course, chains of different lengths may well be present, and also phosphate rings can form. To investigate the possibility of ring formation we compare to the nearby 50/50 crystalline phase, $Na_3P_3O_9$, which consists solely of phosphate rings [8], and to the 50/50 glass. We find, first of all, that the 50/50 glass shows only $Q^{[2]}$ groups and no $Q^{[1]}$ groups, and thus also contains only phosphate rings. Next, to make a detailed comparison of the $Q^{[2]}$ groups of the 56/44 glass to those in the 50/50 compositions, we make use of the second dimension of the VACSY spectrum. The powder pattern representative of each site ($Q^{[1]}$ and $Q^{[2]}$) can be extracted by taking slices at the appropriate isotropic shifts, and from these patterns the principal chemical shift tensor elements can be determined. These are summarized in Table 1. The obvious difference between the $Q^{[1]}$ and $Q^{[2]}$ powder patterns is summarized by their anisotropies, $\Delta\delta = \delta_{33} - \frac{1}{2}(\delta_{11} + \delta_{22})$. This quantity is positive for the $Q^{[1]}$ site, but negative and much larger for the $Q^{[2]}$ site. The difference in magnitude arises from the lower local symmetry (cf. $Q^{[0]}$ sites, which are nearly tetrahedral and have $\Delta\delta \approx 0$). Comparing the $Q^{[2]}$ site of the 56/44 glass with the 50/50 glass and crystal, we see that, while δ_{iso} is essentially the same, $\Delta\delta$ is significantly larger in the 50/50 compositions. This difference is due mostly to the increase in magnitude of δ_{33}. We expect that δ_{33} correlates closely with the O-P-O chain bond angles, and thus is affected by the constraints of the ring geometry, as opposed to chains. In the VACSY spectrum of the 56/44 glass we see no evidence of a $Q^{[2]}$ pattern with $\Delta\delta$ or δ_{33} as large as those observed in the 50/50 glass and crystal; the conclusion is that rings are *not* a significant component of the structure of this glass composition. Note that this is not apparent from the isotropic chemical shifts alone, as these are the same in all three cases.

2.2. QUADRUPOLAR SPINS IN ONE-DIMENSIONAL NMR: DOUBLE ROTATION

The direct analog of MAS for quadrupolar nuclei is Double Rotation [4]. In DOR the sample is spun about both the normal magic angle, and also the fourth-rank magic angle, thereby yielding a one-dimensional spectrum with both first- and second-order anisotropies suppressed. It can be used to determine the number and populations of distinct chemical sites of quadrupolar nuclei like ^{17}O and ^{11}B [9]. The experiment is very difficult to execute, due to the stringent mechanical balance of the rotors that must be achieved. The large size of the rotor assembly also limits its rotation frequency to about 1 kHz or less (MAS rotors routinely exceed 10 kHz).

In spite of these difficulties, DOR can be an effective experiment. We have used DOR to study the oxygen sites in glassy B_2O_3, by making a sample enriched in ^{17}O and recording spectra in two magnetic field strengths [10]. We were able to resolve three different oxygen resonances, which we could assign to oxygen in boroxol rings, oxygen bridging boroxol rings, and oxygen bridging non-ring BO_3 groups. A prime difficulty in this work was the low signal-to-noise ratio, and the aforementioned slow spinning speeds. To extract intensities from the spectra we performed time-dependent simulations of the spectra, with finite large-rotor spinning frequency, and then used the resulting site populations to simulate MAS data of the same sample. The MAS data is obtainable with good signal-to-noise ratio, but has lower resolution. Because the parameters extracted from the DOR data also fit the MAS data we could be confident in our assignments.

The lack of observation of a ring-non-ring bridging oxygen site suggests that the rings in B_2O_3 glass tend to cluster. By combining the oxygen site populations with previous boron site populations we could derive a model which is consistent with the boron and oxygen NMR data and which predicts that the boroxol rings cluster in groups with size 20–30 Å. This length scale for intermediate range order in B_2O_3 glass has been estimated previously from light-scattering data [11, 12]; the NMR has been valuable in providing an estimate of what chemical structures give rise to the order.

2.3. TWO-DIMENSIONAL NMR OF QUADRUPOLAR SPINS: DYNAMIC ANGLE SPINNING

Dynamic Angle Spinning (DAS) has potentially higher information content than DOR, and is easier to implement mechanically [4]. In DAS the sample rotor axis is flipped between two angles, chosen to refocus both the first and second-order anisotropies. This experiment typically takes longer to execute, because as a two-dimensional method it consists essentially of a series of one-dimensional spectra. We have found DAS to be a particularly powerful probe of glass structure, and have applied it to [11]B in borate glasses and [23]Na in tellurite glasses [13, 14, 15, 16]. Its primary limitation is that it cannot be used in cases where spin-diffusion occurs rapidly on the time-scale of the spinning axis reorientation, which is typically 30 msec; this is why we could not use it to study oxygen sites in B_2O_3 glass.

2.3.1. *Modifier Environments in Tellurite Glasses*

In the sodium tellurite system DAS is an effective probe of the average sodium environments not because of the high spectral resolution it affords, but because it removes all anisotropic broadenings and yields precise values of the sodium NMR parameters. In the $(Na_2O)_x(TeO_2)_{1-x}$ system ^{23}Na NMR shows that there is only one average sodium site, to be sure with a fairly broad distribution of quadrupole and chemical shift parameters. The isotropic shift measured by the DAS NMR experiment includes both the chemical shift and the 2nd-order quadrupole shift. By recording the isotropic shift in two field strengths, both the isotropic chemical shift and the quadrupole interaction strength can be determined. These parameters can be related empirically to the local environment around the sodium ions, using the distance and valence of the coordinating oxygen [17, 18].

We know from crystallography [15, 19] that the structures of crystalline sodium tellurites that occur within the glass-forming range ($Na_2Te_4O_9$ and $Na_2Te_2O_5$) include 5- and 6-coordinate sodium ions. Based on the DAS measurements, the glasses do as well [16]. This result is inferred from the chemical shifts of the sodium ions, which can through the above-noted empirical correlations be related to the coordination number (Figure 2). We find a decreasing average coordination number for sodium in $(Na_2O)_x(TeO_2)_{1-x}$, as a function of composition. Note that the onset of the decrease occurs at nearly the $Na_2Te_4O_9$ composition, which has been suggested previously to be an especially good glass-forming composition in this system [20].

2.3.2. *Network Modification in Borate Glasses*

The high resolution provided by DAS NMR has allowed us to follow structural changes in borate glasses in more detail than has been possible previously. It is well-known that addition of a modifier to B_2O_3 increases the average coordination number of the boron as BO_4^- units are produced. The tetrahedral geometry of these units reduces the anisotropy of their quadrupole interactions nearly to zero, making quantification of four-fold coordinate versus three-fold coordinate boron possible. This has been done

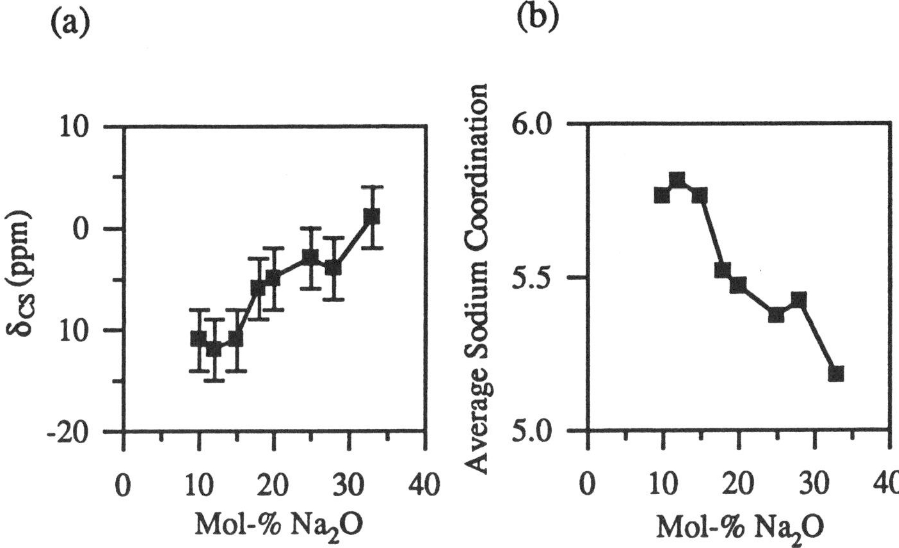

Figure 2. (a) Chemical shift of ^{23}Na in $(Na_2O)_x(TeO_2)_{1-x}$, as a function of composition, as determined from DAS NMR experiments in two different magnetic fields. (b) The coordination number of sodium for these glasses as inferred from the chemical shift and empirical correlations.

for a number of modifiers [1].

The origin of these four-fold boron could be either boroxol ring boron, or BO_3 groups that are not members of rings, as both exist in B_2O_3 glasses and melts. With DAS NMR we can resolve not only the three- and four-coordinate boron, but also both classes of three-coordinate boron [13, 14]. This gives us a unique opportunity to follow the structural changes due to modification, as earlier techniques (wide-line NMR, Raman spectroscopy) cannot resolve all three of these sites.

Figure 3 shows ^{11}B DAS NMR spectra of a series of $(K_2O)_x(B_2O_3)_{1-x}$ glasses [21]. In this figure resolution of three boron sites is observed: ring sites at 4.7 ppm, the non-ring BO_3 unit at 1.0 ppm, and the four-coordinate boron at -0.3 ppm. Note that these are total isotropic shifts, and thus contain the second-order quadrupole shift in addition to the usual chemical shift. This assignment is made based on both comparison with NMR spectra of crystalline potassium borates [21], and from the lineshapes of slices (not shown) of the 2-dimensional DAS spectra at each of the isotropic shifts listed. Figure 3(b) summarizes the spectra, showing the concentration of each of the three sites as a function of added modifier.

Figure 3 shows that the ring boron concentration decreases only slowly as modifier is added, while the non-ring BO_3 fraction decreases much more

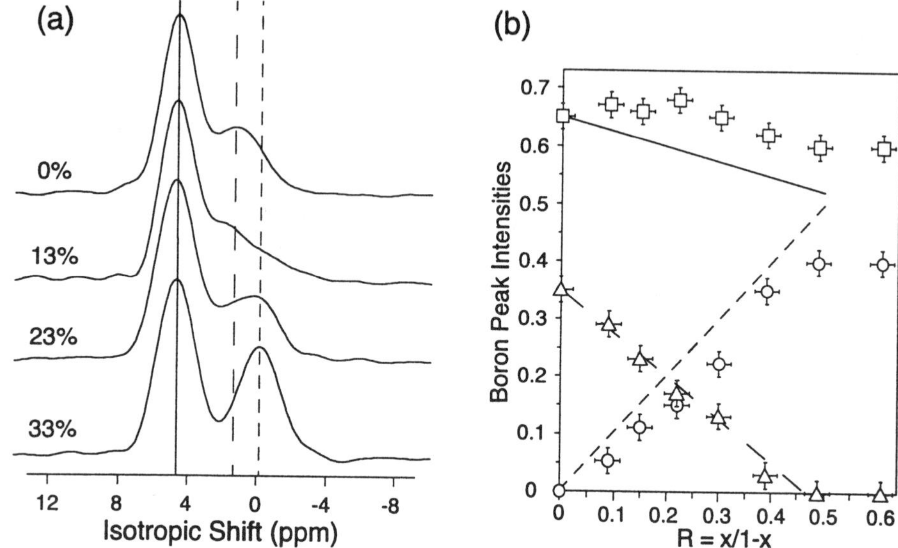

Figure 3. (a) ^{11}B DAS NMR spectra of $(K_2O)_x(B_2O_3)_{1-x}$, showing resolution of three sites, for four compositions. The resonance at isotropic shift of 4.7 ppm is due to ring boron; at 1.0 ppm, to three-coordinate boron outside of rings; and at -0.3 ppm, to four coordinate boron. (b) The fractions of the sites identified in (a), as a function of the modifier/boron ratio. □: rings, △: non-ring BO_3; ○: BO_4^-.

quickly as the four-fold sites appear. Raman spectra we have recorded on these samples [21], as well as those presented in numerous other studies, show that the boroxol rings originally present in the unmodified glass are replaced with new structures, and that this replacement occurs at a high rate as a function of added modifier. Indeed, earlier Raman studies have been interpreted in terms of models that posit *only* modification of the boroxol rings, and none at the non-ring sites.

We have reconciled the apparent discrepancies between the Raman data and the DAS NMR data with the following simple model [21]. We assume four building blocks: boroxol rings, triborate rings (two three-coordinate and one four-coordinate boron in a ring), non-ring three-coordinate boron, and non-ring four-coordinate boron. These are chosen as the minimal set of structures that result from modification at the two basic boron sites. We also assume that the three-coordinate boron in both boroxol and triborate rings have the same isotropic shift—our studies of crystalline potassium borates show that this is reasonable. Then, through normalization and comparison with the data (Figure 3(b) and the Raman spectra) the populations of all four structures can be derived, as a function of composition. The model results are shown (with the two populations of the two ring structures combined) on Figure 3(b), in addition to the experimental data.

The most significant results of this model are as follows. First, in contrast to earlier models, non-ring BO_3 groups are modified preferentially over the boroxol rings. Next, triborate groups are formed faster than boroxol rings are modified, showing that some triborate groups are formed from loose three- and four-coordinate boron. This suggests that the rings have a special stability in the melt, which is perhaps not surprising. Finally, the complementary use of Raman and NMR is crucial to these findings, because of the information these methods yield about structures on different length scales.

3. Summary

In this contribution we have tried to show by example how high-resolution and multidimensional NMR methods can probe glass structure, directly at short length scales and indirectly at intermediate length scales. We showed with VACSY how the powder patterns of different sites may be separated, and demonstrated how these patterns give detailed structural information on a glass, using $(Na_2O)_x(P_2O_5)_{1-x}$ as examples. We discussed use of DOR to study oxygen sites in B_2O_3 glass, and how from these results a model of the intermediate range structure could be inferred. Finally, using DAS we showed how modifier cation environments (in $(Na_2O)_x(TeO_2)_{1-x}$) could be determined, and how structural changes of the modified network itself (in $(K_2O)_x(B_2O_3)_{1-x}$) could be followed.

4. Acknowledgements

We are pleased to acknowledge helpful interactions with Dr. Ulrike Werner-Zwanziger and Prof. Mark Hollingsworth (Indiana University), and Prof. Brad Chmelka and Dr. Mike Janicke (University of California-Santa Barbara). This work was supported by the National Science Foundation by grants DMR-9115787 and DMR-9508625.

References

1. Bray, P. J., Gravina, S. J., Hintenlang, D. H., and Mulkern, R. V. (1988) Nuclear Magnetic Resonance of Glasses, *Magn. Reson. Rev.* **13**, 263–353.
2. Eckert, H. (1992) Structural characterization of noncrystalline solids and glasses using solid state NMR, *Prog. NMR Spec.* **24**, 159–293.
3. Haeberlen, U. (1976) *High Resolution NMR in Solids: Selective Averaging*, vol. Supplement 1 of *Advances in Magnetic Resonance*. Academic Press, New York.
4. Chmelka, B. F. and Zwanziger, J. W. (1994) Solid-State NMR Line Narrowing Methods for Quadrupolar Nuclei: Double Rotation and Dynamic-Angle Spinning,

in B. Blümich and R. Kosfeld (eds.) *NMR Basic Principles and Progress*, vol. 33, pp. 79–124. Springer, Berlin.

5. Frydman, L., Chingas, G. C., Lee, Y. K., Gradinetti, P. J., Eastman, M. A., Barrall, G. A., and Pines, A. (1992) Variable-angle correlation spectroscopy in solid-state nuclear magnetic resonance, *J. Chem. Phys.* **97**, 4800–4808.

6. Zwanziger, J. W., Olsen, K. K., and Tagg, S. L. (1993) Structure and Disorder of Phosphates in Ag_2O-AgI-P_2O_5 Glasses, *Phys. Rev. B* **47**, 14618–14621.

7. Olsen, K. K. and Zwanziger, J. W. (1995) Multinuclear and Multidimensional NMR Investigation of Silver Iodide-Silver Phosphate Fast Ion Conducting Glasses, *Solid State NMR* **5**, 123–132.

8. Ondik, H. M. (1965) The Structures of Anhydrous Sodium Trimetaphosphate, $Na_3P_3O_9$, and the Monohydrate, $Na_3P_3O_9 \cdot H_2O$, *Acta. Cryst.* **18**, 226–232.

9. Youngman, R. E., U. Werner-Zwanziger, and Zwanziger, J. W. (1996) A Comparison of Strategies for Obtaining High-Resolution NMR Spectra of Quadrupolar Nuclei, *Zeitschrift für Naturforschung A* in press.

10. Youngman, R. E., Haubrich, S. T., Zwanziger, J. W., Janicke, M. T., and Chmelka, B. F. (1995) Short- and Intermediate-Range Structural Ordering in Glassy Boron Oxide, *Science* **269**, 1416–1419.

11. Bokov, N. (1994) Light Scattering Studies of Glasses in the Glass Transition region, *J. Non-Crystalline Solids* **177**, 74–80.

12. Moynihan, C. T. and Schroeder, J. (1993) Non-exponential structural relaxation, anomalous light-scattering and nanoscale inhomogeneities in glass-forming liquids, *J. Non-Crystalline Solids* **160**, 52–59.

13. Youngman, R. E. and Zwanziger, J. W. (1994) Multiple Boron Sites in Borate Glass Detected with Dynamic Angle Spinning Nuclear Magnetic Resonance, *J. Non-Crystalline Solids* **168**, 293–298.

14. Youngman, R. E. and Zwanziger, J. W. (1995) On the formation of tetra-coordinate boron in rubidium borate glasses, *J. Am. Chem. Soc.* **117**, 1397–1402.

15. Tagg, S. L., Huffman, J. C., and Zwanziger, J. W. (1994) Crystal Structure and Sodium Environments in Sodium Tetratellurite, $Na_2Te_4O_9$, and Sodium Tellurite, Na_2TeO_3, by X-ray Crystallography and Sodium-23 NMR, *Chem. Mater.* **6**, 1884–1889.

16. Tagg, S. L., Youngman, R. E., and Zwanziger, J. W. (1995) The Structure of Sodium Tellurite Glasses: Sodium Cation Environments from Sodium-23 NMR, *J. Phys. Chem.* **99**, 5111–5116.

17. Xue, X. and Stebbins, J. F. (1993) Na-23 NMR Chemical Shifts and Local Na Coordination Environments in Silicate Crystals, Melts, and Glasses, *Phys. Chem. Minerals* **20**, 297–307.

18. Koller, H., Engelhardt, G., Kentgens, A. P. M., and Sauer, J. (1994) ^{23}Na NMR Spectroscopy of Solids: Interpretation of Quadrupole Interaction Parameters and Chemical Shifts, *J. Phys. Chem.* **98**, 1544–1551.

19. Tagg, S. L., Huffman, J. C., and Zwanziger, J. W. (1996) Sodium Ditellurite $Na_4Te_4O_{10}$ Crystal Structure, in press at *Acta Chem. Scand.*

20. Zhang, M., Mancini, S., Bresser, W., and Boolchand, P. (1992) Variation of glass transition temperature, T_g, with average coordination number, $\langle m \rangle$, in network glasses: evidence of a threshold behavior in the slope $|dT_g/d\langle m \rangle|$ at the rigidity percolation threshold ($\langle m \rangle = 2.4$) *J. Non-Crystalline Solids* **151**, 149–154.

21. Youngman, R. E. and Zwanziger, J. W. (1996) Network Modification in Potassium Borate Glasses: Studies with NMR and Raman Spectroscopies, submitted.

POROUS SILICA
Model fractal materials, their structures and their vibrations

ERIC COURTENS AND RENÉ VACHER,

Laboratoire des Verres, UMR N° 5587, Université de Montpellier II,

F-34095 Montpellier Cedex 5, France

Abstract - The preparation, properties and uses of some porous silicas, leached glasses and primarily silica aerogels, are reviewed. Emphasis is placed on their fractal properties. So far, aerogels form the best material realizations of random mass fractals studied both for their structure and their acoustic vibrations. In these lectures theoretical ideas are informally exposed and illustrated with the simplest possible simulations designed at supporting intuition. These ideas are then applied to aerogels, for which a remarkable degree of consistency was achieved. Inelastic scattering results (Brillouin, Raman, and neutron) in the proper frequency range are explained in terms of fractons and coupling coefficients. Particularly, the spectral shape that accounts for the phonon-fracton crossover, or the localization of acoustic waves, has been tested in considerable details. It can now be extended to other situations, such as glasses, where related phenomena might occur.

1. Introduction

The chemistry of silica, SiO_2, is complex.[1] The material exists in many natural forms such as crystalline quartz, tridymite, or cristobalite, amorphous opal, and volcanic glass. Compounded with other oxides it constitutes the major part of the solid surface of the Earth. A large variety of silica based artificial materials has been synthesized, many of them porous. The purpose of these lectures is not to review them all, but rather to focus on forms of porous silica exhibiting fractal properties, discussing these

255

M. F. Thorpe and M. I. Mitkova (eds.), Amorphous Insulators and Semiconductors, 255–288.

256

properties. From this view point, prominent roles are played by leached glasses (*e.g.* Vycor™), and especially by aerogels.

We begin with an introduction to fractals exposing concepts necessary to subsequent understanding. The accent is not on formal rigor, but on developing intuitive feelings and physical ideas. It is assumed that all readers have somehow been exposed to fractal pictures, which are so common nowadays. If this were not the case, or if a deeper knowledge were desired, excellent books by B. Mandelbrot [2,3] or by others [*e.g.* 4,5] should be consulted. A recent review closely related to our subject is also very useful.[6] Those familiar with the field can skip most of Section 2. We recommend, however, reading the concept of mutually self-similar series which is key to the interpretation of later experimental results, as well as some comments on random models. The percolation models introduced in Section 2 are the simplest possible ones providing intuitive pictures of disordered fractals supporting acoustic vibrations. It must be emphasized that these models are, in many ways, *considerably different* from real materials. Hence, one must caution that comparisons, and particularly extrapolations of quantitative results, can easily be misleading.

Section 3 is devoted to the materials and their structures. First, leached glasses are discussed. These are usually not fractal in themselves (although very porous), but the filling of their pores, if incomplete, can exhibit fractal properties. The remainder and larger part of Section 3 is devoted to aerogels, their preparations, uses, and structures. Depending on the exact reaction conditions, very different materials with distinct microstructures and fractal dimensions are obtained. It will be shown that well characterized mutually self-similar series can be produced. In Section 5, this property is seen to be crucial to a complete study of the phonon-fracton crossover.

In Section 4, the vibrational eigenmodes of fractal solids, or «fractons», are introduced. Mental pictures are created based on scalar vibrations of percolation models. A hypothesis allowing dynamical scaling, the single-length-scale postulate of Alexander,[7] is explained and shown to work for these models.

Experimental results on the vibrations of aerogels are explained in Section 5. First, Brillouin scattering investigations of the phonon-fracton crossover are presented. Long-wave acoustical phonons experience Rayleigh scattering from structural disorder, so that their line width Γ increases with the fourth power of frequency ω, this until the Ioffe-Regel limit [8] $\Gamma = \omega$. In that vicinity they cross over to fractons, which are localized excitations. A spectral shape will be presented that describes perfectly the spectra in the phonon, crossover, and fracton regimes, in many different situations. At higher frequencies, the scattering decays rapidly with ω, so that other light-vibration coupling mechanisms come into play. This will be called the Raman scattering regime, for lack of a better word. Neutron scattering was also used to investigate the fracton density of states. This revealed a crossover in the dominant elasticity, from bond bending at low frequencies to bond stretching at high frequencies. Finally the relation of these vibrations to thermal properties at low temperatures will be discussed.

Section 6 summarizes the main points, indicates some open questions, and discusses possible extensions of some of these results to related situations.

2. Fractal Primer

2.1. SELF-SIMILARITY IN MATHEMATICS, GEOMETRY, AND NATURE

What properties do Koch curves (Fig. 1a), Julia sets (Fig. 1b), clouds, or mountain landscapes possibly share? It is that a small part of each object, when properly scaled up, somehow looks like the entire object. In the words of Mandelbrot [3] these are *shapes made of parts similar to the whole in some way,* which is an intuitive definition of fractals. The essential point is *self-similarity.* For the geometric iteration processes (such as Cantor dust, Koch curves, Sierpinski gaskets, and the like, illustrated in Fig.

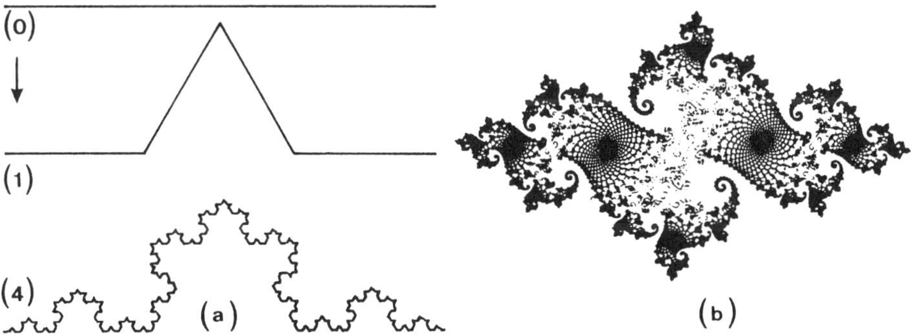

Figure 1. Deterministic fractals: (a). The Koch curve is obtained by successive applications of the generic transformation shown on top, to all straight segments forming the curve. The curve (4) is obtained after the fourth iteration. The zeroth generation, or initiator, is the straight segment (0), and the first generation (1) is here the generator itself. A true fractal results from infinite iteration.

(b). Julia sets are boundaries of domains of attraction for a complex variable z iterated indefinitely with a nonlinear law, here $z \rightarrow z^2 + c$, with $c = -0.74543 + i\,0.11301$; infinity is the only attractor (from [5]).

258

1a), the fractals are *deterministic* and self-similarity is built in by construction. Similarly, Julia sets and other similar sets are deterministic (although their behavior can become chaotic) but the reason for the self-similarity is not immediate. We shall not go into this, as the fractals that interest us are *statistical* ones, just like clouds or mountain landscapes. We take it as an experimental fact that many types of aggregation processes (including simulated ones) lead to hierarchical structures that are fractal. The deep reason for the remarkable self-similarity of some of these processes is a subject of current research and has remained partly a mystery.[9] Probably the simplest illustration of random fractal networks is provided by percolation. For this process, an intuitive feeling for the origin of fractality is easily acquired.

2.1.1. *A simple Model: Percolation Networks*

Take a sheet of squared paper. Let each square be the position for a *site*. Each site can be connected by *bonds* to its four neighboring sites (we assume that no bonds are possible along diagonals). First consider that all sites are empty, and with a pencil start filling them randomly with probability p. As neighboring sites are filled, let the bond between them automatically exist. As p is increased towards a critical value p_c, a *percolation threshold* will be reached (near and mostly above the thermodynamic p_c for small systems) where a continuous path is established between one side and the other side of the squared paper. That situation is illustrated in Fig. 2(a). At that point there exists one large cluster, which contains the connecting path, and which is called the «*infinite*» cluster, this by reference to the thermodynamic limit where this cluster really contains an infinite number of occupied sites. There are also many finite disconnected clusters, of different sizes. If one continues filling in sites, the space occupied by the infinite cluster continues to grow, this at the expense of the finite clusters that progressively attach to it, starting primarily with the largest ones. If we imagine *removing the finite clusters*, the mass distributed in the remaining infinite cluster (if large enough) is fractal. This process is called *site percolation*.[11]

One could proceed in another manner. Imagine that all sites are initially occupied, but without any bonds between them. With a pencil, start filling randomly the bonds with probability p. At a certain point, $p \approx p_c$ (this gives another value for p_c) the first continuous path is established. Again an infinite cluster is formed, and there are many finite clusters not attached to it by any bond. *Removing these finite clusters*,

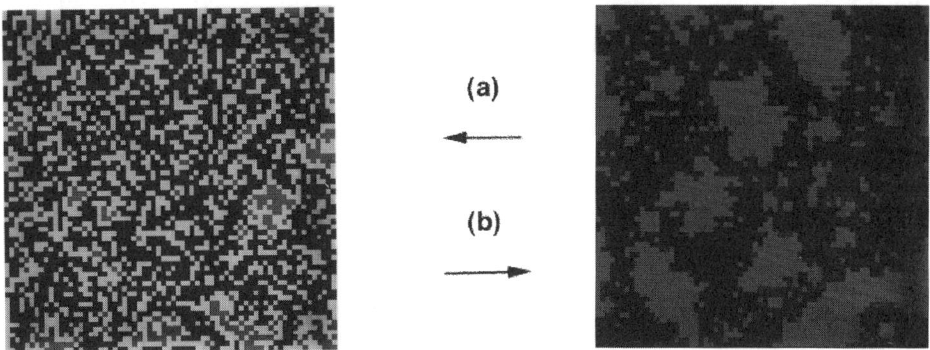

Figure 2. **Examples of site (a) and bond (b)** percolation for 2-dimensional 68×68 realisations. The «infinite» cluster is shown in dark, the finite clusters in grey in (a). One notices the empty sites in (a). The unoccupied bonds in (b) are not represented (Illustration provided by E. Stoll, after [10]).

the remaining mass shows a fractal distribution. This is called *bond percolation*,[11] and it is shown in Fig. 2b.

Comparing the two pictures, we note that site percolation appears to have many small voids. The reason is the initial random filling of sites. This leaves holes having *nothing* to do with fractality, only with *random white noise*. The argument can be made fully quantitative,[10] as shown in Section 2.2.3. The origin of fractality lies in the *removal of finite clusters at all scales*. Of course, for very large systems both processes eventually lead to equally good fractals. However, for dynamical studies, one is strongly restricted in the sizes that can be handled numerically. In that case, it is much preferable to use bond percolation models,[10] as site percolation can even lead to apparent failure of scaling (see *e.g* [12]) for the above reasons.

2.1.2. *Hausdorff-Besicovitch Dimension or the Length of a Coastline*

Consider the coast of Norway, [4] sketched in Fig. 3a. Can one determine its length? This might seem at first a trivial question. However, the practical problem is one of triangulation, and the answer depends on the number of triangulation points that are taken, or in simpler words on the length of the measuring stick! A convenient method is to use a process known as *box-counting*. The picture is covered with squares of sides L, and one counts the number of squares $N(L)$ containing parts of the coastline. If the line were not fractal, one would find $N(L) \times L \approx$ constant, independent of L for small enough L. This constant would be *a measure* for the length of the line. If instead of a

260

line it were a surface that one had to cover, one would obtain $N(L) \times L^2 \approx$ constant, and this constant would again measure the surface in question, while the product $N(L) \times L$ would tend to infinity, and the product $N(L) \times L^3$ would tend to zero as L decreased. We see that it is $N(L) \times L^D$, where D is the dimension of the object, that gives the appropriate measure M_D. Now, in the case of the coast of Norway, one finds that $N(L) \times L^{1.52}$ is a constant, which means that the object is fractal with dimension $D = 1.52$. This is what is called the *Hausdorff-Besicovitch* dimension. To find D, one can simply plot the "box-counting *length*", which is the product $N(L) \times L$, in function of L. The log-log presentation (Fig. 3b) is a line of slope 1-D, which indeed shows that $N(L) \times L$ diverges at small L. The *topological* dimension of a coastline is of course 1, and therefore we speak of a line. A fractal is an object whose *box-counting* dimension is larger than its topological dimension.

So, what will on the average be the length $\Lambda_a(L)$ of the coastline in a box of size L if we decide to use a measuring stick of length $a < L$? This can be considered as the *«length»* of the fractal in its embedding space. We have $M_D = N(a) \times a^D = N(L) \times L^D$. Now, considering only a single box of size L, one has $N(L) = 1$, and

$$\Lambda_a(L) \equiv a \times N(a) = a \times (L/a)^D . \tag{2.1}$$

This is the scaling relation for the length $\Lambda_a(L)$ measured between two points, a distance L apart in the embedding space, when surveying the fractal with a stick of

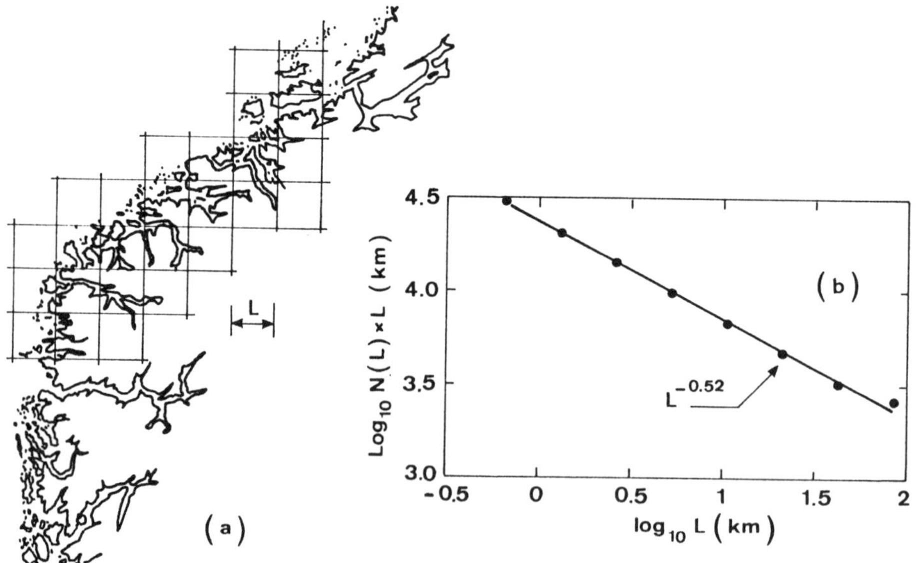

Figure 3. The outline of the coast of Norway (a), and its box-counting length (b), after J. Feder [4].

length a. For the Koch curve of Fig. 1a, the length increases by 4/3 each times the stick is reduced to 1/3 of its previous value: $a \rightarrow b = a/3$ gives $\Lambda_b = (4/3) \times \Lambda_a$. Introducing these two relations in (2.1), one obtains $4/3 = 3^{D-1}$, or $D = ln\ 4\ /\ ln\ 3 = 1.262$.

2.2. RANDOM MASS FRACTALS

2.2.1 *Definition, the Particle Size, and the Persistence Length*

We shall be dealing with real objects, made of particles or grains of typical size (radius) a, of mass m_a, and of density within grains ρ_a. One can define mass fractals as objects assembled from such connected grains, whose mass within a sphere of radius $L > a$, *centered on a grain*, scales on the average like

$$M(L) \approx m_a \times (L\ /\ a)^D .\tag{2.2}$$

Here, $D < 3$ is the fractal dimension of the object. As the volume V of the sphere scales like L^3 (we do not bother about factors like $4\pi/3$), the average density of the fractal within the sphere of radius L *centered on a grain* is given by

$$\rho_f(L) = M(L)/V(L) \approx \rho_a \times (L\ /\ a)^{D-3} .\tag{2.3}$$

Here, we used $\rho_a \approx m_a/\ a^3$. By definition, since Eqs. (2.2) and (2.3) require that there is a grain at the center of the sphere, $\rho_f(L)$ must relate to the correlation function. This will be shown in detail below.

As $D < 3$, one sees from (2.3) that $\rho_f(L)$ decreases with increasing L. If L could tend to infinity the object would have zero average density ! In practice, the biggest L will result from the preparation recipe, but there will be an upper limit to the possible values of L imposed by the mechanical stability of the structure. It is known that thermal stability sets that limit if the grains are small in terms of atomic diameters, whereas the gravitational stability sets the limit for large grains.[13] We call ξ the upper value of L imposed by the preparation recipe. It is the *fractal persistence length* (or also the fractal correlation length). A real material is assembled from fractal blobs of upper size ξ. The macroscopic density ρ of such a material just equals $\rho_f(L)$ at ξ,

$$\rho = \rho_a \times (\xi/\ a)^{D-3} .\tag{2.4}$$

2.2.2. *The Density Correlation Function and the Pair Distribution Function*

Let us first approximate the grains by points at positions $\mathbf{r}_i$ ($i = 1$ to N). Their equal-time (coherent) density-correlation function can then be written [14]

$$G(\mathbf{r}) = \frac{1}{N}\sum_{i,j}\left\langle \delta(\mathbf{r} - \mathbf{r}_{ij})\right\rangle \; , \tag{2.5}$$

where $\mathbf{r}_{ij} = \mathbf{r}_i - \mathbf{r}_j$, the $< >$ means average over realizations, and δ is the Dirac delta function. The pair distribution function is the probability of finding a grain in a unit volume at position $\mathbf{r}$ from a central grain, normalized to their number density. It is

$$g(\mathbf{r}) = \frac{V}{N^2}\sum_i{}'\left\langle \delta(\mathbf{r}-\mathbf{r}_{ij})\right\rangle = \frac{V}{N}\left[G(\mathbf{r})-\delta(\mathbf{r})\right] , \tag{2.6}$$

where the prime on the summation sign means that $i = j$ is excluded from the sum, and V is the Euclidean volume occupied by the N grains (we assume that $V >> \xi^3$). Note that $N/V = \rho/m_a$ is the mean number density of grains. For large r and in a random medium, $G(r)$ tends to N/V and $g(r)$ to 1. To perform the sum in (2.6), we take grains j at the centers of fractal blobs in the sense of (2.2). Then, $\sum_i{}'\left\langle \delta(\mathbf{r}-\mathbf{r}_{ij})\right\rangle$ represents the mean number density of grains at distance r from the centers. In terms of (2.3), this is $d(M/m_a)/dV$. Introducing this in (2.6) one finds

$$g(r) = \frac{m_a}{\rho}\frac{d(M/m_a)}{dV} \approx \rho_f(r)/\rho \; , \tag{2.7}$$

where we neglected a factor $D/3 \approx 1$. The final result, $g(r) = \rho_f(r)/\rho$, has the proper asymptotic limit, $g(r) = 1$ at $r >> \xi$.

In practice two problems arise. Eq. (2.7) is only valid in the fractal region. As r approaches the grain size a, it becomes incorrect to neglect the particle size and shape. If the shape can be approximated by a sphere, and if the polydispersity is moderate, then that problem is easily handled in Fourier space as shown in the next subsection. The second issue, for r approaching ξ, is trickier. Many models were made to describe this crossover, as reviewed in [15]. In the case of diffusion-limited cluster-cluster aggregation, so far only simulations are able to provide the form of $g(r)$.[15]

2.2.3. *Structure and Form Factors*

The static structure factor, $S(\mathbf{q})$, which is the quantity measured in scattering experiments integrated over ω, is the Fourier transform of $G(\mathbf{r})$,

$$S(\mathbf{q}) = \int_V G(\mathbf{r})\, e^{-i\mathbf{q}\cdot\mathbf{r}}\, d\mathbf{r} \; , \tag{2.8}$$

where $\mathbf{q}$ is the scattering vector. Since $G(r)$ tends to a constant for large r, (2.8) contains a term in $\delta(\mathbf{q})$. Subtracting this term, and using (2.6), (2.8) becomes

$$S(\mathbf{q}) = 1 + \frac{N}{V}\int [g(\mathbf{r})-1]\, e^{-i\mathbf{q}\cdot\mathbf{r}}\, d\mathbf{r}, \qquad (2.9a)$$

$$S(q) = 1 + \frac{N}{V}\int_0^\infty \frac{\sin qr}{qr}\,[g(r)-1]\,4\pi r^2 dr. \qquad (2.9b)$$

In the last expression, use was made of the isotropy.

Integrating only over the fractal regime, it is not appropriate to subtract the 1 from $g(r)$ in (2.9). Introducing $g(r)$ from (2.7), one obtains

$$S(q) - 1 \propto q^{-D}. \qquad (2.10)$$

For example, for percolation models such as in Fig. 2, the calculated $S(q)$ is shown in Fig. 4. The exact value, $D = 91/48$, is known in that case.[11] One sees from Fig. 4 the enormous effect of the white noise in site percolation. This precludes using small site-percolation models to derive scaling relations.

More precisely, the scattering intensity per unit volume can be written

$$I(\mathbf{q}) = \frac{1}{V}\langle \rho(\mathbf{q})\rho(-\mathbf{q})\rangle, \qquad (2.10a)$$

where

$$\rho(\mathbf{q}) = \sum_i A_i(\mathbf{q})e^{-i\mathbf{q}\cdot\mathbf{r}_i}. \qquad (2.10b)$$

Here, we introduced the scattering amplitude for grain i centered at $\mathbf{r}_i$, $A_i(\mathbf{q})$. It accounts for the structure of the grains. A_i (sometimes called the form factor) equals

$$A_i(\mathbf{q}) = \int_{V_a} \rho_a e^{-i\mathbf{q}\cdot\mathbf{r}} d\mathbf{r} \qquad (2.10c)$$

V_a is the volume of the grain, $\mathbf{r}$ is a vector from its center, and ρ_a could depend on $\mathbf{r}$ if the structure within the particles were important. Note that the exact meaning of ρ_a will depend on the experiment, $e.g.$ it is the electron density for X-ray scattering, or the scattering-length density for neutron scattering. The double sum implicit in (2.10a) can be decomposed into one-particle terms ($i=j$) and two particle terms ($i \neq j$):

$$I(\mathbf{q}) = \frac{N}{V}\langle |A(\mathbf{q})|^2\rangle + \frac{1}{V}\left\langle \sum_{i,j}' A_i(\mathbf{q})A_j^*(\mathbf{q})e^{-i\mathbf{q}\cdot\mathbf{r}_{ij}}\right\rangle. \qquad (2.11)$$

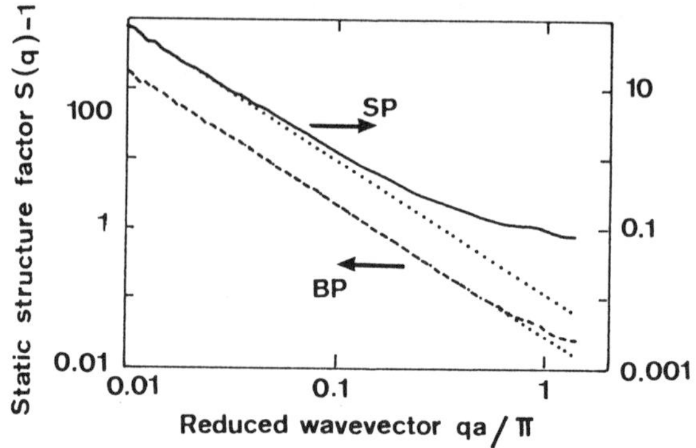

Figure 4. The structure factor $S(q) - 1$ for 2-d site (SP) and bond (BP) percolation models of size 6800× 6800. The dotted lines have slope $-D = -91/48$. The wave vector is normalized by π/a, where a is the size of an elementary square.(After [10]).

For a medium with no correlations between the scattering amplitudes and the relative positions of the grains, the averages under the summation sign in (2.11) are decoupled. Then,

$$I(\mathbf{q}) = \frac{N}{V}\left\langle |A(\mathbf{q})|^2 \right\rangle + \left| \langle A(\mathbf{q}) \rangle \right|^2 \frac{N^2}{V^2} \int_V g(r) e^{-i\mathbf{q}\cdot\mathbf{r}} \, d\mathbf{r} \ . \quad (2.12)$$

For fairly identical centrosymetric particles (for which A is real), we neglect the difference between the two averages of A in (2.12). Subtracting the $\delta(\mathbf{q})$ term still contained in that expression, and writing $P(q) = \langle |A(q)|^2 \rangle$, one gets

$$I(q) \approx \frac{N}{V} P(q) \left\{ 1 + \frac{N}{V} \int_0^\infty [g(r) - 1] \frac{\sin qr}{qr} 4\pi r^2 dr \right\} = \frac{N}{V} P(q) S(q) \ . (2.13)$$

Comparing this to (2.9b), we see that the only effect of the particle structure is the multiplication by $P(q)$, usually called the *form factor*.

If the particles are homogeneous spheres of radius a, the integrals in (2.10c) can be performed in closed form.[16] For $qa \ll 1$, one finds $P(q) = \exp\left[-(qa)^2/5\right]$. For $qa \gg 1$, $P(q)$ shows oscillations of period $2\pi/a$ superimposed to a decay in q^{-4}. The oscillations are hard to see in practice, as these interferences are already washed out by a slight polydispersity. The decay in q^{-4} is the Porod law,[16] which is seen to arise from single particle scattering. More generally, if the surface of the particles is itself fractal with a dimension $D_S > 2$, the law becomes [17,18]

$$I(q) \propto q^{D_s - 6} ,\qquad\qquad (2.14)$$

which for $D_s = 2$ takes the usual Porod value.

2.2.4. *Mutually Self-Similar Series or MSSS* [19]

The scaling of physical properties in the fractal range, $a < L < \xi$, implies that these properties obey power-law dependences on L. This is the case for example for the density, by its very definition (2.2). We saw in the previous section how this can actually be measured in a scattering experiment. The power law q^{-D} of (2.10) directly results from the mass scaling (2.2). However, many other properties should also scale, for example the bulk modulus K. If one had only a single sample, a measurement of this scaling would require a microscopic method to determine the bulk modulus in function of L, $K(L)$. Although, depending on the property of interest, this might sometimes be possible, clearly it would not generally be an easy experimental task.

A way around is to use what we have called a mutually self-similar series of samples. [19] These are materials that are identical in the fractal regime, and differ only by their persistence length ξ. To take the above example, the *macroscopic* bulk modulus for one such sample is $K(\xi)$. So it suffices to study the ξ-dependence of $K(\xi)$ on many samples to determine the scaling $K(L)$.

This assumes that the samples can *really* be made identical, except for ξ. It is not sufficient that different samples have the same fractal dimension D, as illustrated in Fig. 5a which shows schematically the fractal density $\rho_f(L)$. The two samples (1) and (2) cannot be expected to have comparable properties at ξ_1 and ξ_2 . Fig 5b shows

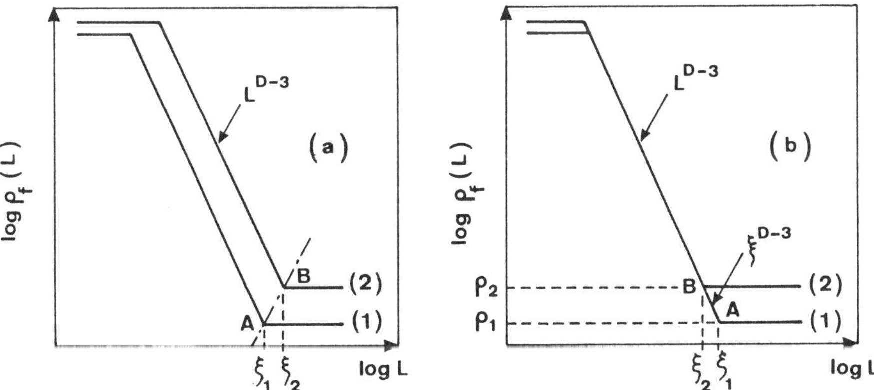

Figure 5. Schematic presentation of $\rho_f(L)$ for (a) fractal samples of same D but not forming an MSSS; (b) the same for an MSSS.

a necessary condition for a MSSS: the points A and B must line up with the fractal slope $\rho_f \propto L^{D-3}$. Hence, if we have an independent way to determine ξ on the different samples, one must find $\xi \propto \rho^{1/(D-3)}$. The curves $S(q)$ must also superimpose. Depending on the actual property of interest, it might further be important, contrary to what is sketched in Fig. 5b, to have identical particle sizes and densities. Finally, if the property depends on the connection between the grains (as would the bulk modulus, for example), it is also important to have similar connections, which cannot be checked simply in an elastic experiment.

It turns out that aerogels, when prepared under closely controlled conditions, can actually be made into what appears good MSSS. This will be shown in sections 3 and 5.

2.2.5. *Surface and Mass Fractals in Nature and in the Laboratory*
Fractals abound in Nature [*e.g.* 3,20], at all scales. The Universe itself, up to the largest catalogued length scales, appears to be a mass fractal with $D = 1.2$. [21] At the other end, tiny colloidal aggregates exhibit remarkable fractal properties.[*e.g.* 22] At intermediate sizes, the surface of the Earth for example has been called a «fractal orange» [2,3].

The solid-state physicist finds mass fractals (or structural fractals) in essentially three types of systems: (1) aggregates or similar systems resulting from a growth and separation process, (2) randomly substituted compounds, in particular magnets, and (3) artificially made structures. The first category comprises the porous silicas described here, but also many other systems that have been studied, *e.g.* silica «smoke» [23], sinters [24], etc. The site-substituted magnets belong to the percolation family. They provide excellent systems to study fractal structure and dynamics,[25,6] in particular the Heisenberg antiferromagnet $Mn_xZn_{1-x}F_2$. The magnon dynamics of disordered antiferromagnets is significantly different from the vibrational dynamics of elastic random media. It will not be discussed here. Artificially made structures are often based on geometrical contructions (*e.g.* Sierpinski gaskets produced by lithography) but also on percolation. They are used, for example, to study phase relationships in superconductors [26] or acoustic localization in Cantor stacks [27].

Surface fractals also abound. The scattering from lignite coals suggests these contain surfaces that are fractal over more than two orders of magnitude in length.[17] Some aggregation models, such as the reaction-limited Eden growth (which describes molds in jelly pots) or the ballistic Vold growth, produce compact objects ($D = 3$ in 3-d) with fractal surfaces ($D_s > 2$).[20] In the case of aerogels, the surface of the grains can be fractal, and can be made smooth by heat treatment in air.[19] Fractal drums [28] provide nice examples of dynamical investigations of surface fractals, a subject which is picking up active attention.

3. Varieties of Porous Silica and their Structures

3.1. LEACHED GLASSES

Leached glasses are porous materials obtained by dissolving away, in a percolation-like manner, the soluble components of a phase-separated multi-component glass. In the final product, only the insoluble component remains, with a large surface area. The most common material is Vycor™, brand N° 7930 of Corning Glass Works,[29] but materials of very different pore sizes can be made by the same principle.[30]

3.1.1. *The Preparation of Vycor and similar Materials*

The preparation is a three-step process: (1) one forms a borosilicate melt of typical composition SiO_2 (75%), B_2O_3 (20%), and Na_2O (5%); (2) this melt is quenched just below its consolute temperature where it slowly separates into a silica-rich phase (with ~96% SiO_2 and 3% B_2O_3) and a borate phase. The system is let to solidify, and it is carefully annealed; (3) the boron-rich phase is leached out with acid at ~100°C. Key to the preparation is the existence of a miscibility gap and the heat treatment near that temperature. Spinodal decomposition presumably takes place. By varying the above conditions, the mean pore radius can be adjusted from 20 to 2000 Å.[30]

3.1.2. *Structure and Properties of Leached Glasses*

There has been a long debate concerning the structure of Vycor (*e.g.* [31] and Refs. therein). A source of difficulty might have been that one must carefully distinguish between the structure of the glass, that of the pores, and that of the glass-pore interfaces. Fig. 6 shows an image of a leached glass of large pore radius, R_p ~ 0.2 μm. [30] It reveals a highly connected network, without apparent hierarchy in the glass. This is clearly very different from the fractal structures of aerogels shown in Fig. 8.

268

Figure 6. Scanning electron microscope image of a leached glass of large pore radius, $R_p \sim 0.2$ μm. (from [30]).

One should note that Fig. 6 is somewhat atypical in that the pores are very large and that the ratio of pore volume to total volume also seems large.

For commercial Vycor, the density is $\rho = 1.45$ g/cm^3, corresponding to a porosity $\phi = 28$ %. The surface area A depends on lot number,[31] which probably reflects the high sensitivity of spinodal decomposition to thermal and compositional conditions. Values of ~ 100 to ~ 200 m^2/g are obtained. If V is the volume of a sample, R_p the mean pore radius, and L the total pore length, approximating the pores by circular tubes, one expects $A = 2\pi R_p L$ and $V\phi = \pi R_p^2 L$, from which $R_p = 2\phi/\rho A$. With $A = 200$ m^2/g this gives $R_p = 20$ Å, a value specified by the manufacturer and confirmed by microscopy. Other values are also found in the literature, *e.g.* [31], consistent with the above formuli if the surface is taken from nitrogen adsorption measurements. The good agreement with the above Euclidean expressions already suggests that the system in not fractal.

Many scattering determinations of the structure of Vycor can be found in the literature. We discuss here a fairly recent one [32]. Returning to (2.11), and if the approximations leading to (2.12) are not valid, one can introduce probability densities describing the distribution of the $A_i(\mathbf{q})$'s, and the pair distribution of $A_i(\mathbf{q}), A_j(\mathbf{q})$ ($i \neq j$). As shown in [32], the scattering can be written as the sum of three terms: (1) a contribution from individual grains, just as the first term on the right-hand side (r.h.s.) of (2.12); (2) a term related to the pair distribution function, quite similar to the second one on the r.h.s. of (2.12); and, (3) a third term containing corrections related to the size distribution and spatial correlations of the grains. The first term gives the Porod scattering. Li and Ross [32] calculate the second term for the Cahn-Hilliard theory of spinodal decomposition [33]. The expression they obtain compares very well with their

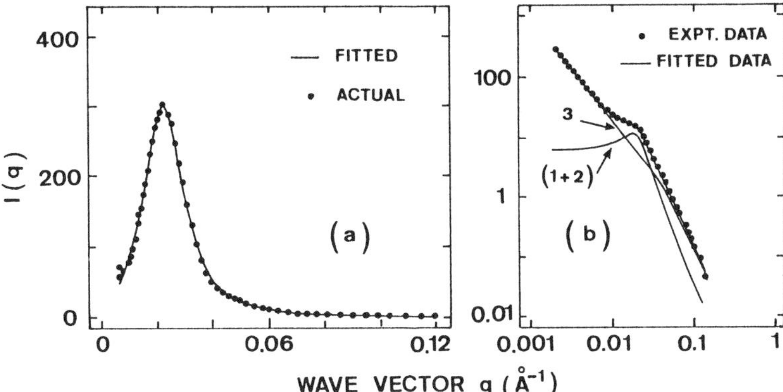

Figure 7. Small angle neutron-scattering results on Vycor (Corning 7930): (a) is obtained with dry Vycor, (b) with Vycor containing a H_2O/D_2O mixture adjusted to match the coherent scattering-length density of SiO_2, and at ~70% humidity while the water is being pumped out. (From [32]).

experimental neutron-scattering measurement of $I(q)$ as shown in Fig. 7a. A quantitative test of the model on systems with different quenching times remains to be performed, as pointed out in [32]. Similar conclusions were also reached in earlier studies, *e.g.* in [34].

3.1.3. *Properties of the Pores*

In the same study, Li and Ross measured water-impregnated Vycor, with the result shown in Fig. 7b. In addition to a signal of the same shape as in Fig. 7a, there is a new component exhibiting a power law q^{-D} over a large region of q, with slope $D \approx 1.7$. The interpretation of this signal as coming exclusively from the correction term (3) might be questioned, as the term (2) should also contain new contributions in this truly different sample. The experimental fact however is rather clear: the partial filling of the pores produces a fractal distribution whose persistence length is larger than 1000 Å. Fig. 7b also reveals that in the Porod region the power-law decay is not in q^{-4} but rather in $q^{-3.6 \dots 3.7}$, suggesting a fractal surface of the grains with $D_S = 2.4 \dots 2.3$. The same value for D_S was obtained in other studies, *e.g.* [31].

3.1.4. *Some Uses of Leached Glasses*

Owing to its large internal surface, and to the relatively good mechanical and chemical stability of its skeleton, Vycor finds many uses in chemistry and biology, wherever adsorption-desorption processes are of interest, *e.g.* in separation. It is also used for gas

270

adsorption and for catalysis. In more fundamental studies, Vycor is often selected to investigate the effect of restricted geometries on various phenomena of physico-chemical [30] or physical importance. For the latter, the pore size adds an extra length scale to systems with intrinsic length scales, such as for phase transitions. One can cite the superfluid transition of He^4, the formation of micelles, the supercooling of liquids, transitions in liquid crystals, phase separation in binary liquid mixtures, glass transitions in model systems, etc. To give just one example of such studies, we consider the phase separation in binary mixtures.

In the bulk, binary mixtures exhibit sharp second order transitions at the consolute point, with critical slowing down and opalescence. In the confined system, one observes metastability and history effects, deep in the two-phase region, with no macroscopic phase separation and extreme slowness. The original experiments were performed with lutidine-water.[35] As lutidine wets the silica, the lutidine-rich phase feels a field that is random, and thus theoretical ideas were first developed similar to the random field Ising model.[36] Another effect of the wetting is however to promote the formation of elongated capsules in cylindrical pores, and if one goes deeper in the two-phase region these capsules can become real plugs.[37] The important role played by these structures in phase separation was confirmed in computer simulations, *e.g.* [38]. Recent measurements of the structure factor $S(q)$ on the same system [39] show that the binary fluid separation is indeed controlled by the formation of capsules.

3.2. GELS AND AEROGELS

Gels are obtained from a wide variety of substances. For silica, they provide an appropriate route towards porous materials. If the particles in the gel are formed by reaction in solution, one speaks of a *polymeric* gel, whereas if they pre-exist in the sol and the gel is obtained by destabilization of this sol, one names it *colloidal*. The wet gels must be dried to obtain the porous material. This is a crucial step owing to the fragility of their solid skeleton. When dried under usual conditions, in which case the liquid forms a meniscus, the product is called *xerogel*. The skeleton is then strongly damaged by the forces originating in the surface tension at the moving meniscus. *Aerogels* are dried above the critical point of the liquid phase, a process called *supercritical drying*.[40] In this case there is no meniscus. This leads to monolithic, highly porous solids.

3.2.1. *Varieties of Aerogels*

Colloidal aerogels can be prepared by acid destabilization of Ludox™ sols.[41,42] These are alkaline suspensions of fairly monodisperse and spherical silica particles. The particle diameter can be selected between ~8 and ~30 nm. The destabilization of Ludox by acid leads to diffusion limited cluster-cluster aggregation (DLCA).[43,44] This process produces mass fractals with $D \approx 1.8$, in agreement with observation if account is taken of finite size effects [15]. A transmission-electron-microscope (TEM) image of such an aerogel is shown in Fig. 8a. One recognizes individual particles and their hierarchical aggregation. The dynamics of such gels was little studied. It shows an early departure from the low frequency phonon regime.[42]

Another method is to grow the particles by reaction in the sol. Typically, one starts from an alkoxysilane, $Si(OR)_4$ with $R- = CH_3-$, C_2H_5-,... which is mixed with water, alcohol, and a catalyst (pH modifier). The hydrolysis of $Si(OR)_4$ forms silicon hydroxide in several steps, and the hydroxide groups polycondense into siloxane bonds, $\equiv Si - O - Si \equiv$. The pH modifies the various reaction rates,[45] and refs. therein. Under neutral conditions (no catalyst added, materials labelled N hereafter) the hydrolysis is very slow and incomplete, and the condensation rapid. Hence, N-aerogels grow like branched polymers, forming eventually small, polydisperse and fluffy particles. Good TEM images are difficult to obtain because of various problems related to charging, coating, or replication of these delicate structures.[46,47]

The addition of a base catalyst increases the hydrolysis rate and the solubility of SiO_2. Larger grains grow, while smaller ones tend to redissolve. The aggregation of grains is initially prevented by their charge. Once the particles are sufficiently large they nevertheless aggregate. For these reasons, base-catalyzed aerogels (hereafter labelled B) are sometimes called colloidal. A TEM picture is shown in Fig. 8b. One

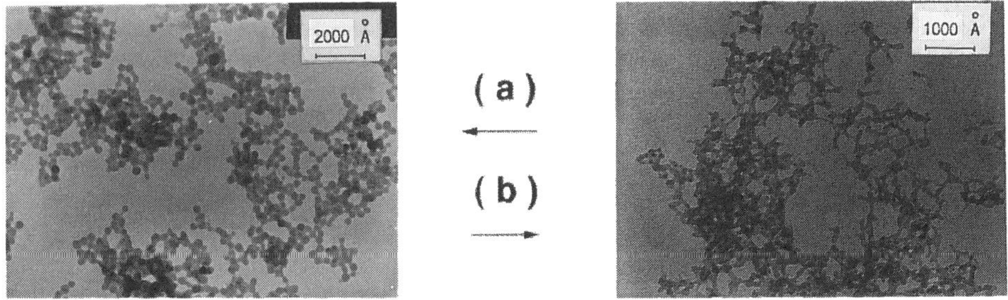

Figure 8. TEM images of small fragments of colloidal aerogels: (a) an aerogel grown from Ludox™ of grain size $a \approx 80$ Å; (b) a B-aerogel with $a \approx 25$ Å. (From [42]).

notices that most grains in (b) belong to closed rings. There exist many other variants to the preparation of aerogels, such as acid catalysis, two-step reaction processes, drying with solvent exchange, etc. In each case the structure might be influenced by the preparation so that a rather careful structural study seems to be a prerequisite to other physical investigations.

3.2.2. *Some Uses of Silica Gels*

Silica gels are light, porous materials with a very high internal surface. They have low dielectric constant and refractive index. They are translucent, and have very low thermal conductivity. In the case of aerogels, large pieces can be made in the form of bricks, plates, or even hollow spheres. These exceptional properties are the reasons for the wide applicability of aerogels. Considerable information can be found in proceedings of conferences on Aerogels,[48,49] or those on Glasses and Ceramics from Gels.[50-51]

The largest quantity of silica aerogels is used for thermal insulation in buildings and appliances. The development of their solar-energy-conversion potential is being actively pursued. In physical chemistry, the high open porosity of silica gels has been know for a long time, and has been the reason for their applications as desiccant and as catalyst support. Aerogels are now used in chemical engineering, not only for catalyzers, but also in devices for gas condensation, chemical separation, and filtering. Aerogels support the gas fuel of inertial-confinement-fusion targets, or laser-fusion pellets.[52] Their low and tunable densities find applications in Cerenkov detectors and particle collectors, in particular those orbiting the Earth. Their low dielectric constants make them useful as fillers in integrated circuit technologies. In another register, aerogels are used as consistency modifiers in food and cosmetics, or as safe pesticides. Silica aerogels can be transformed into glasses by densification at high temperature. When impregnated with refractive-index modifiers, this allows to prepare the rods from which graded optical fibers are pulled. They can also be impregnated by nuclear waste which is then encapsulated by sintering for long term disposal. In fundamental research, aerogels allow to study the properties of matter under fractal confinement, and to investigate the vibrational dynamics of fractals, our purpose here.

3.2.3. *Fractal Structure of Silica Aerogels*

As seen in section 2.2, D and ξ can be determined from $S(q)$ measured in small-angle scattering. The results shown below were obtained with cold neutrons (SANS), giving a q-range typically from $\sim2\times10^{-3}$ to ~0.3 Å^{-1}. This allows to observe fractal behavior between ~1000 Å (upper limit on ξ) and ~10 Å (lower limit on a). As explained above, N- and B-aerogels are expected to be quite different.

Small Angle Scattering from Neutrally Reacted Aerogels   Results on N-aerogels are illustrated in Fig. 9a.[19] The samples are labelled by their macroscopic densities, ρ, in kg/m^3. The intensities are per unit volume of sample. In the homogeneous ($q<\xi^{-1}$) and fractal regions excellent fits are obtained with a smooth exponential crossover [53]

$$g(r) - 1 \propto r^{D-3} \exp{(-r/\xi)} \qquad . \qquad (3.1)$$

The Fourier transform of (3.1) gives the lines of Fig. 9a. These fits have three parameters, D, ξ, and the intensity $I(q=0)$. Not only *each sample independently* has $D = 2.40\pm0.03$, but *for the whole series*, $\xi \propto \rho^{-1.67} = \rho^{-1/(3-D)}$, and also $I(0)\propto \rho^{-3.05} = \rho^{-(2D-3)/(3-D)}$. Both scalings support the MSSS property. The scaling of $I(0)$ implies that the curves $I(q)/\rho$ superpose in the fractal region. As-grown material also has fractal particle surfaces with $D_s \approx 3$, while the Porod law is obtained after oxidation, as shown for the sample marked 356-OX, for which $a \approx 8$ Å.

Small Angle Scattering and Simulations of Base Catalyzed Aerogels   Typical results are illustrated in Fig. 9b. Here the intensities have been normalized per unit sample

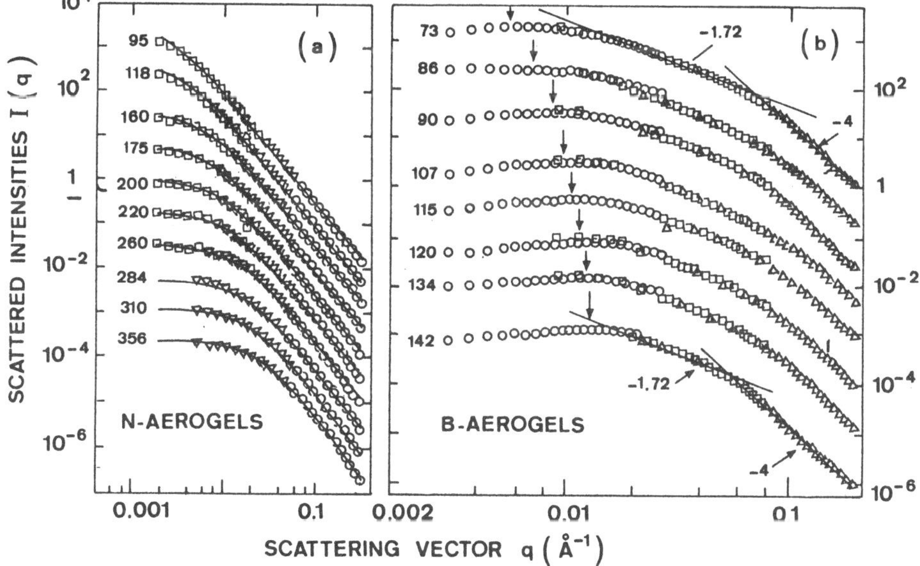

Figure 9. The structure factor $S(q)$ measured: (a) on N-aerogels, from [19]; (b) on B-aerogels, from [54]. The curves are marked by the densities in kg/m^3. They are displaced vertically for visibility, by factors of 4 for (a) and factors of 2 for (b).

274

mass.[54] When undisplaced, the curves *superpose in the fractal region*. Now, contrary to Fig. 9a, the curves show a maximum in the region $q \sim \xi^{-1}$. The physical origin of this maximum is the existence of fractal blobs resulting from the growth. This gives a pseudo-periodicity at the most probable size of the blobs.[19,42] These scattering curves cannot be fitted with the Fourier transform of (3.1), or for that matter with any available analytical expression, as discussed in [15]. It is shown there that *excellent* fits are obtained by DLCA-simulations, and that the maximum in $S(q)$, at the value q_{max} shown by arrows in Fig. 9b, corresponds to $\xi = 2.75/q_{max}$. One finds $\xi \propto \rho^{-1/(3-D)}$ with $D = 1.72$,[54] a value that is also in agreement with simulations [15]. This indicates that the B-aerogels form a MSSS. Excellent fits are also obtained for the size and polydispersity of the grains.[15,55]

4. Fractons: the Eigenmodes of Fractal Structures

4.1. SCALING LAWS ILLUSTRATED BY A SIMPLE MODEL

Consider a very simple model of N masses ($i = 1$ to N) connected by springs. Assume the masses can only displace in one direction, by quantities u_i. Assume further that the restoring forces are simply proportional to the strains with direct neighbors, $u_j - u_i$. This is called *scalar* elasticity.[56] Usual elasticity is tensorial of course, displacements are vectors, and nearest neighbor forces include not only stretching but also at least bending. In spite of its simplicity, the scalar elastic model captures some of the essential ingredients of the dynamics of fractal systems. To fix the ideas, the masses could be the occupied sites of a large BP-system of the type shown in Fig. 2b, with periodic boundary conditions. The support is then fractal from the grain size to a persistence length of the order of the box size. To find the *harmonic* vibrations of such a model one diagonalizes its dynamical matrix. For a fully connected system, one finds N modes, of which only one is uniform with $\omega = 0$, and N−1 are at frequencies ω^{α} with eigenvectors u_i^{α}, orthogonal to the uniform mode, hence $\sum_i u_i^{\alpha} = 0$, for all α.

It helps to inspect some of these modes to find out what happens. At very small ω one has acoustic phonons. However, as soon as the wavelength becomes of the order of the persistence length, the modes localize. Hence, above that (small) frequency, the modes have appreciable intensity only over a restricted region of space, of size λ^{α} for mode α. Images of eigenvectors are found in [57]. These modes cease to be acoustic phonons and are called *fractons*.[58,59] The size of a fracton can be precisely defined by an averaging *over sites* in which the $(u_i^{\alpha})^2$ are taken as weights.[57] Similarly, one can define *local* averages, for example $\overline{\omega_i^2}$, by averaging with the same weights over all fractons. Finally, if the system is sufficiently large, or by taking a sufficient number of realizations, one can average over fractons in a small *frequency interval* $\Delta\omega$ about ω to define an ω-dependant average, which for λ^{α} is $\lambda(\omega) = \langle \lambda^{\alpha} \rangle_{\Delta\omega}$. From the simulations, one finds that: (1) fractons of all frequencies are felt everywhere; (2) more connected regions have higher $\overline{\omega_i^2}$; and (3) the average $\lambda(\omega)$ follows a scaling law, which is written [7,58,59]

$$\lambda(\omega) \propto \omega^{-\tilde{d}/D} \tag{4.1}$$

The new index $\tilde{d}$ is called the *fracton* or *spectral dimension*. The reason for this name is that the density of fracton states in an interval $d\omega$ is given by

$$N(\omega)\, d\omega \propto \omega^{\tilde{d}-1}\, d\omega \tag{4.2}$$

which is just like the density of acoustic phonon states, except that the Euclidean dimension d is replaced by $\tilde{d}$. It is important to understand how (4.2) follows from (4.1). Many proofs exist [7,58] and the relation is vindicated by simulations. Maybe the simplest way is to remark that λ^{-1} plays the role of a wavevector q (as shown below) and that fracton states are on a fractal support. Counting them in the usual way, their density must obey $N(\omega)\, d\omega \propto dq^D \propto d\lambda^{-D} \propto d\omega^{\tilde{d}}$, where (4.1) was used for the last relation. If the support is tightly connected at all scales, one expects $\tilde{d} = D$. However, if the connections increase as λ decreases, then ω increases faster than q and $\tilde{d} < D$. Thus, in general, $\tilde{d} \leq D$. This is why the effective $\tilde{d}$ depends so strongly on the range of forces, even if these remain always short range.[60]

4.2. THE SINGLE-LENGTH-SCALE POSTULATE (SLSP)

This postulate states that for fractons all dynamically relevant length scales, whether wavelength, scattering length, localization length, or else, must simply be proportional to the fracton size $\lambda(\omega)$.[7] This powerful statement allows to predict scaling relations. Its validity was demonstrated by simulations in the case of the BP-model,[10] as

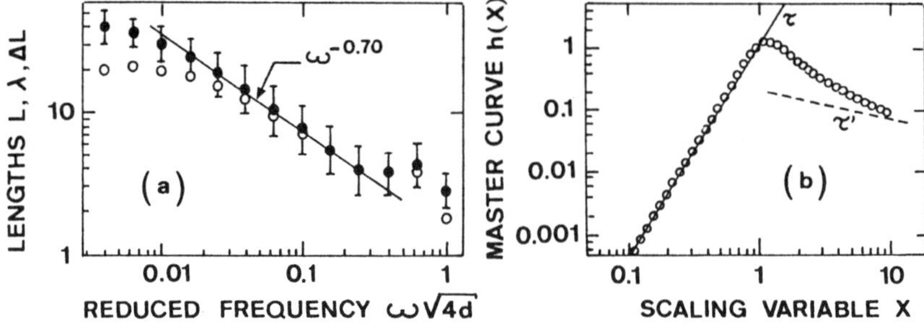

Figure 10. Testing the SLSP on a 2d-BP model. (a) the scaling of dynamical lengths: ●$L(\omega)$; O $\lambda(\omega)$; $\Delta L(\omega)$ is given by the size of the bars. That it is nearly constant in this presentation means that ΔL scales with the same exponent. (b) the master curve $h(x)$ for the dynamical structure factor. (From [10]).

illustrated in Fig. 10. In 10a, the Thouless localization length L, defined by $[L(\omega)]^D \equiv \sum_i (|u_i^\alpha|^2)^2 / \sum_i (|u_i^\alpha|^4)$, [61] its root-mean-square fluctuations $\Delta L(\omega)$, and the size λ itself, all scale with $\omega^{-\tilde{d}/D}$. The exponent for the 2d-BP model is $-64/91$. Scaling also implies that the dynamic structure factor, $S(q,\omega)$, follows a master curve. For this, strain is relevant. The theory calls for another scaling index $\sigma \geq 1$ connecting the coherent strain $\tilde{e}(\omega)$ of fractal blobs to their displacement amplitude $u(\omega)$, [7,62]

$$\tilde{e}(\omega) = u(\omega)/\lambda^\sigma(\omega) \qquad . \qquad (4.3)$$

From (4.1), it is obvious that the scaling variable is $x \equiv q\omega^{-\tilde{d}/D}$. One finds [62] that $S(q,\omega) \propto \omega^{-2} q^y h(x)$, where $y = 2\sigma - D/\tilde{d}$. The master curve $h(x)$ is shown for the 2d-BP model in Fig. 10b. The asymptotic exponents τ and τ' are drawn with the values predicted by the theory.[62] It appears that the simulations might not have reached sufficiently far into the large fracton (low frequency limit) on the τ' side.

5. Inelastic Scattering and the Vibrations of Aerogels

Aerogels are *excellent* systems to investigate fracton dynamics. Their acoustic persistence length can be made of the order of the wavelength of visible light. Hence, the phonon-fracton crossover at ω_{co} can be investigated by Brillouin scattering. With incoherent neutron scattering, one can measure $N(\omega)$. Light scattering is also seen much above ω_{co}, in which case we use the term Raman scattering. Finally, thermal properties provide additional information on fracton dynamics.

5.1. BRILLOUIN SCATTERING NEAR THE PHONON-FRACTON CROSSOVER

Brillouin scattering was measured on MSSS of N- and B-aerogels,[63,64,54] in function of their densities ρ, and of the scattering angle. The latter determines the momentum exchange q in the experiments. For small q, and for heavy samples, the spectra are narrow lines, while they broaden rapidly as q is increased or as the density ρ is decreased. This is due to the crossover from acoustic phonons to localized fractons. The experiment probes excitations having a Fourier component at the *plane* wave vector q. As the excitations localize, many of them have such a component, which leads to *inhomogeneous* broadening, unrelated to the lifetime of the excitations. The fracton linewidths are thus essentially independent of the temperature T.[63]

5.1.1. *A simple minded approach to the interpretation of the spectra and to scaling*
In a first (naïve) approach, the line profiles of N-aerogels were fitted to lorentzians, or to power of lorentzians.[63] This allowed to extract the peak positions ω_p and the widths Γ, as shown in Fig. 11 for various densities. It is seen that ω_p saturates more rapidly for lighter gels, and that $\Gamma \propto \omega^4$.

From these measurements we extracted the exponents of scaling laws for the velocity in the phonon regime, $c_0 \propto \rho^x$, the crossover frequency, $\omega_{co} \propto \rho^y$, and a

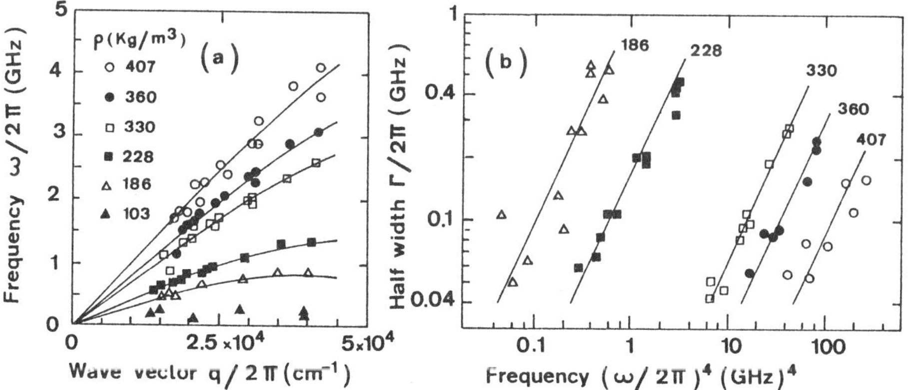

Figure 11. The peak positions ω_p (a) and the linewidths Γ (b) derived from a simple minded analysis of the Brillouin spectra of N-aerogels.[63] Different samples are labelled by their densities in kg/m³.

278

crossover length, $L \propto \rho^{-\zeta}$. It is clear from the above that $L = 1 / q_{co} = \xi_{ac} \propto \xi$, where ξ_{ac} is a persistence length appropriated to the elastic waves [47]. Thus, $\zeta = 1/(3-D)$. The point (q_{co}, ω_{co}) is defined as the intercept of the extrapolated phonon regime, $\omega = c_0 q$, with the fracton regime, $\omega \propto q^{D/\tilde{d}}$, in the (q,ω)-plane. For a MSSS, c_0 scales like $q_{co}^{D/\tilde{d}-1}$, from which $x = (D - \tilde{d}) / \tilde{d}(3 - D)$. Finally, by similar arguments, one finds $y = x + \zeta$. From these scalings and the data of Fig. 11, we estimated $D = 2.36$ and $\tilde{d} = 1.25$. This approach does not use the richness of the spectral lineshapes. It also makes believe that the dispersion that appears in Fig. 11a is similar to the dispersion at the edge of a Brillouin zone, although it is a completely different effect having to do with the scattering from localized excitations.

5.1.2. *Accounting for the spectral shape*

Theories of the spectral shapes were developed in parallel to our scattering measurements.[65,66] The physical picture is that, as ω increases towards ω_{co}, the definition in q of the vibrations becomes less and less precise. Thus vibrations of different frequencies have an appreciable Fourier component at the scattering vector q. This leads to a dependence in frequency (rather than in q, which *for the vibrations* is not well defined) of the parameters entering the spectral function $F(q,\omega)$. These are the linewidth, $\Gamma(\omega)$, and the velocity, $c(\omega)$. From a Green function analysis, one finds [66]

$$F(q,\omega) = \frac{c^2 q^2}{\omega} \frac{\Gamma}{(\omega^2 + \Gamma^2 - c^2 q^2)^2 + 4\Gamma^2 c^2 q^2} . \qquad (5.1)$$

At low $\omega \ll \omega_{co}$, c is equal to c_0 and $\Gamma \propto \omega^4$ owing to Rayleigh scattering of the phonons by the structural disorder. At high frequency, $\omega \gg \omega_{co}$, one has reached the Ioffe-Regel limit $\Gamma \cong \omega$ (for plane waves),[8,67] while $c = \omega/q \propto \omega^z$, with $z = 1 - \tilde{d}/D$ from the fracton dispersion curve (4.1). Complications arise in the intermediate regime. Using the effective medium approximation (EMA) on a percolation model,[66] discontinuities are found at ω_{co}. Such discontinuities seem to arise from the EMA, and they are not apparent in aerogel spectra. Therefore, we devised empirical crossover expressions, [64]

$$c = c_0 \left[1 + (\omega / \omega_{co})^m \right]^{z/m} , \qquad (5.2)$$

$$\Gamma = W\omega \; \frac{(\omega/\omega_{co})^3}{\left[1+(\omega/\omega_{co})^m\right]^{3/m}} \qquad . \qquad (5.3)$$

These have the proper asymptotic behavior. There is only one additional crossover parameter, m. It fixes how rapidly the phonon regime changes into the fracton regime. Although spectral shapes for $q \approx q_{co}$ depend significantly on m, the values obtained for c_0 and ω_{co} are not strongly correlated with m. The other parameter entering (5.3), W, can be set equal to 1, in which case the Ioffe-Regel limit is strictly $\Gamma=\omega$. If this parameter is slightly smaller than 1, it practically amounts to setting the Ioffe-Regel crossover, ω_{IR}, slightly above the strong localization crossover, ω_{co}, a feature which is expected and could be felt if the crossover were sharp (large value of m).[67]

The scattered intensity is related to the spectral shape by

$$I(q,\omega)=A\,q^2\,n_B\,F(q,\omega) \qquad . \qquad (5.4)$$

A is an amplitude coefficient independent of q and ω, the factor q^2 arises from the expansion of the correlation function in the displacements,[62] and n_B is the Bose factor equal to $k_B T/\hbar\,\omega$ for the experiments shown here.

Measurements at 8 values of q were performed on a new high quality MSSS of N-aerogels with 10 different densities.[64] The entire data are adjusted with the same values of z and m, but to each spectrum there are different q_{co}, ω_{co}. The parameter A also varies, as it is difficult to control the scattering volume in these experiments. The

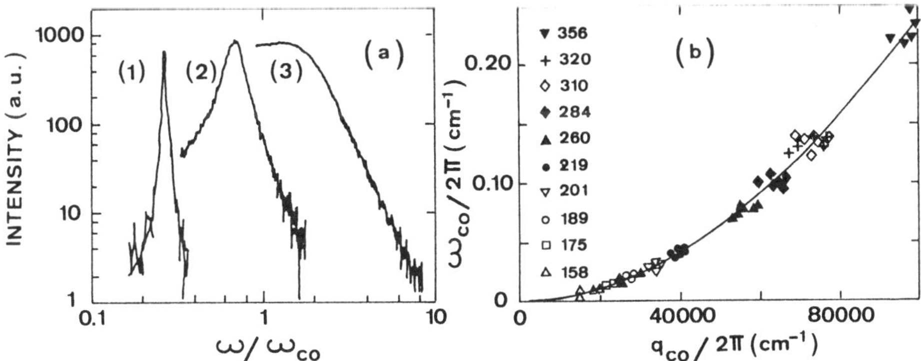

Figure 12. Scaling Brillouin results on N-aerogels: (a) the universal lineshapes for different values of q/q_{co} and $m = 2$, $z = 0.47$, $W = 1$;[68] (b) the fracton dispersion curve traced from the fit parameters of each individual spectrum, q_{co} and ω_{co}, relying on the MSSS-property. (From [69]).

fits are performed iteratively, searching for consistency between the selected z and the fracton dispersion curve (4.1) derived from a plot such as that in Fig. 12b. This relies on the MSSS-property. Excellent fits are obtained with $z = 0.47$ and $m = 2$. An acoustic value for D, written D_{ac}, is then extracted from the scaling of q_{co} with ρ. We find $D_{ac} = 2.46\pm0.03$, in good agreement with the structural determination $D = 2.40\pm0.03$. Finally, from D_{ac} and z, one finds $\tilde{d} = 1.3\pm0.1$. Spectra from different samples can be superposed, as they depend only on q/q_{co} and on ω/ω_{co}. Such universal lineshapes are illustrated in Fig. 12a. The validity of this approach was further confirmed by measurements on uniaxially strained N-aerogels.[70]

Similar measurements on a MSSS of B-aerogels with $D = 1.72$, give $m = 4$ and $z = 0.35$, from which $\tilde{d} = 1.1\pm0.1$. The fits are improved by setting $W = 0.77$. One finds $D_{ac} \cong D$. The higher value of m indicates a sharper crossover in the fracton regime, consistent with the existence of fractal blobs that produce the hump at $\sim 1/\xi$ in Fig. 9. It is this sharpness that allows to distinguish somewhat the localization crossover from the Ioffe-Regel one [67], leading to $W < 1$.

5.2. NEUTRON DETERMINATION OF THE DENSITY OF STATES

The determination of $\tilde{d}$ presented above is restricted to the phonon-fracton crossover region and relies on the MSSS-property. One would like to obtain an independent measurement of $\tilde{d}$ on *single samples,* and one which extends throughout the fracton regime. Neutron scattering allows to determine the density of vibrational states and thus, according to (4.2), the value of $\tilde{d}$. The dynamical structure factor for incoherent neutron scattering is [71]

$$S(Q,\omega) \propto n_B Q^2 \exp\left[-\tfrac{1}{3}Q^2 \langle u^2 \rangle\right] N(\omega)/\omega \qquad . \qquad (5.5)$$

Here, Q is the scattering vector. This expression also applies to coherent scattering when $Q \gg 1/\lambda$, where λ is either a typical fracton size or a phonon wavelength.

Since the excitations of interest cover more than four orders of magnitude in ω, several neutron spectrometers must be employed. At low ω, we utilized spin-echo,[72] at intermediate ω backscattering,[73] and at higher ω time-of-flight spectroscopies. [73,74] The spectra at intermediate frequencies, as illustrated in Fig. 13a, do not agree with Eq. (5.5). On the log-scale of Fig. 13a, one should observe parallel curves at different Q-values at a given T. The simplest way to account for these results is to

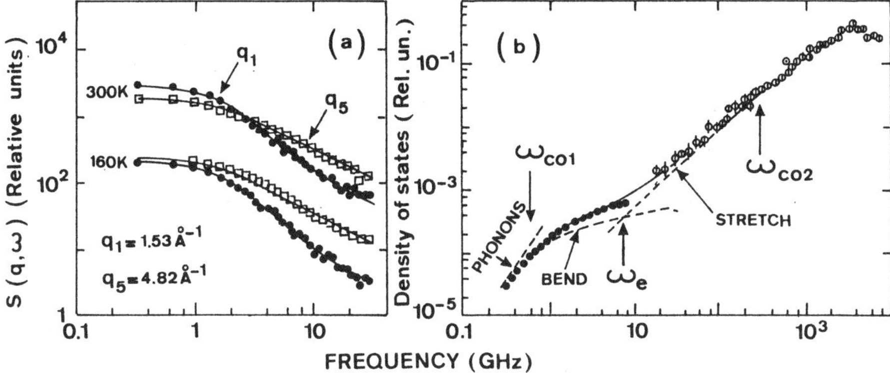

Figure 13. Neutron scattering results obtained on a N-aerogel with ρ = 210 kg/m^3. (a) backscattering spectra at two values of Q and two temperatures; (b) the density of states: dotted line from spin-echo, continuous line from backscattering; points from time-of-flight spetroscopies. (From [73]).

postulate two additive contributions of the type (5.5), having different N and $\langle u^2 \rangle$. Excellent fits to the entire data, as illustrated by the solid lines on Fig. 13a, are then obtained.[73] The existence of two distinct vibrational regimes with different $\tilde{d}$, one dominated by bending at low ω, and the other by stretching at high ω, was predicted by Feng.[75] It has also been confirmed by simulations.[76] For the bending component, our results are consistent with $\tilde{d}_b = 1.3$, and $\langle \frac{1}{3} u_b^2 \rangle = 0.33$ Å^2 at room T. For the stretching, we find $\tilde{d}_s = 2.2$, and $\langle \frac{1}{3} u_s^2 \rangle = 0.06$ Å^2 at room T. These values of $\tilde{d}$ agree with the asymptotic behaviors observed with the other spectroscopies, as shown in Fig. 13b. In addition to the phonon-fracton crossover at ω_{co1}, and the fracton-particle crossover at ω_{co2}, there is also an elasticity crossover ω_e where the vibrations change from bending dominated to stretching dominated.

5.3. LIGHT SCATTERING DEEP INTO THE FRACTON REGIME

As ω increases beyond the fracton crossover, the light scattering is produced by fractons of decreasing size, according to Eq. (4.1). In our experiments these are always much smaller than the light wavelength. Fig. 14a illustrates *polarized* backscattering spectra observed on N-aerogels of low ρ. For N175, ω_{co} is around 0.8 GHz, and the signal at $\omega > \omega_{co}$ falls off with the power law $I(\omega) \propto \omega^{-3-2\tilde{d}/D_{ac}}$ predicted by Eqs. (5.1-3). For smaller ρ, ω_{co} decreases, and at a given ω one observes a progressive departure from the above power law, suggesting that another scattering mechanism becomes operative.

To observe this effect free of acoustic-phonon-like contributions, we measured *depolarized* backscattering.[77] Results are illustrated in Fig. 14b, where the signal is divided by n_B (*i.e.* multiplied by ω). There is first an increase for $\omega > \omega_{co}$, then one slowly enters a new power law regime, $\omega^{-\psi}$, with $\psi \approx 0.37$ for the N-aerogels, and 0.75 for the B-aerogels. The fracton regime ends with the appearance of a hump whose position and shape is relatively independent of the sample density as it is produced by the scattering from particle modes.[77] In other words, to observe the power law over a significant range of ω, one needs to reach deep into a sufficiently extended fracton regime. This might explain the difficulty initially met in reproducing such scaling in numerical simulations,[78] while recent calculations on a large and good fractal model were successful.[79]

Several scattering mechanisms, besides the direct coupling to density, can come into play.[62] In all generality, $I(\omega)$ should depend on the strain exponent σ introduced in (4.3). However, for well connected objects, one expects strong self-similarity and $\sigma = 1$. Eqs. (5.1-3) implicitly made that assumption, which is also used in the following expressions. On the basis of rather straightforward scaling arguments, one finds [62]

$$I_{\text{DIR}} \propto \omega^{-3-2\tilde{d}/D} \quad , \tag{5.5}$$

$$I_{\text{DID}} \propto \omega^{-3-2\tilde{d}+8\tilde{d}/D} \quad , \tag{5.6}$$

$$I_{\text{POC}} \propto \omega^{-3+2\tilde{d}/D} \quad . \tag{5.7}$$

The subscripts correspond to the direct (DIR), dipole-induced-dipole (DID), and acousto-optic or Pockels (POC) mechanisms. One expects that bending dominated

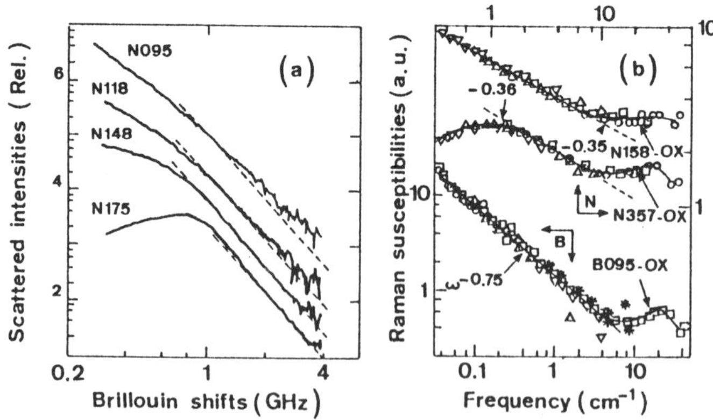

Figure 14.(a) Polarized backscattering from N-aerogels in the fracton regime (from [69]); (b) depolarized backscattering from N- and B-aerogels. The curves were shifted vertically for legibility (From [77] and [73]).

modes should depolarize the light much more effectively than stretching ones. Hence, the value of $\tilde{d}$ in these expressions should be $\tilde{d}_b$ for depolarized backscattering. This view is supported by the absence of any sign of crossover at ω_e in Fig. 14b.

Table I shows the values of ψ calculated from (5.5-7) for the two different MSSS and the three mechanisms. One finds that the slopes in the Raman regime are in agreement with DID for the N-aerogels, and with the acousto-optic mechanism for the B-ones. Presumably, DID-scattering is favored by the small grain size of the N-aerogels. The tightly connected structure of the B-aerogels (confirmed by $\tilde{d}_s \cong D$) and their larger grains enhance the relative importance of the acousto-optic coupling.

Table I. Experimental and calculated indices for the N- and B-aerogels

	D	$\tilde{d}_b$	$\tilde{d}_s$	Ψ_{DIR}	Ψ_{DID}	Ψ_{POC}
N	2.46	1.3	2.2	3.06	0.37	0.94
B	1.72	1.1	1.7	3.28	−0.92	0.72

5.4. THERMAL PROPERTIES

The low-temperature thermal properties of aerogels reflect their fracton dynamics. The specific heat is directly related to the density of vibrational states by

$$C_V = \int_0^\infty \hbar\omega\, N(\omega)\left[dn_B / dT\right] d\omega \propto T^{\tilde{d}}. \tag{5.8}$$

The first equality is general, while the proportionality refers to the contribution of fractons alone. For acoustic phonons, $\tilde{d}$ in (5.8) is simply replaced by $d = 3$. The phonon-fracton crossover is expected at $T \approx \hbar\omega_{col}/3k_B \approx 10$ mK. This is so low that it could not yet be observed. In C_V/T^3 plots,[80-82] the low temperature data is expected to follow a power law, $\propto T^{\tilde{d}-3}$. This has been seen over about two orders of magnitude in T.[82] Owing to the integral in (5.8), all crossovers are fairly broad. The behavior actually observed on materials similar to those used here (N and B) is consistent with the $\tilde{d}$-values in Table I.[81] It should be noted that the low-T specific heat of aerogels, per unit weight, is more than an order of magnitude larger than that of vitreous silica, v-SiO$_2$. Internal friction measurements demonstrated that the contribution of two-level states (TLS) is nearly the same in aerogels and in v-SiO$_2$.[81] Hence, the observed low-T specific heat of aerogels is really produced by the fractons rather than by TLS.

The thermal conductivity, κ, of aerogels shows a plateau at temperatures well below 1 K. This temperature is an order of magnitude lower than for v-SiO$_2$.[80-82] The height of the plateau is typically three to four orders of magnitude lower than in v-SiO$_2$. Aerogels seem to have the lowest known thermal conductivity of all solids, and it thus very difficult to measure. The plateau is consistent with fracton localization, and the increase of κ above 1 K can be explained by fracton hopping.[83-84]

6. Conclusions and Outlook

Porous silica are exceptional materials. Leached glasses are of interest for their large internal surfaces, the possibility to perform physics and chemistry in confined geometries, and their relative strength. Silica aerogels have even more extreme properties. They can be mass fractals up to two to three orders of magnitude in length. In this introductory review, we have shown that a high degree of internal consistency was achieved in measurements of structural and vibrational properties of aerogels over the entire fractal range, including changes of regimes and crossover regions. Fractal and fracton dimensions can be obtained in various ways, and although they take *effective* values, the theoretical relations between them are obviously quite robust in view of the observed consistency.

The reader should not gain the impression that everything is completed and well understood. In fact, new formal developments and/or realistic very large simulations are highly needed. The exact nature of the bending- and stretching-dominated fracton modes, the origin of depolarized scattering and of the slow crossover into the asymptotic Raman regime (Fig. 14b), the cause for the ω-dependence of the measured Debye-Waller factor, [85] are just a few of the many open questions. The different Raman mechanisms and their scaling is a subject of active current debate.[86] The measurements of thermal properties are also far from closed. It would be of great interest to reach down to the lowest possible temperatures and to measure more fractal (*i.e.* lighter) samples that are fully characterized spectroscopically

Finally, one should mention that the purpose of this research goes well beyond the mere understanding of fractals. Some of the results show a manner to treat *strongly disordered systems*, in particular glasses at intermediate length scales. We take as example the scattering near the phonon-fracton crossover described in Section 5.1.2. Eqs. (5.1-4) can also be used to describe the crossover of acoustic phonons into the Ioffe-Regel limit.[87] Although the usual way of handling such situations is that shown in Fig. 11, the approach of Section 5.1.2 reflects more closely the physical phenomena.

Important parts of this work resulted from collaborations with J. Pelous, or were the subjects of Ph.D. theses by E. Anglaret, M. Foret, and A. Hasmy. The excellent materials were elaborated and characterized by T. Woignier and J. Phalippou. Some light-scattering measurements were obtained together with Y. Tsujimi and P. Xhonneux. The neutron scattering studies benefited from the collaboration of G. Coddens, and from the participation of A. Heidemann, C. Lartigue, F. Mezei, and J. Teixeira. The theoretical support of S. Alexander has been essential for insight as well as for the development of the original formalism. We also benefited greatly from exchanges with O. Entin-Wohlman, R. Orbach, and G. Polatsek. Crucial simulations of the structure of aerogels were performed under the direction of R. Jullien. Enlightening simulations on percolation clusters were obtained with M. Kolb, T. Nakayama, E. Stoll, and K. Yakubo. Let them all be warmly thanked here.

7. References

1. Iler, R.K. (1979) *The Chemistry of Silica*, Wiley, New York.
2. Mandelbrot, B.B. (1977) *Fractals: Form, Chance, and Dimension*, Freeman, New York.
3. Mandelbrot, B.B. (1982) *The Fractal Geometry of Nature*, Freeman, New York.
4. Feder, J. (1988) *Fractals*, Plenum Press, New York.
5. Peitgen, O.H. and Richter, P.J.(1986) *The beauty of Fractals*, Springer, Berlin.
6. Nakayama, T.,Yakubo, K., and Orbach, R.L. (1994) Dynamical properties of fractal networks: Scaling, numerical simulations, and physical realizations, *Rev. Mod. Phys.* **66**, 381-444.
7. Alexander, S. (1989) Vibrations of fractals and scattering of light from aerogels, *Phys. Rev.* B **40**, 7953-7965.
8. Ioffe, A.F. and Regel, A.R. (1960) Non-crystalline, amorphous, and liquid electronic semiconductors, *Prog. Semicond.* **4**, 237-291.
9. Pietronero, L. (ed.) (1988) *Fractals' Physical Origin and Properties*, Plenum, New York.
10. Stoll, E., Kolb, M., and Courtens, E. (1992) Numerical verification of scaling for scattering from fractons, *Phys. Rev. Lett.* **68**, 2472-2475.
11. Stauffer, D. (1985) *Introduction to Percolation Theory*,Taylor & Francis, London.
12. Mazzacurati, V., Montagna, M., Pilla, O., Viliani, G., Ruocco, G., and Signorelli, G. (1992) Vibrational dynamics and Raman Scattering in fractals: A numerical study, *Phys. Rev.* B **45**, 2126-2137.
13. Kantor, Y. and Witten, T.A. (1984) Mechanical stability of tenuous objects, *J. Physique Lett.* **45**, 675-679.
14. Lovesey, S.W. (1984) *Theory of Neutron Scattering from Condensed Matter*, Clarendon, Oxford.
15. Hasmy, A., Anglaret, E., Foret, M., Pelous, J., and Jullien, R. (1994) Small-angle neutron-scattering investigation of long-range correlations in silica aerogels: simulations and experiments, *Phys. Rev.* B **50**, 6006-6016.
16. Guinier, A. (1963) *X-Ray Diffraction in Crystals, imperfect Crystals, and Amorphous Bodies*, Freeman, San Fransisco.

17. Bale, H.D. and Schmidt, P.W. (1984) Small-angle X-ray-scattering investigation of submicroscopic porosity with fractal properties, *Phys. Rev. Lett.* **53**, 596-599.

18. Wong, Po-zen and Bray, A.J. (1988) Porod scattering from fractal surfaces, *Phys. Rev. Lett.* **60**, 1344.

19. Vacher, R. Woignier, T., Pelous, J., and Courtens, E. (1988) Structure and self-similarity of silica aerogels, *Phys. Rev. B* **37**, 6500-6503.

20. Vicsek, T. (1989) *Fractal Growth Phenomena*, World Scientific, Singapore.

21. Coleman, P.H. (1988) The fractal nature of galaxy distributions, in [9], pp. 349-363.

22. Weitz, D.A. and Oliveria, M. (1984) Fractal structures formed by kinetic aggregation of gold colloids, *Phys. Rev. Lett.* **52**, 1433-1436.

23. Freltoft, T., Kjems, J., and Richter, D. (1987) Density of states in fractal silica smoke-particle aggregates, *Phys. Rev. Lett.* **59**, 1212-1215.

24. Page, J.H. and McCulloch, R.D. (1986) Ultrasound propagation in sintered metal powder: evidence for a crossover from phonons to fractons, *Phys. Rev. Lett.* **57**, 1324-1327.

25. Uemura, Y.J. and Birgeneau, R.J. (1987) Magnons and fractons in the diluted antiferromagnet $Mn_xZn_{1-x}F_2$, *Phys. Rev. B* **36**, 7024-7035.

26. Gordon, J.M., Goldman, A.M., and Whitehead, B. (1987) Dimensionality crossover in superconducting wire network, *Phys. Rev. Lett.* **59**, 2311-2314.

27. Craciun, F., Bettuci, A., Molinari, E., Petri, A., and Alippi, A. (1992) Direct experimental observation of fracton mode patterns in one-dimensional Cantor composites, *Phys. Rev. Lett.* **68**, 1555-1558.

28. Sapoval, B. (1990) *Les Fractals*, Aditech, Paris.

29. Hod, H.P. and Nordberd, M.E., U.S. Patents N°2,106,744 and N°2,286,275.

30. Drake, J.M. and Klafter, J. (1990) Dynamics of confined molecular systems, *Physics Today*, May 1990, pp. 46-55.

31. Levitz, P., Ehret, G., Sinha, S.K., and Drake, J.M. (1991) Porous Vycor glass: the microstructure probed by electron microscopy, direct energy transfer, small-angle scattering, and molecular adsorption, *J. Chem. Phys.* **95**, 6151-6161.

32. Li, J.C. and Ross, D.K. (1994) Dynamical scaling for spinodal decomposition - a small-angle neutron scattering study of porous Vycor glass with fractal properties, *J. Phys.: Condens. Matter* **6**, 351-362.

33. Cahn, J.W. and Hilliard, J.E. (1958) Free energy of a nonuniform system. I. Interfacial free energy, *J. Chem. Phys.*, **28**, 258-267; Cahn, J.W. (1959) *id.* II. Thermodynamic basis, *ibid*; **30**, 1121-1124; Cahn, J.W. and Hilliard, J.E. (1959) *id.* III. Nucleation in a two-component incompressible fluid; *ibid.* **31**, 688-699.

34. Wiltzius, P., Bates, F.S., Dierker, S.B., and Wignall, G.D. (1987) Structure of porous Vycor glass, *Phys. Rev. A* **36**, 2991-2994.

35. Maher, J.V., Goldburg, W.I., Pohl, D.W., and Lanz, M. (1984) Critical behavior in gels saturated with binary liquid mixtures, *Phys. Rev. Lett.* **53**, 60-63.

36. de Gennes, P.G. (1984) Liquid-liquid demixing inside a rigid network. Qualitative features, *J. Phys. Chem.* **88**, 6469-6472.

37. Liu, A.J., Duriam, D.J., Herbolzheimer, E., and Safran, S.A., (1990) Wetting transitions in a cylindrical pore, *Phys. Rev. Lett.* **65**, 1897-1900.

38. Chakrabarti, A. (1992) Kinetics of domain growth and wetting in a model porous medium, *Phys. Rev. Lett.* **69**, 1548-1551.

39. Lin, M.Y., Sinha, S.K., Drake, J.M., Wu, X.-I., Thiyagarajan, P., and Stanley, H.B. (1994) Study of phase separation of a binary fluid mixture in confined geometry, *Phys. Rev. Lett.* **72**, 2207-2210.

40. Kistler, S.S. (1932) Coherent expanded aerogels, *J. Phys. Chem.*, **36**, 52-64.

41. Foret, M., Pelous, J., Vacher, R., and Marignan, J. (1992) An investigation of the structure of colloidal aerogels, *J. Non-Cryst. Solids,* **147&148**, 382-385.

42. Foret, M., Pelous, J., and Vacher, R. (1992) Small-angle neutron scattering in model porous system: a study of fractal aggregates of silica spheres, *J. Physique 1 France* 2, 791-799; Leverrier-Foret, M. (1992) *Etude par Diffusion élastique et inélastique de Neutrons de la Structure et de la Dynamique vibrationelle d'Aérogels de Silice*, Ph. D. Thesis, Université de Montpellier II (unpublished).

43. Meakin, P. (1983) Formation of fractal clusters and networks by irreversible diffusion-limited aggregation, *Phys. Rev. Lett.*, **51**, 1119-1122.

44. Kolb, M., Botet, R., and Jullien, R. Scaling of kinetically growing clusters, *Phys. Rev. Lett.*, **51**, 1123-1126.

45. Chang, S.Y. and Ring, T.A. (1992) Map of gel times for three phase region tetraethoxysilane, ethanol and water, *J. Non-Cryst. Solids*, **147&148**, 56-61.

46. Bourret, A. (1988) Low-density silica aerogels observed by high-resolution electron microscopy, *Europhys. Lett.* **6**, 731-737.

47. Vacher, R., Courtens, E., Stoll, E., Böffgen, M., and Rothuizen, H. (1991) Pores in fractal aerogels and their incidence on scaling, *J. Phys.; Condens. Matter* **3**, 6531-6545.

48. Fricke, J. (ed.) (1992) *Aerogels 3, J. Non-Cryst. Solids*, **145**.

49. Pekala, R.W. and Hrubesh, L.W. (eds.) (1994) *Aerogels 4, J. Non-Cryst. Solids*, **186**.

50. Esquivias, L.(ed.) (1992) *Advanced Materials from Gels, J. Non-Cryst. Solids*, **147&148**.

51. Livage, J. (1994) *7th Int. Workshop on Glasses and Ceramics from Gels, J. of Sol-Gel Sci. Tech.*, **2**.

52. Jang, K.Y. and Kim, K. (1992) Evaluation of sol-gel processing as a method for fabricating spherical-shell silica aerogel inertial confinement fusion targets, *J. Vac. Sci. Technol. A*, **10**, 1152-1157.

53. Frelthoft, T., Kjems, K.J., and Sinha, S.K. (1986) Power-law correlations and finite-size effects in silica particle aggregates studied by small-angle neutron scattering, *Phys. Rev. B* **33**, 269-275.

54. Anglaret, E., Hasmy, A., Courtens, E., Pelous, J., and Vacher, R. (1994) Investigation of dynamical scaling in a mutually self-similar series of base-catalyzed aerogels, *Europhys. Lett.* **28**, 591-596; Anglaret, E. (1995) *Corrélations Structure-Propriétés vibrationnelles de Matériaux Fractals, les Aérogels de Silice*, Ph. D. Thesis, Université de Montpellier II (unpublished).

55. Hasmy, A., Vacher, R., and Jullien, R. (1994) Small-angle scattering by fractal aggregates: A numerical investigation of the crossover between the fractal regime and the Porod regime, *Phys. Rev. B* **50**, 1305-1308.

56. Alexander, S. (1984) Is the elastic energy of amorphous materials rotationally invariant ? *J. Physique* **45**, 1939-1945.

57. Courtens, E., Vacher, R., and Stoll, E. (1989) Fractons observed, *Physica D* **38**, 41-55.

58. Alexander, S. and Orbach, R. (1982) Density of states of fractals: «fractons», *J. Physique Lett.* **43**, 625-631.

59. Rammal, R. and Toulouse, G. (1983) Random walk on fractal structures and percolations clusters, *J. Physique Lett.* **44**, 13-22.

60. Stoll, E. and Courtens, E. (1990) Connectivity and the fracton dimension of percolation clusters, *Z. Phys. B* **81**, 1-2.

61. Thouless, D.J. (1974) Electrons in disordered systems and the theory of localization, *Phys. Rep.* **C13**, 94-142.

62. Alexander, S., Courtens, E., and Vacher, R. (1993) Vibrations of fractals: dynamic scaling, correlation functions and inelastic light scattering, *Physica A* **195**, 286-318.

63. Courtens, E., Pelous, J. Phalippou, J., Vacher, R., and Woignier, T. (1987) Brillouin-scattering measurements of phonon-fracton crossover in silica aerogels, *Phys. Rev. Lett.* **58**, 128-131.

64. Courtens, E., Vacher, R., Pelous, J., and Woignier, T. (1988) Observation of fractons in silica aerogels, *Europhys. Lett.* **6**, 245-250.

65. Aharony, A., Entin-Wohlman, O., Alexander, S., and Orbach, R. (1987) Cross-over from phonons to fractons *Phil. Mag. B* **56**, 949-955.

66. Polatsek, G. and Entin-Wohlman, O. (1988) Effective-medium approximation for a percolation network: The structure factor and the Ioffe-Regel criterion, *Phys. Rev. B* **37**, 7726-7730.

67. Aharony, A., Alexander, S., Entin-Wohlman, O., and Orbach, R. (1987) Scattering of fractons, the Ioffe-Regel criterion, and the 4/3 conjecture, *Phys. Rev. Lett.* **58**, 132-135.

68. Courtens, E. and Vacher, R. (1988) Fractons in real fractals, in H.E. Stanley and N. Ostrowski (eds.), *Random Fluctuations and Pattern Growth*, Kluwer Academic Publishers, Dordrecht, pp. 20-26.

69. Courtens, E. and Vacher, R. (1989) Experiments on the structure and vibrations of fractal solids, *Proc. R. Soc. Lond. A* **423**, 55-69.

70. Xhonneux, P., Courtens, E., Pelous, J., and Vacher, R. (1989) Scaling non-linear elasticity in silica aerogels, *Europhys. Lett.* **10**, 733-738.

71. Lovesey, S.W. (1984) *Theory of Neutron Scattering from Condensed Matter*, Clarendon, Oxford.

72. Courtens, E., Lartigue, C., Mezei, F., Vacher, R., Coddens, G., Foret, M., Pelous, J., and Woignier, T. (1990) Measurement of the phonon-fracton crossover in the density of states of silica aerogels, *Z. Phys. B* **79**, 1-2.

73. Vacher, R., Courtens, E., Coddens, G. Heidemann, A., Tsujimi, Y., Pelous, J., and Foret, M. (1990) Crossovers in the density of states of fractal silica aerogels, *Phys. Rev. Lett.* **65**, 1008-1011.

74. Vacher, R., Woignier, T., Pelous, J., Coddens, G., and Courtens, E. (1989) The density of vibrational states of silica aerogels, *Europhys. Lett.* **8**, 161-169.

75. Feng, S. (1985) Crossover in spectral dimensionality of elastic percolation systems, *Phys. Rev. B*, **32**, 5793-5797.

76. Yakubo, K. Takasugi, K., and Nakayama, T. (1990) Crossover behavior from vector to scalar fractons in the density of states of percolating nertworks, *J. Phys. Soc. Japan,* **59**, 1909-1912.

77. Tsujimi, Y., Courtens, E., Pelous, J., and Vacher, R. (1988) Raman-scattering measurements of acoustic superlocalization in silica aerogels, *Phys. Rev. Lett.* **60**, 2757-2760.

78. Montagna, M., Pilla, O., Viliani, G., Mazzacurati, V., Ruocco, G., and Signorelli, G. (1990) Numerical study of Raman scattering from fractals, *Phys. Rev. Lett.* **65**, 1136-1139.

79. Terao, T. and Nakayama, T. (1996) Power-law dependence on frequency of the Raman-scattering intensity of percolating networks, *Phys. Rev. B*, **53**, 2918-2921.

80. Calemczuk, R., de Goer, A.M., Salce, B., Maynard, R., and Zarembowitch, A. (1987) Low-temperature properties of silica aerogels, *Europhys. Lett.* **3**, 1205-1211.

81. de Goer, A.M., Calemczuk, R., Salce, B., Bon, J., Bonjour, E., and Maynard, R. (1989) Low-temperature energy excitations and thermal properties of silica aerogels, *Phys. Rev. B*, **40**, 8327-8335.

82. Bernasconi, A., Sleator, T., Posselt, D., Kjems, J.K., and Ott, H.R. (1992) Dynamic properties of silica aerogels as deduced from specific-heat and thermal-conductivity measurements, *Phys. Rev. B* **45**, 10363-10376.

83. Alexander, S., Entin-Wohlman, O., and Orbach, R. (1986) Phonon-fracton anharmonic interactions: The thermal conductivity of amorphous materials, *Phys. Rev. B* **34**, 2726-2734.

84. Jagannathan, A., Orbach, R., and Entin-Wohlman, O. (1989) Thermal conductivity of amorphous materials above the plateau, *Phys. Rev. B* **39**, 13465-13476.

85. Vacher, R., Courtens, E., Coddens, G., Pelous, J., and Woignier, T. (1989) Neutron-spectroscopy measurement of a fracton density of states, *Phys. Rev. B* **39**, 7384-7387.

86. Viliani, G., Dell'Anna, R., Pilla, O., Montagna, M., Ruocco, G., Signorelli, G., Mazzacurati, V., (1995) Raman scattering from fractals: Simulation of large structures by the method of moments, *Phys. Rev. B* **52**, 3346-3355.

87. Foret, M., Courtens, E., Vacher, R., Suck, J.-B. (1996) Scattering investigation of acoustic localization in strong glasses, preprint.

THE STRUCTURE AND MECHANICAL PROPERTIES OF NETWORKS

M. F. THORPE, B. R. DJORDJEVIĆ AND D. J. JACOBS
*Department of Physics & Astronomy
and Center for Fundamental Materials Research
Michigan State University, East Lansing, MI 48824, USA*

1. Introduction

In these lectures, I will talk about continuous random networks and their mechanical properties. I have collected together most of the ideas that have been put forward over the past ten years on constraint counting and the resultant floppy modes in random networks. I show how model systems can help us better understand glasses via *rigidity percolation*. These ideas involving rigidity percolation have been tested experimentally in bulk glasses. I also examine marginal cases where the instability is caused by the surface, leading to the number of floppy modes scaling with the surface area. These effects may be important in porous silica, zeolites and possibly biological systems.

The study of network structures has fascinated scientists in many areas - engineering, mechanics and biology going back more than a century [1]. Maxwell was intrigued with the conditions under which mechanical structures made out of struts, joined together at their ends, would be stable (or unstable) [1]. To determine the stability, without doing any detailed calculations (that would have been impossible then except for the simplest structures), Maxwell used the method of *constraint counting*. This counting is an approximate method that proves to be accurate for structures where the density (of struts or joints) is approximately uniform. Maxwell's constraint counting method is exact for some geometries that I will discuss in this paper. The idea of a constraint in a mechanical system goes back to Lagrange [2] who used the concept of holonomic constraints to reduce the effective dimensionality of the space. The difficult part is to determine which constraints are *linearly independent*. If the linearly independent constraints can be identified, then the problem is solved – however in most large systems this identification is not possible except using some numerical procedure on an actual realization.

The problem under consideration is a *static* one – given a mechanical system, how many independent deformations are possible without any cost in energy? These are the zero frequency modes, which I prefer to refer to as *floppy modes* because in any real system there will usually be some weak restoring force associated with the motion.

Often it is more convenient to look at the system as a *dynamical* one, and assign potentials or spring constants to deformations of the various struts (bonds) and angles.

M. F. Thorpe and M. I. Mitkova (eds.), Amorphous Insulators and Semiconductors, 289–328.
© 1997 *Kluwer Academic Publishers. Printed in the Netherlands.*

290

It does not matter whether these potentials are harmonic or not, as the displacements are virtual. However it is convenient to use harmonic potentials so that the system is linear. It is then possible to set up a Lagrangian for the system and hence define a dynamical matrix which is a real symmetric matrix and as a consequence has real eigenvalues. These eigenvalues are either positive or zero. The number of finite (non–zero) eigenvalues defines the *rank* of the matrix. Thus our counting problem is rigorously reduced to finding the rank of the dynamical matrix. The rank of a matrix is also the number of linearly independent rows or columns in the matrix. Neither of these definitions is of much formal help, and a numerical determination of the rank of a large matrix is difficult and of course requires a particular realization of the network to be constructed in the computer. Nevertheless the rank is a useful notion as it represents the proper mathematics for this problem, which now becomes well posed.

The genius of Maxwell [1] was to devise the simple constraint counting method that allows us to *estimate* the rank of the dynamical matrix and hence the number of floppy modes. In the next section, I discuss the application of these ideas to bulk covalent network glasses. In section 5, I expand this work to look at the marginal case when surface floppy modes become important.

Until recently it has not been possible to improve on the approximate Maxwell constraint counting method, except by using numerical simulations on small systems. Even these have had problems as will be described later. In the last year a new powerful integer algorithm, called the Pebble Game [3] has become available. This allows very large systems to be analyzed in 2d and is currently in the process of being extended to 3d systems.

The layout of these notes is as follows. In the next section we give a brief history of the construction of random network models for covalent glasses. In section 3 we describe the powerful and simple ideas of constraint counting, and in section 4 we discuss the exact algorithm; the pebble game and the powerful insights that it has provided into rigidity percolation. In section 5 we discuss *surface floppy modes*. In section 6 we describe the current state of experiments on glasses that are relevant to this topic.

2. Continuous Random Networks

A perfect crystal is a structure in which the atoms, or groups of atoms, are arranged in a pattern that repeats itself periodically in three dimensions to form an infinite solid. For instance, crystalline diamond makes a lattice which is produced by the periodic repetition of the 8 atom diamond cubic cell (or the smaller 2 atom primitive unit cell) in all three spatial directions. A piece of the crystalline diamond structure is shown in Figure 1.

Amorphous materials do not have the periodicity (long range order) characteristic of a crystal. Starting from a crystalline structure and slightly displacing every atom in a random manner, the periodicity is certainly destroyed, but the random structure obtained is not yet amorphous. An amorphous structure is *topologically* different from a crystalline one. Thus, to obtain an amorphous structure from a crystalline one, it is necessary not only to introduce randomness in the atomic positions, but also to change the *topology* of the original perfect lattice. We can schematically sketch atomic arrangements in three typical cases: a crystal, an amorphous solid, and a gas, as in Figure 2.

Figure 1. A piece of a crystalline diamond structure. The characteristic sixfold rings in a crystalline diamond structure can be seen.

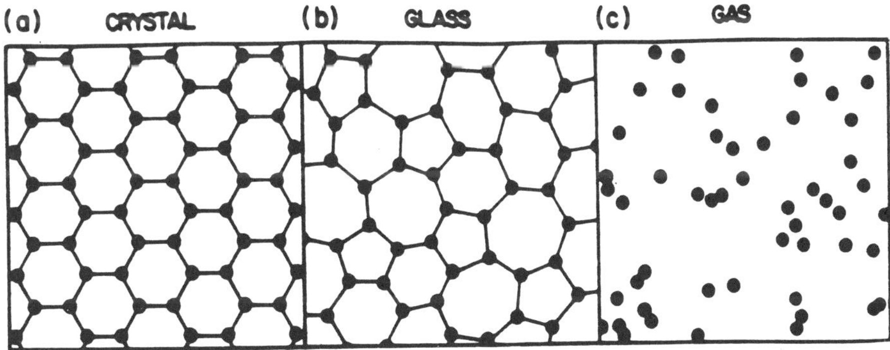

Figure 2. Schematic sketches of the atomic arrangements in (a) a crystalline solid, (b) an amorphous solid, and (c) a gas. (Zallen [4])

In Figure 3 we show a piece of the 3d *amorphous* diamond structure which was obtained in our computer simulations [5]. By amorphous diamond, we mean a structure that is 4 - fold coordinated everywhere and therefore similar to a-Si and a-Ge, except for the relatively stronger bond bending forces.

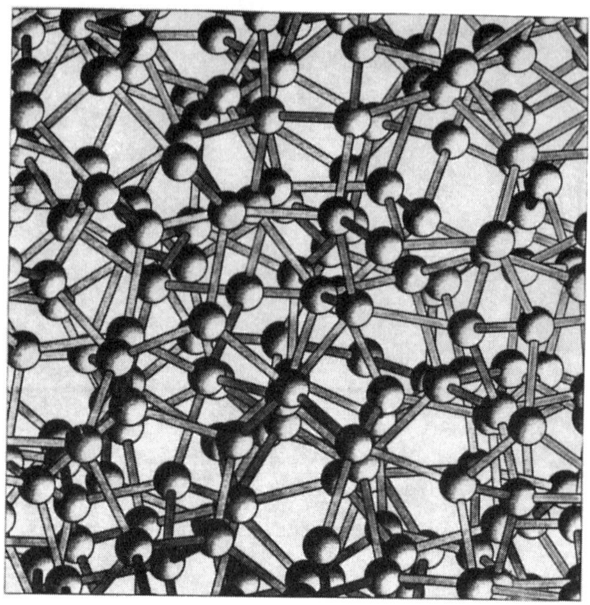

Figure3. A piece of an amorphous diamond structure.

Binary compounds like SiO_2, can be found both in the crystalline and in the amorphous state, the later being topologically different from crystalline SiO_2. Schematic two-dimensional representations of the crystalline and amorphous structures for a hypothetical A_2B_3 compound are shown in Figure 4. Similarly, the crystalline diamond network possess only 6-fold rings of atoms, while the amorphous network also has 5-fold, 7-fold, and higher order rings, and consequently the number of 6-fold rings is reduced from its crystalline value.

The main experimental tool for verifying the structure of glasses is via the radial distribution function (RDF) that can be obtained by Fourier transforming X-ray or neutron diffraction data. This is illustrated in Figure 5.

Modeling of amorphous structures can be divided in two major classes. The first one is the *Dense Random Packing* of hard spheres, (DRP), in all its varieties, and the second is based on the *Continuous Random Network*, (CRN), concept. The DRP model was quite successful in explaining the amorphous structure of some liquids and also amorphous metals. Atoms in these systems are connected with each other predominantly by *non-directional* forces, which makes DRP a suitable model for the description of these materials. DRP modeling has been pioneered by Bernal [8,9], and subsequently developed by Finney [10], Bennett [11] and others. For a review of modeling amorphous metals with DRP models see an excellent and comprehensive article of Cargill [12].

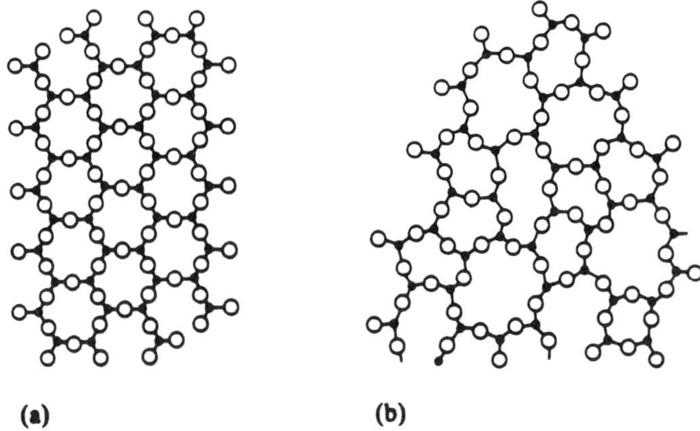

Figure 4. Crystalline and amorphous form of a binary compound. (a) Hypothetical crystalline compound A_2B_3. (b) Amorphous form of the same compound. (Elliott 1984 [6])

In these lectures, however, we shall focus our attention only on the *Continuous Random Network* models which are suitable for describing the structures of materials with predominantly covalent *directed* bonds, namely, covalent glasses. Covalent glasses have a much higher degree of short-range order than metallic glasses and are dominated by covalent bonding with a definite number of nearest neighbors; with distinct bond lengths and bond angles. The prototypical covalent amorphous solids are amorphous silicon, (a-Si), and amorphous germanium, (a-Ge) which belong to the semiconductors from Group IV. Every atom in these structures has four nearest neighbors, i.e. it is tetrahedrally coordinated.

We note here that the use of the term *continuous* in connection with random networks is somewhat unfortunate, because random networks are, in fact *discrete* in their structure. By *continuous* it is meant that one can build an amorphous network continuously, starting from the short-range order unit, up to an indefinite size, without including unsatisfied bonds or breaking the structure. In other words, *continuous* means that there are no identifiable boundaries in the structure, separating regions of distinctly different type of structure.

2.1. HAND-BUILT CRN MODELS

The first attempts to explain the structure of covalent glasses, around 1930, were based on, the then very natural hypothesis, that amorphous materials consist of very large number of elemental microcrystals, randomly arranged into a very fine polycrystalline structure, which appears as amorphous. All attempts to fit the experimental data with such microcrystalline models failed, and the early idea of a *continuous random network* of Zachariasen [13] has become the only viable alternative. It was proposed by Zachariasen that "...*the atomic arrangement in glass is characterized by an extended three-dimensional network which lacks symmetry and periodicity.*" In Figure 6 we show

294

the historical schematic diagram of Zachariasen which represents his CRN model of amorphous silica.

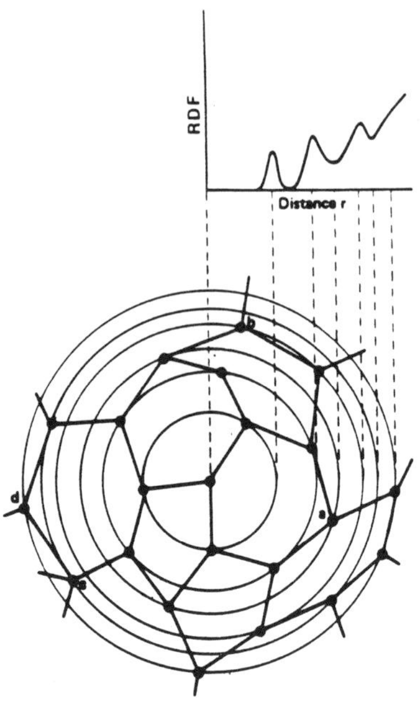

Figure 5. Schematic illustration of the origin of the structural features in the RDF of an amorphous solid. Atoms are shown as lying on sharply defined rings for simplicity (Wooten and Weaire [7]).

The first models based on the Zachariasen's ideas were built much later. One of those was the one for SiO_2, constructed by Bell and Dean [14,15]. This was a hand-built model. This model was quite successful in explaining the diffraction data. In building the model, Bell and Dean used basic units consisting of a central Si atom and four rigid wires in a perfect tetrahedral arrangement. The O atoms were attached by inserting the wires into pre-drilled holes in the O atoms. The model was built to have a mean Si-O-Si bond angle close to 160°, but without any particular bond angle distribution. It maintained full coordination without dangling bonds in its interior. The model consisted of 614 atoms. Atomic coordinates were determined with the help of photographic techniques. From those coordinates the *rms* bond-length deviation was small in agreement with the experiment, while the bond-angle deviation was approximately ±30° around a mean value of 153°. The Radial Distribution Function was computed and was in fairly good agreement with the diffraction data of Mozzi and Warren [16].

Very soon, amorphous elemental semiconductors were intensively studied. The amorphous Group-IV semiconductors, a-Si and a-Ge, came to be regarded as prototypical covalent amorphous solids. It was not clear in the beginning whether tetrahedral bonds in such systems can be randomly connected *ad infinitum* in a disordered manner, or not. It was believed by Phillips [17,18], that for instance, amorphous silicon, due to its *overconstrained* structure (a notion that will be explained in detail later), must necessarily have a discontinuous structure, in which regions of continuous random network are limited in size. But it is obvious that, in principle, there is a very close connection between a-Si and amorphous silica. A random network model for one can be easily converted into a corresponding model for the other, by adding or subtracting the oxygen atoms which link the silicon atoms and introducing a weak bond-bending force at oxygen sites.

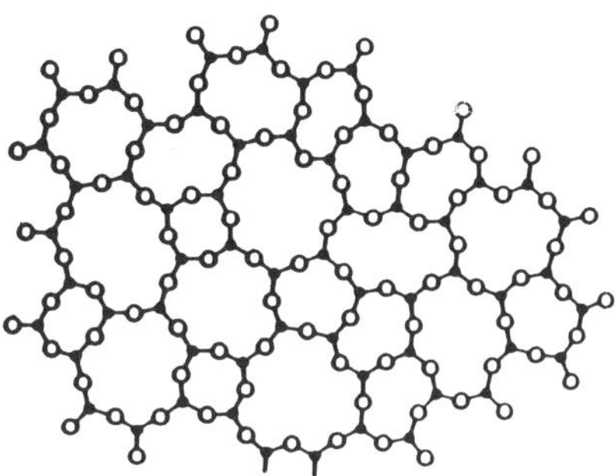

Figure 6. Zachariasen's schematic 2*d* diagram represents his CRN model for amorphous silica. Open circles are two-fold coordinated oxygen atoms, and black dots are Si atoms which are four-fold coordinated in 3*d* (Zachariasen [13]).

The first CRN model for a tetrahedrally bonded amorphous solid was built by Polk [19]. There were no particular rules used in building Polk's model, except that there were no dangling bonds left within the interior of the structure. Whenever a choice about the bonding of a new atom should be decided, simply an arbitrary, random choice was made, just as in the work of Bell and Dean. The model contained a substantial number of five-, six-and seven-membered rings which ensured generation of a non-crystalline structure. There were no dangling bonds in the interior of the structure. Bond-length deviations were restricted within 1%, and bond-angle deviations were within ±10° about the tetrahedral angle of 109°. The model consisted of 440 atoms. The original Polk model was physically extended to 519 atoms and refined by Polk and Boudreaux [20]. The refinement was in reducing the deviation of nearest-neighbor bond lengths from 1% to 0.2% using precise laser-beam measurements of the atomic coordinates and computer techniques. Steinhardt *et al.* [21] and Duffy *et al.* [22] used the 519 model of Polk and Boudreaux [20] and refined it further. This refinement consisted of introducing the Keating [23] potential (see later discussion) in the structure and relaxing the system toward the minimum of the elastic energy in terms of the bond lengths and bond angles.

The method of relaxation adopted by Steinhardt *et al.* [21], is to move each atom separately and sequentially, while keeping the coordinates of the other atoms fixed, until there is no force on it. This process is repeated many times until convergence is achieved. In this way Steinhardt and coworkers found new, relaxed coordinates of 519-

atom model of Polk and Boudreaux. On the other hand, they also built a new 201-atom model starting from a 21-atom cluster with new atoms serially added. Whenever a small group of atoms was added to the growing cluster their positions were relaxed by computer to minimize the forces on them. Thus, the connectivity of this network is determined by the relaxation procedure. They did not find any significant structural difference between the two models. Duffy *et al.* [22] used the same 519-atom model of Polk and Boudreaux but adopted a different relaxation procedure. Each atom was simultaneously moved in the direction of the force on it; a distance proportional to that force. After each move the forces are recalculated, the atoms moved again, and the process iterated until the forces and displacements become small. In this way they obtained results which were pretty much the same as those of Steinhardt *et al.* [21].

Provoked by the controversy involving the presence of five-membered rings in amorphous III-V compounds like GaAs, which would, of course imply the occurrence of energetically unfavorable, *wrong* bonds between like atoms, and to check if different ring statistics would produce different RDF's, Connell and Temkin [24] built a model of a-Ge with no odd-membered rings. The model consisted of 238 plastic and metal tetrahedra. This model did not produce any improvement of the agreement with experiment. Our own calculations of RDF for a-Si and a-C show that RDF is very similar for different ring statistics [5].

All these early hand-built models suffered from the their finite sizes, or in other words, from the lack of periodic boundary conditions. In these relatively small models, a large fraction of the atoms were necessarily located close to the surface. This complicates the analysis of the bulk properties, and the calculation of the RDF. The first hand-built model which had the periodic boundary conditions implemented was made by Henderson [25]. The model consisted of only 61 atoms.

2.2. COMPUTER-BUILT CRN MODELS

Perhaps the very first computer modeling of this kind was done by Kaplow *et al.* [26]. They tried to simulate the structure of amorphous selenium and to compare it to the experimental results. The method Kaplow *et al.* adopted in creating computer model is a good example of a misconception that one can make an amorphous structure just by randomizing the atomic coordinates in a crystalline structure. This is exactly what the authors did in this early work. Their computer-generated coordinates of crystalline arrangement of about of 100 atoms, and then used Monté Carlo procedure to randomize these coordinates. Only those which improved RDF were accepted. After about 10^5 moves a *reasonable* agreement with experiment was achieved. The problem with this procedure, as we have already mentioned before, is that obtained structure is topologically identical to the crystalline one, and thus can not be expected to give good description of the amorphous selenium. This inverse Monté Carlo method has been considerably refined and is discussed at length in these proceedings [27].

Henderson *et al.* [25] experimented with computer algorithms for creating structures with periodic boundary conditions, but they did not succeed in getting good agreement with experiment. Nevertheless, the approach of Henderson, with generating periodic structures and with using a simple computer algorithm, inspired later work on computer modeling. In contrast to this approach, Shevchik [28] and Polk and Boudreaux [20] tried to imitate in their algorithm the hand-building process but that proved to be very cumbersome procedure.

As we have already stressed, a truly amorphous structure created by a successful randomization of the crystalline atomic positions, must be topologically different from the structure of its crystalline parent, and it must have perfect connectivity. One obvious technique to ensure these two requirements, for instance in creating amorphous silicon, is to start from the diamond cubic structure, remove one atom at random, and rejoin the four dangling bonds in pairs. In this way the number of atoms in the structure clearly decreases, but the topology is changed due to the fact that six-membered rings of the crystalline structure are destroyed, while five-, and seven-membered rings are introduced. Four-fold connectivity of the network is preserved. This method was used by Alben *et al.* [29] who used 64-atoms crystalline supercell as a starting point, and ended up with a 58-atoms model of amorphous silicon. In an unpublished work of Wooten and Weaire (see Wooten and Weaire [7]) similar studies were done on models that contained 700-1000 atoms. Results were less then impressive. Calculated RDF's are in poor agreement with experiment. The structures obtained in this way have large bond-angle distortions, and it was concluded that they are not satisfactory as models of amorphous silicon.

The other technique which illustrates the topological connection between the structure of vitreous silica and amorphous silicon, was employed by Evans *et al.* [30,31]. They used a decoration transformation (see Wright *et al.* [32]) of the vitreous silica random network model of Evans and King [33] which involved removing the oxygen atoms, rejoining the two dangling bonds and adjusting the new network to have the mean bond length appropriate for amorphous silicon. The model achieved relatively good agreement with experiment in the first two peaks, but the model pair correlation function exhibits too much structure at higher separations.

2.2.1. *Guttman Model*

The first approach to computer generation of periodic random networks was extensively developed by Guttman [34,35]. His aim was to construct an ideal random network with four-fold coordination which would describe the structure of amorphous germanium. Guttman does not use a single computer algorithm, but otherwise his approach is quite similar to the Wooten and Weaire method [7]. Guttman starts from the crystalline super-cell which contains many unit cells, and imposes periodic boundary conditions on the supercell. There are no bonds in the model initially. Bonds are introduced in a stochastic way: four neighbors are selected at random from the given set of atoms which are assigned to each atom in the lattice. Thus four bonds are constructed at a time, and the process is repeated until, hopefully, the desired network is created. Guttman noticed that the probability of constructing a fully four-fold coordinated network in this way was *decreasing* with the size of the model. The size of the model varied from 64 atoms to 512 atoms. The structure was relaxed using a simple Hooke's-law bond-stretching, and bond-bending term, similar to that in Eq. (1). Guttman also varied the super-cell size to find the optimal density which minimizes the total strain energy. He found that

298

the optimal model density was 15% greater than the observed crystalline density, while the density of a-Ge experimentally was found to be within the few percent of the crystalline density.

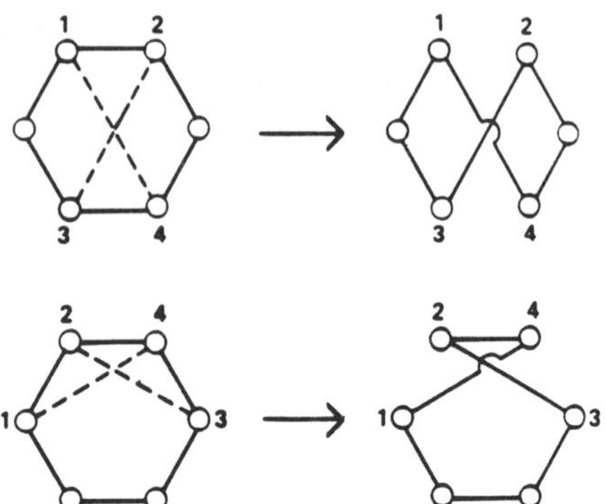

Figure 7. Two examples of bond switches used by Guttman [35]. Six-membered rings of the crystalline lattice are shown, and four atoms are selected. A path 1-2-3-4 is defined and bonds 1-2 and 3-4 are exchanged with non-bonds 2-3 and 4-1 respectively.

Furthermore, a large compressive stress was introduced by random bonding assignments, and the *rms* angular deviation from the perfect tetrahedral angle (109.47°) was found to be about 17°; much greater than the experimental value of abut 10° (see Etherington *et al.*[36]). Guttman then used a specific topological rearrangement to generate the random network. In Figure 7 we show two examples of Guttman's bond rearrangements, or bond switches.

The switch is accepted if it lowers the energy of the network, and rejected if it does not. In this way the RDF was slightly improved, but the models still did not agree very well with experiments, because all Guttman's models contained many dangling bonds.

2.2.2. *Wooten-Weaire Method*

In the Wooten and Weaire model of a-Si and a-Ge, the Keating potential is adopted to describe the interaction between the atoms. The Keating potential (1) was introduced earlier to fit the elastic and vibrational properties of Group-IV elements [23]. It represents a semiempirical description of bond-stretching and bond-bending forces and involves only few parameters. It consists of two terms, the first being the bond-stretching one, and the second the bond-bending.

$$V = \frac{3}{16} \frac{\alpha}{d^2} \sum_{l,i} (\vec{r}_{li} \cdot \vec{r}_{li} - d^2)^2 + \frac{3}{8} \frac{\beta}{d^2} \sum_{l\{i,i'\}} (\vec{r}_{li} \cdot \vec{r}_{li'} + \tfrac{1}{3} d^2)^2 \tag{1}$$

where α and β are bond-stretching and bond-bending force constants, respectively, and d is the strain-free equilibrium bond length in the diamond structure. For Si its value is $d = 2.35$Å, and for Ge, we have $d = 2.46$Å.

Central bond-stretching forces are not enough to stabilize the structure, because the number of constraints imposed by maintaining fixed bond lengths is smaller than the number of degrees of freedom. This will be explained in more detail later. At the moment it is enough to stress the importance of the bond-bending forces for creating a stable network. Wooten and Weaire followed Martin [37] by taking $\beta/\alpha = 0.285$ in building their model of a-Si and a-Ge. Due to the dominance of the α term, when the model is relaxed to minimize energy, all bonds relax to within a few percent to their ideal unstretched length, but because this is not enough to stabilize the structure, the β term defines the stable structure which minimizes the bond-bending energy. Thus, the total energy is roughly proportional to β, and the relaxed structure is independent of β/α (Wooten, Winer and Weaire [38]; Wooten and Weaire [7]). The Keating potential is schematically show in Figure 8.

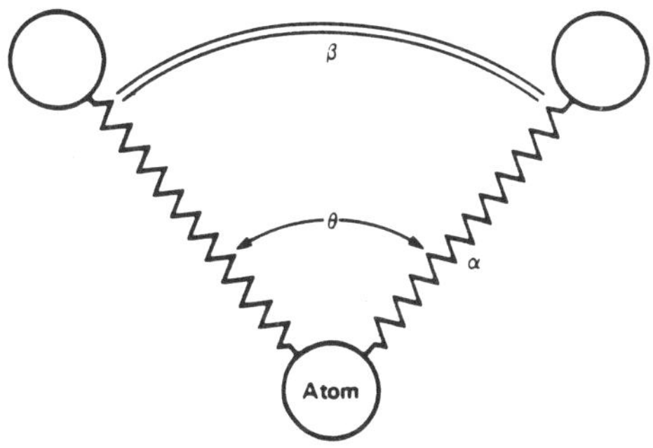

Figure 8. The Keating potential bond-stretching α, and bond-bending β terms (Wooten and Weaire [7]).

Wooten, Winer and Weaire [38] created a new model of amorphous silicon and germanium which significantly improved the agreement with experiment. It consisted of 216 atoms. A new bond-switch was invented which does not produce large bond-length and bond-angle distortions, and is simple and the only one used in the two-stage process which creates the model. This new bond switch is shown in Figure 9.

The bond switch shown in Figure 9 is characterized by interchanging two second-neighbor bonds that are parallel to each other in the perfect diamond structure. This is a local rearrangement of bonds that creates minimal strain into an originally strain-free perfect diamond structure, which was not the case for Guttman's switch (see Figure 7). Atoms are numbered, their coordinates are listed, and a neighbor-table is created. The switch consists of breaking those two bonds [1-2 and 5-6] in Figure 9, and creating two other bonds, [1-5 and 2-6]. The switch is not acceptable if new bonds are longer than 1.7 times the unstretched nearest-neighbor bond, which is the length slightly larger than the second-neighbor distance, as it should be if a successful bond switch is to be made in the original crystalline structure.

300

Figure 9. Wooten-Weaire bond switch. (Left panel) Configuration of bonds in the diamond cubic structure; (Right panel) Relaxed configuration after switching two bonds.

One can define a *ring* as any closed path of bonds. The number of n-membered rings defined in this way increases with n. An *irreducible ring* is defined as one that has no shortcuts across it. That is, given any two atoms (vertices) on the ring, there is no shorter path between any two atoms (as measured by the number of bonds along the path) than a path on the ring itself. One advantage of such rings for topological purposes is that the number of *n*-membered rings is zero for large *n*, but no topologically important rings are omitted, and the complete table of ring statistics remains finite (Wooten and Weaire [39]).

The perfect diamond structure has only 6- membered irreducible rings. The introduction of a bond switch changes the ring statistics by destroying even-membered rings and creating odd-membered rings, 5-, 7-, 9-, etc. membered rings. This process is schematically shown in the 2d case in Figure 10. When the structure is already randomized, bonds that are switched are chosen to be approximately parallel by ensuring that they do not belong to the same 5-, 6-, or 7-membered ring.

One bond switch introduced into the crystalline diamond super-cell affects the topology of the structure by destroying 12 six-membered rings, and simultaneously creating 4 five-membered, and 16 seven-membered rings. During the process of initial randomization, the RDF becomes gradually less structured, reflecting the amorphous nature of the created structure.

After the initial randomization produces a disordered, but highly strained model, it is necessary to continue with creating new bond switches searching for those which will lower the strain energy. Of course, the lowest energy possible is zero, i.e. that of crystal. So one has to be careful to avoid returning to the crystalline state during the process of annealing. Wooten and Weaire [7] showed that the system indeed returns to the original crystalline diamond structure if in the initial process of randomization, the number of introduced bond switches was not large enough. If the initial temperature is too high the subsequent annealing is not able to bring the system down to an acceptably low energy, and the obtained model has too much strain in it. In other words, with too high an initial temperature, the system is trapped in one of the numerous metastable states of relatively high energy. From the other side, the number of bond switches has also to be chosen optimally; if there are too many of them in the beginning, then again, the system will not be able to release a sufficient amount of energy during the annealing process, and the RDF of the final structure will show that the system has too large bond-length and bond-angle distortions. Wooten and Weaire found that for creating an a-Si model the optimal melting temperature for the initial process of randomization has to be kT=1eV. As we explained in the previous Section, this (10,000 K) high temperature is meaningful only within the simulated annealing environment.

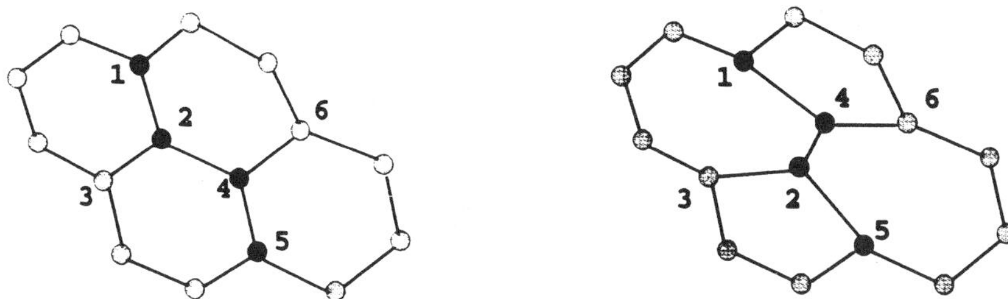

Figure 10. Schematic representation of a Wooten-Weaire topological bond switch. There are four 6-membered rings initially (diamond structure), and after a bond switch is made, two 5- and two 7-membered rings are created.

The annealing process itself is based on Metropolis Monté Carlo algorithm [40] already applied in the literature to general problems of optimization by simulated annealing (Kirkpatrick *et al* [41]; Vanderbilt and Louie [42]). The Metropolis algorithm is generally used as an efficient simulation of a collection of atoms in equilibrium at a given temperature. In the process of creating a model of an amorphous structure, the introduction of topological rearrangements (bond switches) change the total strain energy of the lattice at a given temperature. If that change, ΔE, is negative, i.e. if the introduction of a bond switch lowers the total energy, this bond switch is accepted. On the other hand, if the ΔE is positive (increases the total energy), that bond switch may be accepted with probability $P(\Delta E) = \exp(-\Delta E / kT)$. In practice, a random number uniformly distributed in the interval (0,1) is chosen each time when a bond switch is introduced, and ΔE is calculated. If the picked random number is less than $P(\Delta E)$ than the new configuration of atoms is accepted. If the random number is greater than $P(\Delta E)$ the new configuration of atoms is rejected and the original one (before the bond switch was introduced) is used to start the next step. Of course, the next step always consists of introducing the new bond switch, i.e., in repeating the described basic step. When this is done many times, the procedure simulates the behavior of the system at given temperature T, as it reaches the thermal equilibrium.

The introduction of the Boltzmann factor is essential in the process of annealing because it is able to lift the structure out of a metastable state of higher energy, thus enabling it to find another path in the configurational space toward the lower-energy states. Without it the system would not be permitted to ever increase its energy, and thus it would be easily trapped in a metastable state with no way out. It has been noted (Wooten and Weaire [7]) that allowing the presence of the 4 membered rings in the annealing process facilitated the convergence toward the energy minimum, and helped to avoid trapping into high-energy metastable states.

Thus, after the structure has been initially randomized by the introduction of a large number of bond switches, the temperature is then reduced in small steps, and at each temperature the system is kept long enough by introducing new bond switches and using the Metropolis algorithm until the system is in the equilibrium. As we already mentioned this is followed by the Keating relaxation of the local regions affected by a bond switch. A sequence of the temperatures is chosen, and at the end, when the temperature is zero, one hopes to obtain the structure with all characteristics of an amorphous material. Wooten, Winer and Weaire [38] obtained in this way a good model for a-Si and a-Ge. Their final 216-atom structure had an *rms* bond-angle deviation of about 12.6°. Consequently, the energy of their amorphous models was comparatively low. Our 4096-atom model of a-Si exhibits even lower *rms* bond-angle deviation (10.5° - 11°) which corresponds to a lower strain energy per atom than in the original Wooten and Weaire models. Nevertheless, some physicists object that there is still too much strain present in the amorphous structures obtained by using Wooten-Weaire method, and that a modification of it is needed if one wants to produce more realistic, less strained amorphous structures (Mousseau [43]).

The Wooten, Winer and Weaire model of a-Ge and a-Si [38] has a slight remnant of the original crystal visible in the structure factor, but nevertheless, there was no serious discrepancy with experiment (Wooten and Weaire [7]). The slight memory of the diamond cubic structure that remained in the amorphous structure is a consequence of the insufficient initial randomization and not from the annealing process. It should be noted here that the only essential difference between the a-Si and a-Ge is a simple scaling in proportion to the bond lengths (d=2.35Å for Si, and d=2.46Å for Ge). Otherwise, the experimental data are practically identical for a-Si and a-Ge. In Figure 12 we show the pair correlation function obtained in Wooten and Weaire model for a-Ge compared with experiment.

The discrepancy in the RDF at high r are related to the presence of inhomogeneities in the real sample which are not taken in account in fully tetrahedrally connected models, which always have slightly higher density than samples produced experimentally.

Since McKenzie and coworkers [44] have obtained 85-90% four-fold coordinated amorphous carbon by plasma-arc deposition, there has been an increasing interest in amorphous diamond-like carbon [45,46,47,48]. These diamond-like carbon films are found to be hard, optically transparent, and chemically inert, which makes them potentially important for applications in coating technology and also for use as wide band gap semiconductors. It is known that amorphous silicon and amorphous germanium exhibit a striking similarity in the shape of their radial distribution functions [7,36]. In both of these materials the bond stretching forces are dominant compared to the weaker bond-bending forces. Amorphous carbon is different with respect to both a-Si and a-Ge because of very strong chemical bonding, and because the bond-bending forces are much larger relative to the bond-stretching forces. The ratio of bond-bending to bond-stretching forces β/α is 2.4 times larger for carbon than for silicon. One might have expected a different amorphous structure for carbon (i.e. different ring statistics etc.) compared with silicon and germanium, because of the greater bond stiffness of carbon. Our results [5] show that the RDF of tetrahedral amorphous diamond is remarkably similar to the RDF's of amorphous silicon and amorphous germanium. The carbon first peak is broader than the silicon first peak, which is correlated with the larger *rms* bond-length deviation in carbon. This represents a different apportioning of the strain energy between bond length strain and bond angle strain in carbon and silicon. Nevertheless the overall features of RDF's are essentially unchanged between carbon and silicon. Using a Fourier transformation, the structure factor can be written as

$$I(k) = 1 + \frac{\rho_0}{k} \int_0^\infty \left[\frac{P(r)}{r} - 4\pi r \right] \sin(kr)dr \qquad (2)$$

where $P(r)$ is the RDF and ρ_0 is the average density. In Figure 12. we show a comparison of our calculated structure factor for a.C with that observed in experiments of Gilkes *et al.* [47].

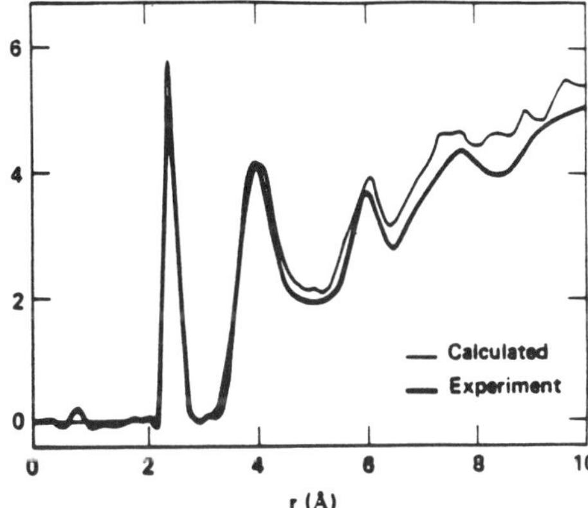

Figure 11. Thin line is the RDF divided by r calculated from the Wooten and Weaire model (1985), and thick line is experimental result for a-Ge.

Our very large (4096 atom), and realistic structural model of amorphous diamond [5] is potentially very appealing for addressing some of the other important issues of the physics of amorphous solids. One of those is the nature of the band tails in the electronic density of states in amorphous materials (Mott and Davis [49]). Recently, Dong and Drabold [50] used our structural model of amorphous diamond [5] to study the nature of the electronic localization in three dimensions in the presence of topological disorder.

Another application of the Wooten-Weaire method was done by Mousseau and Lewis [51] in their computer simulations of amorphous silicon hydrides. The starting point is a model of amorphous silicon obtained by Wooten-Weaire method. Instead of the Keating potential the authors use the Stillinger-Weber potential (Stillinger and

304

Weber [52]) to describe the interaction between atoms. They introduce pairs of H atoms in the a-Si structure in such a way that the fourfold coordination is preserved. The nearest-neighbor Si atoms were allowed to have a bond to each of the two H atoms. After each introduction of hydrogen pairs, the structure is Monté Carlo relaxed using a finite number of Wooten-Weaire bond switches, and relaxed using the Stillinger-Weber potential. The process is repeated until the desired concentration of H atoms is achieved. Mousseau and Lewis calculate Si-Si, Si-H, and H-H partial pair distribution functions, as well as corresponding static factors, and find that the agreement between the model and experiment improves with H concentration.

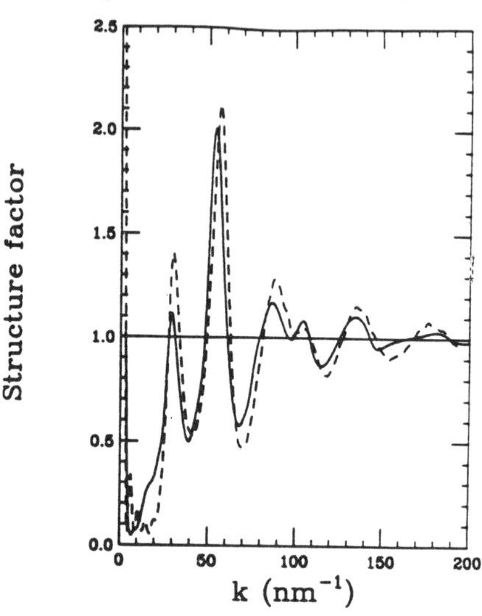

Figure 12. A comparison of our calculated structure factor for a-C [5] (dashed line) with that experimentally observed [47]. Note that our model is 100% sp^3 whereas the experimental sample has from 10-15% *sp^2* carbon in it.

3. Constraint Counting

We start by examining a large covalent network that contains no dangling bonds. I can describe such a network by the chemical formula Ge$_x$As$_y$Se$_{1-x-y}$ where the chemical element, Ge, stands for *any* fourfold bonded atom, As for *any* threefold bonded atom and Se for *any* twofold bonded atom. Every atom has its full complement of nearest neighbors and I consider the system in the thermodynamic limit where the number of atoms $N \to \infty$. There are no surfaces or voids and the chemical distribution of the elements is not relevant, except that I assume there are no isolated pieces, like a ring of Se atoms. The total number of atoms is N and there are n_r atoms with coordination r (r = 2, 3 or 4), then

$$N = \sum_{r=2}^{4} n_r \tag{3}$$

and I can define the mean coordination

$$\langle r \rangle = \frac{\sum_{r=2}^{4} r n_r}{\sum_{r=2}^{4} n_r} = 2 + 2x + y . \tag{4}$$

We note that $\langle r \rangle$ (where $2 < \langle r \rangle < 4$) gives a partial but very important description of the network. Indeed when questions of connectivity are involved, it is the key quantity as we shall see.

In covalent networks like $Ge_x As_y Se_{1-x-y}$, the bond lengths and angles are well defined and small displacements from the equilibrium structure can be described by the Keating potential written in Eq. (1). The bond-bending force is essential to the constraint counting approach. The other terms in the potential are assumed to be much smaller and can be neglected at this stage. If floppy modes are present in the system, then these smaller terms in the potential will give the floppy modes a small finite frequency. For more details see Thorpe [53]. If the modes already have a finite frequency these extra small terms will produce a small uninteresting shift in the frequency. This division into large and small forces is absolutely essential if the constraint counting approach is to be of any use. It is for this reason that it is of little, if any, use in metals and ionic solids. It is fortunate that this approach provides a very reasonable starting point in many covalent glasses.

We will regard the solution of the eigenmodes of the potential of Eq.(1) as a problem in classical mechanics [1,53,54]. The dynamical matrix has a dimensionality $3N$ which corresponds to the $3N$ degrees of freedom. In a stable rigid network we would expect all the squared eigenfrequencies $\omega^2 > 0$ with six modes being at zero frequency. These six modes are just the three rigid translations and the three rigid rotations. I am assuming that our (large) network has free boundary conditions. Of course these 6 modes have no weight in the thermodynamic limit. The total number of zero frequency modes can be estimated by the Maxwell counting algorithm [1]. This approach was first done for glasses by J. C. Phillips [17,18].

The constraint counting proceeds as follows. There is a single constraint associated with each bond which we can assign as $r/2$ constraints associated with each $r-$ coordinated atom. In addition there are constraints associated with the angular forces in Eq. (1). For a twofold coordinated atom there is a single angular constraint; for an $r-$ fold coordinated atom there are a total of $2r - 3$ angular constraints. The total number of constraints is

306

$$\sum_{r=2}^{4} n_r \left[r/2 + (2r-3) \right] . \tag{5}$$

The fraction f of zero–frequency modes is given by

$$f = \left[3N - \sum_{r=2}^{4} n_r \left[r/2 + (2r-3) \right] \right] / 3N \tag{6}$$

which can be conveniently rewritten in the compact form

$$f = 2 - \frac{5}{6} \langle r \rangle , \tag{7}$$

where $\langle r \rangle$ is defined in Eq. (4). Note that this result only depends upon the combination $2x + y$ which is the relevant variable. When $\langle r \rangle = 2$ (e.g. Se chains), then $f = 1/3$; that is one third of all the modes are floppy. As atoms with higher coordination than two are added to the network as crosslinks, f drops and goes to zero at $\langle r \rangle = 2.4$. and the network becomes rigid, as it goes through the phase transition. This mean field approach has been quite successful in covalent glasses and helps explain a number of experiments as will be described later. Also in section 5, we show the results of computer experiments and show that they are rather well described by the results of this section.

4. The Pebble Game

The elastic properties of random networks of Hooke springs has been studied over the past 12 years [53-58]. One of the most interesting findings has been that effective medium theory describes the behavior of the elastic constants and the number of floppy modes remarkably well [56,58] except very close to the phase transition from a rigid to a floppy structure.

Unfortunately, attempts to study the critical behavior in central-force networks have not been very satisfactory and the question of the universality class of the rigidity transition [55,57-61] has remained unresolved. This question is fundamental in understanding the nature of the rigidity transition, and may have important implications as to how the character of the glass transition is affected by the mean coordination, as has been discussed recently via fragile and strong glass formers [62]. We show here how substantial progress can be made in understanding the geometrical nature of generic rigidity percolation [3,63].

There are two important differences between rigidity and connectivity percolation. The first difference is that rigidity percolation is a vector (not a scalar) problem, and secondly, there is an inherent long range aspect to rigidity percolation. For example,

Figure 13(a) shows four distinct rigid clusters consisting of two rigid bodies attached together by two rods connecting at pivot joints. Now the placement of one additional rod, as shown in Figure 13(b), locks the previous four clusters into a single rigid cluster. This non-local character allows a single rod (or bond) on one end of the network to affect the rigidity all across the network from one side to the other.

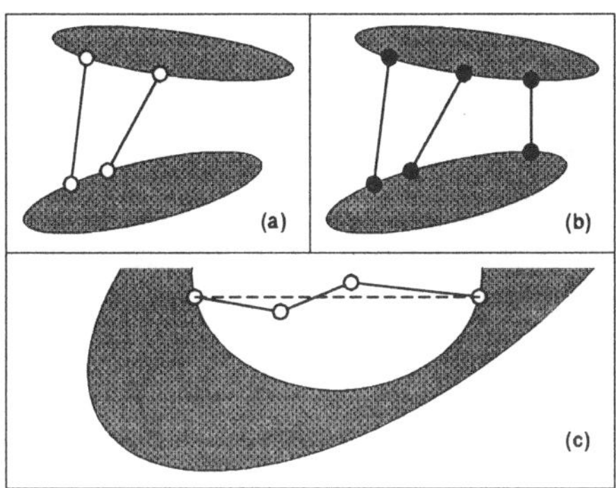

Figure 13. The shaded regions represent 2d rigid bodies. The (closed, open) circles denote pivot-joints that are members of (one, more than one) rigid body. (a) A floppy piece of network with four distinct rigid clusters. (b) Three generic cross links between two rigid bodies make the whole structure rigid. If the bonds were parallel, the structure would not be rigid to shear[64]. (c) Three non-collinear connected rods connecting across a rigid body is generic and contains one internal floppy mode. If they were collinear (along the dotted line), then there would be two infinitesimal (not finite) floppy motions, and under a horizontal compression buckling would occur.

Using concepts from graph theory, we set up generic networks where the connectivity or topology is uniquely defined but the bond lengths and bond angles are arbitrary. A generic network does not contain any geometric singularities [61] which occur when certain geometries lead to null projections of reaction forces. Null projections are caused by special symmetries, such as, the presence of parallel bonds or connected collinear bonds. Rather than these atypical cases their generic counterparts will be present as shown in Figures 13(a) and (b). This ensures that all infinitesimal floppy motions carry over to finite motions [61,65].

All previous studies on rigidity percolation have been on regular (non-generic) lattices which as we now know [3,63] has inadvertently delayed a proper understanding of the rigidity transition. In non-generic (referred to as atypical) networks many geometrical singularities occur which lead to non-linear effects. For example, a diode-like problem frequently occurs in atypical networks where a string of collinear bonds can only be extended with a cost in energy but can be compressed with no cost in energy due to buckling [e.g. Figure 13(c)]. The diode effect complicates studies because it leads to the breakdown of linear elasticity theory which must be reversible. A simple way to view a generic network is to take a regular lattice structure and randomly displace each site location by a small amount. This introduces local distortions throughout the lattice and is in itself a good physical model for amorphous and glassy materials. All prior studies which inevitably involve the non-linear effects arising from geometrical singularities [53-61]. should be considered as a separate problem.

By considering generic networks, the diode effect and the problematic geometric singularities are completely eliminated. Therefore, the problem of rigidity percolation on generic networks leads to many conceptual advantages because all geometrical properties are robust, not depending on a multitude of special cases. Moreover, real glasses are modeled better by generic networks rather than regular atypical networks because of local distortions. In two dimensions, there exist efficient, exact, combinatorial algorithms allowing for the possibility of an in depth study of rigidity percolation.

The rigidity of a network glass is related to how amenable the glass is to continuous deformations that require very little cost in energy. A small energy cost will always arise from weak forces which are present in addition to the hard covalent forces that involve bond-lengths and bond-angles. These small energies can be ignored because the degree to which the network is deformable is well quantified by just the number of floppy modes [53] within the system. A mental picture of floppy and rigid regions within the network has led to the idea of rigidity percolation [53,55]. An interesting model to consider is one with only central forces because its properties are most different from connectivity percolation which can be recovered if all pivot joints are welded fixed [66].

Much understanding of the general phenomena of central-force rigidity percolation can be obtained by studying a random network of Hooke springs. To be specific, we begin by considering a network of Hooke springs characterized by the potential

$$V = \tfrac{1}{2} \sum_{\langle ij \rangle} \alpha_{ij} \eta_{ij} \left(l_{ij} - l_{ij}^{o} \right)^2 \tag{8}$$

where the sum is over all bonds $\langle ij \rangle$ connecting sites i and j in the network. A bond connecting sites i and j is present if $\eta_{ij} = 1$ with probability p and absent if $\eta_{ij} = 0$ with probability $1-p$. The spring constants, α_{ij}, and the equilibrium bond lengths, l_{ij}^{0}, are positive real numbers but are left arbitrary. In addition, the site locations are also left arbitrary as the network is generic. Note that rigidity is a static concept, involving virtual displacements, so that while it is convenient to use harmonic potentials as done in Eq. (8) any set of pair potentials would give the same results for the geometric aspects of rigidity that is of interest here. A collection of sites form a rigid cluster when no relative motion within that cluster can be achieved without a cost in energy. Conversely, the floppy modes correspond to finite motions of the sample which do not cost energy. Therefore, the geometrical properties and the number of floppy modes can be determined by an equivalent bar and joint structure [61]. Note that a d-dimensional system always has at least $d(d+1)/2$ floppy modes due to d global translations and $(d-1)d/2$ global rotations.

The number of floppy modes in d dimensions is given by the total number of degrees of freedom for N sites minus the number of independent constraints. A redundant bond can only add additional reinforcement and/or cause internal stress in an existing rigid body. A key quantity is the number of floppy modes, F, in the network, or normalized per degree of freedom, $f=F/dN$. By defining the number of redundant bonds per degree of freedom as n_r, we can write quite generally,

$$f = \frac{dN - \left(\frac{1}{2} Nzp - dNn_r\right)}{dN} = 1 - \frac{p}{p*} + n_r \qquad (9)$$

where $p* = 2d/z$ and z is the lattice coordination. Neglecting the redundant bonds, as first done by Maxwell [1], we find that f is linear in the bond concentration, p, and goes to zero at the Maxwell approximation, $p*$, for the threshold. The Maxwell approximation gives a very good account of the location of the phase transition and the number of floppy modes, but it ultimately fails since the number of independent constraints is not just the total number of bonds as some bonds are dependent.

Short of the Maxwell constraint counting method, other ways that have been used for calculating the number of floppy modes exactly include rank determination of the dynamical matrix [53], relaxation methods [57,59], transfer matrix methods [67], Gaussian elimination [60] and the equation of motion technique [58]. Hence, many numerical methods have been explored.

Here we focus on the geometrical aspects of rigidity percolation which have not been directly addressed before. Previous studies have used costly relaxation type methods [which use $O(N^3)$ floating point operations] so that networks containing $N \approx 10^4$ sites already present a difficult numerical challenge. Relaxation methods are well suited for calculating elastic constants, but not for characterizing the geometric structure. This is because numerically one cannot identify which bond has exactly zero stress or if a bond accidentally has zero stress. However, until now, this was the only approach available in determining the stress carrying backbone.

Many basic questions have remained open regarding the nature of the rigidity transition in spite of many years of research by many groups. We mention a few points regarding the triangular lattice. Hansen and Roux [59,67] and later supported by Arbabi and Sahimi [60] have indicated that the fractal dimension of the stress carrying backbone is about $D_0 \approx 1.63$ which is very close to the current carrying backbone in connectivity percolation. Furthermore, the elastic moduli critical exponent (also denoted by f), or more precisely the ratio $f/v \approx 1.12$ was obtained for the random bond-dilution problem. Curiously, this value is the same as the full bond-bending model (all angular forces are present) which has an identical geometry to connectivity percolation. These two results gave strong evidence that the geometrical properties of central-force rigidity percolation is identical to connectivity percolation.

At first, the idea that the two types of percolation problems could share the same universality class is surprising since the vector and scalar character have different symmetry properties and because rigidity is a highly non-local phenomenon much different from connectivity. However, Hansen and Roux [59,67] have argued that the reason why this is possible, is because at large length scales, the central forces are able to yield effective angular forces via lever arms. Thus, the central-force problem may renormalize into the full bond-bending model. Knackstedt and Sahimi Knackstedt [60] have suggested using a real space renormalization group calculation, the geometrical properties of site and bond rigidity percolation share the same universality class as far as geometry is concerned. However, the elastic moduli exponents were shown not to be the same. This has been confirmed by simulation [60] as well, namely it has been found that $f / v \approx 1.12$ for random site dilution. If the idea of *lever arms* is correct, it should equally apply to random site-dilution.

4.1. THE PEBBLE GAME ALGORITHM

We have been able to study networks containing more than 10^6 sites, using an integer algorithm which gives exact and unique answers to the geometric properties of generic rigidity percolation. Because of the non-local characteristic of rigidity percolation, [e.g. Figures 13 (a) and 1(b)] burning-type algorithms [68] commonly used in connectivity percolation are useless. This implies that the entire structure needs to be specified (stored in memory) since the rigidity of a given region may depend on bonds far away.

A very efficient combinatorial algorithm, as suggested by Hendrickson [65], has been implemented to (i) calculate the number of floppy modes, (ii) locate over-constrained regions and (iii) identify all rigid clusters for $2d$ generic bar-joint networks. The crux of the algorithm is based on a theorem by Laman [69] from graph theory.

Theorem: *A generic network in two dimensions with N sites and B bonds (defining a graph) does not have a redundant bond iff no subset of the network containing n sites and b bonds (defining a subgraph) violates $b \leq 2n - 3$.*

By simple constraint counting it can be seen that there must be a redundant bond when Laman's condition is violated. This necessary part generalizes to all dimensions such that if $b > dn - d(d+1)/2$ then there is a redundant bond for $n \geq d$. For $n < d$ it follows that if $b > n(n-1)/2$ then there is a redundant bond. Note that $n=1$ is an excluded case in Laman's theorem since two sites are required for a bond to be present. The essence of Laman's theorem is that in two dimensions finding $b > 2n-3$ is the only way redundant bonds can arise. This sufficient part does not generalize to higher dimensions [65].

The basic structure of the algorithm is to apply Laman's theorem recursively by building the network up one bond at a time. Only the topology of the network is specified, not the geometry. Because of the recursion, only the subgraphs which contain the newly added bond need to be checked. If each of these subgraphs satisfy the Laman condition, $b \leq 2n - 3$, then the last bond placed is independent, otherwise it is redundant. By counting the number of redundant bonds, the exact number of floppy modes is determined.

Searching over the subgraphs is accomplished by constructing a pebble game. Each site in the network has two pebbles tethered to it. A pebble is either free when it is on a site or anchored when it is covering a bond. A free pebble represents a single motion that a site can undertake. Consider a single site having two free pebbles, representing two translations. If two additional free pebbles can be found at a different site, then the distance between this pair of sites is not fixed. Placing a bond between this pair of sites will constrain their distance of separation. To record this constraint, one of the four free pebbles is anchored to the bond. Once the bond is covered, only three free pebbles can be shared between that pair of sites. After a bond is determined to be independent, it will always remain independent and covered.

We begin with a network of N isolated sites each having two free pebbles. The system will always have $2N$ pebbles; initially two free pebbles per site. We place one bond at a time in the network connecting pairs of sites. The topological placement of either the sites or bonds will depend on the model under study such as the site or bond diluted generic triangular lattice as done here. The independent bonds must be covered by a pebble, therefore, before a bond can be covered it must be tested for independence. For each bond placed in the network, four pebbles (two on each site at the ends of the bond) must be free for the bond to be independent. When a bond is determined to be independent, any one of the four pebbles can be anchored to that bond. In general, all four pebbles across an added bond will not be free because they are already anchored to other bonds. These anchored pebbles may possibly become free at the expense of anchoring a neighboring free pebble while keeping a particular independent bond covered. In other words, pebbles may be shuffled around the network provided all independent bonds remain covered.

It is always possible to free up three pebbles across a bond, since they correspond to its rigid body motion. When a fourth pebble across a bond cannot be found, then that bond is redundant and it is not covered. In Figure 14 an example of how pebbles are shuffled is schematically shown on a small generic structure. Two distinct pebbles are associated with each site for which each pebble can either be used to cover a bond or is free to cover a bond. The two pebbles closest to a given site as schematically drawn in Figure 14 are the pebbles that are tethered to that site. Thus a pebble may either be on a site (free pebble) or on a bond (anchored pebble) but it always remains tethered to a given site regardless of how the pebbles are shuffled. Note that a bond may be covered by a pebble from either of its end sites. Therefore, free pebbles can be moved across the network by exchanging the site from which a pebble is used to cover a bond.

Over-constrained regions are recorded each time a dependent bond is found. These regions correspond to the set of bonds that were searched in trying to free the fourth pebble but failed. These regions, called Laman subgraphs, violate the condition $b \leq 2n - 3$. An added bond onto a Laman subgraph will be redundant.

We identify all the rigid clusters after the network is completely built. First, we identify isolated sites. Then the rigidity of all other sites is tested with respect to a reference bond. If a test-bond between either one of the pair of sites forming the reference bond and the site in question is found to be (dependent, independent) then that site (is, is not) rigid with respect to the reference bond. The test-bond is actually never added to the network. Since a bond can only belong to one cluster (unlike sites), all the bonds within a rigid cluster are ascribed to a particular reference bond. A systematic search is made to map out all rigid clusters.

312

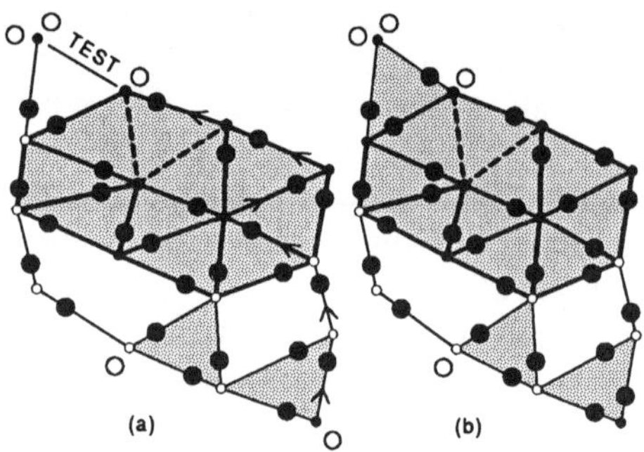

(a)   (b)

Figure 14. A demonstration of the pebble game on a generic network. Independent (redundant) bonds are shown with solid (dashed) lines which are (are not) covered by a pebble. Large (filled, open) circles denote (anchored, free) pebbles on (bonds, sites). The two closest pebbles to a given site are tethered to that site. Small (filled, open) circles denote sites belonging to (one, more than one) rigid cluster. Over-constrained bonds are shown with heavy dark lines. Shaded regions denote 2d rigid bodies. (a) Five free pebbles indicate 5 floppy modes until a new bond is added and tested for independence. A fourth free pebble is found via the path traced by arrows. (b) The added bond is independent and thus covered. There are now six rigid clusters and four floppy modes.

We show in Figure 14(b) the end result of the pebble game applied to a simple structure. Many aspects of rigidity are displayed. It can be seen that: (1) The exact number of floppy modes is determined by the number of free pebbles remaining. A depletion or excess of pebbles to cover a set of bonds distinguishes the over-constrained regions from the floppy regions; unlike the approximate global counting of Maxwell. (2) This network is uniquely decomposed into a set of six distinct rigid clusters, although the clusters are not disconnected. (3) The free pebble along the bottom edge cannot be shuffled over to the rigid body at the top which already has three free pebbles. This free pebble is shared among three bars and two triangles. Generally, free pebbles get trapped in floppy regions consisting of many rigid clusters giving arise to complex collective floppy motion. (4) The number of redundant bonds is unique, whereas their locations are not unique since this depends on the order of placing the bonds. Nevertheless, each redundant bond belongs to a unique over-constrained region (Laman subgraph). For example, there are 19 over-constrained bonds in the rigid cluster at the top of the structure in Figure 14(b), while having only two redundant bonds. (5) A rigid cluster will generally have sub-regions that are over-constrained. If any bond that is over constrained is removed, the rigidity of the network is unchanged.

A section of a large network on the bond-diluted generic triangular lattice at p_{cen} is shown in Figure 15 after the pebble game was applied. Here we see a typical topology and associated geometry of the set of rigid clusters in this section of network. Observe that nearly all individual rigid clusters form connected paths via pivots which are free joints shared by two or more rigid bodies. It can be seen that floppy regions form where the pivot sites cluster. Notice that within this section of the network, there is a spanning rigid cluster where rigidity forms a connected path from top to bottom and left to right.

There are also clusters of over-constrained regions in the network which are not percolating in this section.

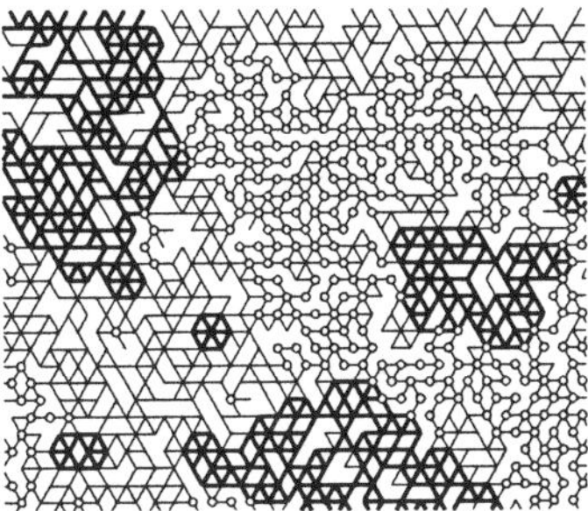

Figure 15. The topology of a typical cut-out region from a bond-diluted generic network at p=0.6603. A particular realization would have local distortions (not shown) similar to Fig. 14. The heavy dark lines correspond to over-constrained regions. The open circles correspond to sites which are acting as pivots between two or more rigid bodies.

4.2. NEW RESULTS FOR BOND AND SITE PERCOLATION

In this section, we present some new results for central-force generic rigidity percolation on both the bond and site diluted triangular net. We begin by working out a better estimate for the bond and site rigidity threshold. The Maxwell counting prediction for the site rigidity threshold gives $q^* = 2/3$ which is the same as $p^* = 2/3$ for bond dilution. This comes about because the number of sites in a site diluted system is only qL^2 and the number of bonds is $q^2 z L^2 / 2$. When the number of floppy modes is normalized per degree of freedom, the final result for f can be expressed in the same way as the left-most side of Eq. (9) with p replaced by q. To get a more accurate estimate for the rigidity threshold, the presence of redundant bonds and floppy inclusions must be accounted for.

In a low concentration expansion the first diagram to contribute redundant bonds is shown in Figure 16, where only 11 of the 12 bonds are independent. This diagram leads to a correction for the number of floppy modes as $n_r = \tfrac{1}{2} p^{12} + O(p^{18})$ and $n_r = \tfrac{1}{2} q^7 + O(q^{10})$ for bond and site dilution respectively. In a high concentration expansion, the first two dominant contributing diagrams are shown in Figures 16(b) and (c) corresponding to dangling ends and isolated sites.

From these two leading diagrams the fraction of floppy modes, f, at high concentrations are given by

$$f = 3(1-p)^5 - 2(1-p)^6 + O\left((1-p)^8\right) \tag{10a}$$

and
$$f = 3(1-q)^5 - 5(1-q)^6 + 2(1-q)^7 + O\left((1-q)^8\right) \tag{10b}$$

314

for bond and site dilution respectively.

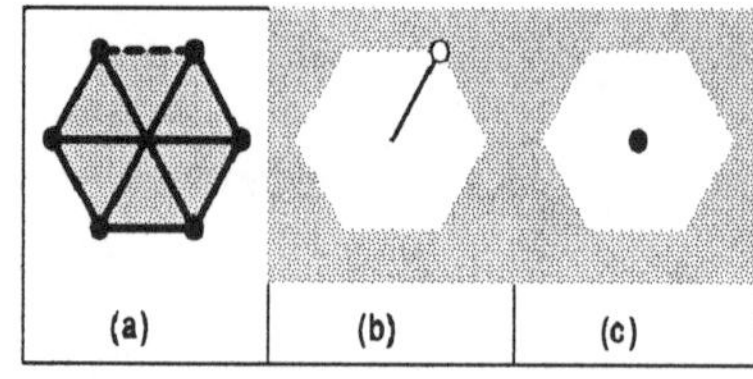

Figure 16. All shaded areas represent a 2d rigid region. The topology is shown for: (a) The lowest order diagram on a triangular network to have a dependent bond (dashed line). All 12 bonds are over-constrained. (b) The lowest order diagram for a floppy inclusion corresponds to a dangling bond. (c) An issolated site is counted as a floppy inclusion and contributes two floppy modes at the next lowest order.

It is these type of corrections that will shift the transition from the Maxwell threshold of 2/3 and be responsible for a non mean-field-like critical behavior. We equate the number of floppy modes from the truncated low and high concentration expansions. For bond and site dilution we estimate the thresholds to be 0.6622 and 0.6877 respectively. Thus we find a small downward shift for bond dilution and a somewhat larger upward shift for site dilution. The pebble game reveals that the rigidity thresholds shift about 50% more than the above estimates to $p_{cen} = 0.66020 \pm 0.0003$; and $q_{cen} = 0.69755 \pm 0.0003$; for bond and site dilution respectively. The site diluted threshold is also in good agreement with that obtained by Duxbury and Moukarzel [70].

A sharp peak in the 2nd derivative of the number of floppy modes shown in Figure 17 appears without any signs of a discontinuity. The peak most resembles a simple cusp. As can be seen there is virtually no difference between the data for linear system sizes L=680,960 and 1150. Only very slight system size dependence has been observed. The trend is for smaller systems to show a cusp-like singularity in $f^{(2)}$ as well, but with the peak slightly shifted to the left with a smaller amplitude.

The behavior of the second derivative suggests that the number of floppy modes is analogous for rigidity and connectivity percolation. In the case of connectivity percolation, the number of floppy modes is simply equal to the total number of clusters, which corresponds to the free energy [71]. It would be nice if a similar result holds for rigidity percolation. We find that the second derivative of the total number of clusters changes sign across the transition, thus violating convexity requirements. Noting that typically rigid clusters are not disconnected, we suggest that the number of floppy modes generalizes as an appropriate free energy. With this assumption, we have estimated the exponent α in the usual context of a *heat capacity* critical exponent. More work needs to be done to see how a free energy can be defined for rigidity percolation.

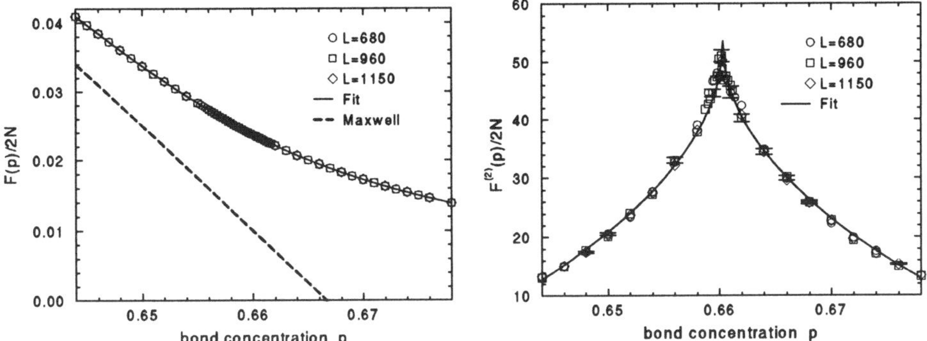

Figure 17. (a) Simulation results for the fraction of floppy modes, f=F/2N, for a bond diluted generic network compared to the Maxwell prediction. All error bars are smaller than the symbols. Note that the non-generic threshold[60] is at 0.641. (b) The second derivative of the fraction of floppy modes for a bond diluted generic network as calculated from Monte Carlo sampling. The fitting results for the cusp yields p_{cen}=0.6603 $\pm$ 0.0003 and an exponent of 0.48 $\pm$ 0.05. Typical error bars are shown which reflect both the statistical errors in the Monté Carlo sampling and the ensemble averaging.

In addition to calculating the number of floppy modes, we analyized the rigid cluster statistics [63] for bond dilution with free and periodic boundary conditions and site dilution with periodic boundary conditions. The free boundary condition data was generated mainly to check boundary effects. Finite size scaling techniques as used in percolation theory [68] are applied here assuming only a single relevant length scale exists.

We also look at the geometrical properties of the over-constrained regions within the network. These over-constrained regions are not necessary to sustain rigidity. An isostatic framework [61], for example, has just the right positioning of bonds to form a rigid cluster without a redundant constraint. However, when external forces are applied to a rigid isostatic framework, using rigid bus bars for example, over-constrained regions will be induced across it, which forms a percolating stressed region. Within a random environment there will be redundant bonds scattered throughout the network. Here the redundant bonds are essentially acting as external forces on an underlying isostatic framework. Physically, the resulting over-constrained regions characterize internal stress caused by bond mismatch. Thus, the over-constrained regions propagate stress (without externally applied forces), and are most closely analogous to the current carrying backbone in connectivity percolation. We find that monitoring the probability for a network of linear size L to contain either a spanning rigid cluster or spanning stressed backbone, leads to the same correlation length exponent, ν, and critical threshold.

From extrapolation using finite size scaling techniques [63] we find $p_{cen} = 0.66020 \pm$ 0.0003 for the bond diluted triangular lattice which is in excellent agreement with that obtained from the cusp singularity in the second derivative of f as shown in Figure 17.

Similarly, for site dilution we find $q_{cen} = 0.69755 \pm 0.0003$. Moreover, the rigidity transition is second order, but in a different universality class than connectivity percolation, with the exponents; $\alpha = -0.48\pm0.05$, $\beta = 0.175\pm0.02$ and $\nu = 1.21\pm0.06$. The fractal dimension of the spanning clusters and the spanning stressed regions at the critical threshold are found to be $d_f = 1.86\pm0.02$ and $d_{BB} = 1.80\pm0.03$ respectively.

Unfortunately at the moment it is not possible to extend the pebble game analysis to $3d$ networks, as this algorithm is based on Laman's theorem which does not generalize to $3d$. This is because of the existence of banana graphs as shown in Figure 18.

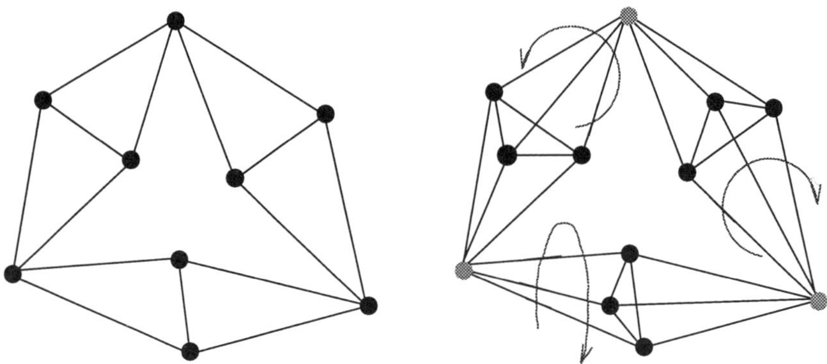

Figure 18. (Left panel) A *single* rigid cluster in two dimensions. (Right panel) Showing 3 bananas in a $3d$ network. There are *four* rigid clusters here - one in each arm plus an additional rigid cluster consisting of the three sites at the corners which are connected by implied bonds. Such non-contiguous rigid clusters cannot occur in $2d$ networks. Such clusters are responsible for the breakdown of Laman's theorem in $3d$.

While the pebble game used in this paper is only applicable in 2d, we are currently extending the rules for the pebble game to 3d. Although the Laman condition is not generally sufficient [65,69,72] in three dimensions, we recently have shown that it can be generalized within certain classes of networks. The problem of the Laman bananas as shown in Figure 18 is conveniently eliminated if angular forces as in the Keating potential Eq. (1) are included. In particular, we have shown that for the bond-bending model, having nearest and next nearest neighbor forces, Laman's theorem can be generalized. Furthermore, a 3d pebble game has been constructed for this special case, for which we our currently verifying the rules. Fortunately, the bond-bending model is precisely the class of models that is applicable to the study of covalent glass-networks!

The recent progress in generalizing the pebble game to 3d for the bond bending model is an important first step to extend the powerful techniques described above to real covalent glasses. This is a fortunate situation, where the situation of physical interest is the one that simplifies the mathematics. Armed with such an algorithm we will be able to access how accurate the mean field prediction for a phase transition at a mean coordination of 2.4 is, and to determine the cluster statistics and universality class of the transition [73].

5. Surface Floppy Modes

The analysis in the previous section has all been in the thermodynamic limit ($N \to \infty$), and the number of floppy modes, F, was a thermodynamic quantity proportional to the number of atoms, N. Thus the floppy modes can be thought of as extended bulk modes. Because there are N such modes, surface effects are normally irrelevant leading to totally negligible corrections. However when the term proportional to N is zero, surface effects can become important and dominant. I will discuss this situation here.

5.1. BASIC COUNTING TECHNIQUES

We use as an illustration the square net with $n_1 n_2$ sites shown in Figure 19. I can apply the Maxwell constraint counting method here also, taking careful account of the sides and corners. The basic algorithm is unchanged; the number of floppy modes F equals the number of degrees of freedom (*twice the number of sites*) minus the constraints (*the number of bonds*).

$$F_2 = 2n_1 n_2 - \left[\tfrac{4}{2}(n_1 - 2)(n_2 - 2) + \tfrac{3}{2}2(n_1 - 2) + \tfrac{3}{2}2(n_2 - 2) + \tfrac{2}{2}4 \right]$$
$$= n_1 + n_2 \tag{11}$$

where the subscript 2 on F denotes two–dimensions. I have checked this computationally for a range of values of n_1 and n_2 and find that the formula (11) works *exactly*. The formula, Eq. (11) is based on topological considerations and has been found to work whether the network is generic or atypical. I have checked this computationally.

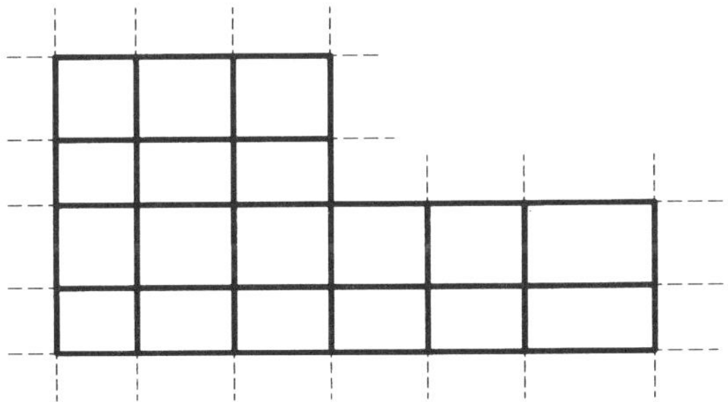

Figure 19. A piece of a square net with 29 sites and 24 surface bonds shown by dashed lines. It is explained in the text that there are 24/2 = 12 floppy modes of which 2 are the macroscopic rigid translations and 1 is the macroscopic rigid rotation.

These planes can now be stacked up to form a simple cubic array and the result for F becomes

$$F_3 = n_3 F_2 + n_3(n_1 n_2) - (n_1 n_2)(n_3 - 1) \tag{12}$$

318

where the second term comes from the extra degrees of freedom and the last term is the additional constraints. Simplifying I find that

$$F_3 = n_1 n_2 + n_2 n_3 + n_3 n_1 . \tag{13}$$

This is clearly a surface term, as it is proportional to the number of surface atoms in a large sample.

We can now generalize to d dimensions

$$F_d = n_d F_{d-1} + (n_d)\prod_{i=1}^{d} n_i - (n_d - 1)\prod_{i=1}^{d-1} n_i \tag{14}$$

which gives

$$F_d = n_d F_{d-1} + \prod_{i=1}^{d-1} n_i . \tag{15}$$

We can solve this equation, using induction and the known answers for $d = 2$ and $d = 3$ to give

$$F_d = \left(\prod_{i=1}^{d} n_i\right)\left(\sum_{i=1}^{d} \frac{1}{n_i}\right). \tag{16}$$

This equation has been checked for various values of n_i for $d = 2$ and $d = 3$ and always works exactly for *both* generic and atypical networks. The reason why this counting is exact is because there are no redundant bonds in the network and thus Maxwell constraint counting is exact.

A nice way of obtaining the equation for F_2 is to count the number of *missing bonds* at the surface and divide by 2. An example is shown in Figure 19. Here there are 29 sites and 24 surface bonds shown by dashed lines. A little thought will convince the reader that the number of floppy modes is 24/2 = 12. This number clearly reduces to the previous answer $n_1 + n_2$ for a rectangular shape, but generalizes the answer to other less regular shapes. Note that the site in the indentation has all four bonds present, and so does not contribute, despite being a surface atom in some sense. This kind of construction also works if there are internal voids. Another way to look at this construction, is to dissect the network into distinct linear chains which decouple. Each linear chain has two (dashed) ends and a single zero frequency mode, leading immediately to the result above. This latter approach is equivalent to using Eq.(15) to solve for F_2 knowing that $F_1 = 1$.

5.2. PROBLEMS WITH PERIODIC BOUNDARY CONDITIONS

The periodic case is more difficult to understand, as symmetry now seems to play a more important role, with clear differences appearing between distorted and undistorted networks [74]. The reader may complain that the periodic case is of no more than academic interest. This complaint may be justified, but we cannot claim to understand this whole area until we do understand the periodic case also. Constraint counting always give exactly zero for any d dimensional *hypercubic* lattice. This result is not exactly correct as there are always the d Goldstone modes, corresponding to the acoustic phonons in the long wavelength limit. For the *distorted or generic* periodic case, I find numerically that this is the total - there are indeed always exactly d modes. The *undistorted or atypical* periodic case can be solved using Bloch's theorem. For a periodic hypercubic lattice, the eigenfrequencies are given by

$$\omega_r^i = \sqrt{\frac{k}{M}} \left| \sin\left(\frac{\pi r_i}{n_r} \right) \right| \tag{17}$$

where $r_i = 0, 1, 2 \ldots n_r - 1$. Because of the decoupling, there are $\dfrac{1}{n_r} \left(\prod_{i=1}^{d} n_i \right)$ floppy modes associated with the direction r, so that the total number of floppy modes is again given by

$$F_d = \left(\prod_{i=1}^{d} n_i \right) \left(\sum_{i=1}^{d} \frac{1}{n_i} \right) \tag{18}$$

which curiously is the same answer I obtained for the square with free boundary conditions [74]. Again this is because the square can be dissected into periodic linear chains, each with a single floppy mode, as in the case with free boundary conditions.

The difference is that the distortions reduce the number of floppy modes from F_d to d in the periodic case, but the number of floppy modes for the non-distorted case is the same as F_d for free boundary conditions. Topologically, the periodic boundaries is equivalent to having long range bonds from end to end along the network. In the case of no distortions, the colinearity of these long range bonds with the 1d chains results in them not contributing as independent constraints.

6. Experiments

6.1. BULK MATERIALS

The above findings have been confirmed using computations on model networks. Some of these results are shown in Figure 20. The computed number of floppy modes (shown in the insert) closely follows Eq. (7) and the elastic constants approach zero from the

320

high coordination side, also at $\langle r \rangle = 2.4$, which is referred to as the point at which *rigidity percolation* occurs. This is a phase transition from rigid to floppy, driven not by temperature, but rather by the mean coordination.

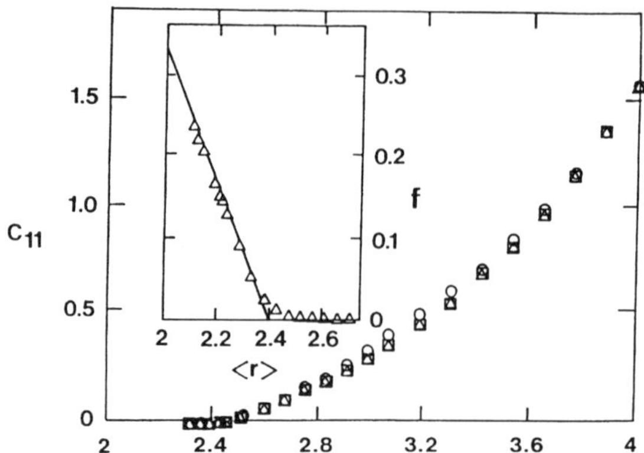

Figure 20   Showing the elastic constant c_{11} for a model network as a function of the mean coordination $\langle r \rangle$ for three different series of random networks. In the insert the fraction of floppy modes, f , is shown. The points are from computer simulation [58], and the solid line in the insert is the straight line given by Eq. (7).

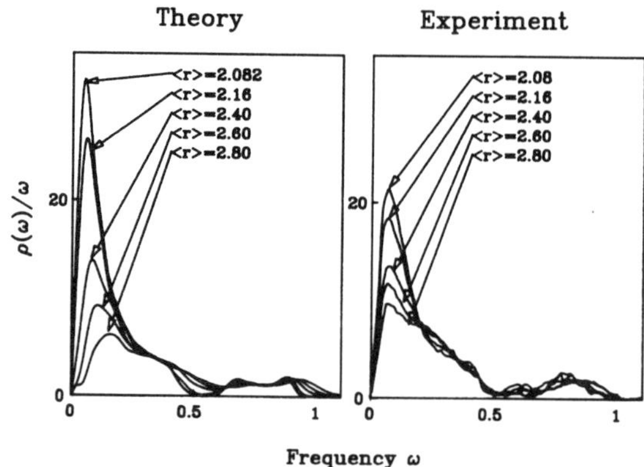

Figure21.   The density of states divided by the frequency, $\rho(\omega)/\omega$, for various values of the mean coordination, $\langle r \rangle$. The left hand set of graphs are theoretical [54] and the right set of graphs are from inelastic neutron scattering experiments [74] on Ge_xSe_{1-x}. The frequency is in units where the maximum frequency is unity.

Measurements of the elastic constants [74] appear to be influenced considerably by the weak forces, and the phase transition is washed out as shown in Figure 21. The best experimental confirmation of these ideas to date comes from inelastic neutron scattering measurements of the density of states [74], shown in Fig. 21. The agreement between theory and experiment is excellent, even when the weak forces are included in a very simple way and adjusted to bring the zero–frequency modes to the correct (low) frequency. Note that the weight in the floppy modes, given by constraint counting, is unaffected by the weak forces, even though there is some background response to contend with.

6.2. CORRECTION FOR DANGLING BONDS

The previous section describes the situation when there are *no* dangling bonds present. Constraint counting fails when dangling bonds are present because the expression for the number of angular forces $2r-3$ gives -1 when $r=1$, instead of the correct answer of zero. Thus Eq. (7) needs correction. This has recently been done in a compact way by a number of authors [75-77] following earlier efforts [78,53].

It is rather straightforward to extend Eq. (7) to include the summation over the dangling bonds ($r=1$) and correct for the miscounting of the angular constraints to give

$$ f = \left[3N - \sum_{r=1}^{4} n_r \left[r/2 + (2r-3) \right] - n_1 \right] / 3N \tag{19} $$

which now leads to the form

$$ f = 2 - \frac{5}{6}\langle r \rangle - \frac{n_1}{3N} \tag{20} $$

where the definition of $\langle r \rangle$ is extended from (2) to include the dangling bonds. The transition now takes place at a lower mean coordination $\langle r \rangle$ which is given by

$$ \langle r \rangle = 2.4 - 0.4 \frac{n_1}{N}. \tag{21} $$

It is not surprising that the transition takes place at a lower mean coordination $\langle r \rangle$, because the dangling ends play no role in the network connectivity. Indeed another conceptual approach is to strip the dangling ends away and define a *skeleton network* that has only 2, 3 and 4 coordinated atoms. The theory described in this section can then be applied to the skeleton network. Equation (21) has recently been applied to networks containing iodine, which forms a dangling end [76,77].

These ideas have also been applied to amorphous carbon networks by Tamor and are described in [74]. Amorphous carbon networks can be thought of as consisting of three kinds of atoms: fourfold (diamond–like) carbon, threefold (graphitic) carbon, and often considerable amounts of atomic hydrogen that ties off dangling ends and so is singly

322

coordinated. Suppose that there are N atoms in the network, with a fraction x_4 of fourfold coordinated carbon, a fraction x_3 of threefold coordinated carbon and x_1 of singly boned atomic hydrogen. Then by definition, we have

$$x_4 + x_3 + x_1 = 1. \tag{22}$$

To illustrate the use of the skeleton network, I apply it to this case. The mean coordination of the skeleton network, with the hydrogen removed is

$$\langle r \rangle = \frac{2(Number\ of\ bonds)}{Number\ of\ sites} = \frac{(4Nx_4 + 3Nx_3 + Nx_1) - 2Nx_1}{Nx_4 + Nx_3} \tag{23}$$

which gives

$$\langle r \rangle = \frac{4x_4 + 3x_3}{1 - x_1} - \frac{x_1}{1 - x_1} \tag{24}$$

The important new term is the $2Nx_1$ in the numerator of Eq. (23), which removes the bonds between hydrogen and the rest of the network that are not present in the skeleton network. In deriving Eq. (24), I have assumed that there are no molecular fragments, like for example methane CH_4, that get detached from the network. If this were the case, then these fragments should also be eliminated for purposes of counting. It appears that the elasticity and especially the hardness properties of carbon networks [77] follows Eq. (24) closely. Indeed the hardness goes all the way from close to that of crystalline diamond at the one extreme, to mush when too much hydrogen is present.

For the *skeleton network*, the number of floppy modes is still given by Eq. (7) as before. The total number of sites in the skeleton lattice is reduced to $N(1 - x_1)$, so that the total number of floppy modes F is given by

$$F = 3N(1 - x_1)(2 - \tfrac{5}{6}\langle r \rangle). \tag{25}$$

We now insert $\langle r \rangle$ from Eq. (24) into Eq. (25) and find that

$$\frac{F}{3N} = 2 - \frac{10}{3}x_4 - \frac{5}{2}x_3 - \frac{7}{6}x_1 \tag{26}$$

which of course reduces to the old result of Eq. (7) if there is no hydrogen present so that $x_1 = 0$. The total number of floppy modes is not changed when the hydrogen is recombined with the skeleton network to reconstruct the original network that existed before the hydrogen was stripped off. This occurs because adding back a single hydrogen atom, adds 3 extra degrees of freedom, but also adds 3 constraints (1 for the central force associated with the bond, and 2 for the angular forces associated with attaching this bond to an atom in the network). Thus there is no change and all the floppy modes are associated with the skeleton network and not specifically with the hydrogen. No direct measurements of these modes have yet been done in carbon networks. Inelastic neutron scattering experiments in the low frequency region, similar to those done for chalcogenide glasses and shown in Figure 21, would be very interesting.

The transition from rigid to floppy occurs when the number of floppy modes, F, goes to zero. The most interesting experiment to date measures the hardness of networks containing carbon as a function of the mean coordination. The results seem to be in quite dramatic agreement with the ideas here.

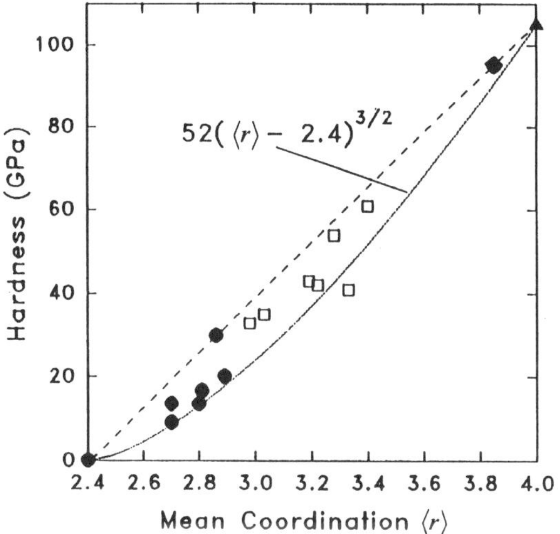

Figure 22. The hardness (measured by nano-indentation) of various diamond films, some containing some hydrogen, as a function of the mean coordination as defined by Eq. (24). The data were complied and presented in this form by Tamor [79] based on previous work by Tamor and others [74], using various preparation techniques. The solid triangle is the result for crystalline diamond.

It has recently been suggested that it is better to plot physical properties, like hardness, not against the mean coordination of the skeleton network, but rather to use the original network [77], which contains dangling ends. This may be connected with the fact that it takes energy to move the dangling bonds apart in an experiment that measures hardness. It would be interesting to have measurements of other physical quantities to see if this observation still persists.

6.3. SILICATE NETWORKS

Although the floppy modes that I have been discussing emanate from the surface, they are in no sense surface modes. The amplitude is not damped away from the surface. They are bulk in extent and involve the whole solid. It is only the total number that scales like the surface area. This concept is important in other marginal structures.

Another 2d example is the kagome lattice which consists of triangles joined at the corners. The most important three dimensional example is provided by silicates $[SiO_2]$ where the SiO_4 tetrahedra are corner–sharing. This case has been examined extensively, both theoretically and experimentally recently [80]. These authors refer to these modes as rigid unit modes [RUM's] in which the SiO_4 tetrahedra are not distorted. These modes can be seen using inelastic neutron scattering. This experiment is not so straightforward as in the chalcogenide glasses, because the number of RUM's is proportional to the surface area $[N^{\frac{2}{3}}]$, rather than the bulk $[N]$, where N is the number of atoms.

It is not yet clear if the exact results given for the number of surface floppy modes in this section, are exactly or only approximately generalisable to the kagome and silicate structures.

7. Summary

Lord Kelvin has said "*I am never content until I have constructed a mechanical model of the subject I am studying. If I succeed in making one, I understand; otherwise I do not*" [81]. This has been the approach that I have found most useful to follow in trying to understand the mechanical aspects of the behavior of glasses.

In this paper, I have tried to collect together some of the more important results regarding floppy modes and constraint counting in glasses that have emerged in the past decade. This work has provided a useful conceptual framework within which to discuss some of the physical properties of glasses. Some of the arguments in this area are subtle and still controversial. For example using a reduced dimensionality of 2 rather than 3 for layered materials like As_xSe_{1-x} does not seem to be correct to me, as these atoms are still embedded in a 3 dimensional space [82]. It is true that using this reduced dimensionality does seem to improve agreement with experiment where the discontinuity seems to be at $\langle r \rangle = 2.67$ rather than $\langle r \rangle = 2.4$, but perhaps there are other explanations for this.

Recent work on surface floppy modes is beginning to look interesting as a way of understanding open structures. So far the only serious application has been to bulk silicate networks [80], but this approach looks promising for other more complex structures like porous silica, clays and zeolites, that contain internal surfaces or voids. Internal floppy modes may permit voids to alter their shape so as to facilitate chemical reactions and catalysis. This remains to be seen. An even more distant hope is that this work may find applications in biological systems, where the ability of molecules to move at little cost in energy probably has important implications for enzyme activity and other biologically important interactions.

8. Acknowledgments

This research was supported by the NSF under grants No. DMR–9024955 and CHE–9224102. A copy of the FORTRAN program *the pebble game* for analyzing two dimensional generic networks is available upon request from M. F. Thorpe.

References

1. Maxwell, J.C. (1864) On the calculation of the equilibrium and stiffness of frames, *Philos. Mag.* **27**, 294- 299.
2. Lagrange, J.L. (1788) *Mécanique Analytique*, Paris.
3. Jacobs, D.J. and Thorpe, M.F. Generic rigidity percolation, *Phys. Rev. Lett.* **75**, 4051-4054.
4. Zallen, R. (1983) *The Physics of Amorphous Solids*, John Wiley & Sons, New York.
5. Djordjević, B.R., Thorpe, M.F. and Wooten, F. (1995) Computer model of tetrahedral amorphous diamond, *Phys. Rev. B* **52**, 5685-5689; Djordjević, B.R., (1996) Ph. D thesis, Michigan State University.
6. Elliott, S.R. (1984) *Physics of Amorphous Materials*, Longman, London and New York.
7. Wooten, F. and Weaire, D. (1987) Modeling tetrahedrally bonded random networks by computer, in H. Ehrenreich, F. Seitz and D. Turnbull (eds.), *Solid State Physics* , **40**, 1-42, Academic, New York.
8. Bernal, J.D. (1959) A geometrical approach to the structure of liquids, *Nature* **183**, 141-147.
9. Bernal, J.D. (1964) The Bakerian lecture, 1962. The structure of liquids, *Proc. Roy. Soc. London A* **280**, 299-322.
10. Finney, J.L. (1970) Random packings and the structure of simple liquids, *Proc. Roy. Soc. A* **319**, 479-547.
11. Bennett, C.H. (1972) Serially deposited amorphous aggregates of hard spheres, *J. Appl. Phys.* **43**, 2727-2734.
12. Cargill, G.S. (1975) Structure of metallic alloy glasses, in H. Ehrenreich, F. Seitz and D. Turnbull (eds.), *Solid State Physics* **30**, 227-320, Academic, New York.
13. Zachariasen, W.H. (1932) Atomic arrangement in glass, *J. Am. Chem. Soc.* 54, 3841-3851.
14. Bell, R.J. and Dean, P. (1966) Properties of vitreous silica: analysis of random network models, *Nature (London)* **212**, 1354-1356.
15. Bell, R.J. and Dean, P. (1972) The structure of vitreous silica: validity of the random network theory, *Philos. Mag.* **25**, 1381-1398.
16. Mozzi, R.L. and Warren, B.E. (1969) The structure of vitreous silica, *J. Appl. Cryst.* **2**, 164-172.
17. Phillips, J.C. (1979) Topology of covalent non-crystalline solids. I. Short-range order in chalcogenide alloys, *J. Non-Cryst. Solids* **34**, 153-181.
18. Phillips, J.C. (1981) Topology of covalent non-crystalline solids. II. Medium-range order in chalcogenide alloys and A-Si(Ge), *J. Non-Cryst. Solids* **43**, 37-77.
19. Polk, D.E. (1971) Structural model for amorphous silicon and germanium, *J. Non-Cryst. Solids* **5**, 365-376.
20. Polk, D.E. and Boudreaux, D.S. (1973) Tetrahedrally coordinated random network structure, *Phys. Rev. Lett.* **31**, 92-95.
21. Steinhardt, P., Alben, R. and Weaire, D. (1974) Relaxed continuous random network models. I. Structural characteristics, *J. Non-Cryst. Solids* **15**, 199-214.

22. Duffy, M.G., Boudreaux, D.S. and polk, D.E. (1974) Systematic generation of random networks, *J. Non-Cryst. Solids* **15**, 435-454.
23. Keating, P.N. (1966) Effect of invariance requirements on the elastic strain energy of crystals with application to the diamond structure, *Phys. Rev.* **145**, 637-645.
24. Connell, G.A.N. and Temkin, R.J. (1974) Modeling the structure of amorphous tetrahedrally coordinated semiconductors, *Phys. Rev. B* **9**, 5323-5326.
25. Henderson, D. (1974) Random tetrahedral network with periodic boundary conditions, *J. Non-Cryst. Solids* **16**, 317-320.
26. Kaplow, R., Rowe, T.A. and Averbach, B.L. (1968) Atomic arrangement in vitreous silica, *Phys. Rev.* **168**, 1068-1079.
27. For more information on the Inverse Monté Carlo method, see the talks by L. Pusztai and N. Zotov in these proceedings.
28. Shevchik, N.J. (1973) Computer-generated structures of amorphous Ge, *Phys. Status Solidi B* **58**, 111-120.
29. Alben, R., Weaire, D., Smith, Jr. J.E. and Brodsky, M.H. (1975) Vibrational properties of amorphous Si and Ge, *Phys. Rev. B* **11**, 2271-2296.
30. Evans, D.L., Teter, M.P. and Borrelli N.F. (1974) Vitreous silica minus oxygen → amorphous silicon: a model study, *A.I.P. Conf. Proc.* **20**, 218-223.
31. Evans, D.L., Teter, M.P. and Borrelli N.F. (1975) The range and kind of order in random tetrahedral structures, *J. Non-Cryst Solids* **17**, 245-258.
32. Wright, A.C., Connell, G.A.N. and Allen, J.W. (1980) Amorphography and the modeling of amorphous solid structures by geometric transformations, *J. Non-Cryst. Solids* **42**, 69-86.
33. Evans, D.L. and King, S.V. (1966) Random network model of vitreous silica, *Nature* **212**, 1353-1354.
34. Guttman, L. (1974) Simulation of continuous random network models with periodic boundary conditions, *A.I.P. Conf. Proc.* **20**, 224-228.
35. Guttman, L. (1975) Vibrational spectra of four-coordinated random networks with periodic boundary conditions, *A.I.P. Conf. Proc.* **31**, 268-272.
36. Etherington, G., Wright, A.C., Wemzel, J.T., Dore, J.C.,Clarke, J.H. and Sinclair, R.N. (1982) A neutron diffraction study of the structure of evaporated amorphous germanium, *J. Non-Crystalline Solids* **48**, 265-289.
37. Martin, R.M. (1970) Elastic properties of ZnS structure semiconductors, *Phys. Rev. B* **1**, 4005-4011.
38. Wooten, F., Winer, K. and Weaire, D. (1985) Computer generation of structural models of amorphous Si and Ge, *Phys. Rev. Lett.* **54**, 1392-1395.
39. Wooten, F. and Weaire, D. (1996), in J. Kalivas (ed.), *Adaptation of Simulated Annealing to Chemical Problems*, Elsevier Science, (in press).
40. Metropolis, N., Rosenbluth, A., Rosenbluth, M., Teller, A. and Teller, E. (1953) Equation of state calculations by fast computing machines, *J. Chem. Phys.* **21**, 1087-1092.
41. Kirkpatrick, S., Gelatt, Jr. C.D. and Vecchi, M.P. (1983) Optimization by simulated annealing, *Science* **220**, 671-680.
42. Vanderbilt, D. and Louie, S.G. (1984) A Monte Carlo simulated annealing approach to optimization over comtinuous variables, *J. Comp. Phys.* **56**, 259.
43. Mousseau, N. (1996), private communication, and these proceedings.
44. McKenzie, D.R., Muller, D.A. and Pailthorpe, B.A. (1991) Compressive-stress-induced formation of thin-film tetrahedral amorphous carbon, *Phys. Rev. Lett.* **67**, 773-776.
45. Drabold, D.A., Fedders, P.A. and Stumm Petra (1994) Theory of diamondlike amorphous carbon, *Phys. Rev. B* **49**, 16415-16422.

46. Hirai, H., Tabira, Y., Kondo, K., Oikawa, T. and Ishizawa, N. (1995) radial distribution function of a new form of amorphous diamond shock induced from C_{60} fullerene, *Phys. Rev. B* **52**, 6162-6165.

47. Gilkes, K.W.R., Gaskell, P.H. and Robertson, J. (1995) Comparison of neutron-scattering data for tetrahedral amorphous carbon with structural models, *Phys. Rev. B* **51**, 12303-12312.

48. Marks, N.A., McKenzie, D.R., Pailthorpe, B.A., Bernasconi, M. and Parrinello, M. (1996) Microscopic structure of tetrahedral amorphous carbon, *Phys. Rev. Lett.* **76**, 768-771.

49. Mott, N.F. and Davis, E.A. (1979) *Electronic Processes in Non-Crystalline Materials*, Clarendon, Oxford.

50. Dong, J. and Drabold, D.A. (1996) Band tail states and the localized to extended transition in amorphous diamond, (submitted to *Phys. Rev. Lett.*).

51. Mousseau, N. and Lewis, L.J. (1990) Computer models for amorphous silicon hydrides, *Phys. Rev. B* **41**, 3702-3707.

52. Stillinger, F.H. and Weber, T.A. (1985) Computer simulation of local order in condensed phase of silicon, *Phys. Rev. B* **31**, 5262-5271.

53. Thorpe, M. F. (1983) Continuous deformations in random networks, *J. Non–Cryst. Solids*, **57**, 355-370.

54. Cai, Y. and Thorpe, M. F. (1989) Floppy modes in network glasses, *Phys. Rev. B* **40**, 10535-10542.

55. Feng, S. and Sen, P. (1984) Percolation on Elastic Networks: New Exponent and Threshold , *Phys. Rev. Letts.* **52**, 216-219.

56. Feng, S., Thorpe, M. F. and Garboczi, E. J. (1985) Effective-medium theory of percolation on central-force elastic networks, *Phys. Rev. B* **31**, 276-280.

57. Day, A. R., Tremblay, R. R. and Tremblay, A–M. S. (1986) Rigid backbone: A new geometry for percolation, *Phys. Rev. Lett.* **56**, 2501-2504.

58. He, H. and Thorpe, M. F. (1985) The Elastic Properties of Glasses, *Phys. Rev. Lett.*, **54**, 2107-2110.

59. Hansen, A. and Roux, S. (1989) Universality class of central-force percolation, *Phys. Rev. B* **40**, 749-752. see especially Figs. 1 and 3.

60. Knackstedt, M. A. and Sahimi, M. (1992) On the universality of geometrical and transport exponents of rigidity percolation, *J. Stat. Phys.* **69**, 887-895; Arbabi, S. and Sahimi, M. (1993) Mechanics of disordered solids. I. Percolation on elastic networks with central forces, *Phys. Rev. B* **47**, 695-702.

61. Guyon, E., Roux, S., Hansen, A., Bideau, D., Trodec, J.-P. and Crapo, H. (1990) Non-local and non-linear problems in the mechanics of disordered systems: application to granular media and rigidity problems, *Rep. Prog. Phys.* **53**, 373-419.

62. Tatsumisago, M., Halfpap, B. L., Green, J. L., Lindsay, S. M. and Angell, C. A., (1990) Fragility of GeAsSe glass-forming liquids in relation to rigidity percolation, and the Kauzmann paradox *Phys. Rev. Lett* **64**, 1549-1552; Böhmer R. and Angell C. A., (1992) Correlations of the nonexponentiality and state dependence of mechanical relaxations with bond connectivity in Ge-As-Se cooled liquids *Phys. Rev. B* **45**, 10091-10094.

63. Jacobs, D. and Thorpe, M. F. (1996) Rigidity percolation in two dimensions: The pebble game, *Phys. Rev. E* **53**, 3682-3693.

64. Heine,V. private communication.

65. Hendrickson, B. (1992) Conditions for unique graph realizations, *SIAM J. Comput.* **21**, 65-84 and private communications.

66. Kantor, Y. and Webman, I. (1984) Elastic properties of random percolation systems, *Phys. Rev. Lett.* **52**, 1891-1894; see also Bergman, D. (1985) Elastic moduli near percolation: universal ratio and critical exponent, *Phys. Rev. B* **31**, 1696-1698.

67. Roux, S. and Hansen, A. (1988) Transfer-matrix study of the elastic properties of central-force percolation, *Europhys. Lett.* **6**, 301-306.

68. Stauffer, D. (1985) *Indroduction to Percolation Theory*, (Taylor and Francis, London).

69. Laman, G. (1970) On graphs and rigidity of plane skeletal structures, *J. Engrg. Math.* **4**, 331-340; see also Lovasz, L.and Yemini, Y. (1982) On generic rigidity in the plane, *SIAM J. Alg. Disc. Meth.* **3**, 91-98.

70. Moukarzel, C. and Duxbury, P. M. (1995) Stressed backbone and elasticity of random central-force systems, *Phys. Rev. Lett.* **75**, 4055-4058.

71. Fortuin, C.M. and Kasteleyn, P.W. (1972) On the random cluster model, *Physica* **57**, 536-564; P. Kasteleyn, W. and Fortuin, C. M. (1969) Phase transitions in lattice systems with random local properties, *J. Phys Soc. Japan,* **26**, 11-14. See also Essam, J.W. (1980) Percolation theory, *Rep. Prog. Phys.* **43**, 833-912.

72. Franzblau, D. S. (1995) Combinatorial algorithm for a lower bound on frame rigidity, *Siam J. on Discrete Math,* **8**, 388-400; Franzblau, D. S. and Tersoff, J. (1992) Elastic properties of a network model of glasses, *Phys. Rev. Lett.* **68**, 2172-2175; D.S. Franzblau, private communications.

73. Jacobs, D.J. and Thorpe, M.F., unpublished.

74. Thorpe, M. F. (1995) Bulk and surface floppy modes, *J. Non–Cryst. Solids*, **182**, 355-142. This mini-review contains references to many experimental results.

75. Angus, J. C. and Jansen, F.(1988) Dense 'dimondlike' hydrocarbons as random covalent networks, *J. Vac. Sci. Technol. A* **6**, 1778-1782.

76. Boolchand, P. and Thorpe M. F. (1994) Glass Forming Tendency, Percolation of Rigidity and 1-Fold Coordinated Atoms in Covalent Networks *Phys. Rev. B* **50**, 10366–10368.

77. Boolchand, P., Zhang, M. and Goodman, B. (1996) Influence of one-fold-coordinated atoms on mechanical properties of covalent networks (1996) *Phys Rev B* **53**, 11488-11494.

78. Döhler, G. H., Dandaloff, R. and Bilz, H. (1981) A topological-dynamical model of amorphycity, *J. Non–Cryst. Solids.* **42**, 87-95.

79. Tamor, M., private communication.

80. Dove, M. T., Giddy, A. P. and V. Heine, Rigid unit mode model of displacive phase transitions in framework silicates, (1993) *Amer Crystal Assoc.* **27**, 65; Giddy, A.P., Dove, M.T., Pawley, G.S. and Heine, V. (1993) The determination of rigid unit modes as potential soft modes for displacive phase transitions in framwork crystal structures, *Acta Crystallographica A* **49**, 697; See also the lecture notes by M. T. Dove, in these proceedings.

81. Quoted in *A Dictionary of scientific Quotations* by A. Mckay IOP Publishing (Bristol and Philadelphia)

82. Tanaka, K. (1988) Structural phase transitions in chalcogenide glasses, *Phys. Rev. B* **39**, 1270-1279.

THE ELASTIC MODULI OF RANDOM NETWORKS: CALCULATIONS

ANTHONY ROY DAY

Physics Department, Marquette University
Milwaukee, WI 53233 USA

1. Introduction

This seminar will be a review of some of the more technical aspects of computer simulations of the elastic properties of the type of random networks discussed in Professor Thorpe's lectures [1][2].

2. Review of Elasticity Theory

We use microscopic models of the structure and energy of an elastic network to evaluate bulk elastic moduli [3] that are valid over length scales much larger than the typical atomic length. If we apply a stress, σ_{ij}, to an elastic body it will be deformed so that a point in the material, initially at position $\mathbf{r}$, will be displaced to a position $\mathbf{r} + \mathbf{u}$ where we assume $\mathbf{u}$ is a small displacement. The stress tensor, σ_{ij}, is defined as the force in direction i applied to a unit area with the normal in direction j. The strain tensor ϵ_{ij} is defined as

$$\epsilon_{ij} = \frac{1}{2}\left(\frac{\partial u_i}{\partial x_j} + \frac{\partial u_j}{\partial x_i}\right). \tag{1}$$

In linear elasticity theory the stress is proportional to the strain and the defining equation is Hooke's law,

$$\sigma_{ij} = C_{ijkl}\epsilon_{kl}, \tag{2}$$

where C_{ijkl} is the elasticity tensor. The elastic energy stored in the material is

$$U = \frac{1}{2}\int_V \epsilon_{ij}C_{ijkl}\epsilon_{kl}dV, \tag{3}$$

M. F. Thorpe and M. I. Mitkova (eds.), Amorphous Insulators and Semiconductors, 329–338.

330

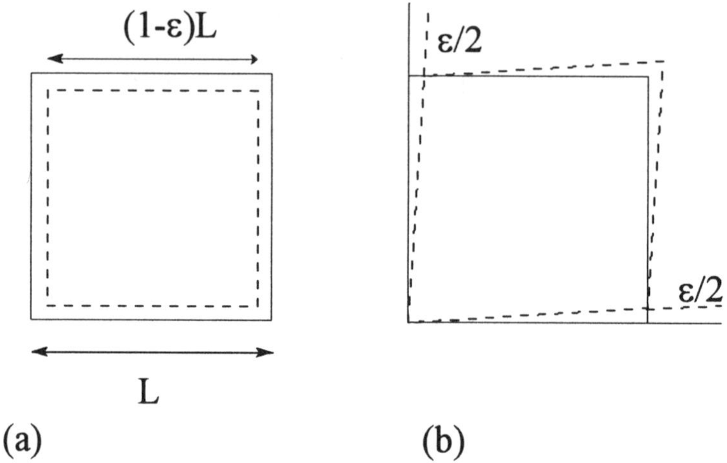

Figure 1.   Showing a hydrostatic compression (a) and a shear deformation (b)

where the integral is over the volume of the material.

The elasticity tensor is a fourth rank tensor but even the most general form only has 21 independent components [4]. In a crystal this is reduced even further, for example, crystals with cubic symmetry have $C_{1111} = C_{2222} = C_{3333}$. Although glasses have no local symmetry they are globally isotropic and there are only two independent elastic constants, which we can choose as the bulk modulus κ and the shear modulus μ. The easiest way to calculate the elastic moduli from a microscopic model is to use the energy equation (3). For example, to calculate the bulk modulus we apply a hydrostatic compression ($\epsilon_{11} = \epsilon_{22} = \epsilon_{33} = \epsilon$, as shown in figure 1a) to the sample, evaluate the energy at equilibrium and then use the relation

$$U = \frac{1}{2}\kappa \left(3\epsilon\right)^2 V. \tag{4}$$

To calculate the shear modulus we apply a shear deformation ($\epsilon_{12} = \epsilon_{21} = \epsilon/2$, as shown in figure 1b) to the sample, evaluate the energy at equilibrium and then use the relation

$$U = \frac{1}{2}\mu \epsilon^2 V. \tag{5}$$

From these two equations we see that the computational problem is to calculate the minimum elastic energy of a network under certain prescribed boundary conditions.

3. Microscopic Models

We start with some well defined network with an energy function describing the interactions between nearby atoms. The model has no dynamics and the interactions do not change as the system is deformed: In particular, if during the deformation two atoms that were originally non interacting approach each other, there is still no interaction. A common microscopic model used is the Keating model [5][6]

$$U = \frac{1}{2}\alpha \sum_{<ij>} \left(\mathbf{r_{ij}} \cdot \mathbf{r_{ij}} - \mathbf{r_{ij}^0} \cdot \mathbf{r_{ij}^0} \right)^2 + \frac{1}{2}\beta \sum_{<ijk>} \left(\mathbf{r_{ij}} \cdot \mathbf{r_{jk}} - \mathbf{r_{ij}^0} \cdot \mathbf{r_{jk}^0} \right)^2, \quad (6)$$

where the first sum is over all nearest neighbor pairs, and the second sum is over all next nearest neighbor triplets. The superscript 0 refers to the equilibrium coordinates of the atoms. In this model, the first term is a bond stretching central force term, the second term contains both a bond bending and a bond stretching component. This model is widely used for amorphous networks in three dimensions. For the generic triangular networks discussed in Professor Thorpe's lectures and shown in figure 2, we use only the first term and refer to it as a central force model.

3.1. THE LINEAR PROBLEM

At this point we need to distinguish between two types of networks. In the first type, such as a randomly diluted crystal [7][8] or the random triangular network[9][10] shown in figure 2, the equilibrium positions are well defined, the deformations are small ($u_{ij} \ll r_{ij}$) and it is appropriate to linearize the energy function. We write $\mathbf{r_{ij}} = \mathbf{r_{ij}^0} + \mathbf{u_{ij}}$, and expand the energy function [equation (6)] up to terms quadratic in u to obtain

$$U = \frac{1}{2}\alpha' \sum_{<ij>} (\mathbf{r_{ij}} \cdot \mathbf{u_{ij}})^2 + \frac{1}{2}\beta' \sum_{<ijk>} (\mathbf{r_{ij}} \cdot \mathbf{u_{jk}} + \mathbf{u_{ij}} \cdot \mathbf{r_{jk}})^2, \quad (7)$$

where we have absorbed some constants into α' and β'. The energy function is a now a quadratic function in u that will be zero when there is no applied strain. Suppose we generate a finite random net containing M atoms. We can minimize surface effects by forming a periodic lattice where the M atoms form a very large unit cell. Figure 2 is an example where the basis vectors of the unit cell, $\mathbf{R_1}$, and $\mathbf{R_2}$, are indicated. If we apply a strain ϵ to the sample, equivalent sites in different unit cells, separated by a lattice vector $\mathbf{R}$, will have displacements $\mathbf{u_i}$ and $\mathbf{u_i'}$ satisfying

$$\mathbf{u_i'} = \mathbf{u_i} + \epsilon\mathbf{R}. \quad (8)$$

332

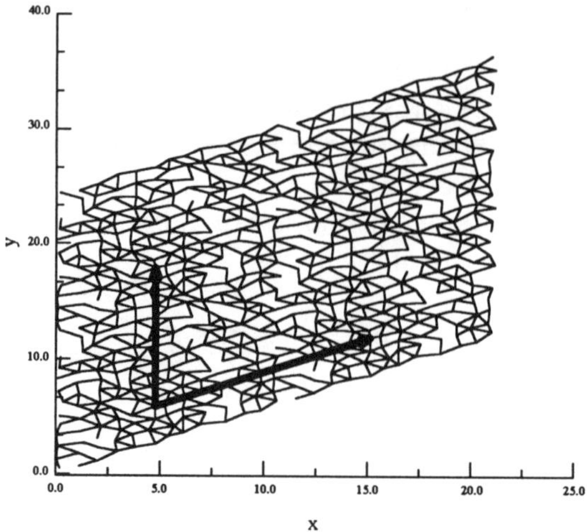

Figure 2. A periodic generic triangular network showing the basis vectors

With these boundary conditions and the energy function given by equation
(7) the energy can be written in the quadratic form

$$U = \frac{1}{2}\mathbf{u}^\mathbf{T}\mathbf{A}\,\mathbf{u} + \mathbf{b}^\mathbf{T}\mathbf{u} \tag{9}$$

where $\mathbf{u}$ and $\mathbf{b}$ are vectors of length $N = dM$ where d is the spatial dimension (i.e. 2 or 3). $\mathbf{A}$ is a $N \times N$ sparse symmetric matrix for which the sum
of the elements of any row is zero. The constant vector $\mathbf{b}$ is proportional
to the applied strain ϵ and its origin is easily understood by considering
a simple one dimensional chain. Determining the elastic constant is now
reduced to the problem of finding the minimum of the energy function U
or solving the matrix equation

$$\mathbf{A}\,\mathbf{u} = -\mathbf{b}, \tag{10}$$

substituting the solution back into equation (9) and using equation (4) or
equation (5) to find the elastic modulus. Although this is a well defined
problem of solving N linear equations (hence the heading *linear problem*),
when N is large it can be computationally very demanding. One must
also be aware that $\mathbf{A}$ has zero eigenvalues associated with the rigid body
translation and rotation, and is thus a singular matrix. We have found that
the conjugate gradient algorithm [11][12] is the most efficient algorithm
available and it works well on both scalar and vector machines. It is an

iterative algorithm that starts from some initial guess for the displacement vector $\mathbf{u}^0$ and iterates to a solution $\mathbf{u}^*$ that satisfies

$$\mathbf{A}\,\mathbf{u}^* = -\mathbf{b}. \tag{11}$$

In principle, the algorithm converges to the solution in exactly N steps, but for tightly bound systems a sufficiently accurate solution is often obtained in many fewer steps. For very weakly bound systems, e.g. percolating networks near threshold, finite precision often results in the problem requiring more than N steps[13]. The details of the conjugate gradient algorithm will be presented in section 4

3.2. THE NONLINEAR PROBLEM

In the second type of network, such as the networks generated by the Wooten and Weaire method [14][15], or a random substitution lattice with large bond length mismatches [16], even finding the equilibrium coordinates is a difficult problem (although it is neccessary that $\mathbf{r}_{ij}^0 \cdot \mathbf{r}_{ij}^0$, the equilibrium bond lengths and $\mathbf{r}_{ij}^0 \cdot \mathbf{r}_{jk}^0$, the equilibrium bond angles, be known). In these networks the equilibrium size is unknown and there is usually elastic energy stored in the network even when there is no applied strain. The energy function is no longer quadratic and finding the energy minimum is not simply the solution of a set of linear equations. For this nonlinear problem there may be several local minima which can cause problems when searching for a solution.

The problem can be set up with periodic boundary conditions in a similar fashion to the linear problem, except now we have

$$\mathbf{r}_i' = \mathbf{r}_i + (1 + \epsilon)\mathbf{R}, \tag{12}$$

and $\mathbf{R}$ is unknown. The most efficient method of solving the problem is to treat the basis vectors $\mathbf{R}_i$ as additional variables and minimize the energy function with $\epsilon = 0$. Then, using these basis vectors apply a strain and calculate the elastic moduli using equation (4) or equation (5), noting that the elastic energy will be

$$U_{el} = U(\epsilon) - U(\epsilon = 0) . \tag{13}$$

A common plan of attack when dealing with nonlinear problems is to solve a linear approximation and repeat. This is the underlying principle of the Newton-Raphson [12] method which we will apply here, extended to a multidimensional problem.

We start with the energy function $U(\mathbf{x})$ where $\mathbf{x}$ is an N vector. We will assume that we have already included the applied strain in $U(\mathbf{x})$ and

334

we wish to find some $\mathbf{x}^*$ such that $U(\mathbf{x}^*)$ is a minimum, for which

$$\mathbf{g}(\mathbf{x}^*) = \nabla U(\mathbf{x})|_{\mathbf{x}^*} = 0 \tag{14}$$

is a necessary condition. Start by doing a Taylor expansion of $U(\mathbf{x})$ about an initial guess $\mathbf{x}^0$

$$U(\mathbf{x}) = U(\mathbf{x}^0) + (\mathbf{x} - \mathbf{x}^0)^{\mathrm{T}} \mathbf{g}(\mathbf{x}^0) + \frac{1}{2}(\mathbf{x} - \mathbf{x}^0)^{\mathrm{T}} \mathbf{H}(\mathbf{x}^0)(\mathbf{x} - \mathbf{x}^0) + O(\mathbf{x} - \mathbf{x}^0)^3 \tag{15}$$

where $\mathbf{H}$ is the Hessian matrix, the second derivative matrix of the function $U(\mathbf{x})$ evaluated at $\mathbf{x}^0$. i.e.

$$H_{ij} = \frac{\partial^2 U}{\partial x_i \partial x_j}. \tag{16}$$

If we truncate the expansion after the quadratic term and try to satisfy equation (14) for this approximation of $U(\mathbf{x})$, the problem is reduced to the solving the set of linear equations

$$\mathbf{H}(\mathbf{x}^0)(\mathbf{x} - \mathbf{x}^0) = -\mathbf{g}(\mathbf{x}^0). \tag{17}$$

The value of $\mathbf{x}$ that solves equation (17) is an approximate solution that can be used as a new starting point for the Taylor expansion. The process is summarized as follows: Given an initial guess $\mathbf{x}^0$, find a sequence of steps $\{\mathbf{d}^k\}$ satisfying

$$\mathbf{H}(\mathbf{x}^k)\,\mathbf{d}^k = -\mathbf{g}(\mathbf{x}^k) \tag{18}$$
$$\mathbf{x}^{k+1} = \mathbf{x}^k + \mathbf{d}^k \tag{19}$$

The efficiency of this process can be dramatically improved if we realize that if $\mathbf{x}^k$ is far from the true solution then the quadratic approximation is poor and there is little point in solving equation (18) exactly. A good strategy is to obtain an approximate solution for equation (18) at each iteration, increasing the accuracy of the solution as the minimum is approached and the quadratic approximation to U improves. The steps in the algorithm are:

Major Iteration:

- compute $U(\mathbf{x}^k)$, $\mathbf{g}(\mathbf{x}^k)$ and $\mathbf{H}(\mathbf{x}^k)$.
- Test for convergence: $|\mathbf{g}(\mathbf{x}^k)| < e$.

Minor Iteration:

- Find $\mathbf{d}^k$ such that $|\mathbf{H}(\mathbf{x}^k)\,\mathbf{d}^k + \mathbf{g}(\mathbf{x}^k)| < \eta$
- Find some λ to minimize $U(\mathbf{x}^k + \lambda \mathbf{d}^k)$

- $\mathbf{x^{k+1} = x^k + \lambda d^k}$
- $k \leftarrow k + 1$ and continue.

The convergence criterion in the major iteration, e, is determined by the machine precision. The convergence criterion in the minor iteration is chosen as $\eta = |\mathbf{g(x^k)}|^{1+t}$ for $t \in (0,1)$. This ensures that when $\mathbf{x^k}$ is far from the exact solution the minor iteration will converge rapidly, and that as $\mathbf{x^k}$ approaches the true solution the minor iteration will provide a more accurate solution to equation (18). The line search at the end of the minor iteration is a simple procedure that can overcome some of the well known disadvantages of Newton's method.

The conjugate gradient method, which can be used to solve the linear problem of section 3.1, is also an appropriate method for solving (18) in the minor iteration. This algorithm is then known as the Truncated Newton Conjugate Gradient Method [17] and has been shown to work well for elasticity problems [18].

4. The Conjugate Gradient Algorithm

The problem is to minimize the function

$$f(x) = \frac{1}{2} x A x + b x + c, \tag{20}$$

where x and b are N vectors, A is an $N \times N$ symmetric matrix and c is a scalar constant. The gradient vector is

$$g(x) = Ax + b, \tag{21}$$

and we define the residual vector $r = -g$. At the solution x^* the residual vector is zero, i.e. $|r| = 0$. The steps in the conjugate gradient algorithm are as follows:

- **Initialization:** Given an initial estimate of the solution, x^0, set $i = 0$ and define the following vectors

$$
\begin{aligned}
g^0 &= Ax^0 + b & (22) \\
p^0 &= -g^0 & (23) \\
r^0 &= p^0 & (24)
\end{aligned}
$$

- **Iteration:** Evaluate

$$
\begin{aligned}
t^i &= Ap^i & (25) \\
\alpha^i &= \frac{p^i r^i}{p^i t^i} & (26)
\end{aligned}
$$

$$x^{i+1} = x^i + \alpha^i p^i \tag{27}$$

$$r^{i+1} = r^i - \alpha^i t^i \tag{28}$$

$$\beta^i = -\frac{r^{i+1} t^i}{p^i t^i} \tag{29}$$

$$p^{i+1} = r^{i+1} + \beta^i p^i \tag{30}$$

− Set $i \leftarrow i+1$ and repeat the iteration.

This algorithm is essentially a Gram-Schmidt process for generating a set of orthogonal vectors $\{r^i \; ; \; r^i r^j = 0 \text{ forall i} \neq \text{j}\}$ and a set of A conjugate vectors, $\{p^i \; ; \; p^i A p^j = 0 \text{ forall i} \neq \text{j}\}$. At each iteration, we update the x in the direction p^i and choose α^i to minimize the function along that direction, i.e. the gradient at x^{i+1} is orthogonal to the direction p^i, $g^{i+1} p^i = 0$. Note that

$$
\begin{aligned}
g^{i+1} &= Ax^{i+1} + b \\
&= Ax^i + b + \alpha^i A p^i \\
&= g^i + \alpha^i t^i
\end{aligned}
\tag{31}
$$

and thus $r^{i+1} = r^i - \alpha^i t^i$ is the residual vector at the new position x^{i+1}.

We see that the initial step direction, p^0 is the steepest descent direction and from equation (30) we see that each subsequent step is in a downhill direction but we choose $p^{i+1} A p^i = 0$ which avoids the well known inefficiency of the steepest descent method.

After the Nth step, we generate $r^N = r^{N-1} - \alpha^{N-1} t^{N-1}$ which is orthogonal to the set of N orthogonal vectors $\{r^0, r^1, \cdots, r^{N-1}\}$. An N dimensional vector space can only have N orthogonal vectors so r^N must be the null vector. i.e. we have converged to the exact solution. This proves that the algorithm converges in exactly N steps. It can also be shown that U decreases with each step.

There are a number of algebraically equivalent forms of the conjugate gradient algorithm which may have some advantages in real calculations. In one of the most useful alternative forms equation (29) is replaced by

$$\beta^i = \frac{r^{i+1} r^{i+1}}{r^i r^i} \tag{32}$$

With this identification it becomes possible to use the conjugate gradient algorithm without any reference to the matrix A as long as we can evaluate the function $f(x)$ and the gradient $g(x)$. If A is difficult to evaluate, or there are long range interactions so A is not sparse, this can be an important consideration. The method is to find α^i by a line search for the minimum along the direction p^i. For a quadratic function this is very efficient and

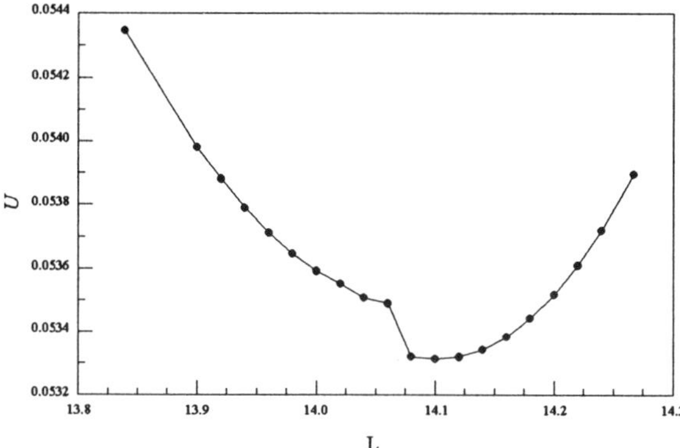

Figure 3.   The elastic energy of a 512 atom periodic amorphous carbon network as a function of the size of the unit cell

only requires knowing the slope at the starting point and one trial point. The line search method does require more floating point operations than if the matrix A is known.

This algorithm works well for the linear problem of section 3.1 and for the minor iteration of section 3.2 where the iterative nature of the solution is particularly well suited to truncating the minimization. For the linear problem calculating α using a line search or using the matrix A usually does not have much effect on the accuracy but for the nonlinear problem using equation (26) to find α is always faster than using a line search and will sometimes work when the line search method fails [18].

5. Concluding Remarks

I have described an efficient algorithm for finding the minimum of a function in a multi-dimensional space. This method is particularly well suited to the energy minimization associated with calculating the elastic constants of an amorphous network. There is a unique solution to the linear problem and in general the method works very well, although if the system is soft (i.e. when the matrix A has some eigenvalues that are very close to zero), the problem becomes ill-conditioned and convergence may be very slow. For the non-linear problem, the situation is more complicated because there may be more than one minimum, the algorithm only finds a local minimum, and the

338

solution x^* will depend on the initial guess x^0. An example of this is shown in figure 3. This curve shows the elastic energy in a amorphous carbon network of 512 atoms with an average coordination number of $< r >= 2.59$ as a function of the size of the unit cell [19]. It is clear that there is an abrupt switch from one local minimum to another at about L=14.07.

This review is the result of my own experience calculating the elastic properties of amorphous materials and composites, and many discussions I have had with Professor Thorpe and his graduate students over the last several years. Their contributions are gratefully acknowledged.

References

1. Thorpe, M.F. (1996) The structure and mechanical properties of networks, this volume.
2. Thorpe, M.F. (1995) Bulk and surface floppy modes, *J. Non-Cryst. Solids* **182**, 135-142.
3. Landau, L.D. and Lifshitz, E.M. (1986) *Theory of Elasticity*, 3rd edition Pergamon Press, New York.
4. Brown, F.C. (1967) *Physics of Solids*, W. A. Benjamin, New York.
5. Keating, P.N. (1966) Effective Invariance Requirements on the Elastic Strain Energy of Crystals with Application to the Diamond Structure, *Physical Review* **145**, 637-645.
6. Keating, P.N. (1964) Relationship between the Macroscopic and Microscopic Theory of Crystal Elasticity I. Primitive Crystals, *Physical Review* **152**, 774-779.
7. Feng, S. and Sen, P.N. (1984) Percolation on Elastic Networks: New Exponent and Threshold, *Phys. Rev. Lett.* **52**, 250-254.
8. He, H. and Thorpe. M.F. (1985) The elastic properties of glasses, *Phys. Rev. Lett.* **54**, 2107-2110.
9. Jacobs, D.J. and Thorpe, M.F. (1995) Generic rigidity percolation: The pebble game, *Phys. Rev. Lett.* **75**, 4051-4054.
10. Moukarzel, C and Duxbury, P.M. (1995) Stressed backbone and elasticity of random central-force systems, *Phys. Rev. Lett.* **75**, 4055-4058.
11. Hestenes, M.R. and Steifel E. (1952) Method of conjugate gradients for solving linear systems, *J. Res. Nat. Bur. Stand.* **49**, 409-436.
12. Press, W.H., Flannery, B.P., Teukolsky, S.A. and Vetterling, W.T.(1986) *Numerical Recipes*, Cambridge University Press, Cambridge.
13. Day, A.R., Tremblay, R.R. and Tremblay, A.-M.S. (1986) Rigid Backbone: A New Geometry for Percolation, *Phys. Rev. Lett.* **56**, 2501-2504.
14. Wooten, F., Winer, K. and Weaire, D. (1985) Computer generation of structural models of amorphous Si and Ge, *Phys. Rev. Lett.* **54**, 1392-1395.
15. Wooten, F. and Weaire, D. (1987) Modeling tetrahedrally bonded random networks by computer. In: Ehrenreich H, Seitz F, Turnbull D (eds) *Solid State Physics, Vol. 40*, Academic, New York, pp. 1-42.
16. Thorpe, M.F. and Garboczi, E.J.(1990) Elastic properties of central force networks with bond-length mismatch, *Phys. Rev. B* **42**, 8405-8417.
17. Dembo, R.S. and Steihaug, T. (1983) Truncated-Newton algorithm for large-scale unconstrained optimization, *Math. Prog.* **26**, 190-212.
18. Day, A.R., Snyder, K.A., Garboczi, E.J. and Thorpe M.F. (1992) Elastic Moduli of a Sheet Containing Circular Holes, *J. Mech. Phys. Solids.* **40**, 1031-1051.
19. Djordjević B. (1996) *Properties of Disordered Discrete and Continuous Solids*, Ph.D. Dissertation, Michigan State University, East Lansing.

ONE-FOLD COORDINATED ATOMS, CONSTRAINT THEORY AND NANO-INDENTATION HARDNESS

P. BOOLCHAND and M. ZHANG
Department of Electrical, Computer Engineering and Computer Science
University of Cincinnati, Cincinnati, OH 45221-0030

B. GOODMAN
Department of Physics
University of Cincinnati, Cincinnati, OH 45221-0011

1. Introduction

Several amorphous and glassy systems contain significant concentrations of monovalent species such as hydrogen, halogens and alkalis. Some of the underlying materials of technological importance include hydrogenated-silicon, diamond-like carbon, chalcohalide glasses, and mixed-oxide glasses. In many instances, the monovalent species are found to be one-fold coordinated (OFC). It appears that a variety of physical properties of these material systems, such as molecular structure, vibrational density of states, glass forming tendency, glass transition temperature, floppy modes per atom or hardness index, all appear to be affected in an intimate fashion by the concentration of the monovalent additives.

Constraint theory has proved to be a particularly useful approach to understand glasses at a basic level. Since its inception, it has been applied to understand [1, 2] the mechanical and vibrational behavior of covalently bonded networks in which atoms bond with a coordination number (r) of two or greater. More recently, the theory has been extended [3] to include one-fold coordinated (OFC) atoms and it has provided a means to understand the observed glass forming region in the Ge-S-I chalcohalide system. OFC-atoms (r=1) have to be treated differently [3,4] from other atoms of the network for which r $\geq$ 2. Atoms possessing a coordination number r $\geq$ 2, contribute [1, 2] both bond stretching (α-) and bond-bending (β)-constraints, but OFC-atoms contribute α-constraints only since no bond angles are involved about such atoms. An interesting consequence of the latter is the fact that the nature of chemical bonding between OFC-atoms and the network, whether covalent or ionic, is no longer important since either interaction can provide a central force (α)-constraint [5,6].

In section 2, we introduce valence force field constraint counting procedures for covalent networks possessing a finite concentration of OFC atoms. In previous discussions, [2,3,7] one has found useful to construct a network formed by *plucking* all OFC-atoms from the starting or *complete* network. The effect of OFC atoms was indirectly taken into account by applying standard constraint counting procedures to the *plucked* network. The notion of using the *plucked* network in contrast to the *complete* network leads to subtle differences in the count of constraints per atom with important consequences on mechanical properties [4]. We shall address these issues in section 2.

M. F. Thorpe and M. I. Mitkova (eds.), Amorphous Insulators and Semiconductors, 339–348.
© 1997 *Kluwer Academic Publishers. Printed in the Netherlands.*

340

The basic constraint counting procedure developed in section 2 provides a useful framework to quantitatively predict nano-indentation hardness of hydrogenated tetrahedral networks of C, SiC and Si. In section 3, we shall compare these predictions with available measurements. A brief summary of our conclusions appears in section 4.

2.   Covalent Networks Containing One-Fold Coordinated Atoms

Consider a covalently bonded network of N atoms containing N_1, N_2, N_3...N_r atoms which are, respectively, 1-fold, 2-fold, 3-fold...r-fold coordinated. We shall examine two cases. In the first case, we will consider a network in which OFC-atoms are bonded to atoms possessing a coordination number $r \geq 3$. In the second case, we will consider a network in which some of the OFC-atoms are bonded to 2-fold coordinated atoms (P_2), in addition some are bonded to atoms with $r \geq 3$. We shall see that in the first case the mechanical behavior of the complete network can be well represented by the plucked network but only near the stiffness threshold. For the second case, the mechanical behavior of the complete network cannot be well represented by the plucked network even near the stiffness threshold. In both cases considered above, away from the stiffness threshold, either in the floppy or the rigid regime, the mechanical properties of the plucked network in general are not a good representation of the complete network.

The average coordination number of the complete network, $\langle r \rangle$, and plucked network $\langle r' \rangle$ are respectively,

$$\langle r \rangle = (1/N) \sum_{r \geq 1} N_r r \tag{1}$$

$$\langle r' \rangle = (1/(N-N_1)) \sum_{r \geq 2} N_r r \tag{2}$$

We shall use unprimed symbols to designate physical quantities for the *complete network* while primed ones for the *plucked network*. One can show [4] that the number of α- and β-constraints per atom in the complete (n_c) and plucked (n_c') networks, respectively, are

$$n_c = 5/2\langle r \rangle + N_1/N - 3 \tag{3}$$

$$n_c' = 5/2\langle r' \rangle - 3 \tag{4}$$

Since rigidity percolates in the plucked network uniquely at $\langle r' \rangle = 2.4$, it is expedient to define [4] over- and under-coordinated networks in terms of the excess *coordination parameter* s, as follows,

$$s = \langle r' \rangle - 2.4 \tag{5}$$

Networks possessing s > 0 are considered to be overcoordinated and rigid, those with s < 0 are thought to be undercoordinated and floppy, while those at s=0 are considered to be optimally coordinated and at a mechanical critical point.

2.1. OFC-ATOMS BONDED TO 3- AND HIGHER-FOLD COORDINATED ATOMS

In the case when OFC-atoms are bonded to 3- and higher fold coordinated atoms, one can show that

$$\langle r' \rangle = (\langle r \rangle - 2N_1/N)/(1-N_1/N) \tag{6}$$

Equations (3) and (4) can then be rewritten in terms of the s parameter as follows,

$$n_c = 3 + 5/2(1-N_1/N)s \tag{7}$$

$$n_c' = 3 + 5/2\ s \tag{8}$$

We shall define the difference $n_c-n_d(=3) = h$, as the *hardness index* per atom. The *hardness index per atom* in the overconstrained regime is the counterpart of *floppy modes per atom*, $f=F/N=n_d-n_c$, in the underconstrained regime: $h= |f| = -f$. Equations (7) and (8) can be rewritten to provide expressions for h and h', respectively,

$$h = n_c - 3 = 5/2(1-N_1/N)s \tag{9}$$

$$h' = n_c' - 3 = 5/2\ s \tag{10}$$

To visualize the role OFC-atoms play on mechanical properties of covalent networks, it is instructive to plot equation (9). In Figure 1, we plot h as a function of s (or $\langle r' \rangle$) for three specific values of N_1/N. In the rigid region (s>0), one finds that h systematically decreases in proportion to N_1/N. Thus presence of OFC-atoms in overcoordinated networks leads to a loss in hardness in proportion to the concentration of such atoms. On the other hand, the plot also shows that for floppy networks, s<0, the count of floppy modes/atom $f=n_d-n_c= -h$, systematically decreases in proportion to N_1/N. This is to say that presence of OFC-atoms in undercoordinated networks leads to a mechanical stiffening of the network. Finally, since each of the plots of h as a function of s, for different values of N_1/N coincide at s=0, we conclude that the stiffness threshold, corresponding to f=h=0, is independent of the concentration of OFC-atoms. It is also obvious from Fig. 1 that for the case when $N_1/N=0$, h=h' and the complete and plucked network become identical. Thus in Fig. 1, the plot of h against s for $N_1/N=0$, also represents the universal prediction of the hardness index h' of the plucked network as a function of $\langle r' \rangle$. In fact, by combining equations (9) and (10) we get

$$h = h'(1-N_1/N) \tag{11}$$

which is perhaps the most transparent statement that OFC-atoms reduce the hardness of the complete network relative to that of the plucked network.

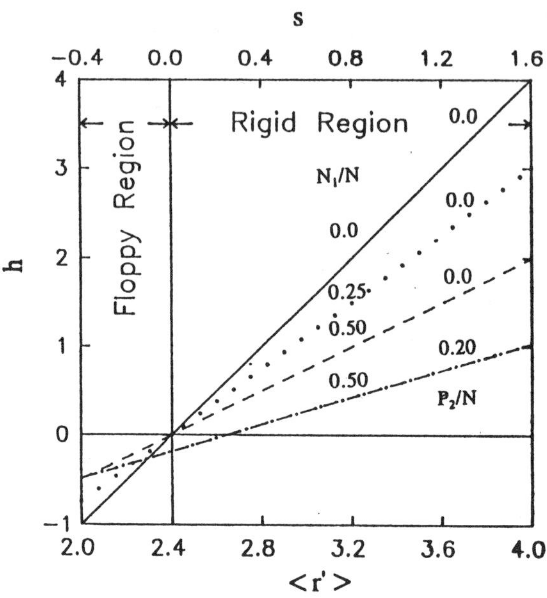

Figure 1. A plot of hardness index h against $\langle r' \rangle$ or s, showing a linear variation. Note that the slope and intercept of the straight lines depend on N_1/N and P_2/N. See text for details. The continuous line corresponding to $N_1=P_2=0$, represents the universal prediction of the hardness index h′ of the plucked network.

2.2. OFC-ATOMS BONDED TO 2- AND HIGHER-FOLD COORDINATED ATOMS

In the previous section, one found that the stiffness threshold condition, s=0, is met simultaneously in the plucked- and complete-networks. We shall now show that the equivalence of the stiffness threshold in the two networks is contingent upon OFC atoms being bonded to atoms of the network which possess a coordination number r of 3 or greater. If this restriction is relaxed, viz, if some OFC atoms (P_2) are bonded to 2-fold coordinated ones, then the stiffness threshold predicted using the plucked network no longer *coincides* with that predicted using the complete network. The two networks become completely inequivalent.

Let P_2 represent the number of 2-fold coordinated atoms in a network that form part of a side chain. To construct a plucked network with no OFC atoms, one would have to remove not only the N_1 atoms that are OFC, but also the P_2 atoms. If $\langle r'' \rangle$ designates the average coordination number of the plucked network, then one can show [8]

$$\langle r \rangle = \langle r'' \rangle - [\langle r'' \rangle - 2](N_1 + P_2)/N \tag{12}$$

In analogy to equations (7) and (8), one can calculate the constraints/ atom in the complete and plucked network for the second case. The results are as follows.

$$n_c = 3 + (5s/2)[1 - (N_1 + P_2)/N] - P_2/N \tag{13}$$

$$n_c'' = 3 + (5s/2) \tag{14}$$

Furthermore, by combining equations (13) and (14) one obtains,

$$h = h''[1 - (N_1 + P_2)/N] - P_2/N \tag{15}$$

which describes the relation between the hardness indices h and h'' of the complete and plucked network. At the outset we note that when $P_2=0$, equation (15) goes over to equation (11). More importantly, equation (15) shows that when rigidity percolates in the plucked network, viz., $h'' = 0$ then $h = -P_2/N$, i.e., the complete network continues to remain floppy as suggested by the negative value of h. Conversely, when rigidity percolates in the complete network, i.e. h=0, then $h'' = P_2/(N-N_1-P_2)$, i.e. the plucked network becomes overconstrained and rigid as can be seen by comparing plot c to plot d in Fig. 1. One has thus demonstrated that the stiffness threshold condition in the two networks no longer coincides in the presence of a finite concentration of P_2 atoms. Fundamentally, the plucked network ceases to be a good representation of the complete network at this stage.

The pictorial example of Fig. 2 helps illustrate the inequivalence of the plucked and complete networks under the above conditions. Consider a complete network including a branch made up of a 5-atom long Se_n chain which terminates in a OFC-I atom. For obvious reasons the branch has to be completely removed in the plucked network. Such removal renders the plucked network rigid when the complete network is floppy.

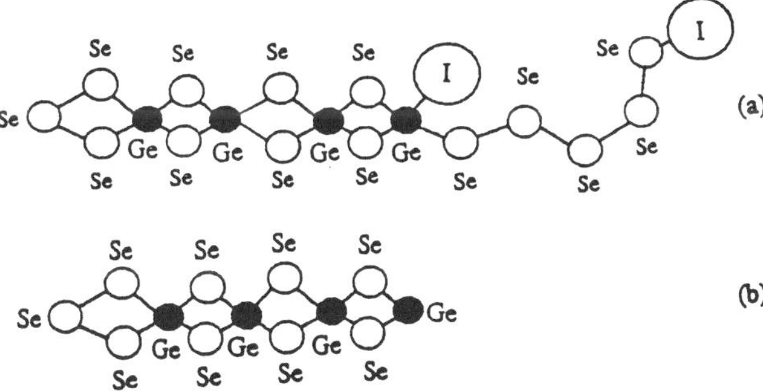

Figure 2. (a) a $Ge_4Se_{14}I_2$ network and (b) its plucked counterpart. The average coordination number in (a) is $\langle r \rangle = 2.10$ and in (b) is $\langle r' \rangle = 2.46$. There is a cross-over from a floppy (a) to a rigid (b) network for the present case when $P_2/N(=0.4)$ is finite.

344

The NIH of crystalline silicon films have been established at 10.4(5) GPa. Jiang et al. [15] prepared hydrogenated amorphous Si films by reactive sputtering of silicon with hydrogen-argon mixtures. They found the NIH of the hydrogenated films to drop by a factor of 2 with only 1 at.% of hydrogen.

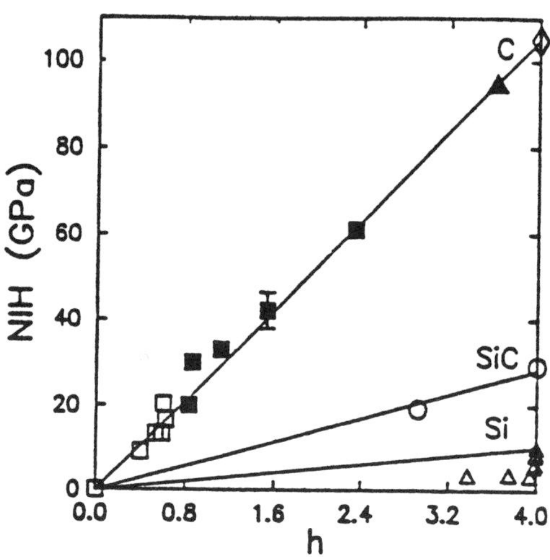

Figure 3. Measured NIH data in GPa of DLC-films, SiC-films and amorphous Si films compared to the predicted hardness (continuous lines) based on constraint counting procedures. The results on DLC-films are taken from those of Tamor et al. [16], Jiang et al. [15] open squares, Weiler et al. [17] filled squares, Lossy et al. [18] filled triangles, those on SiC-films from the work of El Khakani et al. [12] open circles, and those on Si-films from the work of Jiang et al. [15] open triangles.

Such a result cannot be understood even qualitatively in terms of the constraint counting method. The drastic reduction in NIH of hydrogenated a-Si films probably reflects hydrogen passivation of dangling bonds on surfaces of small Si clusters formed on the surface of such films. The single-bond dissociation energies of Si-H, Si-Si and H-H of 373, 310 and 432 kcal/mole do not provide for a strong mixing of Si and H. This chemical factor may be intrinsically responsible for a tendency of hydrogen to segregate from Si in such alloys. Glow discharge deposited amorphous Si samples using silane as a precursor has proved to be a useful photovoltaic material. Such films are known to contain 15 to 20% hydrogen, although we are not aware of NIH measurements on this device quality material.

We have thus shown that the mechanical behavior of the plucked network is, in general, not a good representation of the complete network for the case when the concentration of P_2 atoms is finite. The example of Fig. 2 also illustrates that not only should one reckon the Se atoms next to I as a P_2 atom, but also the remaining four Se atoms in the branch. Thus, in a network, all 2-fold coordinated atoms that form part of a chain that terminates in a OFC-atom contribute to the count of P_2 atoms.

3. Nano-Indentation Hardness of Hydrogenated Group IV Elements

Hardness is an intrinsically plastic property of a material. On a macroscopic level, one physically relates it to plastic flow when work-hardening of a material occurs. On a microscopic level, one relates plastic flow usually to movement of dislocations in a crystalline material. This phenomenon is to be contrasted to elastic deformation of a solid which is an intrinsically reversible process.

Within the constraint theory of glasses, interatomic forces provide a measure of local softness or stiffness of a network, as the case may be. For undercoordinated networks the local softness manifests itself in floppy modes [9], which have been observed [10] in inelastic neutron-scattering and Lamb-Mössbauer factor [11] measurements. For overcoordinated networks such as diamond, silicon and SiC, the local stiffness is connected with plasticity of the materials. Since the average number of constraints/atom increases with average coordination number, one can expect (Nano-Indentation Hardness (NIH) to scale linearly with the average coordination number for each of the tetrahedral solids, as we shall illustrate next.

3.1. CHEMICAL BONDING CONTRIBUTIONS TO NANOINDENTATION HARDNESS (NIH)

Crystalline Si, SiC, C (diamond) consist of ordered networks of tetrahedral building blocks and possess NIH values of $\sim$ 10 GPa [12], 30 GPa [13], and 100 GPa [14], respectively. For a pure tetrahedrally coordinated network, $r=4(n_c-7)$ and $h=4$. The increasing hardnesses measured in the Si $\rightarrow$ SiC $\rightarrow$ C (diamond) sequence reflect the increasing bond strength of the sp^3 covalent bonds as the nearest-neighbor distances decrease. To correlate hardness of the corresponding hydrogenated alloys with h, it is reasonable to postulate the form

$$NIH = \alpha h \tag{16}$$

where α is the chemical-bond scale factor and h represents the dependence of NIH on the network topology or microstructure within the constraint-counting description. Based on the measured NIH of the crystalline tetrahedral semiconductors, and a value of h=4, one obtains the chemical bond scale factor $\alpha=26.25$ GPa for diamond, $\alpha=7.5$ GPa for SiC and $\alpha=2.5$ GPa for Si.

3.2. MICROSTRUCTURE CONTRIBUTIONS TO NIH

Thin-films of diamond-like-carbon (DLC), silicon carbide (SC) and amorphous-silicon (a-Si) have been prepared by plasma decomposition of various hydrocarbons and silane

346

precursors by several groups. The physical properties of the films including NIH are found to depend on the film-microstructure and the amount of bonded hydrogen in the network. Starting from constraint theory, one can predict the hardness index h, given the network microstructure. Using the known value of α, one can then predict the NIH (αh) expected from constraint counting procedures and compare it to the measured value (H) using nanoindenters. In this section we shall illustrate these ideas using published results on DLC-films, SiC-films and a-Si films. We shall see that the NIH results on DLC-films and SiC-films correlate rather well with predicted values based on constraint theory provided one uses the complete network to explicitly include effect of bonded hydrogen [4]. In the case of hydrogenated amorphous-Si films, we have found a considerable deviation between the measured and predicted hardness which may be of a chemical origin.

3.2.1 *Diamond-like Carbon*

Jiang et al. [15] and Tamor et al. [16] have prepared DLC-films using RF Plasma decomposition of methane. Weiler et al. [17] used chemical vapor deposition with plasma beam sources of acetylene as a precursor to deposit DLC-films. Their films displayed a lower hydrogen content and a larger NIH values than those reported by Jiang et al. Graphite like sp^2 and diamond-like sp^3 bonding configurations are prevalent in these films with no detectable evidence of sp-like 2-fold coordinated structures. By establishing the amount of bonded hydrogen in the films, and the ratio of sp^2 to sp^3 reconfiguration, one can uniquely obtain s, the excess coordination parameter and from equation (9), the hardness index of the films from the measured hydrogen content (N_1/N). In Fig. 3, we have plotted the measured NIH in GPa (H) as a function of h. The NIH results on DLC-films scale linearly with h rather well. The correlation of Fig. 3 is suggestive that in these films H is bonded to sp^2- and sp^3-carbon but not to sp-carbon. The latter configurations (P_2), if present, would have produced additional softening of the network as discussed in section 2.2 for which there is no experimental evidence. Furthermore, as shown elsewhere [19,4], attempts to correlate measured hardness H with the predicted hardness ($\alpha h'$) using the plucked network lead to systematic overestimates. These overestimates are the natural consequence of equation (11).

3.2.2 *Hydrogenated Amorphous Silicon-Carbide Films*

El Khakani et al. [12] have prepared stoichiometric amorphous SiC film by pulsed laser ablation and also by triode sputtering and have measured the NIH to be ~ 30 GPa. Plasma enhanced CVD of silane: methane mixtures with argon carrier gas were used to prepare hydrogenated amorphous SiC films containing 27 atomic % of hydrogen. NIH measurements of such films yielded a somewhat lower value of 19.5 (2.0) GPa. Following a procedure similar to the one described earlier for DLC-films, we have calculated the hardness index h=2.3 for such films and have plotted the result in Fig. 3. The predicted hardness results on SiC-films correlate rather well with the measured NIH.

3.2.3. *Hydrogenated Amorphous Silicon Films*

4. Conclusions

Mechanical softening of overconstrained covalently bonded networks by OFC-atoms is analyzed using constraint-counting methods for two cases, one when such atoms are bonded to 3- and higher-fold coordinated atoms of a network and second, when such atoms are bonded to 2- and higher-fold coordinated atoms of a network. Measured Nano Indentation Hardness of DLC- and SiC-films can be quantitatively understood using constraint counting procedures when hydrogen (OFC)-atoms are considered to be bonded to 3- and 4-fold coordinated C in DLC-films, and 4-fold coordinated C- and Si atoms in hydrogenated SiC-films.

5. Acknowledgements

It is a pleasure to acknowledge useful discussions with Jim Phillips, Mike Thorpe, and Stan Ovshinsky. This work was supported by NSF Grants DMR-92-07166 and DMR-93-09061.

6. References

1. Phillips, J.C. (1979) Topology of covalent non-crystalline solids I: short-range order in chalcogenide alloys, *J. Non Cryst. Solids* **34**, 153.

2. Thorpe, M.F. (1983) Continuous deformations in random networks, *J. Non Cryst. Solids* **57**, 355-370.

3. Boolchand, P. and Thorpe, M.F. (1994) Glass forming tendency, percolation of rigidity and onefold-coordinated atoms, *Phys. Rev.* **B50**, 10366-10368.

4. Boolchand, P., Zhang, M. and Goodman, B. (1996) Influence of onefold coordinated atoms on mechanical properties of covalent networks, *Phys. Rev.* **B53**, 11488 (1996).

5. Zhang, M. and Boolchand, P. (1994) The central role of broken bond bending con-straints in promoting glass formation in the oxides, *Science* **266**, 1355-1357.

6. Zwanziger, J.W., Tagg, S.L. and Huffman, J.C. (1995) Broken bond-bending constraints and glass formation in the oxides, *Science* **268**, 1510. Also see Boolchand, P. and Zhang, M. (1995) Response to previous technical comment, *Science* **268**, 1510-1511.

7. Angus, J.C. and Jansen, F. (1988) Dense 'diamond-like' hydrocarbons as random covalent networks, *J. Vac Sci and Technology* **A6**, 1778-1782, Dohler, G.H., Dandoloff, R. and Bilz, H. (1980) A topological-dynamical model of amorphicity, *J. Non Cryst. Solids* **42**, 87-95.

8. Zhang, M. (1995) Constraint theory and broken bond bending constraints in oxide glasses, Ph.D. thesis, University of Cincinnati.

9.   He, H. and Thorpe, M.F. (1985) Elastic Properties of Glasses, *Phys. Rev. Letters* **54**, 2107-2110.

10.   Kamitakahara, W.A., Cappelletti, R.L., Boolchand, P., Halfpap, B., Gompf, F., Neumann, D.A., Mutka, H. (1991) Vibrational density of states and network rigidity in chalcogenide glasses, *Phys. Rev.* **B44**, 94-100.

11.   Boolchand, P., Bresser, W.J., Zhang, M., Wu, Y., Wells, J., Enzweiler, R. (1995) Lamb-Mössbauer factors as a local probe of floppy modes in network glasses, *J. Non Cryst. Solids*, **182**, 143-154.

12.   Kelly, A. and McMillan, N.H. (1986), *Strong Solids*, Oxford University Press, Oxford.

13.   El Khakani, M.A., Chaker, M., Jean, A., Boily, S., Kieffer, J.C., O'Hern, M.E., Ravet, M.F., Rousseaux, F. (1994) Hardness and Youngs modules of amorphous a-SiC thin-films determined by nanoindentation and bulge test, *J. Mat. Research* **9**, 96-103.

14.   Field, J.E., (1979) *Properties of Diamond*, Academic, New York.

15.   Jiang, X., Reichelt, K., and Stritzker (1989) The hardness and Young's modules of amorphous hydrogenated carbon and silicon films measured with an ultralow load indentor, *J. Appl. Phys.* **66**, 5805-5808.

16.   Tamor, M.A., Vassell, W.C. and Carduner, K.R. (1991), Atomic constraints in hydrogenated "diamond-like" carbon, *Appl. Phys. Letts* **58**, 592-594.

17.   Weiler, M., Kleber, R., Sattel, S., Jung, K., Ehrhart, H., Jungnickel, G., Deutschmann, S., Stephan, V., Blandeck, P., Fraunheim, Th. (1994) Structure of amorphous hydrogenated carbon: experiment and computer simulation, *Diamond Related Materials*, **3**, 245-253.

18.   Lossy, R., Pappas, D.L., Roy, R.A., Cuomo, J.J., Sura, M.V. (1992) Filtered arc deposition of amorphous diamond, *Appl. Phys. Lett.* **61**, 171-173.

19.   Thorpe, M.F. (1995), Bulk and Surface Floppy Modes, *J. Non Cryst. Solids* **182**, 135-142.

SILICATES AND SOFT MODES

M. T. DOVE
Department of Earth Sciences, University of Cambridge
Downing Street, Cambridge, CB2 3EQ, UK

1   Introduction

What is the relevance of displacive phase transitions in crystalline silicates to a workshop on glasses? The study of phase transitions brings in the issues of stability, particularly the fine balance that can exist between the relative energies of different phases. It also brings in the issue of how structures can deform, the constraints that operate, and the degree of flexibility. Although the similarities between crystals and glasses must not be emphasised too strongly, there is much to be gained from remembering that the same forces operate in both types of materials. In this chapter I am going to describe recent work that has led to a new understanding of the atomic mechanisms that give rise to phase transitions in crystalline silicates. This new work has led to insights that are of more general applicability, and which may extend to the properties of glasses also.

The type of phase transitions we are concerned with are illustrated in Figures 1 and 2, which show the displacive phase transitions in the silica polymorphs quartz and cristobalite respectively. The displacive phase transition in quartz is reasonably straightforward. The symmetry of the high-temperature phase is broken by small rotations and translations of the SiO_4 tetrahedra. Although not proven by the representation in Figure 1, these rotations and displacements can occur without the tetrahedra needing to distort. The displacive phase transition in cristobalite is similar, yet there is an additional degree of complexity. The representation in Figure 2 indicates that the phase transition can also occur by rotations of tetrahedra that do not require the tetrahedra to distort. However, the cubic symmetry of the high-temperature phase requires that the average positions of the atoms of Si–O–Si linkages must lie along a common straight line. Taken at face value, this might imply the existence of linear Si–O–Si bonds, although it is well known that the stable configuration of two linked SiO_4 tetrahedra has the Si–O–Si bond angle nearer 145°. This indeed is around the angle found in both phases of quartz and in the low-temperature tetragonal phase of cristobalite. Thus it is likely that the cubic phase of cristobalite is rather more complicated than the simple average structure, with local deformations of the structure that give Si–O–Si bond angles somewhat nearer the equilibrium value. However, it is

M. F. Thorpe and M. I. Mitkova (eds.), Amorphous Insulators and Semiconductors, 349–383.

350

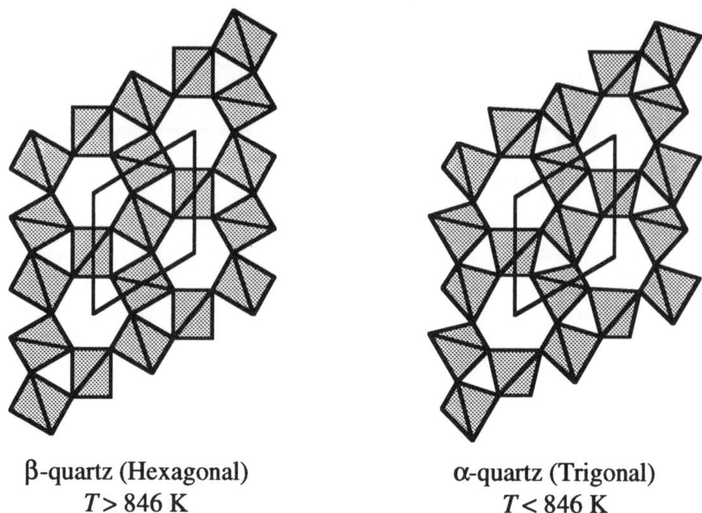

β-quartz (Hexagonal)
$T > 846$ K

α-quartz (Trigonal)
$T < 846$ K

FIGURE 1. Polyhedral representation of the two phases of quartz, showing the small rotations
and displacements of the tetrahedra through the displacive phase transition.

not at all obvious that these local deformations can occur without a substantial cost in energy, because they may necessitate some distortions of the SiO_4 tetrahedra.

These examples represent the type of problem that we are concerned with in this chapter. They involve a number of general issues concerned with the behaviour of an infinite network of linked tetrahedra as found in crystalline and amorphous silicates. Our concern here is with the physics of phase transitions in crystalline silicates. For a phase transition to occur there must be an easy mode of deformation. In the case of silicates this necessarily implies the existence of deformations in which the tetrahedra are not required to distort. The existence of the phase transition also implies a small energy difference between the two phases, and a fine balance between this energy difference and the vibrational entropy.

In this chapter I will discuss an approach to the study of phase transitions in silicates that builds upon the observations sketched in this introduction. At the heart of our approach is the possibility for some of the vibrational modes to propagate without the SiO_4 tetrahedra distorting—these are called "Rigid Unit Modes" (RUMs). This possibility, however, is by no means self-evident, and has required the development of special methods to calculate the RUM spectrum for any material. In order to appreciate the relationship between RUMs and displacive phase transitions I first review the phenomenology and theory of displacive phase transitions. It will later be useful to make comparisons with a simple model that is widely used in the study of phase transitions, which I will introduce in the context of the general theory. The main focus in this discussion will be the concept of the "soft mode". The role of the RUM model will quickly become apparent, and I will spend some time discussing the details of the model and the methods used to determine the existence of RUMs. I will also review the experimental situation. The discussion will then move to the applications of the RUM

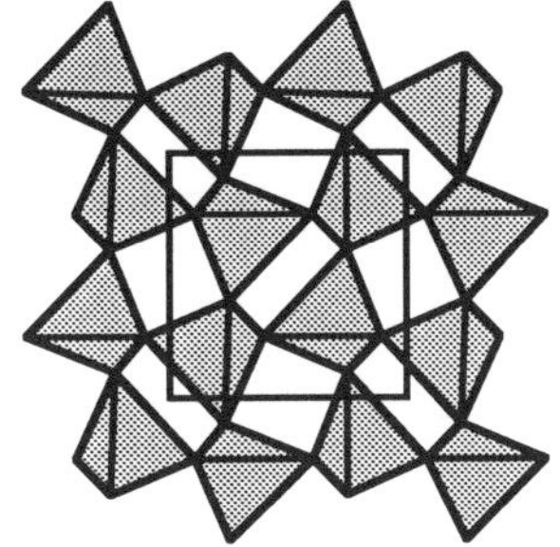

β-cristobalite (Cubic)   α-cristobalite (Tetragonal)

$T > 520$ K $T < 520$ K

FIGURE 2. Polyhedral representation of the two phases of cristobalite, showing the rather larger rotations and displacements of the tetrahedra through the displacive phase transition. The symmetry of the high-temperature phase suggests the possible existence of linear Si–O–Si bonds.

model and its implications for the behaviour of phase transitions. These include the nature of high-temperature phases of materials like cristobalite, the factors that determine the phase transition temperature, and the driving force for a phase transitions. Other applications of the RUM model will encompass thermal expansion and the possibilities of forming local distortions in a structure. Finally I will explore posible applications of the RUM model to glasses.

2   A short history of displacive phase transitions

The study of structural phase transitions extends back over a century [1], and in order to appreciate recent work it is helpful to understand the way the field has developed. The first decades were concerned with characterising the structures of different phases by X-ray crystallography, with other supporting experiments. However, real progress was limited by the lack of a theoretical framework. For example, in 1940 the first soft modes were discovered in quartz using the Raman effect (which had been discovered 12 years previously), and the interpretation of the data was remarkably close to the mark, but ultimately this work did not have a significant impact at the time precisely because it did not fit into a general context [2,3].

The first useful theoretical approach was the phenomenological approach developed by Landau in 1937, and subsequently developed for ferroelectrics by Devonshire. The central concept in this model is the order parameter, η, which gives a quantitative measure of how much the structure has distorted through the phase transition. For example, in quartz η quantifies the symmetry-breaking rotations and translations of the tetrahedra. Whilst there are always some changes in the structure due to changing temperature, it is only those motions that cause a change in the symmetry that are selected in the definition of η. The Landau model can be approached from a

352

number of directions [4–7]. At heart it is a simple Taylor expansion of the change in free energy that comes from the phase transition:

$$\Delta G = \tfrac{1}{2}a(T)\eta^2 + \tfrac{1}{4}b(T)\eta^4 + \cdots \tag{1}$$

For the moment we note that in this polynomial the coefficients $a(T)$ and $b(T)$ may be temperature-dependent, but this point will be made firmer shortly. The polynomial only includes even powers of η because we usually have the requirement that $G(-\eta) = G(\eta)$. At any temperature the equilibrium value of η will correspond to the minimum value of ΔG:

$$\frac{\partial \Delta G}{\partial \eta} = a(T)\eta + b(T)\eta^3 = 0 \tag{2}$$

If both $a(T)$ and $b(T)$ are positive, there will be a single minimum at $\eta = 0$. On the other hand, if $a(T)$ becomes negative there are two minima at

$$\eta = \pm(-a(T)/b(T))^{1/2} \tag{3}$$

So for this polynomial to model the phase transition all we require is that $a(T)$ changes sign at the phase transition, and we can assume that $b(T)$ is a positive constant. The simplest approximate form for $a(T)$ is then

$$a(T) = a(T - T_c) \tag{4}$$

where T_c is the transition temperature. Thus the free energy is written as

$$\Delta G = \tfrac{1}{2}a(T - T_c)\eta^2 + \tfrac{1}{4}b\eta^4 \tag{5}$$

and the minimisation of the free energy yields the temperature dependence of η:

$$\eta(T) = \pm(a(T_c - T)/b)^{1/2} \tag{6}$$

This functional form is found for many second-order phase transitions.

Is this just playing around with numbers? To show that there is more to it than this, let us consider the separate contributions to the free energy difference, the enthalpy and the entropy terms:

$$\Delta G = \Delta H - T\Delta S \tag{7}$$

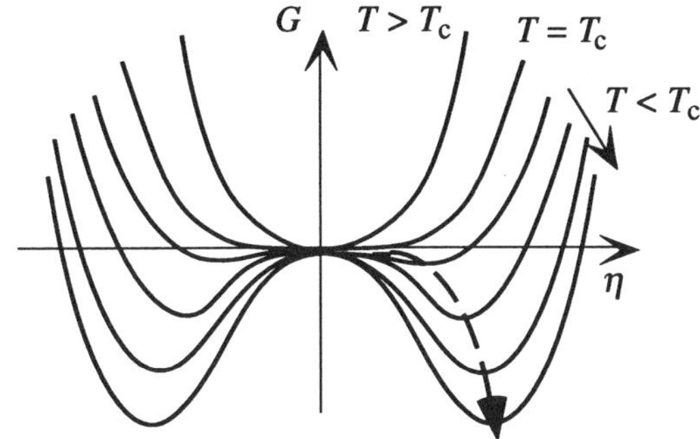

FIGURE 3. Representation of the evolution of the free energy *vs*. order parameter
curves with temperature. The locus of the minima of the curves determines the
temperature dependence of the order parameter.

Since at low temperatures the only significant contribution is from ΔH, this must have a
double-well form to give a non-zero equilibrium value of η:

$$\Delta H = -\tfrac{1}{2} a T_c \eta^2 + \tfrac{1}{4} b \eta^4 \tag{8}$$

where we treat $a T_c$ as a positive constant. Thus ΔH favours the distortion that
characterises the low-temperature phase. The distortions in the low-temperature phase
tend to stiffen the crystal structure, which then cause the vibrational frequencies to
increase slightly. The increase in the frequencies leads to a lowering of the entropy:

$$\Delta S = -\tfrac{1}{2} a \eta^2 \tag{9}$$

Again we must have only even powers of η, and since the entropy difference has a
single maximum at $\eta = 0$ we only need one term. When we combine these two terms
we obtain the earlier result

$$\begin{aligned}
\Delta G = \Delta H - T \Delta S &= -\tfrac{1}{2} a T_c \eta^2 + \tfrac{1}{4} b \eta^4 + \tfrac{1}{2} a T \eta^2 \\
&= \tfrac{1}{2} a (T - T_c) \eta^2 + \tfrac{1}{4} b \eta^4
\end{aligned} \tag{10}$$

The free energy function is shown for a range of temperatures in Figure 3. The evolution
of the free energy curves clearly shows how the minimum of the free energy, which
defines the equilibrium value of η, changes with temperature.

The basic model can be extended in a number of ways. With a negative fourth-
order term, which necessitates the inclusion of a positive sixth-order term, we can model
a first-order phase transition [7]. The model can also include terms containing strain, but
this is now getting away from our main point.

The basic framework of Landau theory has been one of the cornerstones of the study of structural phase transitions. For example, Devonshire adopted a similar approach in his theoretical study of the ferroelectric phase transitions in $BaTiO_3$, which leads us to one of the most significant developments in the study of structural phase transitions. Ferroelectric phase transitions involve displacements of the cations and anions that break the centre of symmetry. The macroscopic effect is the formation of an easily-measured static dielectric polarisation that vanishes above the transition temperature. The big experimental push came in the 1940's (sparked by the need to develop better electrical components for communications in the war) and 1950's. A number of different ferroelectric materials were discovered, many which were members of the perovskite family of structures. By 1960 there was a large accumulation of experimental data [8–10], and a number of theoretical approaches had been developed. Ultimately most of these theories could be cast in the form of a Landau free energy. What was lacking, though, was a general physical (as opposed to phenomenological) theory.

The theoretical breakthrough came from a rather different direction. In the early 1960's the first neutron scattering measurements of phonon dispersion curves were underway. One track in this work was concerned with metals, and a parallel track was concerned with ionic crystals, focusing on the alkali halides. The simplest lattice dynamics models, which included Coulomb interactions and short-range repulsions, were not found to be very successful. Actually this had been anticipated, since it was known that these models could not reasonably reproduce the dielectric properties of ionic materials. It was found that it was necessary to include the effects of ionic polarisability into the lattice dynamics, via the "shell model". In this model the ion is separated into two components, an inner core which contains all the mass, and an outer shell that represents the outer electrons. Both the core and shell are charged, and interact with a simple harmonic potential. More sophisticated variants have since been developed. The application of the shell model was found to be very successful in modelling the lattice dynamics of the alkali halides, giving a physical picture that was consistent with the assumptions of the ionic model. One of the leaders in this work was Cochran. Running his lattice dynamics programs he found that it was easy to produce negative values for the squares of the phonon frequencies if he had not tuned his parameters properly. What he realised was that this problem arose because the equations contained a balance between terms of different sign. This posed the question of whether the balance could be so fine that small changes due to temperature could tip it first one way and then the other, in effect, leading to a stable phase at one temperature and an unstable phase at another. This then suggested a possible mechanism for the origin of the ferroelectric phase transition [11]. The effect of temperature can only come through anharmonic interactions. In the few years following Cochran's work the anharmonic phonon theory was worked into a form that highlighted the role of the interactions that give rise to displacive phase transitions. It became clear in this work that the theory was more general than just for ferroelectric phase transitions [4,9,10].

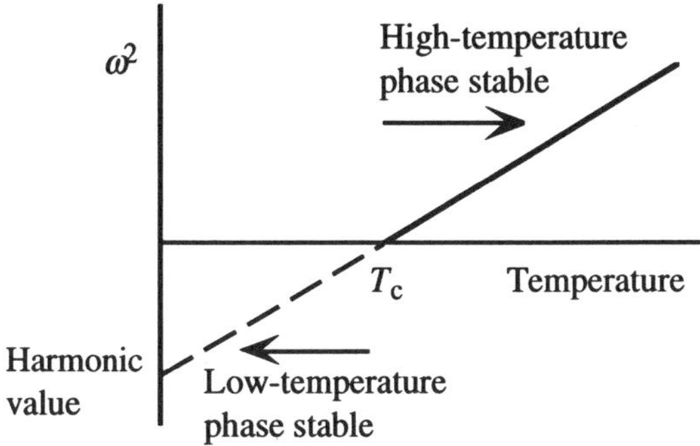

FIGURE 4. Schematic representation of the temperature-dependence of the soft mode. For temperatures above T_c the frequency is real and the structure is stable.

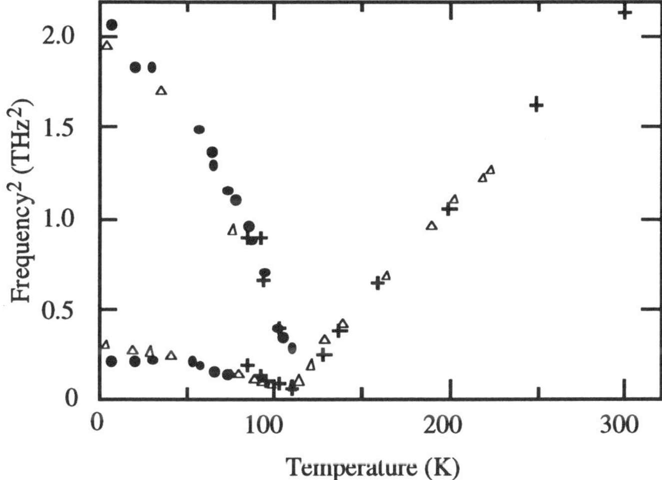

FIGURE 5. Compilation of inelastic neutron scattering data and Raman scattering data for the soft mode in SrTiO$_3$. The crosses [12] and triangles [13] represent data from inelastic neutron scattering experiments, and the squares represent Raman scattering data [14].

The central point in Cochran's theory is that there is a phonon in the high-temperature phase that is unstable at low temperatures. This instability is formally defined by the calculation of a negative ω^2 in harmonic lattice dynamics, appropriate for low temperatures where the anharmonic effects are weak. The effects of anharmonic interactions are to increase the value of ω^2, which means that an unstable ω^2 will become less negative. Eventually, if ω^2 is not too negative, the anharmonic interactions will eventually turn ω^2 positive, so that the phonon becomes stable. The situation is shown schematically in Figure 4. The full theory of the soft mode can eventually be recast into the form of the Landau free energy [4,6–11].

In the years following the development of the soft mode theory, the basic model was confirmed by inelastic neutron scattering experiments and Raman scattering experiments. In Figure 5 experimental data for the displacive phase transition in $SrTiO_3$ are compiled from neutron and Raman scattering experiments [12–14]. This brings us to the early 1970's.

A lot of interest in phase transitions around that time was concerned with the behaviour in the immediate vicinity of the transition temperature, stimulated in part by the ability to accurately and precisely control temperatures for *in situ* diffraction and spectroscopic measurements [4]. An example is the behaviour of the order parameter. Generally this was found to follow a power-law relationship of the form

$$\eta \propto \left(T_c - T\right)^\beta \tag{11}$$

where $\beta = 1/2$ in Landau theory, but may have different values in different models. For many magnetic phase transitions, values of β nearer (although not exactly) $1/3$ have been found. A lot of effort in the 1970's was concerned with measuring β (and corresponding power law exponents for other physical quantities) for different phase transitions. The work was extremely fruitful for many different types of phase transitions, but ultimately for displacive phase transitions, with a few notable exceptions such as $SrTiO_3$, this line of investigation did not lead very far: almost all examples were found to have $\beta = 1/2$. In this respect interest in displacive phase transitions began to decline. The one point of interest concerned measurements of the soft mode at temperatures close to the transition temperature. The basic soft mode model predicted that the soft mode frequency should simply fall to zero at the transition temperature. Experiments showed that it never quite reached zero, and that at the transition temperature a large elastic signal grew in the inelastic neutron spectrum. These features were variously explained by defects, formation of clusters, or inadequacies in the simple anharmonic phonon theory [4].

The point really is that since this time (early 1970's) the progress in understanding the physics of displacive phase transition has not been maintained at the early rate. I think that this has been due partly to the fact that most displacive phase transitions do not have 'interesting' behaviour at temperatures close to the transition temperature, partly to the difficulty in discriminating between different models that give rise to similar behaviour, and partly because (as we will see later) the main equations involve large summations of quantities over reciprocal space. Let us anticipate the details of this last point. The transition temperature is given in one version of the anharmonic phonon theory by

$$k_B T_c = -2\omega_0^2 / \sum_k \alpha_k / \omega_k^2 \tag{12}$$

where ω_0 is the imaginary soft mode frequency at zero temperature, ω_k is any other phonon frequency, and α_k is a fourth-order anharmonic interaction between a phonon and the soft mode (we will derive this equation later). The point here is not to understand the meaning of the equation *per se*, but to appreciate that it requires the summation of all phonons over all wave vectors. At face value it might be thought that there are no physics issues concerned with the calculation of the transition temperature: it may just be a big sum, and what you end up with is the final result from a huge number of different terms. Faced with these sorts of equations, it may not be surprising that more recently the study of displacive phase transitions has not been as vigorous as in the early years.

The rather downbeat picture presented above should not distract the reader from recognising that there are still important issues in the study of displacive phase transitions yet to be resolved, and if one does not feel daunted by large sums over reciprocal space one can still ask relatively simple questions. For example, what really does determine the transition temperature? In magnetism this is an easy question to answer: it is invariably controlled by the nearest-neighbour exchange integral. Could there be something as simple in displacive phase transitions? Even simpler: why do these phase transitions occur? This might have two parts, namely why does the potential energy surface have minima at non-zero values of η, and why is the transition temperature not infinite? These questions concern the existence of easy modes of deformation. We can also ask why one phase transition might be preferred over other possible transitions? Looking back over some of the history, we can ask why Landau theory seems to work so well for many displacive phase transitions, both at temperatures close to T_c and over a wide range of temperatures below T_c? Finally (at least for the moment) we can ask questions about the nature of the high-temperature phases. Do they really look like their average structures, or do they locally resemble more closely the structures of the low-temperature phases?

This article arises from some work my colleagues in Cambridge and I have been engaged on over the past few years, in which we have attempted to look for some answers to these questions using some of the insights offered by crystalline silicates. The resultant model is called the "Rigid Unit Mode" (RUM) model, and has accounted for many features of displacive phase transitions. In fact, although we have not pushed the model very far into applications away from silicates, it has a wider relevance, and we have been able to understand the origins of the phase transitions in a few molecular and ionic crystals as a result. What has proved to be extremely interesting is the wide range of properties that can be understood with the RUM model. For example, we are able to understand properties of the phase transition such as the origin of the actual transition temperature, and also properties such as small and negative thermal expansion, and local distortions of zeolite structures.

3   "Soft Mode" theory of phase transitions

Before I start to describe our recent work, it will be useful to review the general theory of phase transitions, which I will do within the context of the soft-mode theory [6,7].

358

The starting point is to approximate the dependence of the lattice energy on the order parameter η by a simple double-well potential $V(\eta)$:

$$V(\eta) = -\frac{1}{2}\kappa_2\eta^2 + \frac{1}{4}\kappa_4\eta^4 \tag{13}$$

The coefficient $-\kappa_2$ is equivalent to the square of the imaginary harmonic frequency. The model Hamiltonian, containing fourth-order anharmonic terms, can be written as

$$\mathcal{H} = V(\eta) + \frac{1}{2}\sum_k \omega_k^2 Q_k Q_{-k} + \frac{1}{4}\sum_k \alpha_k Q_k Q_{-k}\eta\eta^* \tag{14}$$

where all phonons other than the soft mode are labelled by the subscript k, which subsumes information about the wave vector and the actual phonon branch. Q_k is the normal mode coordinate with corresponding frequency ω_k. α_k is the fourth-order anharmonic coefficient for interactions between the phonon labelled by k and the soft mode. In principle the anharmonic term should include terms with four phonons of quite different wave vector (subject to the selection rule that the sum of wave vectors should equal a reciprocal lattice vector), but in the normal treatment we make an approximation that leads to all these other terms being excluded. In the high-temperature phase the quantity that becomes the order parameter η in the low-temperature phase plays the same role as any other normal mode coordinate. We do not need to include the terms of the form η^2 and η^4 in the last two terms since they are explicitly included in $V(\eta)$, with prefactors $-\kappa_2$ and κ_4 instead of ω^2 and α respectively.

First we consider the behaviour of the soft mode in the high-temperature phase. In the standard treatment of renormalised phonon theory [4,6,7,9,10] we replace pairs of phonon coordinates by their averages, leading to

$$\mathcal{H} = V(\eta) + \frac{1}{2}\sum_k \omega_k^2 Q_k Q_{-k} + \frac{1}{4}\sum_k \alpha_k \langle Q_k Q_{-k}\rangle\eta\eta^* \tag{15}$$

The terms involving η^2 are therefore given as

$$\mathcal{H}(\eta) = -\frac{1}{2}\kappa_2\eta\eta^* + \frac{1}{4}\sum_k \alpha_k \langle Q_k Q_{-k}\rangle\eta\eta^* \tag{16}$$

The thermal average is given by the standard result

$$\langle Q_k Q_{-k}\rangle = \left(n(\omega) + \tfrac{1}{2}\right)\frac{\hbar}{\omega_k} \approx \frac{k_B T}{\omega_k^2} \tag{17}$$

where $n(\omega)$ is the Bose–Einstein distribution [7], and we have used the high-temperature approximation, $k_B T > \hbar\omega_k$, to write the second form. Thus the Hamiltonian can be written as

$$
\begin{aligned}
\mathcal{H}(\eta) &= -\frac{1}{2}\kappa_2\eta\eta^* + \frac{k_B T}{4}\sum_k \alpha_k\eta\eta^* / \omega_k^2 \\
&= \frac{1}{2}\left(-\kappa_2 + \frac{k_B T}{2}\sum_k \alpha_k / \omega_k^2\right)\eta\eta^*
\end{aligned}
\tag{18}
$$

This gives a new effective harmonic frequency for the soft mode, which is called the renormalised frequency:

$$
\omega^2 = -\kappa_2 + \frac{k_B T}{2}\sum_k \alpha_k / \omega_k^2
\tag{19}
$$

This clearly has the form illustrated earlier in Figure 4. The important feature of this equation is that ω^2 is negative (imaginary frequency) at low temperatures, but becomes positive for temperatures above T_c, where

$$
k_B T_c = 2\kappa_2 / \sum_k \alpha_k / \omega_k^2
\tag{20}
$$

This temperature defines the point at which the soft mode frequency is stabilised, and therefore represents the phase transition.

In the low-temperature phase the quantity η is a static deformation rather than a normal mode coordinate. Equation (14) can be rewritten as

$$
\begin{aligned}
\mathcal{H} &= V(\eta) + \frac{1}{2}\sum_k \left(\omega_k^2 + \frac{1}{2}\alpha_k\eta\eta^*\right)Q_k Q_{-k} \\
&= V(\eta) + \frac{1}{2}\sum_k \overline{\omega}_k^2 Q_k Q_{-k}
\end{aligned}
\tag{21}
$$

where

$$
\overline{\omega}_k^2 = \omega_k^2 + \frac{1}{2}\alpha_k\eta\eta^*
\tag{22}
$$

The modified frequency $\overline{\omega}_k$ carries information about the value of the order parameter and provides the basis for the technique of "hard-mode spectroscopy" [15]. From equation (22) we note that measurements of the way a phonon frequency at temperatures below the phase transition deviates from the value extrapolated from the frequency in the high-temperature phase gives direct information on the temperature dependence of η, i.e. from equation (22) we can determine the order parameter from $\eta^2 \propto \left(\overline{\omega}_k^2 - \omega_k^2 \right) \propto \Delta\omega_k$.

With the quasiharmonic approximation we can use the standard expression for the phonon free energy, G_{ph}, of a set of harmonic vibrations (using the nomenclature of the Gibbs free energy in order to emphasise the contact with experiment, but technically we working with constant volume—the differences will not be significant) [7]:

$$G_{\mathrm{ph}} = k_{\mathrm{B}}T \sum_k \ln\left[2\sinh\left(\frac{\hbar\omega_k}{2k_{\mathrm{B}}T} \right) \right] \tag{23}$$

We can expand G_{ph} as a Taylor series in η about the point $\eta = 0$, noting that we use the set of values of $\overline{\omega}_k$ as defined by equation (22) for the frequencies [6,7]. This yields

$$G_{\mathrm{ph}}(\eta) = G_{\mathrm{ph}}(\eta = 0) + \frac{k_{\mathrm{B}}T}{4} \sum_k \frac{\alpha_k}{\omega_k^2}\eta^2 + \cdots \tag{24}$$

In order to simplify the nomenclature can write equation (24) in the form of an Einstein model, although we will not actually be making an Einstein approximation:

$$G_{\mathrm{ph}}(\eta) = G_{\mathrm{ph}}(\eta = 0) + \frac{3RT\tilde{\alpha}}{4\tilde{\omega}^2}\eta^2 + \cdots \tag{25}$$

where

$$\tilde{\alpha} = \frac{1}{3N} \sum_k \alpha_k \tag{26}$$

and

$$\tilde{\omega}^2 = \tilde{\alpha} \,/ \sum_k \left(\alpha_k / \omega_k^2 \right) \tag{27}$$

If we take the part of $G_{\mathrm{ph}}(\eta)$ that depends on η, and add this to $V(\eta)$ as defined in equation (13), we obtain

$$\begin{aligned} G(\eta) &= -\frac{1}{2}\kappa_2\eta^2 + \frac{3\tilde{\alpha}RT}{4\tilde{\omega}^2}\eta^2 + \frac{1}{4}\kappa_4\eta^4 \\ &= \frac{3\tilde{\alpha}R}{4\tilde{\omega}^2}(T - T_{\mathrm{c}})\eta^2 + \frac{1}{4}\kappa_4\eta^4 \end{aligned} \tag{28}$$

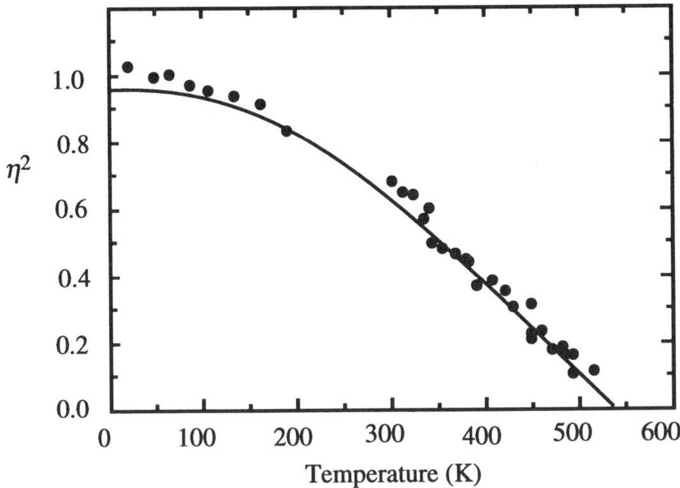

FIGURE 6. Temperature dependence of the square of the order parameter in anorthite [6] with the curve obtained from minimisation of the free energy using two fitted parameters.

where T_c is the transition temperature, now given by

$$k_B T_c = \frac{2\kappa_2 \tilde{\omega}^2}{3\tilde{\alpha}} \tag{29}$$

This is identical with the result given in equation (20). Note that the free energy is now written in a form that is equivalent to the Landau free energy, equation (5).

In principle one can extend the expansion of the phonon free energy, equation (24) to higher order. This would add a temperature-dependent term to the higher-order coefficients in the Landau free energy. The fourth-order coefficient would be given as

$$b = \kappa_4 - \frac{3\tilde{\alpha}^2 k_B T}{\tilde{\omega}^4} \tag{30}$$

For all practical temperatures the temperature-dependent part of equation (30) is negligibly small in value compared to the value of κ_4 [6].

The theory developed in this section has accomplished a number of objectives. It has given a mechanism for a displacive phase transition, in which an unstable phonon can be stabilised by anharmonic interactions. The theory has provided a link between microscopic interactions and quantities such as the phase transition temperature. Finally, the phonon theory has naturally led to a form equivalent to Landau theory, giving the phenomenological parameters some physical interpretation.

As an example of the application of this model, I show in Figure 6 data for the order parameter for the displacive phase transition in the mineral anorthite, $CaAl_2Si_2O_6$ over the range of temperatures from nearly 0 K to the transition temperature [16]. In this

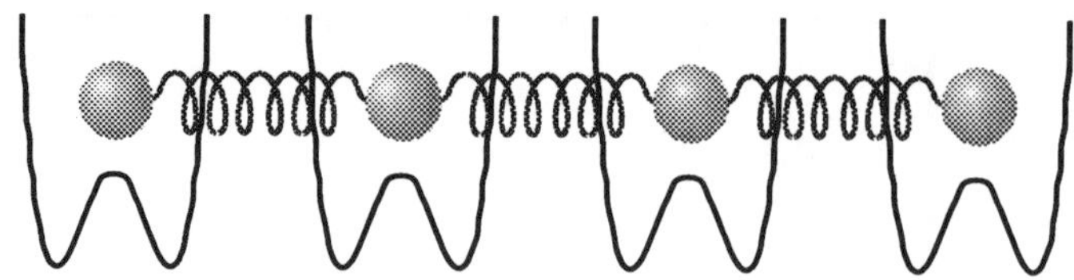

FIGURE 7. One dimensional representation of a simple model that simulates a displacive phase transition. The atoms vibrate in local potential energy wells, and interact with nearest neighbours as represented by the harmonic springs.

application I have used an Einstein approximation to replace all values of $\overline{\omega}_k$ and α_k by average values ω_{ave} and α_{ave} respectively, and have fitted values of η obtained from the minimisation of $G(\eta)$ to the experimental data. Only the values of ω_{ave} and α_{ave} needed to be adjusted. The results gave $\omega_{\text{ave}} = 12$ THz, and $\alpha_{\text{ave}} = 0.02$ THz^2. An interesting point to be aware of is that the value of α_{ave} is substantially below the value you might have expected from measurements of individual phonon frequencies. Usually you measure changes of order 100 GHz over the whole range of temperatures, whereas the average shift given by this value of α_{ave} is only 0.4 GHz. A detailed calculation I once did for quartz showed that individual values of α_k can be rather larger, but will be both positive and negative. Thus the average value, α_{ave}, will actually be relatively small. From the equations it is clear that α_{ave} has to be positive for a phase transition to occur. It is interesting to note that as an average over a whole range of positive and negative quantities it could be so small as to approach zero, giving a transition temperature that will approach an infinite value. This leads to an interesting question as to why many transition temperatures are not in the regime of solar temperatures, or equally why they are not in the cryogenic regime. I will argue below that the RUM model actually points us in the direction of an answer.

4   A simple model of a phase transition

A number of useful insights have followed from the simple model shown in Figure 7 [4,17–20]. The model has atoms vibrating in local double-well potentials, and interacting with their nearest neighbours by harmonic forces of stiffness J. The local double-well potentials are written in the form of equation (13). In this case there is only a single anharmonic force constant: $\tilde{\alpha} = \alpha_k = \kappa_4$ for all k (there is now only one phonon branch). The energy can be written for the d-dimensional model as

$$
\begin{aligned}
E &= \sum_j \left(-\tfrac{1}{2}\kappa_2 u_j^2 + \tfrac{1}{4}\kappa_4 u_j^4\right) + \tfrac{1}{2}J\sum_{j\neq j'}\left(u_j - u_{j'}\right)^2 \\
&= \sum_j \left(\tfrac{1}{2}(2dJ - \kappa_2)u_j^2 + \tfrac{1}{4}\kappa_4 u_j^4\right) + \tfrac{1}{2}J\sum_{j\neq j'}u_j u_{j'}
\end{aligned}
\tag{31}
$$

To calculate the transition temperature we use the values of ω_k^2 at $T = T_c$, given as

$$\omega_k^2 = J \sum_j 2d - \exp(i\mathbf{k} \cdot \mathbf{r}_j) \tag{32}$$

and the sum is over all nearest neighbours of separation $\mathbf{r}_j$. Thus we have

$$\tilde{\omega}^2 = 1 / \sum_k \omega_k^{-2} = gJ \tag{33}$$

where $g = 0.66$ for the three-dimensional version of the model where each atom has six neighbours. This leads to an expression for the transition temperature

$$k_B T_c = \frac{2gJ\kappa_2}{\kappa_4} \tag{34}$$

The interesting thing to note about this model is that at zero temperature, where the atoms will be located in the bottoms of the potential energy wells, the order parameter has the value

$$\eta_0^2 = \frac{\kappa_2}{\kappa_4} \tag{35}$$

so that

$$k_B T_c = 2gJ\eta_0^2 \tag{36}$$

Thus the transition temperature is determined by two quantities, the force constant of the spring, and the maximum displacement possible.

I have introduced this model because I will make use of the simple relationship between T_c, J and η_0 later on. It should also be remarked that the limit $J \gg \kappa_2$, the so-called "displacive limit", gives the limit of the model for which Landau theory is appropriate. In this case the strong springs mean that the movement of one atom will influence the atoms a long distance away. In effect, the displacive limit is characterised by long correlation lengths. The opposite limit gives behaviour more typical of an order–disorder phase transition (in fact the model becomes equivalent to the Ising model) [4].

5   The Rigid Unit Mode model of displacive phase transitions in silicates

The idea that a displacive phase transition in a silicate can involve small rotations and translations of undistorted tetrahedra lies at the heart of the Rigid Unit Mode model. The idea is not new, and was first discussed in relation to the phase transition in quartz [21,22], and with reference to the crystal structures of the feldspars [22], but given in

364

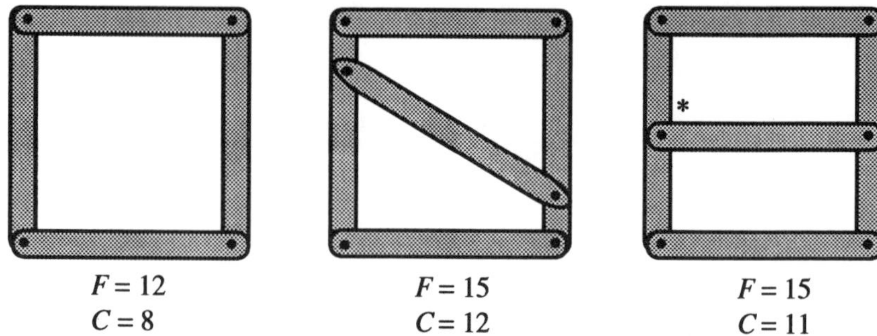

FIGURE 8. A two-dimensional example of the effect of symmetry on the number of independent constraints. The three structures are formed by bars hinged at common corners. Each bar has 3 degrees of freedom, and each linkage has two constraint equations. The number of degrees of freedom for each structure is $F - C$. Two of these are the uniform translations of the structure and a third is the uniform rotation. In the left-hand structure, there is an additional degree of freedom, the possibility to shear the structure. This degree of freedom is lost in the middle structure as a result of the extra constraints imposed by the cross-brace bar. Although the right-hand structure has the same number of constraints as the middle structure, it can clearly also be sheared. The solution to this problem is to note that the two constraints associated with the joint marked * can be replaced by a single constraint that makes the middle bar parallel to the top and bottom bars.

terms of static distortions of the crystal structures. The idea that phonons could involve motions of rigid tetrahedra was first developed for the analysis of neutron scattering data from quartz [23], but at this time the relationship to the phase transition was not explored. The significant advance came from Vallade and co-workers in Grenoble [24–28]. They undertook a detailed study of the intermediate incommensurate phase that occurs over a small temperature interval between the α and β phases. As part of this work they proposed that the soft mode for the phase transitions will be a RUM, and so they calculated (by hand) all the RUMs that could exist in the β-quartz structure. They found that there is a RUM for each wave vector along $\mathbf{a}^*$, and, without going into details here, they were able to account for the complete sequence of phase transitions based on these RUMs.

Our own contribution [20,29–37] came following the inspiration of the Grenoble group. Since displacive phase transitions are common in aluminosilicate minerals, we recognised that the existence of RUMs as a general feature could account for these phase transitions. The point is that for any aluminosilicate crystal structure it is necessary to identify a low-frequency phonon mode that can propagate with the SiO_4 and AlO_4 tetrahedra moving as rigid units with no distortions. In this case the RUM will be the natural soft mode for the phase transition. Our first task was to develop a method to calculate the RUMs for any crystal structure [31]. Following this we recognised that the RUMs do more than act as the soft modes for the phase transitions [33]. But before I discuss these aspects of the model, I should deal with some general points.

It is important to note that the existence of any RUMs at all in a framework silicate is not trivial. In a simple mechanical structure, such as a framework structure made of loosely jointed tetrahedra, the number of possible modes of deformation is simply equal to the difference between the total number of degrees of freedom, F, and the total number of constraints, C. Each tetrahedron has 6 degrees of freedom, and each corner has 3 constraint equations associated with the linkage to the corner of the connected tetrahedron, so the number of constraints per tetrahedron is also 6. Thus the framework structure made of connected tetrahedra appears to be exactly constrained with $F = C$, and hence will have no modes of deformation. However, symmetry can make some of the constraints redundant, so that there is a slightly-lower number of independent constraints than the number of degrees of freedom. This principle is illustrated in Figure 8, where I show a number of two-dimensional arrangements of linked rods with constraints acting at the linkages. For a square of rods linked at corners and cross-braced by a fifth rod, the balance between the constraints and degrees of freedom only allows for uniform translations and rotations of the arrangements. However, if the fifth rod is parallel to two others, the two constraints that act at one end of this rod can be replaced by a single constraint that the rod should remain parallel to two others. This reduction in the total number of constrains then gives back to the structure one more degree of freedom, which in this case is the same shear deformation that would occur if the fifth rod was not in place.

6 Calculation of the number of rigid unit modes in a framework of linked triangles

Having evaluated the role of symmetry, we now ask whether it is possible to count the independent constraints to determine the number of RUMs for a structure. As an example we take a two-dimensional arrangement of triangles, Figure 9. These are arranged to form hexagonal rings. Consider the two darkest triangles in Figure 9a. Each triangle has 3 degrees of freedom (translations in two directions, and one rotational variable). We define the accumulation of the number of degrees of freedom with the variable F. When we link together two triangles there will be a number of constraints. So when we link together the two darkest triangles there will be two constraints that tie together the positions of the two connected corners. We define the accumulation of the number of constraints with the variable C. When we add the four triangles of medium-grey shading, we increase the value of F by $4 \times 3 = 12$. To count the additional number of constraints we need to be a little more careful. Adding the top and bottom triangle increases C by $2 \times 2 = 4$. But we can only add the remaining two triangles to complete the hexagonal ring in one way, since the distance spanned across the corners marked as ℓ is fixed by the way we added the third and fourth triangles. Thus not only do we have constraints on the positions of the corners of the triangles, we also have constraints on their orientations. Thus we have increased C by $2 \times 3 = 6$. We can repeat this process to add more hexagonal rings in a row, each time adding four more triangles. If we make a row containing N hexagonal rings, in total we will have $F_1 = 6 + 12N$ and $C_1 = 2 + 10N$. The number of RUMs in this row is equal to $F_1 - C_1 = 4 + 2N$. We have added the subscripts to denote row number 1: subsequent rows will have different formulae.

366

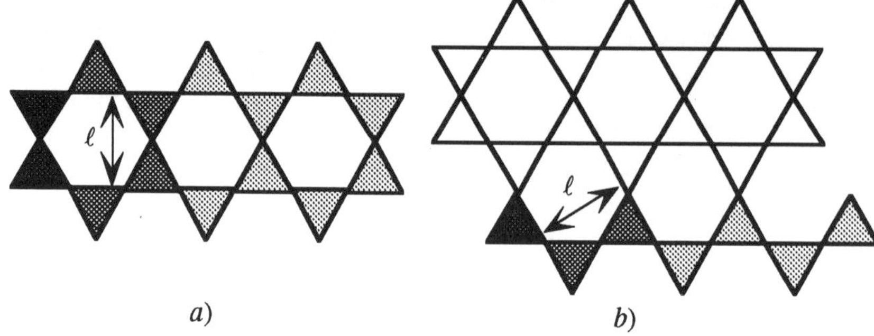

a)   b)

FIGURE 9. The development of the two-dimensional hexagonal array of triangles showing the separate additions of degrees of freedom and constraints. *a)* shows a single row of hexagons. The first pair of triangles are shown as dark grey, with constraints acting only on the linked corners. The hexagon is closed by the four middle-grey triangles. There are additional constraints associated with the need to close the hexagonal as described in the text. Addition of further groups of four triangles to create additional hexagons follow in exactly the same manner as the first group of four triangles. *b)* shows the addition of a second row. The orientation of the first triangle (dark grey) is unconstrained. When we add a pair of triangles (middle grey) to close the hexagonal the positions and orientations are completely constrained. Addition of further pairs to complete the new row of hexagons follows in the same manner as the first pair.

We now build up the structure in the second dimension in the way shown in Figure 9b. Starting with a completed row, we first add the darkest triangle. This adds 3 degrees of freedom and 2 constraints. To complete the ring we add two more triangles (the two shaded with the intermediate grey). Since the distance marked ℓ is now fixed, adding these two triangles increases both F and C by 6. To develop this row we add two triangles as a time, and completing the last ring is the same as forming the first ring. The new row containing N rings will have accumulated $F_2 = 3 + 6N$ and $C_2 = 2 + 6N$, so that the additional number of RUMs is $F_2 - C_2 = 2$. Any subsequent rows will follow exactly as the second row. Thus for M rows of N rings, we will have a total number of RUMs of $(F_1 - C_1) + (M - 1)(F_2 - C_2) = 2 + 2N + 2M$, corresponding to three lines of RUMs in reciprocal space.

A similar calculation can be performed for arrangements of squares, where there are 4 corners per unit. The simple counting (not taking account of degenerate constraints) gives 4 constraints per square, one more than the number of degrees of freedom. The effect of symmetry is to reduce the number of independent constraints to allow one mode of deformation [36].

This analysis has not incorporated any constraint on the volume of the system. Without this constraint all the RUMs have an equivalent status. However, for a deformation to also correspond to a phonon mode it is necessary that the volume change is zero to first-order. This point is illustrated in Figure 10, which shows an array of squares that represents a two-dimensional perovskite structure. In the ordered arrangement any infinitesimal rotations of the octahedra, $\delta\theta$, will lead to a volume change proportional to $(\delta\theta)^2$. On the other hand, in the distorted structure, where the average rotation is non-zero, any infinitesimal rotations of the octahedra will lead to a

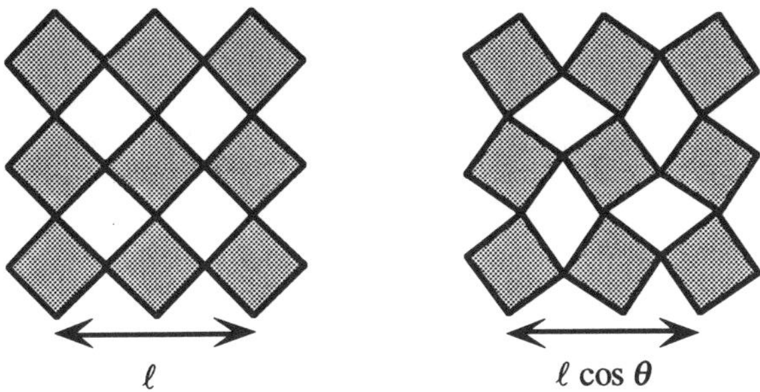

FIGURE 10. Distortion of an array of squares, representative of a two-dimensional perovskite structure. The arrangement on the left has the maximum volume. The squares in the arrangement on the right have rotated by an angle θ. This is the only possible RUM distortion. Note the change in volume caused by the distortion.

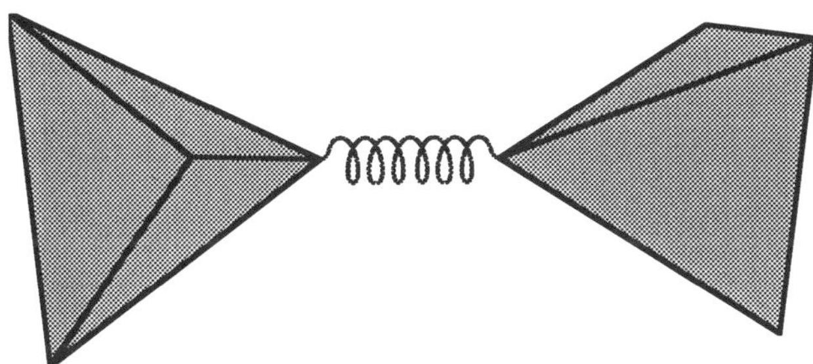

FIGURE 11. Split-atom representation of a pair of linked tetrahedra.

first-order volume change proportional to $\delta\theta$. Only the first rotation can propagate as a vibrational mode. The problem with the counting methods described above is that they do not differentiate between distortions that change volume to first or second order.

7   Practical method for the calculation of RUMs: the "Split Atom" method

The scheme of counting independent constraints does not provide a practical method for the enumeration of the number of deformation patterns possible in a three-dimensional framework structure. Nor, as we have just mentioned, does it differentiate between volume-preserving and volume-changing RUMs. And the simple counting also only gives a final number of RUMs, and only with some effort can it be adapted to provide information about the wave vectors of the RUMs. We have developed a computational approach to this problem [29–31] that in practice involves some degree of compromise. The essence of this approach is to view each tetrahedron as an individual rigid body. The atoms at the corners of tetrahedra are then split into two, one per linked tetrahedron. The model has two types of constraints: the tetrahedra should be perfectly rigid, and the positions of the split atoms should coincide exactly. In our approach we treat the rigid-tetrahedra constraints in a strict sense, but we treat the constraints on the positions of the split atoms in a slacker sense by inventing a harmonic force between the two split atoms that is designed to give an equilibrium separation of zero. This model is illustrated in Figure 11. In principle the spring has an infinite stiffness, but for computational convenience it is given a finite value. Actually, this spring stiffness is formally equivalent to the stiffness of the octahedra or tetrahedra to lowest order.

The split-atom model gives equations of motion that are conveniently solved within the framework of molecular lattice dynamics [7,38], and we have incorporated this model into a standard molecular lattice dynamics computer program [31,32]. The program solves the dynamical matrix for any given wave vector [7], and any RUM at this wave vector is identified as a zero-frequency solution. In this sense the program calculates phonon modes that can propagate without distortions of the tetrahedra or octahedra, and these are the solutions that do not involve any first-order change in volume of the crystal, but the same modes can also be represented as static distortions of the structure. The implementation of the split atom method is called CRUSH [31,32], and is freely available on the WWW [39]. The method works with octahedra as well as tetrahedra. There is also an option to take account of flexing of the Si–O–Si angle. The usefulness of this approach is that it gives explicit information about the wave vectors, and for each RUM at any wave vector it gives the associated normalised rotations and translations of the rigid units.

Complete sets of RUMs have been calculated for a number of framework silicates [35], and representative examples (for the high-temperature phases in each case) are given in Table 1. In general there are always some RUMs, and these frequently lie on lines or planes of wave vectors in reciprocal space. In some cases, as we will see below, RUMs can be found for wave vectors that lie on complicated curved surfaces in reciprocal space [34,35]. Some of the RUMs we have calculated can be identified with the soft modes for displacive phase transitions. The soft mode for the α–β phase transition in quartz (Figure 1) corresponds to the RUM at zero wave vector. The phase transition in cristobalite (Figure 2) has a double-degenerate RUM with wave vector [1,0,0]. In tridymite, another form of silica, there is a complicated sequence of phase transitions, and two of the RUMs at zero wave vector are involved in this sequence. In leucite, $KAlSi_2O_6$, there are two phase transitions which involve the RUM at zero wave vector and a soft transverse acoustic mode, with wave vector along [1,1,0], that also corresponds to a RUM. We found that when there is a displacive phase transition there are fewer RUMs in the lower-symmetry phases. For example, in the low-temperature phase of leucite the only RUMs are acoustic modes. The planes of RUMs in β-cristobalite are reduced to lines along [1,1,0] in α-cristobalite. The case of sanidine, the high-symmetry disordered form of the feldspar structure, is interesting in this respect. Calculations on an ideal structure with a monoclinic angle of 120° give a totally symmetric RUM at zero wave vector [35]. When this RUM is imposed on the structure, together with a RUM shear, only a few of the RUMs are preserved. Some of the phase transitions in the feldspars occur with the help of a soft acoustic RUM. Other phase transitions (such as the one to which the data in Figure 6 refer) involve a RUM

TABLE 1. Numbers of rigid unit modes for symmetry points in the Brillouin zones of some aluminosilicates [35], excluding the trivial acoustic modes at $k = 0$. An asterisk indicates a RUM that acts as the soft mode for a displacive instability in either the parent structure or a related structure. The "—" indicates that the wave vector is not of special symmetry in the particular structure. The numbers in brackets denote the numbers of RUMs that remain in the low-temperature phases.

k	Quartz $P6_2 22$	Cristobalite $Fd3m$	Tridymite $P6_3/mmc$	Sanidine $C2/m$	Leucite $Ia3d$	Cordierite $Cccm$
$0,0,0$	1* (0)	3 (1)	6*	0	5* (0)	6
$0,0,\frac{1}{2}$	3 (1)	—	6	1*	—	6
$\frac{1}{2},0,0$	2 (1)	—	3*	—	—	6
$\frac{1}{3},\frac{1}{3},0$	1 (1)	—	1	—	—	6
$\frac{1}{3},\frac{1}{3},\frac{1}{2}$	1 (1)	—	2	—	—	0
$\frac{1}{2},0,\frac{1}{2}$	1 (1)	—	2	—	4 (0)	2
$0,1,0$	—	2*	—	1	—	—
$\frac{1}{2},\frac{1}{2},\frac{1}{2}$	—	3 (0)	—	0	0	—
$0,1,\frac{1}{2}$	—	—	—	1	—	—
$0,0,\xi$	3 (0)	2 (0)	6	—	0	6
$0,\xi,0$	2* (0)	2 (2)	3	1	0	6
$\xi,\xi,0$	1 (1)	1 (0)	1	—	4 (0)	6
ξ,ξ,ξ	—	3 (0)	—	—	0	—
$\frac{1}{2},0,\xi$	1 (0)	—	2	—	0	2
$\xi,\xi,\frac{1}{2}$	1 (1)	—	0	—	0	0
$\frac{1}{2}-\xi,2\xi,0$	1 (1)	—	1	—	—	6
$\frac{1}{2}-\xi,2\xi,\frac{1}{2}$	1 (1)	—	0	—	—	0
$0,\xi,\frac{1}{2}$	0 (0)	—	1	1	—	—
$\xi,1,\xi$	—	1 (0)	—	—	0	—
$\xi,\zeta,0$	1 (0)	0	1	—	0	6
$\xi,0,\zeta$	0 (0)	0	2	1	0	0
$\zeta,1,\zeta$	—	0	—	1	0	—
ξ,ξ,ζ	0 (0)	1 (0)	0	—	0	0

distortion, but the existence of the RUM is very sensitive to the size and shape of the unit cell [35].

There are other cases that have one or more RUMs for every wave vector, not just for a restricted set of wave vectors. Our first example was the high-symmetry phase of the sodalite structure [35]. This has one RUM for each wave vector. Since the sodalite structure contains the basic structural building blocks for a number of zeolites, it is likely that a high-symmetry zeolite phase may have several RUMs per wave vector. We have found this to be the case [37]. I will discuss some of the implications of this later.

370

8 Experimental verification

8.1 DIFFUSE SCATTERING IN ELECTRON DIFFRACTION

The most striking experimental studies have been measurements of the diffuse scattering seen in electron diffraction patterns of the two silica phases cristobalite [39,40] and tridymite [34,42]. The intensity for scattering of radiation from single phonons for a particular scattering vector $\mathbf{Q}$ and phonon frequency ω varies as

$$S(\mathbf{Q}, \omega) \propto \frac{k_\mathrm{B}T}{\omega^2}|\mathbf{Q} \cdot \mathbf{e}|^2 \delta(\omega) \tag{37}$$

where $\mathbf{e}$ is the phonon eigenvector, and we have assumed the classical high-temperature approximation [7]. The diffuse scattering seen in electron (or X-ray) diffraction includes scattering from all energies $\hbar\omega$:

$$S(\mathbf{Q}) \propto \int S(\mathbf{Q}, \omega)\mathrm{d}\omega \tag{38}$$

The key point is that the intensity of diffuse scattering from phonons varies as ω^{-2}, so it will arise predominantly from the low-frequency phonons. Strong diffuse scattering is therefore likely to be a good signature of the presence of RUMs. In the case of cristobalite and tridymite the RUMs lie on planes in reciprocal space [35], and these are seen as sharp lines in the diffraction patterns as the plane of reciprocal space in the diffraction patterns intersects the RUM planes. The example of tridymite is shown in Figure 12. In all cases the diffuse scattering seen in the electron diffraction patterns are explained by the RUM calculations [35]. For tridymite, however, there are also intricate curved surfaces of RUMs in reciprocal space, which are seen as the curves of diffuse scattering in the electron diffraction patterns in Figure 12 [34,42]. The CRUSH calculations of the detailed sets of lines of diffuse scattering provide a confirmation of the predictions of the RUM calculations. The calculations of the curved surfaces in tridymite have reproduced exactly the measured diffraction patterns, providing a particularly stringent validation of the model [34].

8.2 PHONON DISPERSION CURVES BY INELASTIC NEUTRON SCATTERING

Inelastic neutron scattering has been used to measure some of the RUMs in quartz [23–28] and leucite, $KAlSi_2O_6$ [33,43]. In quartz the RUMs provide the mechanism for the displacive phase transitions, and the RUMs have been measured in some detail. In leucite the measurements of the RUMs are more tentative, but confirm the basic picture. The neutron scattering data show that the RUMs are mostly overdamped, but it was possible to extract the phonon dispersion relations that are shown in Figure 13. The inelastic neutron scattering data show that the RUMs have frequencies in the range 0–1

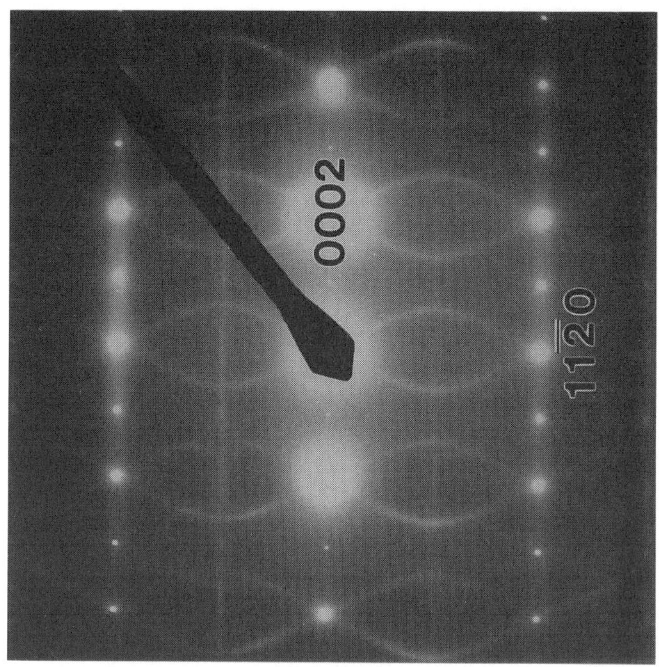

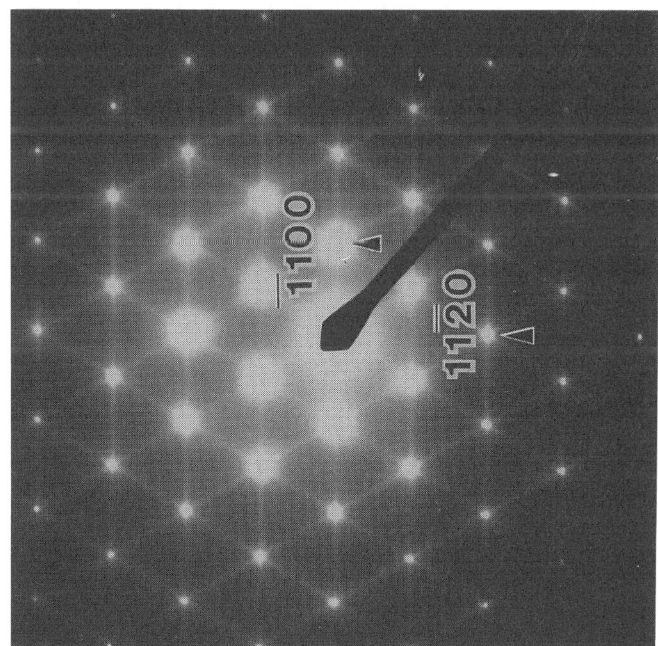

FIGURE 12. Electron diffraction patterns of the high-temperature phase of cristobalite viewed down [001](left) and [1̄10] (right). The streaks and curves correspond to diffuse scattering from RUMs [34,42].

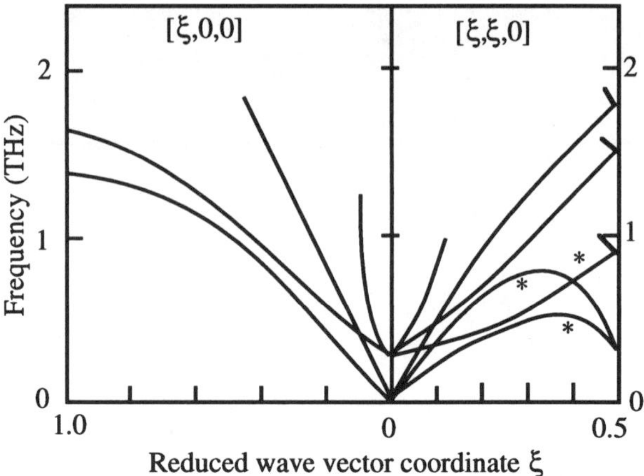

FIGURE 13. The low-frequency phonon dispersion curves of leucite determined by inelastic neutron spectroscopy by Boysen [43]. The phonon branches indicated by * are identified as the RUMs. Not all the phonon branches have been measured, and the interpretation of the measurements that are available are only tentative. Nevertheless, the low-frequency modes along [110] can be interpreted within the RUM model.

THz. The CRUSH calculations give RUMs only for wave vectors along [1,1,0], which is consistent with the experimental data. From Figure 13 we note the existence of a triply-degenerate RUM at zero wave vector. For wave vectors along [1,1,0] the degeneracy is lifted, and only one component remains a RUM. The frequencies of the other components rapidly increase as the wave vector increases, and along [1,0,0], where there are no RUMs, the frequencies of all components increase rapidly. The other point to note from Figure 13 is the existence of a RUM that is an acoustic mode.

8.3 INELASTIC NEUTRON SCATTERING FROM POWDER SAMPLES

A striking experimental verification of the RUM picture came from inelastic neutron scattering measurements on a powdered sample of cristobalite [44,45]. In this case the experiments measure a quantity that is related to the phonon density of states:

$$S(\omega) \propto \int S(\mathbf{Q}, \omega) d\mathbf{Q} \propto k_B T g(\omega) / \omega^2 \tag{39}$$

The experiment was very simple, and involved a measurement of $S(\omega)$ at one temperature in each phase [44]. The low-frequency results are shown in Figure 14. The striking point is that at low energies there is a significant increase in the neutron scattering in the high-temperature phase. In this phase there are planes of RUMs, most of which vanish in the low-temperature structure. The dramatic changes in the low-frequency spectra provide a validation of the RUM model. Also, the experimental data give a frequency scale for the RUMs of 0–1 THz, as found in the single-crystal

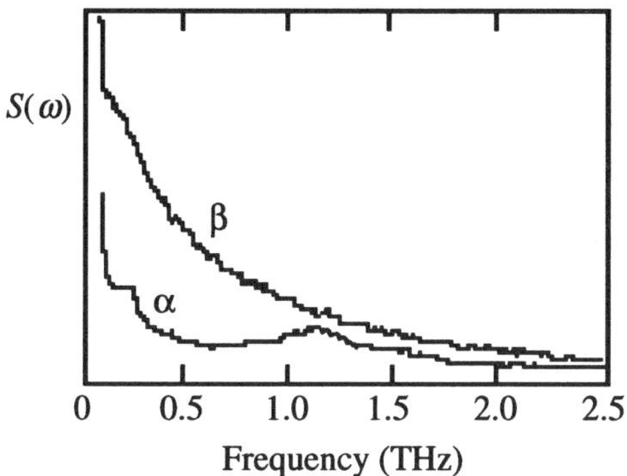

FIGURE 14. Inelastic neutron scattering spectrum from polycrystalline samples of cristobalite in the α and β phases [44,45]. The data were obtained on the TFXA spectrometer at ISIS.

measurements on leucite, Figure 13. These results for cristobalite have been backed up by molecular dynamics simulations [45].

9 Rigid unit mode model and the soft mode theory

The idea of a RUM-driven phase transition is most easily illustrated with the model of a two-dimensional perovskite structure represented in Figure 10. If two neighbouring squares rotate by angles φ_1 and φ_2, we can write the energy to lowest order as

$$E(\varphi_1, \varphi_2) = \frac{1}{2} K(\varphi_1 + \varphi_2)^2 \tag{40}$$

This form expresses the fact that the only rotations that do not involve distortions of the squares are $\varphi_1 = -\varphi_2$. On the other hand, the greatest distortions of the squares are required when $\varphi_1 = \varphi_2$, which will lead to the greatest energy. The force constant K can be interpreted as the force required to distort the squares, the stiffness of the units. In the split-atom representation, the separation of the split atoms will be proportional to $\varphi_1 + \varphi_2$ to lowest order.

Now consider a three-dimensional framework of connected tetrahedra (or octahedra). For simplicity we subsume all the rigid-body motions of a tetrahedron into one parameter φ, and note that when two neighbouring units move together without distorting we can say that $\varphi_1 = -\varphi_2$. The energy of the whole structure is given as

$$E_1 = \frac{1}{2} K \sum_{i,j} (\varphi_i + \varphi_j)^2 = K \sum_i \varphi_i^2 + K \sum_{i,j} \varphi_i \varphi_j \tag{41}$$

374

This function does not contain any forces to either encourage or discourage the rotations of the tetrahedra—it simply provides the energy that will encourage the tetrahedra to move in concert so as to avoid any distortions.

In general the high-temperature phases are fully expanded, so any rotations of the tetrahedra, whether as part of a local distortion or at a phase transition, will lead to a reduction in the volume. This is clearly seen in Figure 10. The long-range dispersive interactions between the highly-polarisable oxygen anions will favour as high a density as possible, and will therefore act as an inward pressure. Since the volume change will scale as $-\varphi^2$, we can express the effects of this inward pressure as the following energy:

$$E_2 = -\mathcal{P} \sum_i \varphi_i^2 \tag{42}$$

Any given structure cannot collapse indefinitely. Either neighboring oxygen atoms begin to get too close, or there may be a collapse of the framework about a cation which eventually inhibits further collapse. These effects lead to short-range repulsive forces. The harmonic parts can be subsumed within the parameters K and $\mathcal{P}$. However, short-range steric forces have a strong anharmonic component, so we can simply write the short-range energy as a lowest-order anharmonic energy:

$$E_3 = \beta \sum_i \varphi_i^4 \tag{43}$$

This term may also include the higher-order terms neglected in the expansion of equation (40). When we add together these three contributions, we obtain

$$E = E_1 + E_2 + E_3 = \sum_i \left\{ (K - \mathcal{P})\varphi_i^2 + \beta\varphi_i^4 \right\} + K \sum_{i,j} \varphi_i\varphi_j \tag{44}$$

The important point to note is that this exactly matches the simple model introduced in Section 4, where K plays the same role as J, and $\mathcal{P}$ and β play the parts of κ_2 and κ_4. Actually the signs of the last terms are different, but this is not important as it simply expresses the fact that in the simple model two neighboring atoms want to move the same way whereas here two neighboring tetrahedra will rotate in opposite senses. By identifying K with the strong stiffness of the tetrahedra, and $\mathcal{P}$ with the weak internal pressure due to the long-range dispersive interactions, we note that the model lies in the limit $K \gg \mathcal{P}$, which is equivalent to the displacive limit described in Section 4. This then provides a natural explanation for why many phase transitions found in silicates appear to behave as in the displacive limit and are then accurately described by Landau theory.

The transition temperature for our simple theory can be obtained by comparison with equation (34). From the analogy we find that the transition temperature is directly determined by K, the stiffness of the tetrahedra:

$$k_{\mathrm{B}}T_{\mathrm{c}} \propto K\mathcal{P} / \beta = K\varphi_0^2 \qquad (45)$$

where φ_0 is the rotation of the tetrahedra at 0 K. We have not included in this equation the prefactor corresponding to g in equation (34) which accounts for the detailed structure topology (it will be of order unity), but the use of a dispersion surface that will be different from (32) will give a different numerical value. The correct dispersion surface will account for the existence of lines and planes of RUMs [20]. We have recently performed a calculation for quartz using a realistic value for K and taking account of all topological factors, and have obtained a value of the transition temperature that is reasonably close to the experimental value [33].

It may seem paradoxical that the transition temperature is determined by the stiffness of the tetrahedra when the soft mode is a RUM that does not involve any distortion of the tetrahedra. The point is that the vibrational entropy comes from all the other phonons, whose frequencies in the simple model are determined by K. The transition temperature in the general case, equation (29), is determined by the ratio of the average ω^2 to the shift in ω^2, which is determined by the single anharmonic coefficient β.

In this simple approach we have considered the behaviour of only one phonon branch, namely that which contains the soft mode, and we have quietly neglected all the other phonon branches. Are these neglected phonon branches important? The general equation for the transition temperature, equations (20) and (29), involves contributions from all phonons, not just those on the branch containing the soft mode. However, lattice dynamics calculations for quartz [33] have shown that the contribution to the thermodynamic functions from the complete set of phonons are largely self-cancelling. It was found that the mean value of α_k is substantially lower than the magnitudes of the individual values of α_k, but with a wide spread of positive and negative values of α_k. Since equation (29) gives $T_{\mathrm{c}} \propto \langle \alpha_k \rangle^{-1}$, a small value of $\langle \alpha_k \rangle$ could lead to an extremely large ('uncontrolled') value of T_{c}. So we could ask why isn't $T_{\mathrm{c}} \sim \infty$? We are rescued by the branch containing the soft mode. Within the quasiharmonic approximation this branch is uncoupled from the other phonon branches. The anharmonic behavior of this branch is determined by the anharmonic coefficient β which operates equally for all wave vectors. Thus for this branch the mean value $\langle \alpha_k \rangle = \beta$ will not vanish, so that T_{c} will have a finite (i.e. controlled) value.

10 The origin of the driving force for displacive phase transitions in silicates

There are two components to the driving force for displacive phase transitions in silicates. First we have the coupling between neighboring atoms that allows the possibility of long-range ordering. This arises from the stiffness of the tetrahedra which leads to a local deformation propagating over large distances. Second we have a longer-range driving force that drives the deformation, which is described by the double-well potential $V(\eta)$. We have identified three contributions to $V(\eta)$ [33]. First is the effect of long-range interactions, which some preliminary calculations have suggested are mostly due to the long-range dispersive interactions between the highly-polarisable oxygen anions. These are attractive, and want to pull the structure in to the highest density possible. Generally high-temperature phases have structures of maximum volume, and any RUM distortion will lead to a lowering of the volume.

The second contribution to $V(\eta)$ is the short-range attraction between a cation such as K^+ or Ca^{2+}, which occupies a large cavity site, and neighboring oxygen anions. These interactions may lead to a collapse of the cavity about the cation, which will propagate over large distance. This collapse is limited by the size of the cation. This appears to be the crucial factor driving the phase transition in leucite [33,35].

The third contribution to $V(\eta)$ arises from the energy associated with the Si–O–Si bond angle (either Si could also be Al). This has an ideal value of ~145°, and bonds with angles that differ from this will have a higher energy. In some materials like cristobalite, the high-temperature phase appears to have a bond angle of 180°. In reality this will lead to a degree of disorder with neighboring tetrahedra rotating to try and reduce this angle, as apparently in cristobalite [44,45], but in order for as many bonds as possible to have the ideal bond angle the structure needs to undergo the displacive phase transition. Thus in cristobalite the energy associated with the Si–O–Si bond angle contributes to $V(\eta)$ as a driving force for ordering. On the other hand, there are cases where the Si–O–Si bond angle is already near its ideal value in the high-temperature phase, so the energy associated with this bond will either oppose the phase transition or select the phase transition given by the RUM distortion with the smallest effect on the Si–O–Si bond angle [33]. For example, in the high-temperature phase of quartz the Si–O–Si bond angle is already close to its ideal value, and only a few of the RUM distortions will give no first-order change in this angle [35]. Leucite is like quartz in having the Si–O–Si bond angle close to its ideal value in the high-temperature phase. In this case there is a first-order change in the bond angle caused by all the RUM distortions, but the RUMs that are associated with the phase transition are those for which the changes in the bond angle are smallest [33].

11 The nature of high-temperature phases

The average structure of the high-temperature phase of cristobalite deduced from diffraction studies has a linear Si–O–Si bond [46]. Since it has been established both by energy calculations [47–49] and by comparison with other silicates that linear bonds are energetically unfavourable compared to bonds with an angle of around 145°, it has been suggested that there must be some degree of disorder in the high-temperature phase. For example, the Si–O bonds may be tilted at an angle to the {111} directions, but their orientations about these directions must be disordered in order to be consistent with the crystal symmetry. This interpretation is consistent with measurements of the lattice parameters [50] at high temperatures, and is supported by a recent measurement of the radial distribution function for the high-temperature phase of cristobalite. The preliminary data are shown in Figure 15. I have marked on the positions of the interatomic distances given by the ideal structure. The actual Si–O and O–O nearest-

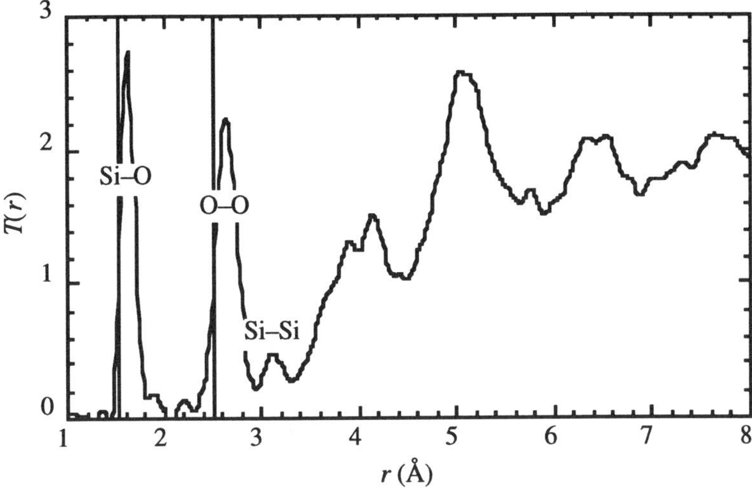

FIGURE 15. Pair distribution function for the high-temperature crystalline phase of cristobalite. The vertical lines indicate the interatomic distances that would be found if the ideal cubic phase was a realistic model for the local structure. The original data were obtained using the LAD spectrometer at ISIS.

neighbour distances are longer than those predicted by the ideal structure, which implies that the local structure is rather different from the average structure, although the Si–Si distance is the same as that given by the average structure. These data are consistent with an average rotation of the SiO_4 tetrahedra of 17° away from their average high-symmetry orientations, indicating the existence of considerable local disorder.

The problems come when attempts are made to produce a model that gives this disorder. For the Si–O bonds are not independent entities, but are part of the nearly-rigid SiO_4 tetrahedra. Therefore the motion of one Si–O bond will be strongly coupled to the motions of the other Si–O bonds in its tetrahedron. Two models have been proposed that are based on the existence of domains of lower symmetry structures [51,52]. Part of the appeal of domain models is that they provide the means of accounting for disorder of the orientations of the Si–O bonds and the correlations between connected Si–O bonds at the same time. Any distortions of the tetrahedra only occur within the walls between domains. The RUM model provides a way of building up a domain wall with only a minimum amount of distortion of the tetrahedra. If RUMs exist over a line of plane of wave vectors, it will be possible to construct a linear combination of RUM distortions as a Fourier summation over all wave vectors, leading to a distortion pattern in real-space [35,44,53]. This distortion pattern could be a set of domains with domain walls. But if there is enough freedom in the Fourier summation, the sizes of the domains and the sizes of the domain walls could be quite arbitrary without significant distortions of the tetrahedra. Thus the RUM model allows the distinct domains to evaporate, leaving a dynamically disordered phase that is generated by a superposition of a large number of large-amplitude RUM phonons [44].

The RUM model as described in this chapter is strictly a harmonic model, appropriate for the limit of infinitesimal rotations and translations. Since rotations cannot be added vectorially, when the RUM amplitudes are large there will be a large anharmonic interaction between different RUMs, and it is not immediately obvious that the RUM model as sketched here will apply when the rotational amplitudes are large as in cristobalite [54]. The experimental data discussed in Section 8, particularly the diffuse electron scattering, indicate that the RUM picture retains its validity when the RUMs have finite amplitudes. Manoj Gambhir (poster presented at this ASI) has been performing some molecular dynamics simulations with the split-atom model, and has reached the same conclusion, although the anharmonic interactions least affect the RUMs with wave vectors along [1,1,1].

12 Thermal expansion

We have made a lot of the fact that a small rotation of a rigid unit will drag in its local environment, leading to a net reduction in volume. In a dynamic situation the tetrahedra will be continuously rotating back and forth due to the propagating RUMs. The equilibrium rotation from one RUM will be given by

$$\left\langle \theta^2 \right\rangle \propto k_{\mathrm{B}} T \, / \, \omega^2 \tag{46}$$

This will lead to a net volume change:

$$\Delta V \propto -\left\langle \theta^2 \right\rangle \propto -k_{\mathrm{B}} T \sum \left| \mathbf{e}_k \right|^2 \omega_k^{-2} \tag{47}$$

where we now sum over all RUMs, and $\mathbf{e}_k$ is the rotational component of the eigenvector of each RUM. The important point is that the RUMs give a net reduction in volume on increasing temperature, thereby giving a negative contribution to the thermal expansion. If this is larger than the positive expansion of the Si–O and Al–O bonds the overall thermal expansion will be negative. Examples where this is the case, albeit for single directions, are quartz at high temperature, and cordierite. Recently it was discovered that ZrW_2O_6 has an isotropic negative thermal expansion at all temperatures, and we were able to suggest a RUM mechanism for this [55].

13 Local deformations in zeolites

We remarked in Section 7 that some zeolites have more than one RUM for each wave vector. This allows the formation of a local RUM distortion

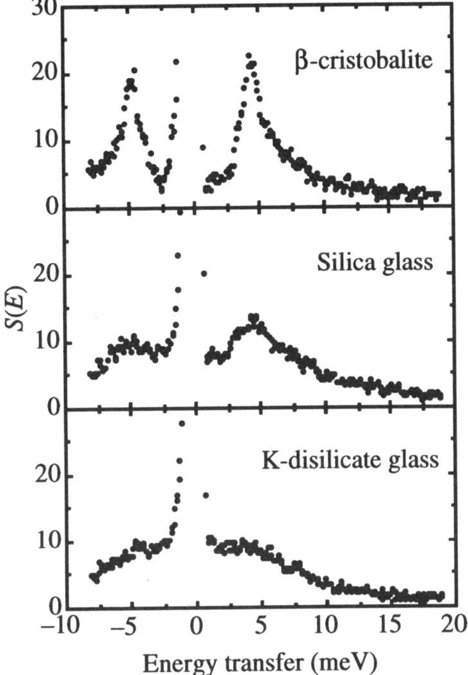

FIGURE 16. Inelastic neutron spectra for polycrystalline cristobalite and two amorphous silicates. The data were collected on the PRISMA spectrometer at ISIS.

$$\theta(\mathbf{r}) = \sum_{\mathbf{k}} c(\mathbf{k})\theta(\mathbf{k})\exp(i\mathbf{k}\cdot\mathbf{r}) \tag{48}$$

where $\theta(\mathbf{k})$ is the rotational component of the RUM with wave vector $\mathbf{k}$, and $c(\mathbf{k})$ is the weighting and phase factor for each RUM. We have constructed local deformations in some standard zeolites and have found that with only 100 RUMs, a RUM distortion can be localised on a single cavity in the zeolite structure [37]. This easy way to develop localised deformations of the structure may provide an understanding of the catalytic properties of zeolites.

14 Glasses

Finally we should link these ideas back to glasses. Let us start with silica glass. The counting of constraints might imply that there will be no RUMs in glasses. This idea gains some support by noting that our CRUSH analysis indicates the absence of any RUMs in some low-temperature phases [35]. However, there will be static fluctuations in the density across a silica glass, giving regions where the local environment may

380

allow some RUMs to propagate. On the other hand, the disorder present in glasses may allow some tetrahedra to have one or two non-bridging bonds, which could lead to a reduction in the net total of constraints.

In Figure 16 I show some recent inelastic neutron scattering data for amorphous silica, the low-temperature phase of cristobalite, and a potassium disilicate glass. The data have been compiled from many individual spectra, each with a different parabolic trajectory in $Q - E$ space. The data have been corrected for factors dependent on wave vector, and for multiple scattering, and the final spectra are proportional to $g(E)/E^2$, where $E = \hbar\omega$. The different spectra are on a common intensity scale. The cristobalite sample sets a baseline, since the CRUSH analysis shows that there are very few RUMs in this phase. Two points emerge from these data. First is that there are clearly more low-energy modes in silica glass than in cristobalite, showing that there is a non-zero number of RUMs allowed in the glass network despite the absence of symmetry that normally might be expected to lower the number of independent constraints. It should be noted that, apart from this difference, the spectra for cristobalite and silica glass are remarkably similar, including the existence of the peak at around 5 meV that is conventionally associated with the 'Boson peak'. The second point is that there are many low-energy modes in the K-disilicate glass, so many in fact that the 5 meV peak is partly masked. Given that there are many non-bridging Si–O bonds in this sample, it is not surprising that there are so many RUMs possible, for there are many more internal degrees of freedom than constraints. What is significant is that this spectrum provides an estimate of the energy scale for the RUMs, in the range of 0–5 meV, which closely corresponds to the frequency range of 0–1 THz for RUMs in the crystalline silicates discussed in Section 8.

What is the origin of the low-energy modes in the silica glass that correspond to the RUMs we have explored in the crystalline silicates? There are two possibilities. One is the existence of non-bridging bonds as mentioned above. Another is the existence of low-density regions in the glass structure where there are large rings of linked tetrahedra. Our original impression was that the lack of symmetry in any glass structure would most likely lead to the lack of degenerate constraints. We cannot yet tell which possibility is most likely. This is an issue that needs to be tackled by careful analysis of neutron scattering data from glasses, probably by use of appropriately-constrained Reverse Monte Carlo methods. However, what is clear is that when the origin of the low-energy excitations is understood, some of the ideas we have sketched in this chapter, particularly (but not only) the ideas on thermal expansion, may have important applications.

15 Acknowledgements

The work described in this chapter owes an awful lot to a number of co-workers. First is Volker Heine who helped initiate the development of the RUM model. Andrew Giddy did a lot of the early development work; in particular he set up the initial implementation of the CRUSH program and developed the analysis programs. Kenton

Hammonds became the custodian of the package, and made many useful improvements. Kenton also pushed through many of the applications to specific materials. Ian Swainson did a lot of the experimental work on cristobalite. Alix Pryde and Manoj Gambhir are presently taking the model further, answering some of the remaining questions and applying the model to a greater quantitative extent. Alex Hannon (ISIS, UK) helped in the collection and analysis of the $S(Q)$ data for cristobalite, and he and Mark Harris (ISIS, UK) collected and analysed the inelastic neutron scattering data for the silicate glasses. I am grateful to Ray Withers (Canberra) for providing the electron diffration photographs. Without any doubt the original inspiration for our work was given by Marcel Vallade (Grenoble). The work on the RUM model has been supported by NERC and EPSRC (UK), with some of the experiments being supported by the ISIS neutron scattering facility (Rutherford Appleton Laboratory, UK). To all these people and organisations I can only say "Thank you". Finally, I also owe a great debt of thanks to my wife Kate and daughters Jennifer-Anne, Emma-Clare and Mary-Ellen, the unseen heroes in this endeavour.

16 References

[1] Dolino, G. (1990) The α–inc–β transitions of quartz: A century of research on displacive phase transitions, *Phase Transitions*, 21, 59–72.

[2] Raman, C. V. and Nedungadi, T. M. K. (1940) The α–β transformation of quartz, *Nature*, **145**, 147.

[3] Saksena, B. D. (1940) Analysis of the Raman and infrared spectra of α-quartz, *Proceedings of the Indian Academy of Science*, **A12**, 93–139.

[4] Bruce, A. D. and Cowley, R. A. (1981) *Structural Phase Transitions*, Taylor and Francis, London.

[5] Salje, E. K. H. (1992) Application of Landau theory for the analysis of phase transitions in minerals, *Physics Reports*, **215**, 49–99.

[6] Dove, M. T. (1997) Theory of displacive phase transitions in minerals, *American Mineralogist* (in press).

[7] Dove, M. T. (1993) *Introduction to Lattice Dynamics*, Cambridge University Press, Cambridge.

[8] Jona, F. and Shirane, G. (1962) *Ferroelectric Crystals*, Pergamon, Oxford

[9] Lines, M. E. and Glass, A. M. (1977) *Principles and Applications of Ferroelectrics and Related Materials*, Clarendon Press, Oxford.

[10] Blinc, R. and Zeks, B. (1974) *Soft Modes in Ferroelectrics and Antiferroelectrics*, North Holland, Amsterdam.

[11] Cochran, W. (1960) Crystal stability and the theory of ferroelectricity, *Advances in Physics*, 9, 387–423.

[12] Cowley, R. A., Buyers, W. L., and Dolling, G. (1969) Relationship of normal modes of vibration of strontium titanate and its antiferroelectric phase transition at 110 K, *Solid State Communications*, 7, 181–184.

[13] Shirane, G. and Yamada, Y. (1969) Lattice-dynamical study of the 110 K phase transition in $SrTiO_3$, *Physical Review*, **177**, 858–863.

[14] Fleury, P. A., Scott, J. F., and Worlock, J. M. (1968) Soft phonon modes and the 110 K phase transition in $SrTiO_3$, *Physical Review Letters*, **21**, 16–19.

[15] Salje, E. K. H. (1992) Hard mode spectroscopy: experimental studies of structural phase transitions, *Phase Transitions*, **37**, 83–110.

[16] Redfern, S. A. T. and Salje, E. (1987) Thermodynamics of plagioclase. 2. Temperature evolution of the spontaneous strain at the $I\bar{1} - P\bar{1}$ phase transition in anorthite, *Physics and Chemistry of Minerals*, **14**, 189–195.

[17]   Giddy, A. P., Dove, M. T., and Heine, V. (1989) What does the Landau free energy really look like for structural phase transitions? *Journal of Physics: Condensed Matter*, **1**, 8327–8335.

[18]   Padlewski, S., Evans, A. K., Ayling, C., and Heine, V. (1992) Crossover between displacive and order-disorder behavior in the Φ^4 model, *Journal of Physics: Condensed Matter*, **4**, 4895–4908.

[19]   Radescu, S., Etxebarria, I., and Perezmato, J. M. (1995) The Landau free-energy of the 3-dimensional Φ^4 model in wide temperature intervals, *Journal of Physics: Condensed Matter*, **7**, 585–595.

[20]   Sollich, P., Heine, V., and Dove, M. T. (1994) The Ginzburg interval in soft mode phase transitions: Consequences of the Rigid Unit Mode picture, *Journal of Physics: Condensed Matter*, **6**, 3171–3196.

[21]   Grimm, H. and Dorner, B. (1975) On the mechanism of the α–β phase transformation of quartz, *Physics and Chemistry of Solids*, **36**, 407–413.

[22]   Megaw, H. D. (1973) *Crystal structures: a working approach*, W. B. Saunders, Philadelphia.

[23]   Boysen, H., Dorner, B., Frey, F., and Grimm, H. (1980) Dynamic structure determination for two interacting modes at the M-point in α- and β-quartz by inelastic neutron scattering, *Journal of Physics C: Solid State Physics*, **13**, 6127–6146.

[24]   Berge, B., Bachheimer, J. P., Dolino, G., Vallade, M., and Zeyen, C. M. E. (1986) Inelastic neutron scattering study of quartz near the incommensurate phase transition, *Ferroelectrics*, **66**, 73–84.

[25]   Bethke, J., Dolino, G., Eckold, G., Berge, B., Vallade, M., Zeyen, C. M. E., Hahn, T., Arnold, H., and Moussa, F. (1987) Phonon dispersion and mode coupling in high-quartz near the incommensurate phase transition, *Europhysics Letters*, **3**, 207–212.

[26]   Dolino, G., Berge, B., Vallade, M., and Moussa, F. (1989) Inelastic neutron scattering study of the origin of the incommensurate phase of quartz, *Physica*, **B156**, 15–16.

[27]   Dolino, G., Berge, B., Vallade, M., and Moussa, F. (1992) Origin of the incommensurate phase of quartz: I. Inelastic neutron scattering study of the high temperature β phase of quartz, *Journal de Physique, I.* **2**, 1461–1480.

[28]   Vallade, M., Berge, B., and Dolino, G. (1992) Origin of the incommensurate phase of quartz: II. Interpretation of inelastic neutron scattering data, *Journal de Physique, I.* **2**, 1481–1495.

[29]   Dove, M. T., Giddy, A. P., and Heine, V. (1992) On the application of mean-field and Landau theory to displacive phase transitions, *Ferroelectrics*, **136**, 33–49.

[30]   Dove, M. T., Giddy, A. P., and Heine, V. (1993) Rigid unit mode model of displacive phase transitions in framework silicates, *Transactions of the American Crystallographic Association*, **27**, 65–74.

[31]   Giddy, A. P., Dove, M. T., Pawley, G. S., and Heine, V. (1993) The determination of rigid unit modes as potential soft modes for displacive phase transitions in framework crystal structures, *Acta Crystallographica*, **A49**, 697–703.

[32]   Hammonds, K. D., Dove, M. T., Giddy, A. P., and Heine, V. (1994) CRUSH: A FORTRAN program for the analysis of the rigid unit mode spectrum of a framework structure, *American Mineralogist*, **79**, 1207–1209.

[33]   Dove, M. T., Heine, V., and Hammonds, K. D. (1995) Rigid unit modes in framework silicates, *Mineralogical Magazine*, **59**, 629–639.

[34]   Dove, M. T., Hammonds, K. D., Heine, V., Withers, R. L., Xiao, Y., and Kirkpatrick, R. J. (1996) Rigid unit modes in the high-temperature phase of SiO_2 tridymite: calculations and electron diffraction, *Physics and Chemistry of Minerals*, **23**, 55–61.

[35]   Hammonds, K. D., Dove, M. T., Giddy, A. P., Winkler, B., and Heine, V. (1996) Rigid unit phonon modes and structural phase transitions in framework silicates, *American Mineralogist*, **81**, 1057–1079.

[36]   Dove, M. T., Gambhir, M., Hammonds, K. D., Heine, V., and Pryde, A. K. A. (1996) Distortions of framework structures, *Phase Transitions* (in press).

[37]   Hammonds, K. D., Deng, H., Heine, V., and Dove, M. T. (1996) How floppy modes give rise to adsorption sites in zeolites, *Physical Review Letters* (submitted).

[38]   Pawley, G. S. (1972) Analytic formulation of molecular lattice dynamics based on pair potentials, *Physica Status Solidi*, **49b**, 475–488.

[39]   The programs and documentation are available from `http://www.esc.cam.ac.uk/`.

[40] Hua, G. L., Welberry, T. R., Withers, R. L., and Thompson, J. G. (1988) An electron diffraction and lattice-dynamical study of the diffuse scattering in β-cristobalite, SiO_2, *Journal of Applied Crystallography*, **21**, 458–465.

[41] Welberry, T. R., Hua, G. L., and Withers, R. L. (1989) An optical transform and Monte Carlo study of the disorder in β-cristobalite SiO_2, *Journal of Applied Crystallography*, **22**, 87–95.

[42] Withers, R. L., Thompson, J. G., Xiao, Y., and Kirkpatrick, R. J. (1995) An electron diffraction study of the polymorphs of SiO_2-tridymite, *Physics and Chemistry of Minerals*, **21**, 421–433.

[43] Boysen, H. (1990) Neutron scattering and phase transitions in leucite, in E. K. H. Salje, *Phase transitions in ferroelastic and co-elastic crystals*, Cambridge University Press, 334–349.

[44] Swainson, I. P. and Dove, M. T. (1993) Low-frequency floppy modes in β-cristobalite, *Physical Review Letters*, 71, 193–196.

[45] Swainson, I. P. and Dove, M. T. (1995) Molecular dynamics simulation of α- and β-cristobalite, *Journal of Physics: Condensed Matter*, **7**, 1771–1788.

[46] Schmahl, W. W., Swainson, I. P., Dove, M. T., and Graeme-Barber, A. (1992) Landau free energy and order parameter behaviour of the α–β phase transition in cristobalite, *Zeitschrift für Kristallographie*, **201**, 125–145.

[47] Lasaga, A. C. and Gibbs, G. V. (1987) Applications of quantum mechanical potential surfaces to mineral physics calculations, *Physics and Chemistry of Minerals*, **14**, 107–117.

[48] Lasaga, A. C. and Gibbs, G. V. (1988) Quantum mechanical potential surfaces and calculations of minerals and molecular clusters, *Physics and Chemistry of Minerals*, **16**, 29–41.

[49] Gibbs, G. V., Downs, J. W., and Boisen, M. B. (1994) The elusive Si–O bond, *Reviews in Mineralogy*, **29**, 331–368.

[50] Swainson, I. P. and Dove, M. T. (1995) On the thermal expansion of β-cristobalite, *Physics and Chemistry of Minerals,* **22**, 61–65.

[51] Wright, A. F. and Leadbetter, A. J. (1975) The structures of the β-cristobalite phases of SiO_2 and $AlPO_4$, *Philosophical Magazine*, **31**, 1391–1401.

[52] Hatch, D. M. and Ghose, S. (1991) The α–β phase transition in cristobalite, SiO_2, *Physics and Chemistry of Minerals*, **17**, 554–562.

[53] Swainson, I. P. and Dove, M. T. (1993) Comment on "First-principles studies on structural properties of β-cristobalite", *Physical Review Letters*, **71**, 3610.

[54] Liu, F, Garofalini, S. H., Kingsmith, R. D., and Vanderbilt, D. (1993) Comment on "First-principles studies on structural-properties of β-cristobalite" – Reply, *Physical Review Letters*, **71**, 3611.

[55] Pryde, A. K. A., Hammonds, K. D., Dove, M. T., Heine, V., Gale, J. D., and Warren, M. C. (1996) Origin of the Negative Thermal Expansion in ZrW_2O_8 and ZrV_2O_7, *Science* (submitted).

VIBRATIONAL DYNAMICS IN GLASSES

C. LEVELUT, F. TERKI, Y. SCHEYER AND J. PELOUS
Laboratoire des Verres
Université Montpellier II
Place Eugène Bataillon, case 069
34095 Montpellier Cedex
France

1. Introduction

The dynamics of glasses and glass transition has received considerable attention in recent years. A particular effort has been directed to the frequency range from 5 to 100 cm^{-1} (0.15 to 3 THz) in the Raman or neutron scattering spectra which is expected to reflect the cooperative motions of atoms over correlation lengths in the nanometer range. In this frequency range the inelastic neutron and Raman scattering data are dominated below the glass transition temperature T_g by a strong broad peak around $20 - 50cm^{-1} \equiv 0.6 - 1.5THz$ called the boson peak (BP) because its intensity varies with the temperature like the Bose factor and above T_g by a strong quasielastic line below $10cm^{-1} \equiv 0.3THz$.

Several models have been proposed for the interpretation of the maximum in the vibrational density of states responsible for the BP. Some authors associated this peak to a characteristic frequency of clusters [1, 2] or to the localization of the phonons by disorder through a mechanism of phonon scattering by density fluctuations or by coupling with low energy excitations [3, 4, 5, 6]. Most of these models associate the position of the boson peak ω_{max} to a length [7, 8] defined as $\ell \propto V/\omega_{max}$ where V is the sound velocity. This length ℓ has been associated with a cluster size, a correlation length over which short- and medium-range order are maintained [6], or a localization length in the Ioffe-Regel definition [6, 5]. ℓ ranges from 1 to 2 nm, this scale corresponds to medium range order which also relates to another quasi universal feature, the so-called First Sharp Diffraction Peak [9, 10, 11]. Many authors discussed the correlation between V and ω_{max} (*i.e.* $\ell = cste$) or between ℓ and the lengths deduced from the position Q_1 or the

385

M. F. Thorpe and M. I. Mitkova (eds.), Amorphous Insulators and Semiconductors, 385–394.
© *1997 Kluwer Academic Publishers. Printed in the Netherlands.*

width ΔQ of the FSDP ($d_1 = 2\pi/Q_1$ and $d_2 = 2\pi/\Delta Q$). We present two studies illustrating a possible way to induce modifications of the medium-range-order without changing the short-range-order i. e. studying the effect of preparation condition for a fixed chemical composition.

Above T_g, the spectra are dominated by the quasielastic contribution. This contribution exhibits a very strong temperature dependence and is assigned to relaxational modes. Several models attempted to consider both relaxational and vibrational contributions [12, 13]. In one of them, the authors assume that the vibrational excitations seen as the boson peak at $T \ll T_g$ are strongly coupled to some relaxing variable which induces a broadening of each frequency of the BP [12]. The "vibration-relaxation" model [13] suggests a transformation from more or less damped vibrations at high frequencies to low-barrier relaxations at low frequencies. This model is a simplified classical version of the soft potential model for glasses [14, 15]. In the framework of the two-level systems or the soft potential model, a proportionality relationship between the attenuation α and the Raman reduced intensity I_R have been predicted if the two physical properties are related. The relationship affects either the total [16] or the relaxational [17] reduced intensity. These models have been introduced to explain the low temperature properties of glasses. In this paper, we present a correlation between I_R and α for several samples at fixed temperature (section 2) and we extend this connection to the high temperature regime in a strong glass (section 3).

On another hand, many discussions also focus on the relaxational processes and their temperature dependence across the liquid-glass transition. New steps in the investigations were stimulated by the achievement of the so-called mode coupling theory (MCT) [18]. Two relaxational processes are expected for a single variable: a slow (α) relaxation with a strong temperature dependence, and a fast (β) relaxation varying slowly with the temperature. A quantitative agreement of the predictions of this theory is fulfilled for fragile (in Angell's classification [19]) glasses. Nevertheless, for strong glass-formers, a quantitative discrepancy between the predictions of the MCT and the experimental results, assigned to the increase of the vibrational contribution not taken into account by the MCT [20], is observed [21]. Depending on the experimental technique, the measurements can be sensitive to the α or the β process or to both. The comparison of the times measured by different techniques as well as the interpretation of the relaxational processes are much debated. Brillouin light scattering is a useful tool to probe these processes in the $10^{-8} - 10^{-10}$ s range. The relaxational process observed in the quasielastic Raman intensity has often been identified with the β relaxation of the MCT [22]. Other authors proposed a different approach where this so-called fast process is due to an increase of the density of states at low energy when the temperature is raised toward the glass transition [23, 24]. This β_{fast} process is then different from the β_{slow} of the MCT.

2. Relation between low-frequency dynamics and medium-range order versus fictive temperature

The two examples of light scattering studies presented here illustrate the effect of preparation conditions or impurities on series of glasses with a fixed chemical composition [25, 26].

The first investigation was carried out on a series of barium-crown glasses of various fictive temperature T_f in the transformation range. Each sample was kept at a temperature T_f in the transformation range during a time sufficiently long to allow its physical properties to stabilize at their equilibrium value for that temperature. The sample was then rapidly quenched to room temperature. Secondly, we investigated samples of vitreous silica of different origins, as described in ref [26] (tetrasil, puropsil, sol-gel, fiber, irradiated quartz and silica). The samples differ either by their OH content or their fictive temperature. For the barium crown glasses, the velocity and attenuation of hypersounds as measured by Brillouin scattering depends strongly upon T_f: the former decreases by about 5%, while the latter increases by 50% when T_f increases over the transformation range $590 - 740^\circ C$ [27]. For the silica samples, the values of transverse and longitudinal hypersonic velocities (V_L and V_T) as well as the longitudinal attenuation (α_L) are reported in table I. V_L and V_T decrease with increasing OH-content whereas α_L increases. In contrast to the barium crown glasses, the velocities decrease with increasing fictive temperature; this variation is related to the well-known anomaly in elastic properties of vitreous silica. The irradiated quartz and silica show a larger sound velocity and a lower sound wave attenuation than all the other samples. The effect of neutron-irradiation has already been partly explained by a very high fictive temperature.

Figure 1 shows the reduced Raman intensity for two barium crown glasses stabilized at the limits of the transformation range. Above ω_{max}, the two curves coincide almost perfectly. At lower frequencies, I_R is larger for the sample of higher T_f. This difference is not temperature dependent and reflects changes in the vibrational properties of the glass. In these samples, ω_{max} increases with increasing sound velocity (both quantities decreases with increasing T_f). The reduced Raman intensities for tetrasil, puropsil and sol-gel samples are very similar except for a small softening of modes for the sol-gel silica in the whole range of measurements. The figure 2a shows a hardening of modes for the fiber compared to the puropsil and a

TABLE 1. Hypersonic velocity and attenuation for the silica samples deduced from the Brillouin data and position of the maximum of the boson peak obtained by fitting the VH polarized Raman spectra. The velocities V_L and V_T are in ms^{-1}, the attenuation α_L and the position of the BP in cm^{-1}, the characteristic lengths in $\mathring{A}$. ℓ is calculated using the Debye sound velocity. The errors bars are $10ms^{-1}$ for V_L, $40ms^{-1}$ for V_T, $300cm^{-1}$ for α, $0.02\mathring{A}$ for d_1, $1\mathring{A}$ for d_2. The errors bars for ω_{max} correspond to the upper and lower limits of the 90 % confidence interval deduced from the fitting procedure.

sample	V_L	α_L	V_T	ω_{max}	ℓ	d_1	d_2
sol-gel	5910	1950	3810	52.3 ± 5	26.1 ± 2.5	4.16	21
tetrasil	5960	1800	3750	58.3 ± 4	23.6 ± 2	4.16	21
puropsil	5970	1550	3780	58.5 ± 3	23.6 ± 1.5	4.16	21
fiber	6015	-	3900	59.2 ± 3	23.5 ± 1.5	4.16	21
irr. quartz	6160	650	3885	72.4 ± 5	19.7 ± 2.5	3.95	16
irr. silica	6185	630	-	75 ± 6	19 ± 3	3.95	16

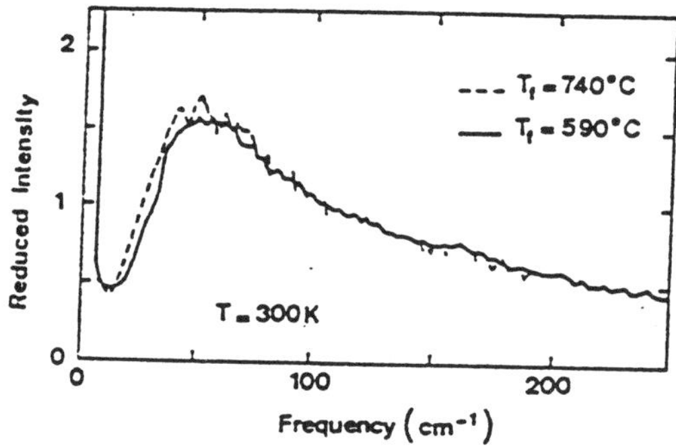

Figure 1. Reduced Raman intensity for two barium-crown dense glasses of different fictive temperature.

strong hardening associated with a decrease of the width of the peak in the $250 - 450cm^{-1}$ range for the irradiated samples. These changes are similar to the effect of high temperature pressure induced compaction [28]. The values of ω_{max} reported in table I were determined by taking into account a lorentzian to describe the relaxation. Therefore, the vibrational and relaxation contributions are described by independent parameters, this is the so-called superposition model [29]. ω_{max} decreases with decreasing sound velocity.

For the barium crown glasses, the measurements of the FSDP demonstrate that the position and width are independent upon T_f. Therefore, there is no correlation between the FSDP and the BP for this glass: d_1

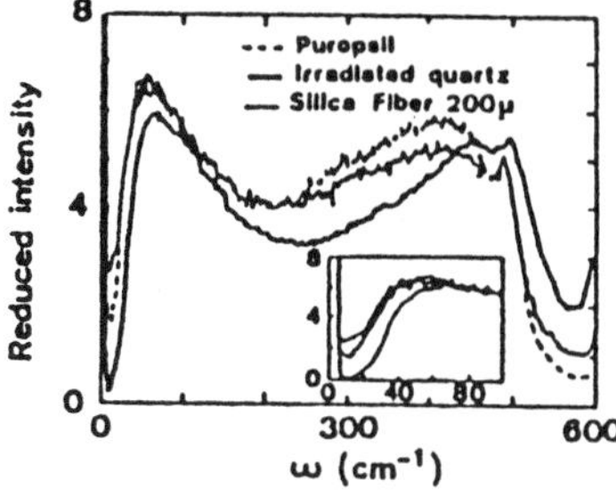

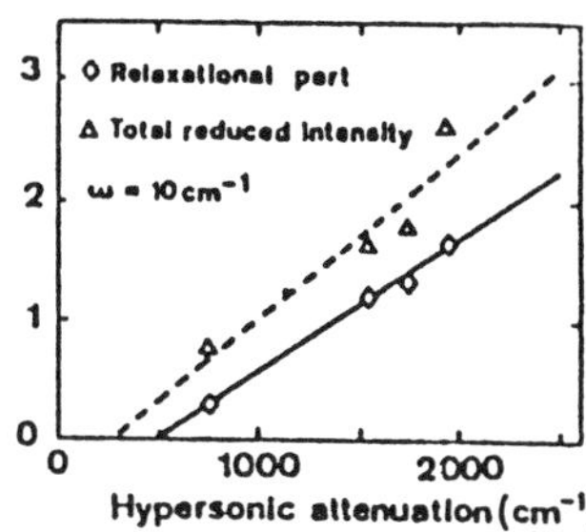

Figure 2. a-Comparison of the VV polarized Raman reduced Stokes intensity $I_R(\omega) = I(\omega)/\omega(n(\omega, T) + 1)$ for puropsil, tetrasil fiber and irradiated quartz. b-Total intensity and relaxational contribution to reduced intensity in the VH polarization at $10cm^{-1}$ for four silica samples versus hypersonic attenuation at about $1cm^{-1}$. The lines are the results of linear regressions.

and d_2 are constant whereas ℓ increases with increasing T_f. For the vitreous silica, the measurements of the FSDP show very little changes. This indicates that, in those two strong glass-formers, the MRO is very similar for glasses formed at different temperatures in the transformation range. The variation of OH content and T_f does not change ℓ more than 10 % (ℓ increases with increasing OH content and decreases with increasing T_f). In contrast, noticeable changes are induced by fast-neutron irradiation: the irradiation produces a broadening of the FSDP, together with a shift to higher q-values. Thus d_1 and d_2 decreases by about 30 %. The characteristic length ℓ also decreases by 30 % under the effect of irradiation, as already observed by Konstantinov [30]. This results leads to a possible correlation between the FSDP and MRO but for silica samples only. However the variation of ω_{max} and V are correlated for none of the samples presented in this paper. Concerning the correlation between hypersonic attenuation and Raman intensity, there are very few experimental evidences using Brillouin measurements of the attenuation in the GHz range, although a recent study establishes that the temperature and frequency dependence of the quasielastic scattering agree qualitatively with the predictions of the soft potential model (i. e. proportionality with the hypersonic attenuation) [31]. Such correlation is quantitatively achieved for the barium crown glasses : both quantities increase in a similar extent with increasing T_f. In the silica samples, the intensities of the different samples are comparable in a large frequency range because there are very few changes in composition. A good correlation is found (figure 2b) using either the total Raman reduced intensity or its relaxational part but it is better if one takes only the relaxational contribution. In both cases, the linear law intercepts the attenuation axis at a non-zero value: there are contributions to the attenuation due to the scattering of acoustic waves by non-propagating defects which

390

do not contribute to the quasielastic intensity.

Finally, in these two series of strong glasses with fixed chemical composition, the BP is not clearly correlated to the medium-range-order characteristic probed by the FSDP nor to the sound velocity. In contrast, the proportionality relationship between the attenuation and the relaxational intensity is fulfilled.

3. Influence of the temperature: competition between relaxation and vibration in a strong optical glass

In this section, we present a light scattering study carried out as a function of the temperature (across the glass-transition) [32]. We emphasize a careful comparison of the models in order to demonstrate that more attention must be devoted to the determination of the position of the BP and of the relaxational intensity which are crucial points in checking the reliability of the theoretical predictions mentioned in the introduction.

The longitudinal and transversal sound velocities decrease with increasing temperature, slowly up to about $735K$ (resp. 695 K for V_T) and then a major break of slope occurs below T_g ($\simeq 800K$). The VH reduced Raman intensities corresponding to various temperature are presented in figure 3. The variations of ω_{max} are similar to that of the velocities. However, the break of slope occurs at significantly lower temperature than that of the velocities. The decrease of the boson peak frequency above the break of slope depends on the approach used for the extraction of vibrational modes. We used two models, in addition to the simple determination of the apparent position of the BP: the superposition model (SM) defined above and the coupling model (CM), which introduces correlations between the vibrational and relaxational contributions. A quantitative comparison was done by comparing the values of the relative slopes (RS) $\frac{dV}{dT} \times \frac{1}{V}$ and $\frac{d\omega_{max}}{dT} \times \frac{1}{\omega_{max}}$ after the break. The RS should be equal if V and ω_{max} are proportional. The absolute value of the RS for ω_{max} decreases by 60% when a relaxational contribution is added and further decreases by 20% if the coupling between the two contributions is taken into account. A qualitative correlation is then achieved between the variation of ω_{max} using the CM and the transversal and longitudinal velocity. Nevertheless, the correlation is better for the transversal one. The VH Raman spectra are more sensitive to depolarized modes which probably originate from transverse acoustic modes. Moreover, we expect that the maximum of the vibrational density of state is dominated by transversal modes by analogy with crystals where it is related to the Debye velocity which reflects mainly the variations of the transversal modes.

The hypersonic longitudinal attenuation is nearly constant up to $770K$

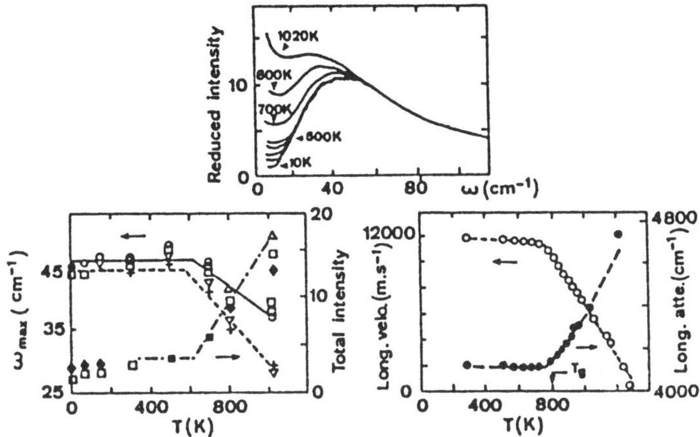

Figure 3. Reduced Raman intensity for BaF4 from 10 to 1020K (a), hypersonic longitudinal velocity and position of the boson peak (b), longitudinal hypersonic attenuation and Raman quasielastic intensity (c) as a function of temperature.

and then increases sharply by a factor six when the temperature increases up to $1000K$. The transverse attenuation, measured less accurately, has similar variations. The relaxational part of the reduced attenuation has been determined from the values of the fit parameters for the Lorentzian simulating the relaxation in the SM or after subtracting an "experimental" vibrational contribution deduced from the low temperature spectra (scaled by the ratio of ω_{max} or velocities to account for the softening with increasing temperature). All the methods produce a temperature-dependence very similar to the one of α_L. The position of the break of slope (770 K) for the attenuation, close to that of ω_{max}, is higher than that of the Raman intensity ($550 - 650K$). The softening of the boson peak occurs as the fast relaxation sets in and below T_g. Such behavior, already observed by Buchenau et al [13] in fragile glasses is explained by a transformation of vibrational modes to relaxational ones as the temperature increases. Above the break of slopes, the comparison of the relative slopes yields a 60% higher value for the attenuation at $1cm^{-1}$ than for the total or relaxational part of the Raman intensity at $10cm^{-1}$. The frequency dependence of the RS for the Raman intensity is too weak to account for this discrepancy. We expect that the transverse attenuation rather than the longitudinal one has the same origin as the VH Raman intensity; thus, the Raman intensity should be better connected with the transverse attenuation.

We have demonstrated a strong correlation between the temperature dependence of V and that of ω_{max} on the one hand, and between the variations of α and I_R on the other hand. Therefore, the Raman features (at a

392

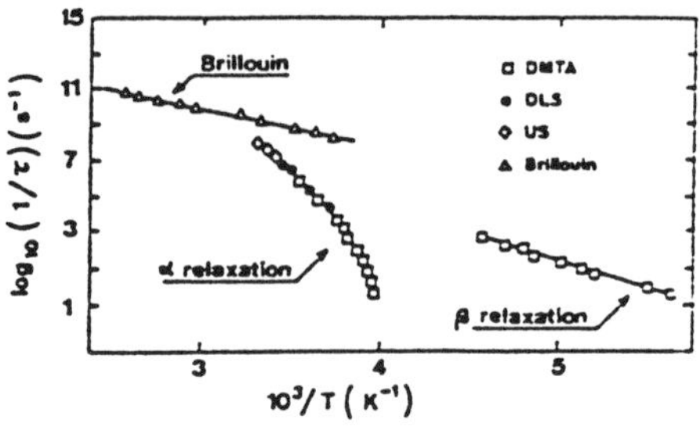

Figure 4. Relaxation times determined by Brillouin scattering compared to lower frequency determinations.

few tens of cm^{-1}) and the Brillouin data (at $1cm^{-1}$) are clearly connected suggesting that the same mechanism is responsible for the vibrational dynamics within the $1 - 50cm^{-1}$ range.

In this section, we present a light scattering study in a fragile glass. We used Brillouin scattering to investigate the relaxational processes above the glass transition on polyurethanes [33]. The specificity of these samples is to present a gelation transition when the ratio of two constituent is changed; this allows to investigate samples corresponding to various crosslink densities. We used two different chain length for one of the constituant to vary also the length between crosslinks. The samples obtained by condensation of polyoxypropylated triol and hexamethylene diisocyanate are characterized by the stoichiometric ratio: $r = \frac{[NCO]}{[OH]} \leq 1$. The glass transition temperature is in the 210-260K range. Brillouin scattering measurements were performed with the VV polarization in the $1.2T_g - 1.5T_g$ range.

The Brillouin spectra can be interpreted in terms of relaxational modes coupled to acoustical modes. The dynamical structure factor of a generalized linear hydrodynamic theory including non-exponential relaxation (through the introduction of a Cole-Davidson relaxation distribution function) is fitted to the complete spectra. For the sample $r = 1$ (fully gelled), the average relaxation time determined by Brillouin scattering (τ_{Br}) is compared in figure 4 with the relaxation time determined for α relaxation at lower frequencies by ultrasonic measurements, dynamic light scattering and dynamical mechanical thermal analysis [34]. τ_{Br} exhibits an Arrhenius behavior, in contrast to the lower frequency determination. This is in agreement with several Brillouin studies in glass-forming liquids [35, 36]. However, the amplitude of the relaxation and the parameter characteristic of the distribution width are the same as for the α relaxation. As in other polymers, the Brillouin probed relaxation splits from the α relaxation, the nature of the Brillouin relaxation is not yet elucidated. One possible explanation was previously suggested by Cummins [37] who interpreted this relaxation as an α process disturbed by the β process. For the samples synthesized from the same triol, the relaxation times are superimposed in a rescaled Arrhenius plot ($\log \tau_{CD}$ versus T_g/T) The influence of the crosslink density is fully accounted for by the changes T_g. If the molecular weight of the triol increases, the relaxation time decreases and does not vary anymore with r (the triol dominates and T_g is constant).

We have recently performed measurements of $S(q,\omega)$ across T_g by inelastic neutron scattering in these samples. The result demonstrates the existence of a vibrational contribution and suggests that the same analysis as used in strong glasses could be transposed to the fragile ones.

References

1. Phillips, J. C. (1981) Microscopic origin of anomalously narrow Raman lines in network glasses, *J. Non-Cryst. Solids* **431**, 347-355.
2. Duval, E., Boukenter A. and Achibat T. (1990) Vibrational dynamics and the structure of glasses, *J. Phys. : Condens. Matter* **2**, 10227-10234.
3. Akkermans E. and Maynard R. (1985) Weak localization and anharmonicity of phonons, *Phys. Rev. B* **32**, 7850-7862.
4. Schirmacher, W. and Wagener, M. (1993) Effective medium theory for phonons in glasses: the mean free path, in S. Hunklinger, W. Ludwig and G. Weiss (eds), *Phonons 89*, World Scientific, Singapore, p531-533.
5. Klinger, M. I. (1992) Correlation between atomic soft configurations (dynamics) and medium-range order in glasses, *Phys. Lett. A* **170**, 222-224.
6. Elliot, S. R. (1992) A Unified Model for the low-Energy Vibrational Behavior of Amorphous Solids, *Europhys. Lett* **19**, 201-207.
7. Martin, A. and Brenig, W. (1974) Model for Brillouin scattering in amorphous solids, *Phys. Status Solidi* **B64**, 163-172.
8. Malinovski, V. K. and Sokolov, A. P. (1986) The nature of the boson peak in Raman scattering in glasses, *Solid State* Commun. **57**, 757-761.
9. Sokolov, A. P., Kisliuk, A., Soltwisch M. and Quitmann,D. (1992) Medium-range-order in glasses: comparison of Raman and diffraction measurements, *Phys. Rev. Lett.* **69**, 1540-1543.
10. Novikov, V. N. and Sokolov, A. P. (1991) A Correlation between low-energy vibrational spectra and first sharp diffraction peak in chalcogenides glasses, *Solid State Commun.* **77**, 243-247.
11. Börjesson, L.,Hassan, A. K., Swenson, J., Torell, L. M. and Fontana, A. (1993) Is there a correlation between the first sharp diffraction peak and the low-frequency vibrational behavior of glasses?, *Phys. Rev. Lett* **70**, 1275-1278
12. Gochiyaev, V. Z., Malinovsky, V. K., Novikov, V. N. and Sokolov, A. P. (1991) Structure of the Rayleigh line wing in highly viscous liquids, *Philosophical Magazine B* **63**, 777-787.
13. Buchenau, U., Schönfield, C., Richter, D., Kanaya, T., Kaji, K. and Wehrmann, R., (1994) Neutron scattering study of the vibration-relaxation crossover in amorphous polycarbonates, *Phys. Rev. Lett* **73**, 2344-2346.
14. Karpov, V. G., Klinger, M. I., Ignat'ev, F. N. (1983) Theory of the low-temperature anomalies in the thermal properties of amorphous structures, *Sov. Phys. JETP* **57**, 439-448.
15. Parshin, D. A. (1994) Soft-potential model and universal properties of glasses (review) *Phys. Solid State* **36**, 991-1024.
16. Gurevich, V. L., Parshin, D. A., Pelous, J. and Schober, H. R. (1993) Theory of low energy Raman scattering in glasses, *Phys. Rev. B* **48**, 16318-16331.
17. Theodorakopoulos, N. and Jäckle, J. (1976) Low-frequency Raman scattering by

defects in glasses, *Phys. Rev.* B **14**, 2637-2641.

18. Götze, W. and Sjögren. L. (1992) Relaxation processes in supercooled liquids, *Rep. Progr. Phys.* **55**, 241-376.

19. Angell, C. A. (1988), Perspectives on the glass transition, *J. Phys. Chem. Solids* **49**, 863-871.

20. Sokolov, A. P., Rössler, E., Kisliuk, A. and Quitmann, D. (1993) Dynamics of strong and fragile glass formers: differences and correlation with low-temperature properties, *Phys. Rev. Lett.* **71**, 2062-2065.

21. Rössler,E., Sokolov, A. P., Kisliuk, A. and Quitmann, D. (1994) Low-frequency Raman scattering on different types of glass formers used to test predictions of mode-coupling theory, *Phys. Rev. B* **49**, 14967-14978.

22. Li, G., Du, M., Chen, X.K., Cummins, H. Z. and Tao, J. T., (1992) Testing mode-coupling predictions for α and β relaxation in $Ca_{0.4}K_{0.6}(NO_3)_{1.4}$ near the liquid-glass transition by light scattering, *Phys. Rev. A* **45**, 3867-3879

23. Buchenau, U., Galperin, Yu. M., Gurevich, V.L. and Schober, H. R., (1991) Interaction of soft modes and sound waves in glasses, *Phys. Rev. B* **43** 5039

24. Zorn, R., Richter, D., Frick, B. and Farago, B. (1993) Neutron scattering experments on the glass transition of polymers, *Physica A* **201** 52-66

25. Levelut, C., Gaimes, N., Terki, F., Cohen-Solal, G., Pelous, J. and Vacher, R. (1995) Glass-transition temperature: Relationship between low-frequency dynamics and medium-range order, *Phys. Rev. B* **51**, 8606-8609.

26. Terki, F., Levelut, C., Boissier, M. and Pelous, J. (1996) Low-frequency dynamics and medium-range order in vitreous silica, *Phys. Rev. B* **53**, 2411-2418.

27. Vacher, R., Delsanti, M., Pelous, J., Cecchi, L., Winter, A. and Zarzycki, J. (1982) A Brillouin scattering study of glasses of various fictive temperature, *J. Mater. Sci.* **9**, 829-834.

28. McMillan, P., Piriou, B. and Couty, R. (1984) A Raman study of pressure-densified vitreous silica, *J. Chem. Phys.* **81**, 4234-4236.

29. Krüger, M., Soltwisch, M., Petscherizin, I. and Quitmann, D. (1992) Light scattering from disorder and glass-transition-dynamics in $GeSBr2$, *J. Chem. Phys.* **96**, 52-63.

30. Konstantinov, A. V., Maksimov, L. V., Silin A. R. and Yanush, O. V. (1990) Medium-range order of radiation modified silica glasses studied by spectroscopic and optical method, *J. Non-Crystallin. Solids* **123**, 286-290

31. Brodin, A., Fontana, A., Börjesson, L., Carini, G. and Torell, L. M. (1994) Low-energy modes in phosphate glasses: a comparison with the soft potential model, *Phys. Rev. Lett.* **73**, 2067-2070.

32. Terki, F., Levelut, C., Prat, J.-L., Boissier, M. and Pelous, J. (1996) Low-frequency dynamics in an optical strong glass: vibrational and relaxational contributions, to be published

33. Levelut, C., Scheyer, Y., Boissier, M., Pelous, J., Durand. D. and Emery, J. R. (1996) A Brillouin scattering investigation of relaxation versus crosslink density in glass- and gel-forming polymers, *J. Phys. : Condens. Matter* **8**, 941-957.

34. Tabellout, M., Baillif, P.-Y., Randrianantoandro, H., Litzinger, F. and Emery, J. R.(1995) Glass-transition dynamics of a polyurethane gel using ultrasonic spectroscopy, dynamic light scattering, and dynamical mechanical thermal analysis, *Phys. Rev. B* **51**, 12995-13007.

35. Fioretto, D., Livi, A., Rolla, P. A., Socino, G. and Verdini, L. (1994) The dynamics of poly(n-butyl acrylate) above the glass transition, *J. Phys: Condensed Matter* **6**, 5295.

36. Wang C. H., Fytas G. and Zhang J.(1985) Rayleigh-Brillouin scattering studies of segmental fluctuations in (poly)siloxanes. *J. Chem. Phys.* **82**, 3405-3412.

37. Cummins H. Z., Li G., Du W., M. Hernandez J. (1994) Structural relaxation dynamics in supercooled liquids *J. Non-Crystalline Solids* **172-174**, 26-36.

OPTICAL STUDIES OF NANOPHASE TITANIA

R. J. GONZALEZ and R. ZALLEN
Department of Physics, Virginia Tech
Blacksburg, VA 24061-0435   USA

1. Introduction

Titania, TiO_2, is a somewhat unlikely insulator to be included in this volume. To see
this, we can compare it to SiO_2, the prototypical glass-forming insulating solid which
features prominently in these proceedings.

Property	SiO_2	TiO_2
bonding	covalent	highly ionic
coordination	low (4,2)	high (6,3)
structural units	tetrahedra	octahedra
linkage	corner-sharing	edge-sharing
Zachariasen rules	OK	not OK
glass former	among the best	among the worst

Although it is a rotten glass-former, amorphous TiO_2 can, in fact, be prepared by sol-gel
techniques. Its atomic-scale structure has been studied [1], and its optical properties
(Raman and infrared) will appear briefly in the present article. So it is not the worst
glass former mentioned in these proceedings; that honor probably belongs to KCl. The
present article will deal mainly with nanophase titania, characterized by particle sizes
under 100 nm in size.

The high ionicity of TiO_2, and its consequent high dielectric constant (roughly 50,
depending on direction and crystal phase), makes it a candidate to replace silicon oxide
and silicon nitride (two very important amorphous insulators) as the dielectric layer in
the next generation of ultrathin thin-film capacitors. This is its most important high-tech
application. Titania already has plenty of low-tech applications. It is the world's most
important white pigment; well over 10^9 kg/year is produced for paint. Among its
material properties are polymorphism (three crystal phases coexist indefinitely at
ordinary conditions), high refractive index, transparency in the visible, and complete
nontoxicity. Applications related to the last two include its use as an ultraviolet-
absorbing optical filter and as a whitener in toothpaste.

M. F. Thorpe and M. I. Mitkova (eds.), Amorphous Insulators and Semiconductors, 395–403.
© *1997 Kluwer Academic Publishers. Printed in the Netherlands.*

TiO_2 also is used as a catalyst support, since it can be prepared in the form of ultrafine nanometer-scale particles. One way of doing this is with sol-gel techniques. In this article, we describe some results recently obtained at Virginia Tech on optical studies of nanophase titania obtained by thermal and other treatments of sol-gel titania particles.

2. Sol-Gel Synthesis

The metal alkoxide used as the precursor is $Ti(OC_2H_5)_4$, called titanium tetraethoxide or tetraethylorthotitanate (TEOT). A simplified statement of the chemistry [2] is:

$$hydrolysis, \qquad Ti(OC_2H_5)_4 + 4H_2O \rightarrow Ti(OH)_4 + 4C_2H_5OH;$$

$$condensation, \qquad Ti(OH)_4 \rightarrow TiO_2 + 2H_2O.$$

An important variable is the water concentration relative to the alkoxide, given by the molar ratio $R=[H_2O]/[TEOT]$. Edelson and Glaeser [3] introduced a water soluble polymeric surfactant, hydroxypropylcellulose (HPC), into the sol-gel procedure in order to stabilize small particle sizes; the polymer adsorbs onto the forming particles and provides steric hindrance that impedes aggregation. In our work, we used variations on this technique (varying R, using HPC or omitting it, etc.); details are given elsewhere [4].

3. Infrared Reflectivity of Bulk and Nanophase Anatase TiO_2

Three crystal forms of titania are found in nature: rutile, anatase, and brookite. Rutile is the "true" stable phase at standard temperature and pressure (in the sense that graphite is the "true" stable phase of carbon at STP), and it is by far the most-studied and best-understood phase. Large single crystals of rutile have long been available, and the lattice dynamics of rutile TiO_2 in the lattice-fundamentals regime has been studied by polarization-dependent far-infrared reflectivity measurements [5]. This is not the case for anatase TiO_2. The polarization-dependent single-crystal far-infrared reflectivity of anatase has only recently been measured [6] by our group. The results are shown in Fig. 1.

The clean polarization-dependent reflectivity spectra of Fig. 1, along with the theoretical fits shown [5], determine the transverse-optical (TO) and longitudinal-optical (LO) zone-center infrared-active phonon frequencies for anatase. For E//c, $\bar{\nu}(TO)$ is 367 cm^{-1} and $\bar{\nu}(LO)$ is 755 cm^{-1}; for E$\perp$c, the two TO/LO pairs are

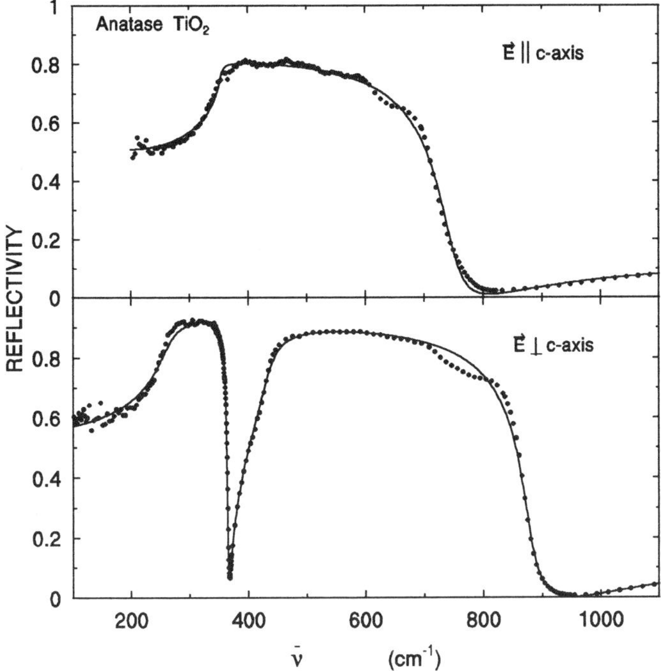

Figure 1. The polarization-dependent far-infrared reflectivity of single-crystal anatase. The curves are fits based on the factorized form of the dielectric function.

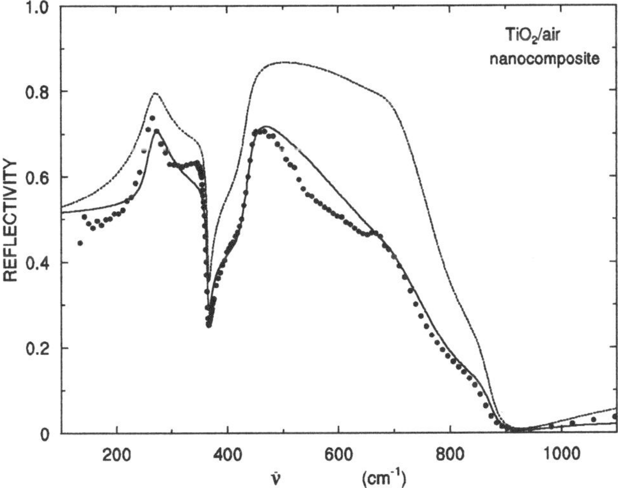

Figure 2. The infrared reflectivity of a pressed pellet prepared from sol-gel titania particles annealed at 600°C, analyzed with a combination of effective medium theory, bulk crystal data, and surface roughness. The dashed curve corresponds to the assumption of an abrupt air/pellet interface, the solid curve corresponds to the assumption of a graded surface layer of thickness 1.5 μm.

262/366 and 435/876 cm^{-1}. The very large TO-LO splittings demonstrate the high ionicity of anatase TiO_2.

In Fig. 2, we show the results of infrared reflectivity measurements on a dense pressed pellet of nanophase anatase. As discussed in the following section, we have studied annealing-induced phase transformations in sol-gel titania. Anatase dominates for anneal temperatures between 300 and 800°C. The pressed pellet of Fig. 2 was prepared from sol-gel particles and then annealed in pellet form at 600°C. X-ray diffraction linewidth measurements determined the average nanocrystallite size to be 50 nm. This spectrum was first analyzed using the crystal dielectric functions (determined from Fig. 1) and the Bruggeman effective medium theory [7]. The model assumes the pellet to be a composite of two fictitious isotropic materials, one having dielectric function $\varepsilon_{//}$ and the other having dielectric function $\varepsilon_{\perp}$. Effective medium theory was then used to estimate ε_{pellet} of the composite, assuming volume fractions of 1/3 (for $\varepsilon_{//}$) and 2/3 (for $\varepsilon_{\perp}$). The result is the dashed curve of Fig. 2 which, though containing the main qualitative features of the observed spectrum, is not quantitatively close for the vertical scale.

The dashed curve discussed above assumes the usual abrupt air/pellet interface. The solid curve shown in Fig. 2 incorporates the effect of surface roughness; a graded surface layer of 1.5 μm thickness was used here. Effective medium theory is again used to optically characterize the graded layer, but now the two components are taken to have dielectric functions ε_{pellet} (for nanocrystalline anatase) and 1.00 (for air). This method is rather successful.

4. Annealing-Induced Phase Transformations

Optical techniques are effective for studying the phase transformations that occur upon annealing sol-gel titania. Figure 3 presents a series of Raman-scattering spectra observed after anneals up to 1000°C. The Raman spectra were taken at room temperature, using the argon green line. Anneals were done in air for 40 minutes.

X-ray diffraction studies [1] have shown that the as-grown titania is amorphous, and this is revealed in the Raman spectra by the broad overlapping bands seen at the lower left of Fig. 3. Three spectra are shown for 250°C. These correspond to data taken with a Raman microprobe at different parts of the sample. The sharp, strong, anatase band at 141 cm^{-1} is already present at $T_{anneal} = 250$°C, indicating that some crystallization (into the anatase phase) has occurred although the material is still primarily amorphous. Conversion to anatase is complete by about 400°C. Anatase dominates for anneals to 700°C. Rutile appears alongside anatase at $T_{anneal} = 800$°C, and conversion to rutile is complete at 1000°C.

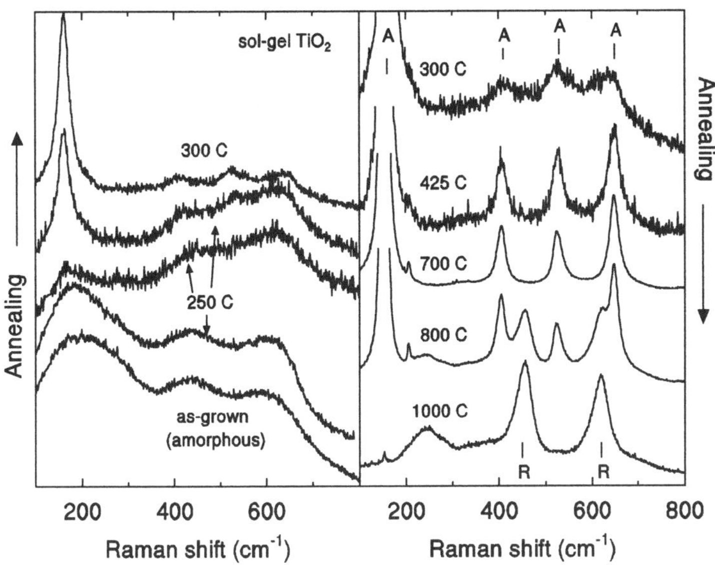

Figure 3. Room-temperature Raman spectra of ultrafine sol-gel titania particles, after annealing in air to different temperatures. The spectra clearly display the progression from amorphous TiO_2 to the anatase (A) and then the rutile (R) crystal forms.

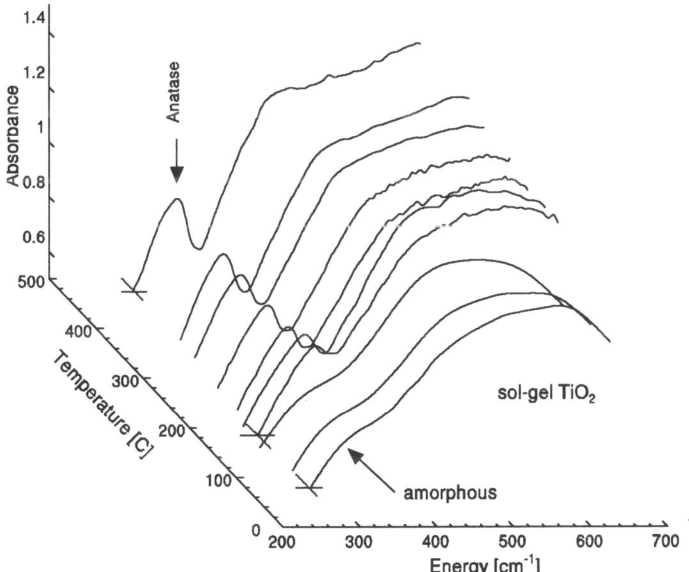

Figure 4. Infrared absorption spectra of sol-gel TiO_2 particles, diluted in KBr and pressed into pellets. These are in-situ temperature-dependent measurements.

Figure 4 displays infrared absorption measurements on the titania particles, taken during annealing. Unlike the Raman results, these are in-situ measurements taken while the sample was at elevated temperature. Powder samples were diluted in KBr (0.5% titania, by weight), pressed into pellets, and mounted in an optically accessible heater located within the evacuated sample space of a Bomem FTIR spectrometer. The heating rate was 1.25°C/min. A characteristic anatase band near 360 cm^{-1} clearly makes its appearance by 200°C, even though the material is still mainly amorphous (from x-ray and Raman) at this temperature.

5. Search for Finite-Size Effects in Anatase Raman Spectra

In semiconductors [8], AlO(OH) [9], and some other crystals, there is a well-documented finite-size effect observed for nanocrystalline samples; Raman bands broaden and shift with decreasing nanocrystal size L. The mechanism is as follows. For a nanocrystal of size L, the strict infinite-crystal k-space selection rule is replaced by a relaxed version characterized by a k-space uncertainty of about (1/L). Instead of k=0 only, a range of k-values from 0 to (1/L) are now Raman-allowed. Since the phonon dispersion curve $v(k)$ is not flat, the observed Raman band shifts and broadens. The smaller the L, the larger is the shift and broadening.

Because the anatase phase was the dominant form over a wide range of anneal temperatures (Fig. 3), and because the average nanocrystal size increases with annealing, we have looked for a finite-size effects in the Raman spectrum of nanocrystalline anatase. Crystallite sizes (L) were estimated from the linewidth of the (101) anatase peak observed in x-ray diffraction.

Figure 5 shows a comparison of the Raman spectrum of "bulk" anatase (micron-size crystals) with that of a nanocrystalline powder obtained by annealing sol-gel particles in air at 300°C. The strong and sharp bulk-anatase line at 141 cm^{-1} broadens and blue-shifts, while the 638 cm^{-1} band broadens and red-shifts. We have followed in detail the peak position and the linewidth (full width at half maximum) of the dominant 141 cm^{-1} Raman band, during the course of annealing.

Two different annealing atmospheres were used, air and argon, because it is known that stoichiometry changes can influence the Raman spectra of annealed TiO_2 particles [10]. Figure 6 shows a compilation of our results in the form of the correlation between the Raman-lineshape parameters (peak position v and linewidth Γ of the 141 cm^{-1} anatase band) and the nanocrystal diameter L. Despite the scatter, it can be seen that the argon-annealed samples lie close to a single curve in each panel of Fig. 6. Similarly, all of the air-annealed samples lie close to a single curve in each panel of Fig. 6. But the argon

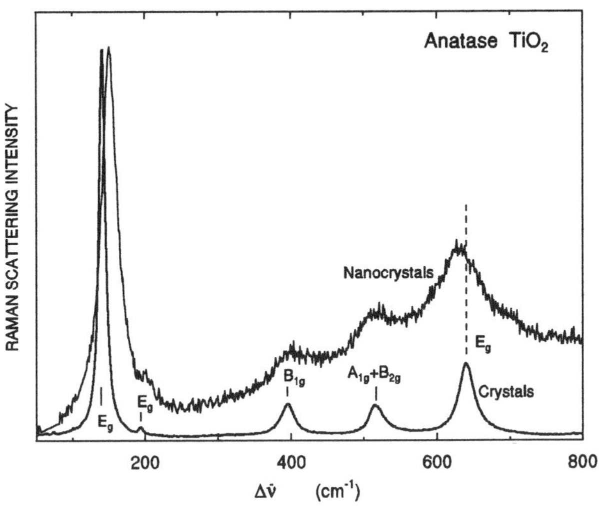

Figure 5. Comparison of the Raman spectrum of a nanocrystalline powder with that obtained from a powder of micron-size anatase crystals. The nanocrystalline sample was obtained by annealing sol-gel particles in air at 300°C.

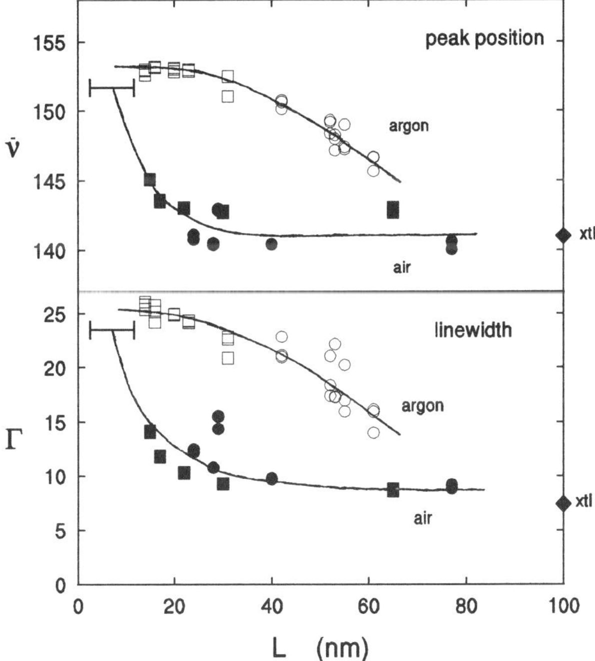

Figure 6. Raman lineshape versus nanocrystal size for the main anatase band of the annealed nanoparticles.

and air curves, for both $\nu(L)$ and $\Gamma(L)$, are distinctly different from each other. The argon curves make it only part way to the bulk-crystal values, while the air curves quickly approach the bulk values (becoming indistinguishable from them by L = 40 nm).

If L was the only physical quantity governing the changes in the Raman spectra, then each panel of Fig. 6 would contain a single curve. But this is not the case. The air/argon differences must be ascribed to the influence of stoichiometry [10].

For the air anneals, the initial rapid approach to the bulk-crystal values is then attributed to uptake of oxygen and the approach to stoichiometry of initially oxygen-deficient material. For the argon anneals, there is little effect because stoichiometry is little affected. From the Raman linewidth of 25 cm^{-1} for our small-L samples and the linewidth-versus-stoichiometry results of Parker and Siegel [10], it appears that the initial stoichiometry is $TiO_{1.98}$.

Since our results are interpretable on the basis of stoichiometry changes, no claim can be made here for the observation of finite-size effects. However, the following coincidence (?) is worth noting. Using the known dispersion curves of corresponding phonon modes in rutile, we have calculated the expected finite-size effects for the 141 and 638 cm^{-1} anatase bands. The calculated effects have the same signs (blue-shift for 141, red-shift for 638) and same relative sizes (large for 141, small for 638) as the observed effects. It therefore seems that phonon dispersion, the sign and size of the $\nu(k)$ curvature near k=0, also plays an important role in the sensitivity to stoichiometry of a Raman band.

6. Summary

A variety of optical and annealing investigations were performed on nanophase titania prepared by sol-gel techniques. The TO and LO frequencies of anatase were obtained from clean polarization-dependent infrared reflectivity measurements (Fig. 1) and, in combination with effective medium theory and the inclusion of surface roughness, were used to account for the infrared reflectivity of polycrystalline anatase (Fig. 2). Raman and infrared spectra were obtained for amorphous titania and were used to monitor amorphous-to-anatase and anatase-to-rutile annealing-induced phase transformations (Figs. 3 and 4). Extensive studies were done on the Raman spectra of anatase nanocrystals (Figs. 5 and 6). These have confirmed the importance of stoichiometry considerations, although interesting similarities to calculated finite-size effects indicate that phonon dispersion may also play a role in the sensitivity to stoichiometry.

7. Acknowledgments

We wish to thank the following individuals, each of whom has either collaborated with us on parts of this work or has helped us in other ways: A. Gaynor, R. M. Davis, H. Berger, R. J. Bodnar, and F. Harrison.

8. References

1. Wang, Q.J., Moss, S.C., Shalz, M.L., Glaeser, A.M., Zandbergen, H.W., and Zschack, P. (1992) in *Physics and Chemistry of Finite Systems: From Clusters to Crystals*, edited by P. Jena et. al., Kluwer Academic Publishers, Dordrecht.
2. Barringer, E.A. and Bowen, H.K. (1982) *J. Am. Ceram. Soc.* **65**, C-199; Barringer, E.A. and Bowen, H.K. (1985) *Langmuir* **1**, 414.
3. Edelson, L.H. and Glaeser, A.M. (1988) *J. Am. Ceram. Soc.* **71**, C-198.
4. Nagpal, V.J., Davis, R.M., and Riffle, S.J. (1994) *Colloids and Surfaces* **A 87**, 25; Gonzalez, R.J., Gaynor, A., Zallen, R., and Davis, R.M., to appear in *J. Mater. Res.*
5. Spitzer, W.G., Miller, R.C., Kleinman, D.A., and Howarth, L.E. (1962) *Phys. Rev.* **126**, 1710; Gervais, F. and Piriou, B. (1974) *J. Phys. C* **7**, 2374.
6. Gonzalez, R.J., Zallen, R., and Berger, H. submitted to *Phys. Rev. B*.
7. Bruggeman, D.A.G. (1935) *Ann. Phys. (Leipzig)* **24**, 636; Aspnes, D.E. (1982) *Am. J. Phys.* **50**, 704.
8. Richter, H., Wang, Z.P., and Ley, L. (1981) *Sol. St. Commun.* **39**, 625; Nemanich, R.J., Solin, S.A., and Martin, R.M. (1981) *Phys. Rev. B* **23**, 6348; Tiong, K.K., Amirtharaj, P.M., Pollak, F.H., and Aspnes, D.E. (1984) *Appl. Phys. Lett.* **44**, 122; Holtz, M., Zallen, R., Brafman, O., and Matteson, S. (1988) *Phys. Rev. B* **37**, 4609; Zallen, R. (1992) *J. Non-Cryst. Solids* **141**, 227.
9. Doss, C.J. and Zallen, R. (1993) *Phys. Rev. B* **48**, 15626.
10. Parker, J.C. and Siegel, R.W. (1990) *J. Mater. Res.* **4**, 1246; Parker, J.C. and Siegel, R.W. (1990) *Appl. Phys. Lett.* **57**, 943.

ELECTRONIC STRUCTURE METHODS WITH APPLICATIONS TO AMORPHOUS SEMICONDUCTORS

D. A. DRABOLD

Department of Physics and Astronomy, Condensed Matter and Surface Science Program, Ohio University, Athens, Ohio, 45701 USA

1. Introduction and Background

The theoretical description of amorphous and glassy materials is a persistent challenge to scientists. The amorphous state possesses many features unknown in other forms of matter. Some unique qualities of glasses include; floppy modes and the associated floppy to rigid transition, other dynamical features such as tunneling modes and two level systems; special electronic character, such as the exponential bandtails (so called Urbach tails in optical measurements), and unusual defect structures and associated electronic signatures associated with the glassy state. A recurring theme of theory work on amorphous systems is the need to handle small energy differences (as between two close but distinct conformations) in an approximate but reliable fashion. This is a challenge even for the most sophisticated methods in current use.

In this paper, I will concentrate on the electronic structure techniques required for modeling amorphous insulators, including electronic structure based molecular dynamics schemes which produce reliable interatomic forces and therefore enable the modeling of glasses with considerable reliability. Naturally electronic structure techniques are required for any study of the electronic states, defect states and optical properties of these materials. The methods which I present in Section 2 range from simple but efficient descriptions of the electronic structure, to state of the art methods which are computationally prohibitive for all but the smallest systems, but are essential for questions involving very high precision or small energy differences.

405

M. F. Thorpe and M. I. Mitkova (eds.), Amorphous Insulators and Semiconductors, 405–436.
© *1997 Kluwer Academic Publishers. Printed in the Netherlands.*

1.1. THE NEED FOR ELECTRONIC STRUCTURE CALCULATIONS

It is well to observe at this point that the essential complexity of atomistic descriptions of amorphous insulators stems from the complex many-body interactions between the constituent atoms, mediated by the electronic structure of the material. Thus, in amorphous Si, the forces on an atom depend in minute detail upon the local environment (sometimes up to several shells from the reference atom), the location of the Fermi level (a global quantity) and other concepts unique to the electronic structure. Thus, any fundamental approach to evaluating the plausibility of the topology of a network requires an electronic structure calculation, or at least a total energy functional which is known to accurately track the electronic structure. A corollary is that credible first attempts at modeling (for which accurate potentials are not yet available) require the flexibility of an electronic structure approach.

1.2. LINKS TO EXPERIMENT

Some of the most revealing experimental probes of structure, bonding and dynamics (vibrational modes) involve the electronic aspects of the system either directly (as in photoemission measurements, or optical measurements) or indirectly (to the extent that the electronic structure determines the forces, dynamical matrix and therefore vibrations, etc). Thus, a measurement of the density of states by photoemission techniques provides specific spectral information about a state, and in conjunction with electronic structure work on realistic models, may identify the specific structures responsible for the experimental signature, thereby providing microscopic information about the sample.

1.3. A LITTLE HISTORY

Probably the first important electronic structure calculation salient to amorphous insulators was the work of Weaire and Thorpe[1], who explained how an optical gap was possible in the absence of long range order in the quaternary column IV systems a-Si or a-Ge. Using the simplest tight-binding model, it was shown that the local bonding was the key to understanding the basic structure of the electronic density of states. This insight has proven to be robust, and much detailed *local* information has been gleaned from electron sensitive measurements, but little if anything has been learned from this type of study about the longer range properties of the network. Another aspect of the work of Weaire and Thorpe is that one can 'turn the argument around' and to note that the existence of a gap *implies* that electronic effects are spatially local, since the existence of such a gap is precisely the condition needed to justify an exponentially localized generalized Wannier (real-space localized) representation of the electrons[2]. Such Wannier functions can be truncated at the range of several Å in real systems with arbitrarily little loss of accuracy. This turns out to be a key for extending electronic structure methods to very large systems. An explicit recipe for the computation of these functions is given in Section 3.5.

A breakthrough occurred in the middle eighties when Car and Parrinello[3] showed how to combine the accuracy of *ab initio* density functional methods with force calculations and molecular dynamics techniques. This work, and subsequent work by Sankey[4] with a local basis representation for the density functional orbitals, has enabled the theorist to compute credible atomistic models of amorphous solids and make links to forms of experiment previously inaccessible. While this class of methods must not be oversold [there are severe limitations on the size of systems which may be addressed, the length of simulations possible, and the local density approximation (LDA) itself], it is clear that this work is one of the most important achievements for the developing understanding of glassy systems in general. It has also had a dramatic impact on empirical potential calculations[5], and empirical tight-binding calculations, providing microscopic information to produce reasonably dependable functional forms for the total energy and thereby contributing to the impressive empirical force calculations reported here by Vashishta[5] and others.

The rest of this paper will be organized as follows: In Section 2 I describe methods in use on amorphous insulators in increasing level of complexity from electronic structure motivated empirical potentials to the current state of the art. In Section 3 I describe some important developing special topics of particular interest to the amorphous insulator community, the so-called "order N" methods (which enable electronic structure calculations with CPU and memory demand scaling *linearly* with the system size), and therefore allows study of larger models and longer time scales.

2. METHODS

2.1. SEMI-EMPIRICAL POTENTIALS AND ELECTRONIC STRUCTURE

Successful empirical potentials in current use have functional forms which have been tuned to mimic to the maximum possible degree the energies and forces predicted by electronic structure techniques. A typical procedure is to "guess" a functional form with free parameters and to then determine the parameters most consistent with a variety of accurate total energy calculations, and in some cases also by comparing to experiment. If the energy functional was chosen with enough of the "underlying physics", then the functional can provide a reasonably transferable (translation: correct in

408

a wide variety of local bonding environments) description of the energetics and interatomic forces. The role of the electronic structure calculation here is simply to provide useful information on energetics and forces.

At a practical level, the success of the empirical approaches has been mixed. If such potentials are used carefully, they can answer questions about model systems that are extremely large (an important ingredient of realism not yet directly attainable with *ab initio* techniques). Normally the reliability is connected with limiting the application of the potential to systems which exhibit bonding reasonably close to conformations for which the potential was fit in the first place. In an extremely disordered environment (such as liquid Si or C), it seems unlikely that an empirical potential will offer a very satisfactory treatment of structure and particularly of *dynamics* in the near future. An excellent review of work in this area has been provided by Carlsson[6].

A recent development of interest is work of Ercolessi[7] who has developed an efficient scheme for computing empirical potentials based upon a "force matching" method. The idea is to take an enormous number of configurations for which the forces are accurately known (from an *ab initio* calculation), (as for example from a thermal simulation representative of the type of study planned for the desired potential), and then minimize the error in the forces predicted potential relative to the forces from the *ab initio* database. If a suitable functional form is selected in the first place, it has been found possible to reduce the error in the potential to acceptable levels. Such an approach has the appealing property that one can partly gauge its accuracy by considering the deviation of the potential from the *ab initio* results. Also there is likely to be "importance sampling" if a thermal simulation is used to form the database. So far, this procedure has been applied to Al[7]; work is in progress for a C potential; a much more challenging area , since the C-C interaction involves both sp^2 and sp^3 bonding, and these states are nearly degenerate.

2.2. EMPIRICAL TIGHT BINDING

The simplest explicit electronic structure calculation is the "empirical tight-binding (ETB) approximation", in which we imagine that the electronic eigenstates can be represented by a linear combination of atomic orbitals: $|\psi_i\rangle = \sum_\mu a_\mu^i |\mu\rangle$ where μ is a site-orbital index and i indexes the state (eigenvector). This method enables the calculation of an approximate one-body Hamiltonian matrix, whose eigenvalues are taken to approximate the allowed electronic energies and the eigenvectors are the states. In the usual implementation of ETB calculations, the basis is taken to be orthonormal: $\langle\mu|\nu\rangle = \delta_{\mu\nu}$. Also, most ETB Hamiltonians include interactions only with nearest neighbors and include only two-center contributions. The sum of the occupied eigenvalues is the (attractive) electronic contribution to the total energy. To compute the system (ions + electrons) energy a repulsive interaction must be added to the electronic part. This is obtained from some fitting procedure. ETB is the simplest approach enabling an estimate of the many-body forces characteristic of covalently bonded materials.

I outline here how an ETB calculation is implemented, since it is simple and is illustrative of many of the concepts of electronic structure calculations in amorphous solids.

(1) We begin by considering a supercell model (large unit cell with periodic boundary conditions) of an amorphous solid with N atoms and atomic coordinates $\{\mathbf{R}_i\}_{i=1}^{N}$ and three lattice vectors specifying the periodic boundary conditions. We can view the $\{\mathbf{R}_i\}_{i=1}^{N}$ as specifying a set of "basis vectors" for a crystal with a very large and topologically complex unit cell. Such a large unit cell possesses a band structure, as does any periodic system, but since the cell is supposed to represent an *amorphous* system, it must be large to have credibility. We will suppose that the $\mathbf{k}$ dispersion is negligible because of the large cell size. In calculations *this point must be checked*. The significance of this is discussed further in Section 3.1. We limit our discussion here to the $\mathbf{k} = \mathbf{0}$ point of the Brillouin zone, valid for a large enough model.

(2) Next, we set up the Hamiltonian matrix $H_{\mu\nu} = \langle\mu|\hat{H}|\nu\rangle$ where $\hat{H}$ is the Hamiltonian operator and $|\mu\rangle$ are a set of orbitals (typically s,p_x,p_y, and p_z for a column IV material) centered on each atom. As the matrix elements in this representation depend in detail on the network topology, it is convenient to work with "molecular coordinates" specifying the interatomic hopping. These are V_{ss}, $V_{sp-\sigma}$, $V_{pp-\sigma}$, $V_{pp-\pi}$, in the usual chemistry nomenclature for an sp^3 model. Explicit forms for the distance dependence of these interactions (see for example Harrison[8]) plus simple rules[9] connecting the molecular and $|\mu\rangle$ representation enables the calculation of $\hat{H}$ in the $|\mu\rangle$ representation. The H matrix eigenvalue problem then reads $H|\psi_i\rangle = \epsilon_i|\psi_i\rangle$, the usual orthogonal eigenvalue problem, where the electronic eigenvalues are supposed to be approximated by ϵ_i. For electronic state density calculations or questions of the spectral signature of a defect, an exact diagonalization of H is sufficient.

(3) The calculation of forces is easy when we possess the exact eigenvalues and eigenvectors; then the Hellman-Feynman theorem can be employed in the form:

$$\mathbf{F}_\alpha = -2\,\partial\sum_{i\,occ}\langle\psi_i|H|\psi_i\rangle/\partial\mathbf{R}_\alpha = 2\sum_{i\,occ}\langle\psi_i|-\partial H/\partial\mathbf{R}_\alpha|\psi_i\rangle \qquad (1)$$

Here, the sum on i is restricted to occupied states. Since total energies, forces, charge densities and other ground state properties depend upon

410

the *occupied* eigenfunctions and eigenvalues, it is natural at this point to introduce the single-particle density operator $\hat{\rho} = \sum_{i\ occ} |\psi_i\rangle\langle\psi_i|$, which in this context is just the projector onto the occupied subspace. In a particular representation $|\mu\rangle$, this is just the usual density matrix:

$$\rho_{\mu\nu} = 2 \sum_{i\ occ} \langle\mu|\psi_i\rangle\langle\psi_i|\nu\rangle \tag{2}$$

giving the usual expression for the electronic energy, $E = \text{Tr}(\rho H)$, and the forces are:

$$\mathbf{F}_\alpha = -\sum_{\mu\nu} \rho_{\mu\nu}\ \partial H_{\nu\mu}/\partial\mathbf{R}_\alpha \tag{3}$$

We will have more to say about density matrix methods in Section 3.5. Note that this form assumes that the overlap matrix remains exactly the unit matrix through any atomic motions, which is suggestive of why forces in particular are sensitive to the assumption of orthogonality. It is also important to remind the reader that Eq. (3) prescribes only the (purely attractive) electronic part of the force; an additional empirical repulsive pair potential must be added to the electronic energy to specify a system energy; the derivative of this term contributes likewise to the forces, and a dynamical simulation is a complicated balancing act between the electronic and repulsive terms. Atomic trajectories are obtained by integrating the (classical) equations of motion for the atoms using one of many schemes for performing numerical integrations.

For the study of bandtails in amorphous Si the tight-binding model is sometimes used in conjunction with Bethe lattice[10] methods. In this approach, a small cluster (involving typically from several to $\approx$ 20 atoms) is studied, and the cluster is terminated artificially by an infinite network saturating each surface bond in a tight- binding sense, and with each saturating bond being connected to an infinite non-interacting "lattice" of "atoms". Because of its peculiar non-crossing topology, each Bethe tree is isomorphic to a one dimensional problem, so that with Green's function and transfer matrix techniques, this problem can be solved exactly. Such calculations provide useful insight into the formation of defect/bandtail states as a function of bond angle distortion[10, 11], and also to indicate the spectral signature of specific defects (like dangling vs floating (five-fold) bonds).

As to the question of specific ETB Hamiltonians "on the market", there are several, and their transferability is variable. In particular, some are designed for producing good band structures[12], but are at the same time quite inappropriate for total energies and forces. As with the empirical potential methods, these Hamiltonians are most reliable for conformations "near" what the Hamiltonian was fit to in the first place. The entire underlying idea of an orthogonal TB model is the notion that there exist some underlying set of generalized Wannier functions with the property that these are mutually orthonormal in real-space. This is a justified point of view, which, however is valid only for a particular topology – if a different structure is considered, another set of Wannier functions emerge, and the justification for an orthogonal basis fails (unless the Hamiltonian is re-fit to the new structure). To be fair, it is probable that the failure is not serious until significant distortions are introduced; such is often the case in amorphous insulators. The most widely used ETB Hamiltonians for force calculations are those of Goodwin-Skinner-Pettifor[13] type. The original [13] was for Si; this was adapted by Xu *et al.*[14] to carbon systems. The difficulties with the orthogonal Hamiltonians seem to emerge most dramatically in *force* calculations [a universal feature of simulations is that the forces are more sensitive to approximations than quantities like the energy]. A specific discussion of this point is given in Ref. [15, 16].

More consistently reliable results can be obtained from a variety of nonorthogonal Hamiltonians. Menon[17] has proposed an empirical, nonorthogonal TB Hamiltonian which has been applied widely to clusters and recently has been generalized to bulk systems. Also in this category is a Hamiltonian whose form is motivated by density functional theory, explicitly using squeezed atomic orbitals as basis functions[18]. This Hamiltonian has been tested extensively on total energies and equilibrium structures of simple structures[19], and neglects three-center integrals. Both of these methods seem to provide reasonably faithful descriptions of the chemistry while paying only a slightly higher price than the orthogonal Hamiltonian (the only computational difference is that the "generalized eigenvalue problem": $H|\psi\rangle = \epsilon S|\psi\rangle$, where H, S are the Hamiltonian and basis positive-definite overlap matrix, respectively, and ϵ is the energy eigenvalue).

2.3. GENERAL COMMENTS ON EMPIRICAL METHODS

As a closing note on the semiempirical methods discussed in this paper, we remind the reader *not* to be impressed by large numbers of fitting parameters for either empirical potentials or tight-binding models. Too often, functional forms without much physical basis are "forced" upon the existing information about a material. Then, by construction, the function reproduces *what it was fit to* very well, but sometimes fails spectacularly for questions that don't seem so different from the fitted database. This is the famous "Ockham's razor" principle, telling us roughly that we should use the simplest representation while accounting for as many facts as reproducibly as possible. Thus, some comparatively simple potentials have been quite successful (for example Vashishta's SiO_2 model potential), while I know of at least one empirical potential for elemental Si which has 36 free parameters, which has difficulty with some of the smallest Si clusters, though this is where much of the fitting data came from. Exactly the same case arises for ETB models which, at the initial step, have much of the underlying physics removed (for example by the assumption of perfect orthogonality). Then an attempt is made to patch this assumption up, which can be difficult. Note that these comments are *not* a general indictment of empirical methods; only a cautionary note that the empirical potentials and Hamiltonians must be carefully and sensibly constructed and applied, and attention must be paid to verifying that the method is not being applied to a system more complex than it can handle. For an interesting *quantitative* measure of Ockham's principle, I refer the reader to the Bayesian probability theory literature[20].

2.4. DENSITY FUNCTIONAL THEORY

As we discussed above, the complexity of electronic structure and force calculations arises from the many-body nature of the interactions between the electrons. Currently, it would seem that direct attacks on the many-electron problem is too difficult to have direct impact on amorphous systems, requiring as they do a large number of atoms to provide a model worth investigating. Thus, all the successful electronic structure calculations salient to amorphous insulators have involved some kind of mapping of the many-body problem into an effective one-electron problem. Historically, the Hartree and Hartree-Fock approaches were the first success in this direction; descendants of these methods are widely used today, particularly in quantum chemistry. The ETB work captures some of the many-body effects, albeit in an approximate fashion, and the exact connection between the ETB model and the real many-body problem is obscure.

Of course, the true many-body Hamiltonian treats both the ions and electrons on a quantum mechanical basis. Because of the large mass difference between the electrons and nuclei, it is standard to decouple the nuclear and electronic degrees of freedom with the adiabatic or Born-Oppenheimer approximation[21], in which the electrons are assumed to respond instantly to motions of the ions (the electrons are taken to be in their ground state for all instantaneous ionic conformations). Moreover, the nuclei are treated as classical particles which move in a potential determined by the electrons in their ground state (computed for the given ionic coordinates). For most studies of amorphous solids, this is a reliable approximation, and I am unaware of any such calculation which has not started with the Born-Oppenheimer approximation.

For atomistic force calculations on solids, the current method of choice is density functional theory, due to Kohn, Hohenberg and Sham[22]. Its name comes from its predicted connection between the total ground state electronic energy of a system and the electronic charge density. The following rigorous statements embody the foundation of zero-temperature DFT:

(1) The ground state energy of a many electron system is a functional of the electron density $\rho(\mathbf{x})$:

$$E[\rho] = \int d^3x V(\mathbf{x})\rho(\mathbf{x}) + F[\rho], \tag{4}$$

where V is an external potential (due for example to interaction with ions, external fields, e.g., *not* with electrons), and $F[\rho]$ is a *universal* functional of the density. The trouble is that $F[\rho]$ is not exactly known, though there is continuing work to determine it[23]. The practical utility of this result stems from:

(2) The functional $E[\rho]$ is minimized by the true ground state density.

It remains to estimate the functional $F[\rho]$, which in conjunction with the variational principle (2), enables real calculations. To estimate $F[\rho]$, the usual procedure is to note that we already know some of the major contributions to $F[\rho]$, and decompose the functional in the form:

$$F[\rho] = e^2/2 \int d^3x d^3x' \rho(\mathbf{x})\rho(\mathbf{x}')/|\mathbf{x} - \mathbf{x}'| + T_{ni}(\rho) + E_{xc}(\rho) \tag{5}$$

Here, the integral is just the electrostatic (Hartree) interaction of the electrons, T_{ni} is the kinetic energy of a *noninteracting* electron gas of density ρ, and $E_{xc}(\rho)$ is yet another unknown functional, called the "exchange-correlation" functional, which includes nonclassical effects of the interacting electrons. Eq. 5 is difficult to evaluate directly in terms of ρ (because of the term T_{ni}). Thus, one introduces single electron orbitals $|\chi_i\rangle$, for which $T_{ni} = \sum_{i\ occ}\langle\chi_i| - \hbar^2/2m\nabla^2|\chi_i\rangle$, and $\rho = \sum_{i\ occ}|\chi_i(\mathbf{x})|^2$ is the charge density of the physically relevant *interacting* system. The value of this decomposition is that $E_{xc}(\rho)$ is a smoothly and reasonably slowly varying functional of the density: we have included the most difficult and rapidly varying parts of F in T_{ni} and the Hartree integral, as can be seen from essentially exact many-body calculations on the homogeneous electron gas[24]. The Hartree and non-interacting kinetic energy terms are easy to compute and if one makes the "local density approximation" (taking the electron density to be *locally* uniform), and using the results for the homogeneous electron gas, functional (Eq. 5) is fully specified.

With noninteracting orbitals $|\chi_i\rangle$, (with $\rho(\mathbf{x}) = 2\sum_{i\ occ}|\langle\mathbf{x}|\chi_i\rangle|^2$), then the minimum principle plus the constraint that $\langle\chi_i|\chi_j\rangle = \delta_{ij}$ can be translated into an eigenvalue problem for the $|\chi_i\rangle$:

$$\{-\hbar^2\nabla^2/2m + V_{\text{eff}}[\rho(\mathbf{x})]\}|\chi_i\rangle = \epsilon_i|\chi_i\rangle, \tag{6}$$

414

where the effective (density) dependent potential V_{eff} (in practical calculations orbital dependent) is:

$$V_{\text{eff}}[\rho(\mathbf{x})] = V(\mathbf{x}) + e^2 \int d^3x' \rho(\mathbf{x}')/|\mathbf{x} - \mathbf{x}'| + \delta\epsilon_{xc}/\delta\rho \qquad (7)$$

In this equation ϵ_{xc} is the parameterized exchange-correlation energy density from the homogeneous electron gas. The quantities to be considered as physical in local density functional calculations are: the total energy (electronic or system), the ground state electronic charge density $\rho(\mathbf{x})$, and related ground state properties like the forces. In particular, it is tempting to interpret the $|\chi_i\rangle$ and ϵ_i as genuine electronic eigenstates and energies (and indeed this can often be useful); such identifications are not rigorous. It is instructive to note that the starting point of density functional theory was to depart from the use of orbitals and formulate the electronic structure problem rigorously in terms of the electron density ρ; yet a practical implementation (which enables an accurate estimate of the electronic kinetic energy) led us immediately back to orbitals! This illustrates why it would be very worthwhile to know $F(\rho)$, or at least the kinetic energy functional since we would then have a theory with a structure close to Thomas-Fermi form and would therefore be able to seek *one* function ρ rather than the cumbersome collection of orthonormal $|\chi_i\rangle$.

According to the usual application of the variational principle, we err as minimally as possible because of incompleteness in the basis, but of course the representation selected to represent the $|\chi_i\rangle$ must be able to reasonably approximate ρ for reliable results. Of course, the more fundamental assumption of the LDA can itself cause trouble (this is discussed further at the beginning of Section 2.5).

The usual implementation of LDA leads to a nonlinear set of coupled integrodifferential equations. The origin of this nonlinearity is that V_{eff} in the Schrödinger-like (Eq. 6) is ρ-dependent, which in turn depends on the eigenvectors $|\chi_i\rangle$, which in turn depend on V_{eff} and so on. This nonlinearity is dealt with in the usual way "iterating to self consistency", an expensive inner loop on an already challenging computational task.

Consistent with chemical intuition, a minimal basis set of s and p functions is primarily responsible for the bonding and ground state electronic properties. The pseudopotentials are introduced so that only the really relevant electrons, namely the valence orbitals are *explicitly* treated in the calculation. Thus the valence electrons in the LDA solid are interacting with each other and with the ions via exotic potentials mimicking the nucleus dressed with the core electrons. This is an excellent approximation for most applications in the area of amorphous insulators. Pseudopotentials themselves have developed a considerable lore, and the classic modern work in the area is the paper of Bachelet, Hamann and Schlüter[25]. The choice of pseudopotential is governed partly by the basis used: for plane-waves special "soft" pseudopotentials are needed[26, 27]; plane-wave calculations also require a special "separable" form, enabling a factorization that saves considerable computational effort[28] to reduce the number of plane-waves needed; real-space basis functions are immune to this problem.

The next two subsections focus on two heavily used practical implementations of LDA; the first with a basis local in real-space, the second local in momentum space.

2.4.1. *Local basis: "Ab initio Tight Binding"*

Because of formal similarity to the ETB method of Section 2.2, I begin by describing a spatially local basis approach to construct the density functional orbitals $|\chi_i\rangle$. This is a program that was developed by Sankey[4], and his coworkers, and has been extensively applied to studies of amorphous materials.

The essential approximations are: (1) the LDA; (2) nonlocal (angular momentum dependent), norm-conserving pseudopotentials; (3) a minimal basis set of valence orbitals per site (this has recently been extended by Yang[29] to include d states in the basis). These basis functions are slightly excited (confined) pseudoatomic orbitals (PAO) (a PAO is a valence eigenfunction for a given atom type in free space, calculated self consistently within the LDA using pseudopotentials to eliminate the core: in the case of Si, the PAO's can be thought of as approximate atomic 3s and 3p levels); (4) A linearized (non self consistent) version of density functional theory, the "Harris functional"[30] is used which enables calculations without self consistent iterations. This approximation is best discussed by Foulkes and Haydock[31] and is quite remarkably good except in the most ionic systems like SiO_2. For additional discussion see the work of Smith[32].

The use of PAO's is appealing, because the chemistry is naturally built into the basis functions. The payoff is that for materials like Si and C, a minimal basis of four orbitals per site can be used, so that in a model with N atoms, the overlap and Hamiltonian matrix has dimension $n_b N$, for $n_b = 4$ or 9, the number of basis functions per atom (sp or spd) whereas $n_b \approx 100 - 1000$ for many plane-wave calculations (depending in detail on how carefully such a plane-wave calculation is implemented, and how 'hard' the pseudopotential is, which determines the number of plane-waves needed). Also the eigenvectors are expressed naturally in terms of the local PAO's, which is helpful in interpreting the physical meaning of the results. Finally, since the method is implemented entirely in *real-space*, no artificial periodicity is ever imposed, making the approach much more amenable to surface and cluster calculations, where desired.

416

Sankey and coworkers constructed this method to optimize performance for *ab initio* molecular dynamics simulations; the details are well beyond the scope of this paper. The method is able to handle systems with up to ≈ 500 atoms with exact diagonalization on a workstation, and has been extended recently to include quantum order N methods (see Section 3 below). The ability to perform exact diagonalizations (as opposed to "iterative minimization" schemes) is valuable for systems with states in the gap and particularly metallic systems. The limitation of non self consistency has been lifted recently by Demkov *et al*[33] and Ordejón *et al.*[34]

While this approach is relatively easy to understand, it is vastly more complex than the ETB method above. In particular, to make the scheme efficient, all the two and three-center matrix elements have to be tabled, so no integrals have to be computed during an electronic structure calculation or MD simulation; rather many calls are made to one and two dimensional interpolators. The calculations of forces, particularly with respect to the exchange correlation interaction, is tedious, and "Pulay corrections" must obviously be included, since the basis functions move with the atoms. The formal expressions for matrix elements, forces etc. are substantially more complex than for plane-wave pseudopotential methods of the next Section. I refer the reader particularly to References [4, 33, 34] for the details of the theory. This scheme has found considerable use in amorphous insulators.

2.4.2. *Plane-waves: Car-Parrinello methods*

The first *ab initio* molecular dynamics simulation was performed in 1985 by Car and Parrinello (CP). I begin with a comment on what "CP" means. Often a "CP simulation" refers to a molecular dynamics calculation using a plane wave basis and an iterative minimization scheme to solve the electronic structure problem (the self consistent LDA equations). More properly, "CP" refers only to a method for coupling the approximation of stationary states of giant basis eigenvalue problems with associated ionic dynamics, even for a basis other the plane-waves.

In the next paragraphs, I will discuss the basic CP equations of LDA on a plane-wave basis. To keep to the essentials I will write the expressions only for the Γ point of the Brillouin zone. For a complete discussion of the method, see Ref. [35]. As usual, the LDA equations take the form (for LDA Hamiltonian $\mathcal{H} = p^2/2m + V_{\text{eff}}$):

$$\mathcal{H}|\chi_i\rangle = \epsilon_i|\chi_i\rangle \tag{8}$$

where

$$|\chi_i\rangle = \sum_{\mathbf{G}} A^i_{\mathbf{G}}|\mathbf{G}\rangle, \tag{9}$$

where $|\mathbf{G}\rangle = 1/\sqrt{\Omega}\,\exp(i\mathbf{G}\cdot\mathbf{x})$, Ω is the cell volume, and $\mathbf{G}$ labels reciprocal lattice vectors (*a plane-wave basis restricts our studies to periodic systems*), so that the usual apparatus of reciprocal lattices is appropriate. The sum is cut off for $G = G_c$ large enough ("wavelength" small enough) that the smallest salient features in the problem are adequately described. An obvious advantage of this representation is that the kinetic energy operator is diagonal and therefore trivial. Matrix elements of V_{eff} are obtained by fast Fourier transforms, the existence of which makes the plane-wave approach tractable. The *problem* with Eq. (8) is that for systems large enough to interest us, the Hamiltonian matrix is too large to diagonalize with "classical" methods.

The key contribution of CP was to note that the diagonalization (Eq. 8) is unnecessary; it is adequate instead to consider the LDA expression for the total energy (a 'partial trace' over the occupied electronic subspace) and vary the χ_i:

$$E(\{\chi_i\}) = \sum_{i\ occ} \langle \chi_i|\mathcal{H}|\chi_i\rangle \tag{10}$$

where two important facts must be emphasized: (1) the sum is over the *occupied* subspace and (2) $\langle \chi_i|\chi_j\rangle = \delta_{ij}$. Thus, using Eq. (9), E can be viewed as a function of an extremely large number of parameters A_G^i ($n_{pw} \times n_e/2$) for n_{pw} the number of reciprocal lattice vectors used and n_e the number of electrons (double occupancy assumed). The idea is that if a set of mutually orthogonal $|\chi_i\rangle$ can be found, adequate in number to accomadate the electrons in the problem, and minimizing the functional (Eq. 10), then the E so obtained is the LDA ground state energy, to the extent that the basis is complete.

To perform the minimization, CP introduces a fictitious Lagrangian[3] which includes both the electronic and ionic degrees of freedom:

$$\mathcal{L} = \sum_i \mu/2\langle \dot{\chi}_i|\dot{\chi}_i\rangle + \sum_\alpha 1/2 M_\alpha \dot{\mathbf{R}}_\alpha^2 - E(\{\chi_i\},\{\mathbf{R}_\alpha\}). \tag{11}$$

In this equation, $E(\{\chi_i, \mathbf{R}_\alpha\})$ is the same as that given in Eq. (10) (the expectation value for the electronic energy, viewed as a functional of the orbitals χ_i), and the quadratic terms involving time derivatives (dots) are for the electrons (χ_i) and ions ($\mathbf{R}_\alpha$). Parameter μ is a fictitious mass assigned to the electron (it has nothing to do with the real mass, and is adjusted to make the calculation proceed efficiently). Note at this point that this method *does not* provide any information about electron dynamics in the usual quantum mechanical sense of time evolution. Rather it is just a trick to find the stationary states of the LDA Hamiltonian (approximate the occupied eigenstates from eigenvalue problem Eq. 8). Thus, the "dynamics" for the electrons has only the utility of helping to solve the

418

eigenvalue problem, and in fact if the fictitious kinetic energy becomes significant compared to the physical ionic kinetic energy, the dynamics for the ions will be suspect. In the absence of the "fictitious electronic kinetic energy", this Lagrangian would just generate the Newton equations of motion for the ions $M_\alpha \ddot{\mathbf{R}}_\alpha = -\partial E/\partial \mathbf{R}_\alpha$, as one would expect. The additional term adds an extra equation however: $\mu|\ddot{\chi}_i\rangle = -\delta E/\delta\langle\chi_i| + \sum_k \Lambda_{ik}|\chi_k\rangle$. Here, Λ_{ik} is a matrix of Lagrange multiplers required to maintain orthogonality $\langle\chi_i|\chi_j\rangle = \delta_{ij}$. In practical methods the Λ_{ik} are obtained by compelling the eigenvectors $|\chi_i\rangle$ to be orthonormal at each time step, and computing Λ_{ik} from $\Lambda_{ik} = \langle k|\mathcal{H}|i\rangle - \mu\langle\dot{\chi}_k|\dot{\chi}_i\rangle$. Then the ionic and fictitious electronic "forces" are completely specified, and the determination of the stationary states can proceed. Note that there are effectively two time scales in the dynamics so generated: one is determined by the artificial "mass" μ of the electrons; the physically salient time scale is set by the ionic masses M_α. This method generates unreliable dynamics for the ions if significant energy is transferred from the ionic (physical) degrees of freedom to the fictitious electronic degrees of freedom. That this must happen for long times is apparent from the equipartition theorem (Eq. (11) is a quadratic form, and asymptotically energy will reach equipartition). It develops that the rate of transfer from the ions to the unphysical electron kinetic energy depends critically on the energy gap (rapid transfer for small gap), which makes the CP method difficult to apply to metals. This is best discussed in Reference [36].

2.5. BEYOND THE LOCAL DENSITY APPROXIMATION

At the present time (1996), the LDA is the universal choice for *ab initio* calculations on amorphous systems. There are, however, well known problems with the approximation: (1) the optical gap is always poorly estimated (normally *underestimated*). This does not affect ground state properties like charge density, total energy and forces (and therefore dynamics), but it can be a serious headache for calculations of conduction states, as for example in the case of transport or optical properties. Moreover the gap is of particular interest for certain types of applications like photovoltaic device studies. While the so called "scissors" approximation is useful for the bulk (this amounts to just introducing an offset of conduction states to match experiment), this is not possible in cases where we do not already know the answer from experiment! (2) In strongly (electronically) inhomogeneous systems such as SiO_2, the basic assumption of weak spatial variation of the charge density is not well fulfilled, and the LDA has difficulty. (3) The LDA implicitly assumes that the system is paramagnetic; the local spin density approximation[37] (LSDA) (in which a separate "spin up" and "spin down" density functional is used) is useful for systems with unpaired spins, as for example a half filled state at the Fermi level. (4) A final example of less interest to the glass community is that weak bonds (as for example of fluctuating dipole type) are poorly described in LDA; and hydrogen bonding is also not very well described in the LDA.

Two general classes of attack on the problem can be identified. Several workers, but especially Perdew[38], have worked on the natural next step to the LDA: inclusion of effects proportional to the gradient of the charge density. Recent improvements along these lines are called "generalized gradient approximations" (GGA), and these seem to have led to significant improvements in SiO_2[39], and intermolecular binding in water is better described with GGA than in the LDA. In some ways the GGA's have been disappointing; on very precise measurements on molecules the results have been mixed. Overall, however, the GGA seems to be an improvement over the conventional LDA, and it is an area with much activity. In the general context of "patching up LDA", the "GW" approximation of Hedin[40] has been much applied to surface problems, particularly by Louie[41].

The other approach is to abandon the LDA altogether, and either develop entirely new exchange-correlation functionals[23], or abandon density functional theory for essentially "exact" calculations from the full manybody Hamiltonian. The most promising work appears to be the Quantum Monte Carlo (QMC) work of Mitas and coworkers. Such calculations allow a statistical sampling of the true many particle wave functions, and enable accuracy much better than LDA calculations. At the present time, it is a stretch to handle more than about 100 electrons. The increasing relevance of QMC to large systems is due in large measure to the introduction of pseudopotentials[42] into the scheme. This has allowed exceedingly accurate calculations on 20-atom clusters of Si[43] and C[44]. An indication of the salience of these methods arises from considering 20 atom C clusters[44] with carefully converged *ab initio* Hartree Fock calculations, LDA calculations, using several GGA approximations and QMC gave different results for the energies of C_{20} isomers, and different GGA's actually changed the energy ordering of some of the isomers[45]. The QMC calculations revealed that in C there are very subtle effects arising from electron correlations that have some structural implications even in a 20-atom C cluster: certainly one can expect that such effects exist in a-C as well, though it is unlikely that their inclusion will change our qualitative understanding of the material substantially.

In addition to molecular applications, QMC has been applied to a few solids, and gives excellent agreement on cohesive energies (LDA tends to overbind by as much as 20%); and also reproduced the band structure of the column IV materials very well (for the case of diamond, see Ref. [46]).

420

At the moment, force calculations have not been implemented in QMC, and it is still limited to systems too small to have direct relevance to amorphous systems, but it seems certain that within the next several years, with methodological developments and improvement in computer technology, accuracies at the level of 0.01 eV/atom will be attainable[47], and will find use in the glass community.

3. Linear system size scaling methods

3.1. THE NEED TO MODEL LARGE SYSTEMS

The common approach to model a glass or amorphous solid is to set up a supercell and impose periodic boundary conditions to eliminate surface artifacts. Of course this construction introduces a physically nonexistent periodicity which can nevertheless strongly influence the results. We can view this in different ways: Most properly we should admit to ourselves that the use of supercells and periodic boundary conditions *is* a calculation on a crystal with a large unit cell. From an electronic point of view then, Bloch's theorem applies, and quantities like the total energy and the forces involve quadratures in reciprocal space. *If it is the case that there is a significant dispersion of the bands, then it is likely that the cell is too small to use as a model for an amorphous material.* At a practical level this should be checked by computing the Hamiltonian (and overlap for a nonorthogonal local basis calculation) at several $\mathbf{k}$ values and checking that the total energy *and especially the forces* are converged to whatever tolerance is set[48]. At a more physical level, one can understand the effects of small cells as causing the formation of impurity bands. In many amorphous semiconductors (like a-Si) defects are really quite rare, occurring less than once per 1000 atoms. The best supercell calculations typically have a few to several *per cell* – there are many calculations that purport to describe a-Si with more than 10% defects. Because of both inter- and intracell interactions (because the defects tend to have similar electronic energies), the states on the defects mix, and this reduces the localization substantially. At its most extreme it is possible even for a specific defect to directly interact with its own image in neighboring cells. This problem is quite obvious when comparisons are made to experiments; the localization is always much higher (as measured by spin resonance experiments) in the real material than in computer models).

It is important to note that these effects extend well beyond inadequately localized electron states: Since the electronic eigenvectors are different, so is the density matrix and therefore the forces. Thus the structure itself, the vibrational modes, and anything else of physical importance is affected.

3.2. SPECTRAL ORDER N METHODS

To begin the discussion of quantum "order N" methods (so called because the computational cost (memory and CPU) scales linearly with N, the number of atoms in the system), I discuss a method for extracting the density of electronic states from giant matrices. I limit the discussion to the orthogonal eigenvalue problem[49], though the ideas have recently been extended to the case of a non orthogonal basis[50].

Let H be a large, sparse, Hermitian matrix of dimension N. Ideally one would like the eigenvalue spectrum ϵ_i, where $H|\psi_i\rangle = \epsilon_i|\psi_i\rangle$. We note that the information carried by the density of states is contained in any one single vector $|\xi\rangle$ in the family of vectors of the form

$$|\xi\rangle = N^{-1/2} \sum_{j=1}^{N} e^{i\phi_j}|\psi_j\rangle, \tag{12}$$

where ϕ_j specifies an arbitrary phase. Note that the expectation value of the DOS operator $\hat{\Delta}(E) = \delta(E - \hat{H})$ between any $|\xi\rangle$ gives the exact density of states; $\rho(E) = \langle\xi|\hat{\Delta}(E)|\xi\rangle$. The vectors $|\xi\rangle$ equally weigh all of the eigenvalues of the spectrum of H. For this reason we call such $|\xi\rangle$ *impartial vectors*. The Hamiltonian H has moments μ_n,

$$\mu_n = \int_{-\infty}^{\infty} dE\, E^n \rho(E) = \frac{1}{N}\mathrm{Tr}H^n. \tag{13}$$

We then see that an impartial vector also generates exact moments through its expectation value,

$$\mu_n = \langle\xi|H^n|\xi\rangle. \tag{14}$$

However, as we discuss below, the expectation value is an $O(N)$ operation, while taking the trace in Eq. 13 is not.

Skilling [51] noted the possibility of extracting moment data from the operation of sparse matrices on random vectors and derived useful error estimates for errors for integrals over $\rho(E)$. Silver and coworkers [52] used random vectors to generate moment data and used an orthogonal polynomial fit for the DOS of the 2D 4x4 Heisenberg model. Here we greatly extend the practical value of this earlier work and investigate its appropriateness for electronic structure applications by applying it to problems so large that they would otherwise be intractable with existing techniques. We should point out that Lanczos[53] understood most of this as early as 1956, and many workers[54, 55] have redeveloped ideas related to the work we present here. It is historically relevant to note that a very early moment

422

expansion (for thermodynamic quantities related to the vibrational DOS) dates prior to the first World War[56]!

The first step of our technique is to show that approximate moments of ρ are readily obtained. Let $|x\rangle$ be an arbitrary random normalized vector in the space of H. Now consider the following sequence ν_k, where

$$\nu_k = \langle x|H^k|x\rangle = \sum_{lm} x_l^* x_m \, (H^k)_{lm}. \tag{15}$$

These objects are the moments of a nonnegative function $\rho_x(E)$ for any x as can be seen by inspecting Eq. 15 with x expressed in the energy representation. Now consider an average over an ensemble of RVs $|x\rangle$:

$$\overline{\nu_k} = \sum_{lm} \overline{x_l^* x_m} \, (H^k)_{lm} \to \mathrm{Tr}\, H^k = \mu_k, \tag{16}$$

where the last result is fulfilled for *any* sampling scheme (random or otherwise) which produces $\overline{x_l^* x_m} \to \delta_{lm}$, and the bar denotes ensemble average. Thus, a random process can lead, without any knowledge of the energy eigenvectors, to arbitrarily accurate estimates for the moments μ_k of the DOS. Some such processes are more efficient than others. A simple approach is to sample the components of $|x\rangle$ independently from the unit normal distribution then rescale to get $\langle x|x\rangle = 1$. In our first report of this method[49], we thought it necessary to average the DOS (instead of the moments). In fact we find that averaging the moments and using the averaged moments to obtain a DOS is adequate, and more efficient. In addition, an appealing aspect of this simple method is that for estimates of the DOS, the number of vectors $|x\rangle$ *decreases* with increasing system size.

The ν_k are easily computed, as observed by Skilling [51] and Silver *et al.* [52] for sparse H, since ν_k may be computed recursively. If we define $|y_k\rangle = H^k|x\rangle$, then $|y_{k+1}\rangle = H|y_k\rangle$, and $\nu_k = \langle x|y_k\rangle$. Thus, we may accumulate the ν_k by repeated operations of a sparse matrix on a vector. These calculations are $O(N)$.

A unique feature of our work, salient for computing integrals over the DOS, is to use special vectors selected to approximate $|\xi\rangle$, to produce improved moment data $\{\nu_k\}$ from H. We show elsewhere[49] that this dramatically improves the method for computing band energies, the Fermi level or any other quadratures involving ρ. To obtain these improved vectors, we construct a penalty function of a vector $|x\rangle$ subjecting it to three constraints $(\mu_0, \mu_1$ and $\mu_2)$,

$$\pi(|x\rangle) = (\langle x|x\rangle - \mu_0)^2 + (\langle x|H|x\rangle - \mu_1)^2 + (\langle x|HH|x\rangle - \mu_2)^2 \tag{17}$$

The exact moments μ_1 and μ_2 are both $O(N)$ calculable, and by normalization, $\mu_0=1$. We minimize the function (Eq. 17) with a conjugate gradient (CG) method [57]. We find that it is always straightforward to generate $|x^*\rangle$ such that $\pi(|x^*\rangle) = 0$ from an initial random vector $|x\rangle$. Note that such $|x^*\rangle$ is 'closer' to an impartial vector $|\xi\rangle$ than an $|x\rangle$ merely chosen at random and normalized [51, 52].

The second key step in our technique is to transform the information contained in $|x^*\rangle$ into $\rho(E)$ through moment data (Eq. 15). As demonstrated by a variety of workers [58, 59, 60, 61], Maxent offers a very rapidly convergent approach to computing the density of states from its moments. Because of its information theoretic origin [62], Maxent introduces no artifacts stemming from *ad hoc* approximations. It is also manifestly nonnegative, a desirable feature for a spectral function. For numerical convenience, we scaled and shifted H so that the DOS has support only on (-1,1), and we used Tchebychev polynomials, T_n, instead of raw powers. It is these shifted and scaled units that are used in the figures of this paper. In practice, one can use low-order Maxent approximations to the DOS to obtain an approximate width, and modify H accordingly. Thus, no highly accurate guess is required *a priori* for the width. Given averaged moment data $\overline{\nu_k(x)}$, the Maxent reconstruction is $\rho_M(E) = \exp[\sum \lambda_n T_n(E)]$, where the Lagrange multipliers λ_n are determined by requiring the Maxent DOS reproduce the input moments $\{\overline{\nu_k(x)}\}$. See Turek [60] for a stable algorithm to solve the Maxent moment problem, and for details on the use of orthogonal polynomials see Ref. [51]. The ease of computing integrals over the DOS is also useful for other order N methods[63] which do not provide DOS (and therefore Fermi level) information. For examples of the spectral resolution obtainable with these methods, see References[64], [65]. Codes to implement these calculations are available from the author[66].

3.3. APPLICATION: BAND TAILING IN AMORPHOUS DIAMOND

I have chosen to discuss one detailed application in this paper involving a large system (order N methods) and electronic structure issues. One of the central issues of the physics of glassy and amorphous solids is the nature of the band tails in the electronic density of states (DOS). Particular issues include: (1) What is the origin of the ubiquitous exponential shape of the tails seen in photoemission and less directly in optical absorption measurements[67] and (2) How does the spatial character of the electronic eigenstates change from the highly local midgap states to the extended states interior to the valence or conduction bands? The nature of the electronic states for electron energies ranging between midgap (localized) to valence or conduction (extended) is of obvious interest to the theory of doping and transport. The tools required to address such questions in an unambiguous way are (1) *very large and realistic structural models of a representative amorphous system*. Small models can give a good account only of the the most highly localized midgap states, and necessarily fail in describing the spatial structure of the states as the volume of the state exceeds the volume of the supercell. Tantalizing hints of the nature of band tailing have been observed in earlier work on small supercells[68] and with elegant calculations using Bethe lattice techniques[10], which cannot provide a useful description of the spatial structure of the disorder-influenced electronic states. The structural model of this paper is a 4096-atom cubic supercell model of a-diamond (a-D) provided by Djordjevic, Thorpe and Wooten.[69] We note that a-D is a hypothetical, entirely fourfold material at a density of 3.52 gm/cm^3 as crystalline diamond, but with topological (primarily bond angle) disorder. It is probably related to tetrahedral amorphous carbon (ta-C), which, however has a lower density (3.0 gm/cm^3) and contains about 15% sp^2 sites[70]. Column IV amorphous semiconductors a-Ge and a-Si are the materials most resembling a-D, with their large proclivity for sp^3 bonding (good quality unhydrogenated a-Si is believed to have less than 0.1% non-sp^3 sites). That the cell of Ref. [69] is a structurally credible model of amorphous diamond can be inferred from our LDA relaxations of smaller versions (216 and 512 atom models) constructed in an analogous fashion with the Wooten-Weaire-Winer (WWW) method[71]. We found that these smaller supercells of a-D were practically unchanged upon relaxation. We reported on this in more detail elsewhere[72]. At present, several thousand atom structural models are unattainable from *ab initio* molecular dynamics simulations. The second tool needed is: (2) *electronic structure methods able to cope with the large Hamiltonian matrices from which spectral information is required.* We use recently developed maximum entropy[62] spectral methods[49, 65] to handle the state density calculations and a shifted Lanczos[73] scheme to compute the electronic states of interest from the sparse 16384×16384 Hamiltonian matrix. We use a block Lanczos method[74] and sparse matrix techniques to implement these calculations. We make the simplest reasonable choice of an electronic Hamiltonian; the orthogonal tight-binding Hamiltonian of Xu et al[14]. For the *spectral* calculations we report here this Hamiltonian is a reasonable choice. Its reliability is also in little doubt for an entirely four-coordinated matrix as we study here. We concentrate on the valence band tail in this paper since the basis of the Xu et al. Hamiltonian is minimal (one s and three p orbitals per site).

A variety of experiments on amorphous systems show the density of band tail states falling exponentially into the bandgap[75]. It is clear that both structural and thermal disorder contribute to the tail[68]: this work focuses entirely on the structural origins of exponential tailing. Where struc-

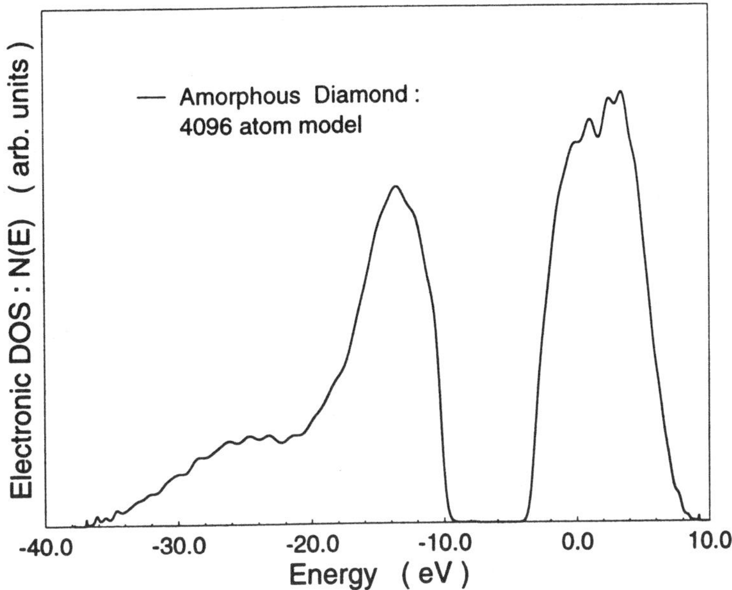

Figure 1. Total electronic density of states (DOS) of amorphous diamond: 80 moments and 50 random vectors were used.

tural broadening is concerned, an early argument of Halperin and Lax, [76] led to a DOS $N(E) \propto \exp(-\gamma |E|^{1/2})$ in three dimensions, and later Soukoulis *et al.* (1984)[77] modified the theory with scaling localization arguments and obtained the correct exponential form of the DOS. As a complement to this work, we give a simple argument below which also leads to exponential tails. The electronic DOS of our amorphous diamond model is computed with the maximum entropy method[49] and illustrated in Fig. 1.

Compactly stated, random vectors and an averaging scheme are used to obtain up to 100 moments of the density of states of the sparse Hamiltonian matrix, and maximum entropy techniques are used to reconstruct the DOS from the moments. Care was taken to properly converge the results with respect to both the number of moments and random vectors[49]. For an illustration of the spectral resolution this method affords, see Ref.[65] and [64]. In Fig. 2a we show the valence band edge region for diamond in a 4096 atom cell and the tails from the Djordjevic cell. The crystalline diamond cell has a defect-free gap with sharp band edge, while the amorphous diamond model has extended band tailing at both valence and conduction bands, as well as a few defect states in the middle of the bandgap. Different numbers

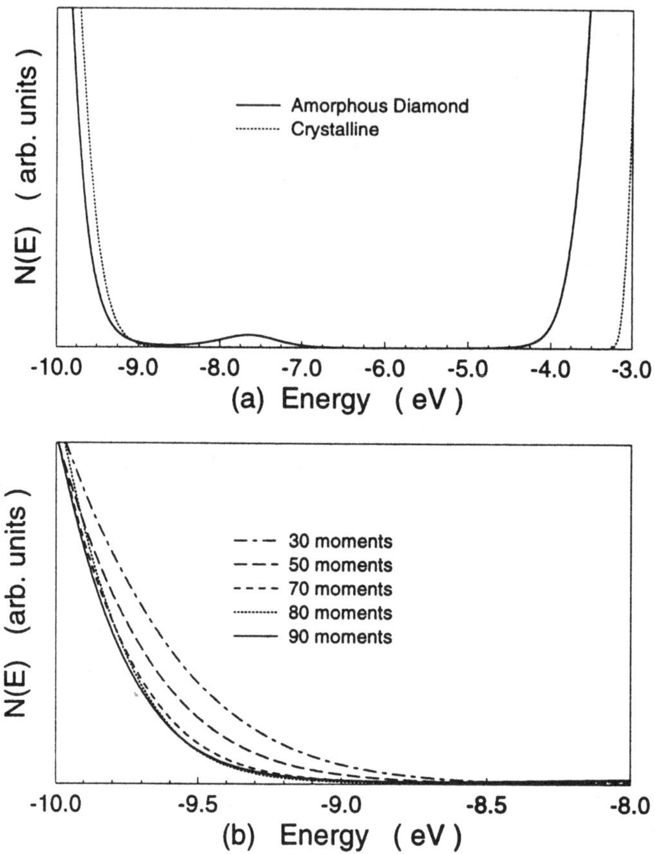

Figure 2. (a) Electronic DOS in the bandgap region. The solid curve depicts a-D; dotted curve is crystalline diamond. (b) Results with different numbers of moments. Convergence of the maximum entropy reconstruction is obtained with 80 moments and 50 vectors.

of moments and random vectors are used to compute DOS. Fig. 2b shows that our result of 80 moments and 50 vectors is well converged. These parameters, particularly the number of random vectors selected, is very conservative.

A semi-log plot in Fig. 3 reveals that the band tail falls exponentially , which agrees with the experimental observation. The tail decay parameter E_0 [such that the valence DOS $\propto \exp(-E/E_0)$] is about 180 meV (versus approximately 60 meV seen in photoemission studies on a-Si)[75]. We also find it useful to present a simple heuristic argument for the origin of structural exponential tailing for the valence edge, based upon the following assumptions. First, as suggested by Bethe lattice calculations[10], we view

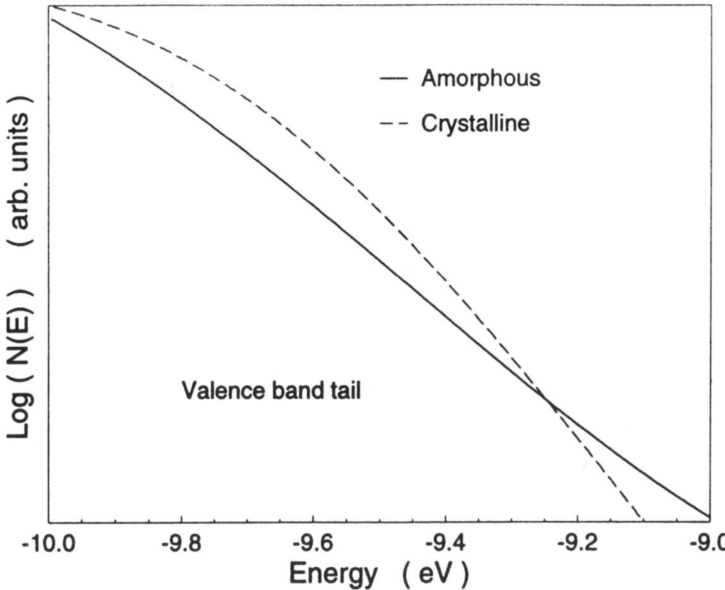

Figure 3. Semi-log plot of electronic DOS in the valence band tail. The linearity of the graph for a-D implies an exponential valence tail. The value of the DOS varies by a factor of ~ 150 over the plotted energy range.

band tailing as originating in bond angle distortions from the tetrahedral angle θ_T; and we further assume that an energy deviation from the diamond valence edge value can be assigned associated with these distortions. This effectively assumes that the states in question are substantially localized. Second, we have observed that the distribution of the cosine of bond angles θ in the cell is very well approximated with a normal distribution $p(\xi) = \exp[-(\xi - \xi_t)^2/2\sigma^2]/(2\pi\sigma^2)^{1/2}$ with $\xi = \cos(\theta)$, $\xi_t = \cos(109.01°)$ (the mean bond angle is near θ_T, as expected), and $\sigma = 0.149$ (corresponding to a dispersion in θ of about $9.0°$). This bond angle distribution was generated by the WWW method, and is seen in the smaller 216- and 512- models as well, and as we pointed out, *it is preserved under an LDA relaxation in 216 and 512 atom models* lending some credence to the view that the normally distributed cosines are realistic at least for dominantly four-coordinated systems. To investigate this further, we would like to consider bond angle distributions in supercell models generated entirely from *ab initio* methods, but there are two problems with this: (1) the statistics are poor because of the small cell size and (2) the unphysically large number of defects obtained in virtually all *ab initio* simulations introduces a complicating factor in interpreting the resulting distributions. With the assumption of normally distributed cosines, we take the crystalline valence band edge to be at energy λ_V. As $\lambda(\xi)$ is presumably a minimum for $\xi_T = \cos(\theta_T) \approx \xi_t$, this function can be approximated for small distor-

428

tions as $\lambda = \lambda_V + K(\xi - \xi_t)^2$, where K is a positive constant[78]. As the probability density function (PDF) of $\xi = \cos\theta$ is normal, one can easily write the PDF for λ, the highest valence band eigenvalue as broadened by the structural disorder of this model, by using the usual rule for changing variables in a PDF. One obtains: $N(\lambda) \propto \exp(-|\gamma(\lambda - \lambda_V)|)$, where $\gamma = (2\ \sigma^2\ K)^{-1}$. An essentially identical argument can be stated for the conduction tail. We note that in this simple picture, *any* mechanism causing approximately Gaussian bond angle disorder leads to exponential tails. Consider a very simplified model for the network dynamics, and assume the lattice is in thermal equilibrium at finite temperature T. In a valence force field model[79], the energy associated with bond bending involves a term of the form $U = \epsilon_0[\delta\cos(\theta)]^2$, where $\delta\cos(\theta)$ is the deviation of the angle between adjacent bonds from $\cos(\theta_T)$. Parameter ϵ_0 may be inferred from Ref. [79] for a variety of materials. If one neglects other distortions, comparison to the canonical distribution function suggests that thermal disorder leads to normally distributed cosines (with the width of the distribution $\sigma^2(T) \propto T$). The resulting linear dependence in the exponential "tail width" is clearly seen in experiments for the conduction tail in Ref. [75] (for a-Si), and less obviously for the valence tailing, which seems to be mostly structural in origin.

To show the nature of these band tail states, we apply a shifted Lanczos method[73] to probe the midgap and valence tail energy region. We computed about 30 eigenstates in the valence band tail. The localization is characterized by the inverse participation ratio (IPR) defined as: $I(\psi_j) = N\sum_{i=1}^{N} a_i^{j^4} / \sum_{i=1}^{N} a_i^{j^2}$ where $\psi_j = \sum_{i=1}^{N} a_i^j \phi_i$ is the jth eigenvector and $\{\phi_i\}$ is the (tight-binding) orthogonal basis and $N = 16384$, the number of basis functions. Note that $I = 1$ for completely uniform extended states and $I = N$ for a state completely localized on a single orbital: results are presented in Fig.4.

The first several defect states in the middle of the gap ($E \approx -8eV$) are strongly localized. As the energy approaches the interior of the valence band, (from about $-9eV$ to about $-10eV$), the degree of localization substantially decreases. This localized to extended transition is more clearly shown in Fig. 5. Atoms are assigned different grey scale renderings (darker more charge localization) according to their contribution to the eigenvectors. Atoms making little contribution (where $< 0.05\%$ of the "charge" is located) are omitted in Figure 5. The origin of the midgap defect states is found to be an atom that has very large bond angle distortion ($\Delta\theta \approx 20°$). Such states are strongly localized on the defect core and its nearest neighbors. As we expect, such large angle distortions are rare in the model, so only a few (3 of 16384) midgap states are found. States deeper into the band tail do not just localize on several atoms, but tend to have weight

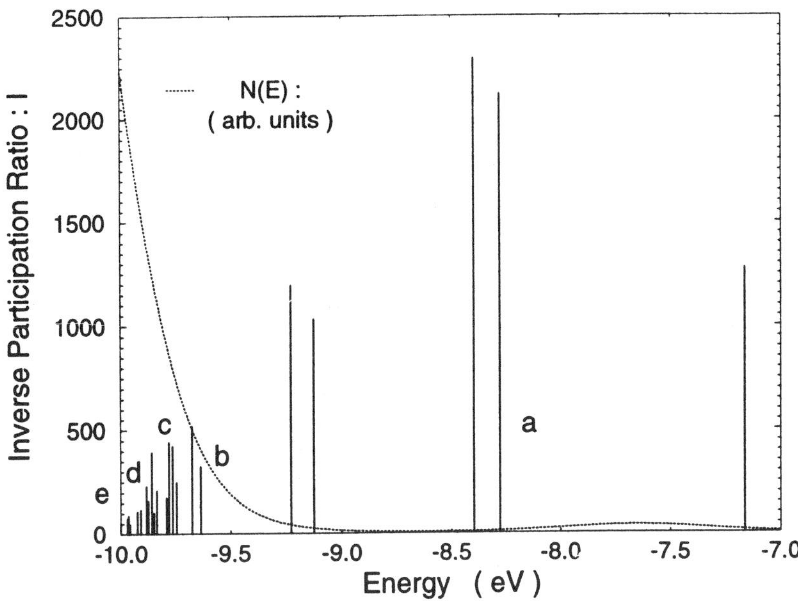

Figure 4. DOS and localization as measured by the inverse participation ratio in the valence band tail region. The letters refer to Fig. 5.

on certain clusters of atoms. The deeper the state is inside the band, the larger of the size of cluster, which implies smaller IPR. There is an interesting pattern, when the size of the cluster increases to certain point, the localized states start to "bifurcate" into two smaller spatially separated pieces (compare Fig. 5c and Fig. 5d). As we have pointed out elsewhere[80], using the simple language of perturbation theory, it is natural to see this as a "resonant phenomenon" in the sense that electronic states can spatially mix "energetically similar" parts of the network that are well separated in the supercell because of "small energy denominators" . We have seen a very similar resonance effect on vibrational eigenstates in glassy $GeSe_2$[81]. As one progresses deeply into the valence tail, the fraction of the network associated with the given energy increases until the states become fully extended; for E close to $-10eV$, there is no clear pattern of clustering and states become quite extended.

3.4. FORCE CALCULATIONS

A more challenging problem is to extract *local* properties in an order N fashion. This includes *forces*, which are of course essential to any MD simulation. Here we outline the main ingredients (see Ref. [63] for a very complete discussion), which are (i) a new band structure energy functional

430

(a)   (b)

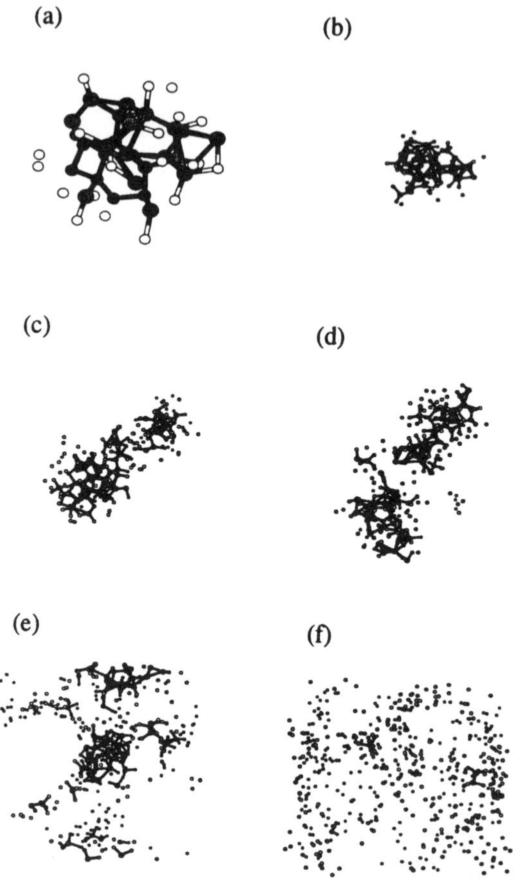

(c)   (d)

(e)   (f)

Figure 5.   Spatial character of the local to extended transition. I is the inverse participation ratio (see text) (a)$E = -8.27eV$: highly localized midgap state (I=2120); (b)$E = -9.63eV$: less compact cluster (I=326); (c)$E = -9.84eV$: larger cluster (I=210); (d)$E = -9.92eV$: weight on two separated clusters (I=110); (e)$E = -10.0eV$: nearly extended state (I=49); (f)$E = -11.0eV$: extended valence state (I=5).

which, when freely minimized with no orthogonalization constraints, yields the exact ground state energy and a set of orthonormal states defining the ground state subspace, and (ii) description of the system in terms of localized, Wannier-like functions, that are truncated beyond a certain cutoff radius R_c. The unconstrained energy functional has the form

$$\tilde{E} = 2 \sum_{i,j=1}^{N_e} H_{ji}(2\delta_{ij} - S_{ij}), \tag{18}$$

where $S_{ij} = \langle \chi_i | \chi_j \rangle$ and $H_{ij} = \langle \chi_i | \hat{H} | \chi_j \rangle$ are the overlap and Hamiltonian matrix elements between occupied orbitals $\{|\chi_i\rangle\}$ ($i = 1, ..., N_e$) and $2N_e$ is the number of electrons. The set of functions defining the ground state is the one that minimizes the functional $\tilde{E}$. This is a key point: the χ_i are *not necessarily eigenvectors*; rather, they are just a set of orthonormal vectors spanning the occupied subspace with no unoccupied component. As the χ_i form an orthonormal set, the transformation connecting the χ_i to the eigenvectors is *unitary*. Therefore *all trace quantities (the band energy, the force, etc.)* are given as well by the χ_i representation as the eigenvectors. The χ_i are easier to obtain, and also allow the freedom of constructing them to be spatially local in insulators, which is of critical importance from a computational point of view. In particular, one of the possible equivalent sets of functions is a set of localized wave functions (LWF's), centered at different positions, similar to the Wannier functions for crystals. In the Order-N method of Ordejón *et al.* these LWF's are truncated beyond a cutoff radius R_c from the center of the LWF. When this is done, all the operations necessary to compute and minimize the energy functional $\tilde{E}$ scale linearly with the size of the system, since for each LWF $|\chi_i\rangle$ only the matrix elements H_{ji} and S_{ij} with orbitals $|\chi_j\rangle$ inside the radius R_c must be calculated (a number that is independent of the system size). It can also be shown that, once the LWF's have been obtained, the forces on *all* the atoms can also be obtained in $O(N)$ operations.

3.5. NON VARIATIONAL (PROJECTION) METHODS

As the preceding discussion reveals, a spatially local representation of electronic states is very desirable, enabling us to partition space into regions involving the overlap of a finite number of Wannier-like functions and accurately compute the electronic structure order N. One can obtain these orbitals by making an initial (local) guess and performing the minimization described above. This is challenging in an arbitrary bonding environment. Thus, we have constructed Tchebychev polynomial approximations (after scaling and shifting H so that its spectrum lies on [-1,1]) of the projection (density) operator[82, 83]:

$$\hat{\rho} = \lim_{\beta \to \infty} (1 + \exp[\beta(\hat{H} - \mu)])^{-1} \approx \sum_{l=0}^{M} \gamma_l(\beta, \mu)\, T_l(\hat{H}) \qquad (19)$$

where the limit really only implies that β should be large enough that $\hat{\rho}$ annihilates unoccupied states, and multiplies occupied states by unity, to a suitable degree of convergence; μ is the chemical potential specifying the demarcation between occupied and unoccupied states. We employ Tchebychev polynomials because of their property of rapid uniform pointwise convergence to the Fermi-Dirac operator[82]. The number of terms (M) required depends upon the gap (which determines β), and the total bandwidth of the spectrum. The γ_l are easy to determine from the form

432

of the Fermi-Dirac distribution[57]. In a typical practical calculation (as for example c-Si or c-GaAs) $M \approx 50$ has been found to be adequate. A larger gap requires a smaller β (which is easier to approximate since the projection operator is smoother for energies near μ). Thus, *as usual*, insulators are easier to handle than metals[84]. As one might expect, if this operator is applied to a spatially local vector, this locality is substantially preserved, *and the resulting state is entirely constructed from eigenvectors* $|\psi\rangle$ *in the occupied subspace.* These projected states are suitably designated as "generalized Wannier functions", and what is best, these calculations can be carried out without any minimization: just a single application of $\hat{\rho}$ on an arbitrary vector, which is O(N). We[85] have adapted the method to making the initial guesses as a complement to the Ordejón *et al.* order N scheme.

The use of a projector leads to a very simple approach to force and dynamics calculations. Note that Equation 19 is a general operator equation. In an arbitrary representation $|\omega\rangle$, the density operator becomes the density matrix, and it is straightforward to compute the band energy

$$E = \sum_{\omega\nu} \langle\omega|\hat{\rho}|\nu\rangle\langle\nu|\hat{H}|\omega\rangle \tag{20}$$

band structure forces

$$\mathbf{F}_\alpha = -\sum_{\omega\nu} \langle\omega|\hat{\rho}|\nu\rangle\langle\nu|\partial\hat{H}/\partial\mathbf{R}_\alpha|\omega\rangle. \tag{21}$$

and anything else related to the occupied subspace. In the foregoing equations, we may directly compute

$$\langle\omega|\hat{\rho}|\nu\rangle \approx \sum_{l=0}^{M} \gamma_l(\beta,\mu)\,\langle\omega|T_l(\hat{H})|\nu\rangle, \tag{22}$$

which is O(N) since (1) use of the Tchebychev recurrence relation avoids computation of matrix powers and (2) for real-space localized orbitals $|\omega\rangle$ and $|\nu\rangle$ sufficiently separated, $\rho_{\omega\nu} = 0$ for an insulator, (thus, such matrix elements need not be computed) a result following again from Kohn's[2] work. Recently, Goedecker[83] has implemented orthogonal tight-binding MD with essentially these approximations, and Stephan and Drabold[85] have generalized to *ab initio* MD with a local basis by including the effects of overlap.

3.6. PHONONS FROM ELECTRONIC STRUCTURE WITH LINEAR SCALING

In this section I point out that vibrational spectra may also be acquired in an order-N fashion[64]. The approach is based upon the finite differences dynamical matrix approach. The dynamical matrix $D_{\alpha I,\beta J}$, defined as $D_{\alpha I,\beta J} = (M_I M_J)^{-1/2} \partial^2 E / \partial u_{\alpha I} \partial u_{\beta J}$, where E is the total energy of the system, α and β are cartesian coordinates on atoms I and J, respectively, and the M's are ionic masses, is computed by finite differences: atom I is displaced an amount $\Delta x_{\alpha I}$ in the α direction, and the forces on all the atoms $\mathbf{F}_{\beta J}$ are computed. Then, $D_{\alpha I,\beta J} \approx (M_I M_J)^{-1/2} \mathbf{F}_{\beta J} / \Delta x_{\alpha I}$, since the force on an atom is the derivative of the total energy with respect to the displacement of that atom: $\mathbf{F}_{\beta J} = -\partial E / \partial x_{\beta J}$. We will assume that at the beginning of the computation we have solved the electronic problem for the system in equilibrium, in which all the atoms experience a zero force. We have therefore minimized the functional $\tilde{E}$ and obtained the localized electronic wave functions $\{\chi_j^0\}$ and the total energy E^0. During the process of computing the dynamical matrix, when atom I is displaced, we must minimize the electronic energy again, and find the new optimized localized functions $\{\chi_j^I\}$. At this point, we introduce the first basic approximation of our method: since the functions $\{\chi_j^I\}$ are localized, only those in the neighborhood of atom I will differ significantly from the solutions of the equilibrium system $\{\chi_j^0\}$. Therefore we can, as an approximation, optimize only those functions χ_j which are located close enough to the atom that is moving, leaving all the rest of the functions unchanged. The criterion used in this work is to allow variations in only those wave functions which include the atom that is being displaced. We denote this set of functions as $\mathcal{S}_I \equiv \{\chi_j \mid |\mathbf{R}_j - \mathbf{R}_I| < R_c\}$, where $\mathbf{R}_j$ is the center of χ_j, $\mathbf{R}_I$ is the position of atom I and R_c is the cutoff radius of the LWF's.

In order to obtain the optimized set of wave functions $\mathcal{S}_I$, we must minimize the energy functional $\tilde{E}$, defined in Eq. (18), with respect to the functions $\chi_j \in \mathcal{S}_I$. This is therefore a *restricted energy minimization*, since the functions that do not belong to set $\mathcal{S}_I$ are not allowed to change. During the minimization, there is a majority of terms in Eq. (18) that are constant: those involving matrix elements between functions χ_i and $\chi_j \notin \mathcal{S}_I$. It is convenient to separate those terms from the ones that do vary in the energy functional:

$$\begin{aligned}
\tilde{E} &= 2 \sum_{i \text{ or } j \in \mathcal{S}_I} H_{ji}(2\delta_{ij} - S_{ij}) + 2 \sum_{i,j \notin \mathcal{S}_I} H_{ji}(2\delta_{ij} - S_{ij}) \\
&= \tilde{E}_1^I + \tilde{E}_2^I
\end{aligned} \tag{23}$$

so that $\tilde{E}_2^I$ is constant during the minimization, and does not need to be computed. Therefore, the functions $\chi_j \in \mathcal{S}_I$ are obtaining minimizing the 'local' functional $\tilde{E}_1^I$. This operation scales as $O(1)$, i.e., does not depend on the size of the system. This is easy to see, since if the system size is increased the number of functions included in the set $\mathcal{S}_I$ will not increase, provided that the system is large enough. Therefore, the solution of the electronic problem for the $3N$ atomic displacements (one per atom per direction in space) takes an $O(N)$ effort. For each atomic displacement, once the electronic wave functions have been obtained in $O(1)$ operations, we need to compute the forces on all the atoms in the system. This would be an $O(N^2)$ task, since for each of the $3N$ atomic displacements we must obtain N forces. Here we introduce the second approximation of our method: only the forces on atoms closer than a certain cutoff R_f from the displaced atom are computed, the rest of the forces being set to zero. This approximation takes into account the fact that the forces on each atom depend ony on its local environment, and not on the details of the structure in distant regions. This also reflects the fact that the dynamical matrix elements between distant atoms decreases rapidly with distance. Our approximation can therefore be put in terms of imposing a cutoff on the dynamical matrix:

$$D_{\alpha I,\beta J} = 0 \quad \text{if} \quad |\mathbf{R}_I - \mathbf{R}_J| > R_f \tag{24}$$

so that the resulting matrix is sparse. Its calculation and storage are both $O(N)$. With the dynamical matrix in hand one can exactly diagonalize D for small systems or use the Maxent spectral technique[49, 65] for large systems. In addition, the shifted Lanczos can be used for computing a limited number of vibrational eigenvectors in a user specified energy range[73]. Of course this method also works for empirical potentials in which case the germain approximations are that D have finite range and one of the spectral techniques described in this paper.

4. Acknowledgements

A number of teachers, colleagues and students have contributed to parts of the work reported here. I would like to thank Peter Fedders, Otto Sankey, Richard Martin, Ed Jaynes, Pablo Ordejón, Ron Cappelletti, Uwe Stephan Jian Jun Dong, Dominic Alfonso, Petra Stumm and Mark Cobb. I would like to thank the National Science Foundation for support under Grant DMR-93-22412.

References

1. Weaire, D. and Thorpe, M. F. Phys. Rev. B 4 2508 (1971).
2. Kohn, W. Phys. Rev. **115**, 809 (1959).
3. Car, R. and Parrinello, M., Phys. Rev. Lett. **55**, 2471 (1985).
4. Sankey, O. F. and Niklewski, D. J. (1989), Phys. Rev. B **40** 3979
5. For example, Nakano, A. *et al.*, (1994) Phys. Rev. B **49** 9441
6. Carlsson, A. E. (1990), in *Solid State Physics, Adv. in research and Applications*, edited by H. Ehrenreich and D. Turnbull (Academic, New York, 1990), Vol. 43, p.1
7. Ercolessi, F. *et al.* (1994), Europhys. Lett. **26** 583
8. Harrison, W. (1980) *Electronic Structure*, Freeman, San Francisco,
9. Slater, J. C. and Koster, G. F. (1954) Phys. Rev. **94** 1498

10. Allan, D. and Joannopoulos, J. (1984) in *Hydrogenated Amorphous Silicon II*, ed. by J. Joannopoulos and G. Lucovsky (Springer, Berlin) p. 5
11. Davidson, B. N. (1992) Ph.D. Thesis, N. C. State University (unpublished).
12. Vogl, P. *et al.* (1983), J. Phys. Chem. Solids **44**,365
13. Goodwin, L. *et al* (1989) Europhys. Lett. **9** 701 see also Mercer, J. and Chou, M. Y. (1993) Phys. Rev. B **47** 9366
14. Xu, C. H. *et al.* (1992), J. Phys. Cond. Matt. **4** 6047.
15. Drabold, D. A., *et al.* (1994), Phys. Rev. Lett. **72** 2666
16. Wang, C. Z. and Ho, K. M. (1994) Phys. Rev. Lett. **72** 2667
17. Menon, M. *et al.* (1993) Phys. Rev. **47** 12754
18. Seifert, G. and Jones, R. (1991), Z. Phys. D **20** 77
19. Porezag, D. *et al.* (1995), Phys. Rev. B **51** 12947
20. Bretthorst, G. L. (1988), *Bayesian Spectrum Analysis and Parameter Estimation*, Springer (New York)
21. Born, M. and Huang, K. (1954) *Dynamical Theory of Crystal Lattices*, Oxford Univ. Press, Clarendon
22. Hohenberg, P. and Kohn, W. (1964) Phys. Rev. **136** B 864; Kohn, W. and Sham, L. J. (1965) Phys. Rev. **140** A 1133
23. Carlsson, A. E., submitted to Phys. Rev. Lett.
24. Ceperley, D. M. and Alder, G. J. (1980), Phys. Rev. Lett. **45** 566
25. Bachelet, G. B. *et al.* (1982) Phys. Rev. B **26** 4199
26. Vanderbilt, D. (1990), Phys. Rev. B **41** 7892
27. Troullier, N. and Martins, J. L., (1992) Phys. Rev. B **41** 1754
28. Kleinman, L. and Bylander, D. M. (1982), Phys. Rev. B **48** 1425
29. Yang, S. (1996), Ph.D. thesis, University of Illinois (unpublished).
30. Harris, J. (1985) Phys. Rev. B**31**, 1770
31. Foulkes, W. M. C. and Haydock, R. (1989) Phys. Rev. B **39** 12 520
32. Smith, J. R. *et al.* (1995), Phys. Rev. Lett. **74** 3084
33. Demkov, A. A. *et al.* (1995) Phys. Rev. B **52** 1618
34. Ordejón, P. *et al.* (1996) Phys. Rev. B **53** R10441
35. Payne, M. *et al.* (1992) Rev. Mod. Phys. **64** 1045
36. Pastore, G. *et al* (1991) Phys. Rev. B **44** 6334
37. Fulde, P., (1993) *Electron Correlations in Molecules and Solids* Springer-Verlag, Berlin.
38. Perdew, J. (1996), Intl. J. Quant. Chem **57** 309
39. Hamann, D. R. (1996), Phys. Rev. Lett. **76** 660
40. Hedin, L. (1965), Phys. Rev. **139** A 796
41. Hybertsen, M. K. and Louie, S. G. (1986), Phys. Rev. B **34** 5390
42. Mitas, L. and Martin, R. M. (1994), Phys. Rev. Lett. **72** 2438
43. Grossman, J. C., Mitas, L. (1994), Phys. Rev. Lett. **74** 1323
44. Grossman, J. C. *et al.* (1995) Phys. Rev. Lett. **75** 3870
45. Ragavachari, K. *et al.* (1993), Chem. Phys. Lett. **214** 357
46. Mitas, L. (1995) in *Electronic Properties of Solids Using Cluster Methods*, ed. by T. A. Kaplan and S. D. Mahanti, Plenum, New York, p. 151.
47. Mitas, L. Computer Phys. Comm. (to be published).
48. Drabold, D. A. *et al.* (1990) Phys. Rev. B **42** 5345
49. Drabold, D. A. and Sankey, O. F. (1993) Phys. Rev. Lett. **70** 3631
50. Röder, H. *et al*, submitted to Phys. Rev. B; Stephan, U. and Drabold, D. A. (unpublished).
51. Skilling, J. (1989) in *Maximum Entropy and Bayesian Methods*, J. Skilling, Ed., Kluwer, Dordrecht p155.
52. Silver, R. N. *et al.* (1994) Intl. J. Mod. Phys. C **5** 735
53. Lanczos, C. (1956) *Applied Analysis* Prentice Hall, New York
54. Strohmaier, B. *et al.* (1985) Phys. Rev. C **32** 1379.
55. Haydock, R. (1980) in *Solid State Physics: Advances in Research and Applications*

Edited by F. Seitz, H. Ehrenreich and D. Turnbull, Academic, New York
56. Thirring, H. (1913) Phys. Zeit. **14** 867. Montroll used moment calculations exploiting trace invariance to compute vibrational state densities in the forties. Further discussion can be found in Born and Huang[21].
57. Press, W. H. *et al.* (1986) *Numerical Recipes, The Art of Scientific Computing*, (Cambridge University Press, Cambridge).
58. Mead, L. R. and Papanicolaou, N. (1984), J. Math. Phys. **27** 2903
59. Carlsson, A. E. and Fedders, P. A. (1986) Phys. Rev. B **34** 3567
60. Turek, I. (1988) J. Phys. C **21** 3251
61. Drabold, D. A. and Jones, G. L. (1991) J. Phys. A **24**, 4705
62. Jaynes, E. T. (1983), *Papers on Probability, Statistics and Statistical Physics*, Kluwer, Dordrecht
63. Ordejón, P. *et al.* (1995) Phys. Rev. B **51** 1456. References to other order N schemes may also be found in this paper.
64. Ordejón, P. *et al.* (1995), Phys. Rev. Lett. **75**, 1324
65. Drabold, D. A. *et al.* (1995), Solid State Commun.,**96**, 833
66. web: http://www.phy.ohiou.edu, email: drabold@roma.phy.ohiou.edu
67. Mott, N. F. and Davis, E. A. (1979) *Electronic Processes in Non-Crystalline Materials* 2nd edn. Oxford Univ. Press, Clarendon
68. Drabold, D. A. *et al.* (1991), Phys. Rev. Lett. **67** 2179
69. Djordjevic, B. R. *et al.* (1995), Phys. Rev. B, **52**, 5685
70. McKenzie, D. R. *et al.* (1991), Phys. Rev. Lett. **67** 773
71. Wooten, F. and Weaire, D. (1987), in Solid State Physics: *Advances in research and Applications*, edited by H. Ehrenreich and D. Turnbull, Academic Press, New York, 1987, vol. 40, p.2.
72. Drabold, D. A. *et al.* (1994), Phys. Rev. B **49** 16415
73. Grosso, G. *et al.* (1993), Il Nuovo Cimento D **15**, 269
74. Golub, G. H. and Van Loan, C. F. (1983) , *Matrix Computations* (Johns Hopkins University Press, Baltimore)
75. For amorphous Si, see particularly Aljishi, S. *et al.*, (1990) Phys. Rev. Lett. **64**, 2811 and references therein.
76. Halperin, B. I. and Lax, M. (1966), Phys. Rev. **148**, 722; (1967) Phys. Rev. **153**, 802
77. Soukoulis, C. M. *et al.* (1984), Phys. Rev. Lett. **53**, 616, and references therein.
78. This assumption means that we are assuming that the energy states are inhomogeneously broadened (an idea intimately connected with localization; see Drabold, D. A. and Fedders, P. A. (1988), Phys. Rev. B **37** 3440). Such an assumption is probably invalid for small distortions in a crystal, but it is plausible for a disordered network.
79. Martin, R. M. (1970) Phys. Rev. B **1** 4005
80. Drabold, D. A. *et al.* Submitted to Phys. Rev. B
81. Cappelletti, R. L. *et al* (1995) Phys. Rev. B **52** 9133
82. Sankey, O. F. *et al.* (1994) Phys. Rev. B **50** 1376.
83. Goedecker, S. and Colombo, L. (1994) Phys. Rev. Lett. **73** 122
84. Equation (19) is still valid for metals, and β becomes the reciprocal temperature (assuming that the electrons and lattice are in equilibrium); however the expansion will require a very large M, and the resulting projected states will not be very spatially local (they can be expected to decay as a power-law).
85. Stephan, U. and Drabold, D. A. (unpublished)

AMORPHOUS SILICON

Defects and Disorder

GUY J. ADRIAENSSENS
Laboratorium voor Halfgeleiderfysica
Katholieke Universiteit Leuven
Celestijnenlaan 200D, B-3001 Heverlee, Belgium
guy.adri@fys.kuleuven.ac.be

1. Introduction

Amorphous silicon (a-Si), and its characteristic defect, the silicon dangling bond, are definitely amongst the most-widely studied topics in the field of amorphous materials, with probably only silicon dioxide and its E' center having generated comparable activity. While the evident reason for this interest is purely utilitarian, and follows from the eminent suitability of these substances for microelectronic applications, a wealth of purely scientific information on the dangling bond defect has been accumulating in the process. In spite of that, many questions relating to a physical understanding of amorphous silicon defect states have remained unresolved till now. The complexity and lack of definition of the amorphous state are the reason. Frequently, the analysis of experimental data requires that assumptions be made about some of the model parameters to compensate for our lack of knowledge; but that in turn often means that models which were designed to account for a given set of experimental observations, are not always able to properly describe the situation under modified circumstances. This intrinsic ambiguity in our understanding of the amorphous silicon defect configurations – a direct consequence of the need to make those assumptions – will be a major theme throughout this chapter.

Contrary to just about every other amorphous material discussed in this volume, amorphous silicon can be prepared as either intrinsic, or p-, or n-type semiconductor, and is sufficiently conductive to allow its use as an active element in electronic devices. Consequently, electrical measurements can be used to obtain information on the distribution of electronic states in a-Si, in addition to the standard optical and X-ray spectroscopies. Since the conductivity is especially sensitive to the localized electronic states in

M. F. Thorpe and M. I. Mitkova (eds.), Amorphous Insulators and Semiconductors, 437–468.
© 1997 *Kluwer Academic Publishers. Printed in the Netherlands.*

438

the band gap of the semiconductor, electrical spectroscopies are in fact the primary ones for the study of the electronic density of states (DOS) in the gap of a-Si. Given that the electrical properties form the main distinction between a-Si and the other traditional amorphous semiconductors, we will discuss the distribution of electronic gap states in detail, and indeed make extensive use of measured conductivities, both in the dark and under (or after) illumination.

As silicon is normally four-fold coordinated in the condensed state, it is not a glass former; amorphous silicon is, therefore, produced directly as a thin solid film, mostly from a vapour. As illustrated schematically in Fig. 1, silicon atomic $3s$ and $3p$ orbitals hybridize to four, tetrahedrally arranged, sp^3 orbitals which then form σ and σ^* bonding and anti-bonding orbitals with neigbouring atoms. In the solid, these levels give rise to the valence

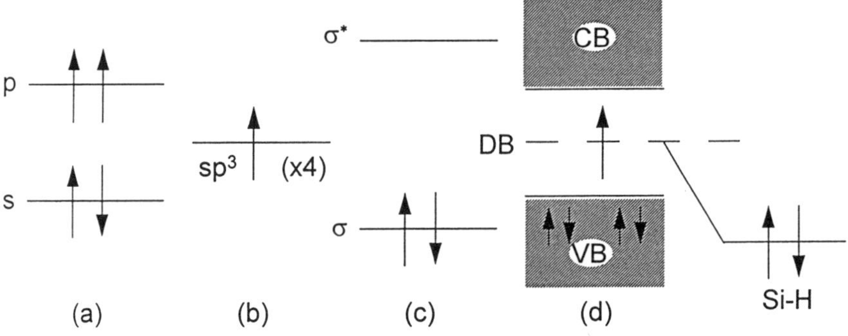

Figure 1. Schematic representation of the energy relations between the Si atomic s and p orbitals (a), the hybrid sp^3 orbitals (b), the isolated σ and σ^* bonding and anti-bonding orbital (c), and the valence and conduction bands in the solid (d). Also shown are the position of a Si dangling bond (DB) and of a Si-H bond.

and conduction bands. Any sp^3 orbital which does not find a partner for bonding will, to first approximation, retain its original energy and become a defect center (the silicon dangling bond) in the band gap. Since each dangling bond contains one electron, such centers can be observed through electron spin resonance (ESR). In a-Si, the absence of a regular lattice leads to the frequent occurrence of isolated dangling bonds, and therefore to a much higher defect density in the gap than is found in crystalline silicon. A neutral dangling bond defect is often indicated by the symbol D^0. If such defect traps an electron it becomes negatively charged, and is therefore given the symbol D^-. The energy of a D^- state is raised above the D^0 level by U, the Hubbard repulsion energy between the two electrons now occcupying the same sp^3 orbital. The energy of the positively charged defect D^+ (obtained when D^0 loses its electron) is not shifted from the D^0 position.

a-Si layers, prepared by standard thermal evaporation or sputtering techniques, were found to have a dangling-bond density on the order of $10^{20} cm^{-3}$. This defect density makes them totally unsuitable for any practical use. The first report on a-Si with a low defect density was published in 1969 by Chittick, Alexander and Sterling [1] who used a radio-frequency glow-discharge decomposition of silane gas (SiH_4) to produce the amorphous layers. It was later realized that this technique entailed the incorporation of hydrogen in the growing film, and that the hydrogen strongly reduced the dangling bond density by forming Si-H bonds. The Chittick *et al.* result led to the demonstration by Spear and Le Comber [2] that a-Si prepared by glow discharge could be doped, and thus to the present use of the material. The glow-discharge technique, now generally referred to as *plasma-enhanced chemical vapour deposition* (PECVD), has remained the standard preparation method for what has become known as *hydrogenated amorphous silicon* (a-Si:H). The introduction of hydrogen during or after preparation of evaporated or sputtered films did succeed in lowering the defect density there as well, but not quite to the levels obtained with the plasma technique. Defect densities as low as $10^{15} cm^{-3}$ are presently being attained.

The amorphization of crystalline silicon (c-Si) by implantation of Si ions [3] is one further way of obtaining a-Si which is worth mentioning. It is not meant for practical applications, but it allows the preparation of homogeneous and impurity-free amorphous layers for studying the basic a-Si lattice. In a recent study [4], subsequent controlled hydrogen implantation was used to investigate the behaviour of hydrogen in a-Si:H.

2. Hydrogenated Amorphous Silicon

A very readable and comprehensive survey of fundamental properties and characteristics of a-Si:H, including all key references, may be found in a recent book by Street [5]. We will, therefore, restrict ourselves to a brief reminder of some essential aspects of the material, to its dangling bond defect, and to the possible role of hydrogen in determining its properties. Intrinsic a-Si:H will be subject, with only an occasional reference to doped material.

As mentioned above, PECVD from SiH_4 has become the standard preparation method for a-Si:H layers. Optimal results are obtained from deposition on substrates heated to 250°C, which makes it an attractive low-temperature process in terms of semiconductor processing. Capacitively-coupled systems are used since they can readily be scaled up for the deposition of large-area layers as are needed for solar-cell or flat-panel-display applications. The technique provides ease of doping through the addition of small amounts of, for instance, phosphine (PH_3) or diborane (B_2H_6) to the silane gas, and also allows the production of silicon alloys through the admixture of gases such as germane (GeH_4), methane (CH_4), or ammonia (NH_3). Dilution of the starting gases with hydrogen (H_2) does allow control of the growth rate and quality of the films and can even lead to the production of microcrystalline rather than amorphous layers.

Hydrogen plays a very important role in virtually every aspect of a-Si:H production and use, and it is a role which is still poorly understood. The passivation of the dangling bond defects by hydrogen is a known fact, just as it is also known that good-quality PECVD a-Si:H contains about 10 at.% hydrogen, much more than is needed to passivate the defects. How and where the 'superfluous' hydrogen is incorporated in the material, and how it influences the material properties remains the subject of many investigations. The recent hydrogen implantation study of Acco *et al.* [4] showed that the solubility limit of hydrogen in the a-Si lattice is 4 at.%, so some inhomogeneous distribution of the remainder is required, e.g. with hydrogen decorating the inner surface of voids that have been detected in PECVD material, or being trapped there in molecular form.

2.1. LIGHT-INDUCED DEFECTS

The extensive interest for the behaviour of hydrogen in a-Si:H is directly related to the notion that hydrogen plays a key role in the creation of a set of metastable light-induced defects that significantly degrade a-Si:H quality. The phenomenon has become known as the Staebler-Wronski effect (SWE) [6], and is a main concern of a-Si:H solar cells research since it causes a considerable drop in cell efficiency during the first hours of operation. Within the experimental accuracy of the ESR or gap-state spectroscopies that have been used for their characterisation, the light-induced defects have revealed the same parameters as the intrinsic dangling bonds. Somewhat unclear still is whether the last statement is valid for essentially all, or just for the larger part of the light-induced defects.

Although the light-induced defects and the SWE have received most of the attention, it should be noted that prolonged submission to a high current density will also create defects in a-Si:H. The process is known as current degradation. Again – to first approximation – the additionally created defects cannot be distinguished from the intrinsic ones.

The defect creation mechanism is a very slow one; for instance, it takes many hours of sunlight before the defect density saturates at some $10^{17} cm^{-3}$. The ensuing defects are stable at room temperature, but can be annealed out by raising the temperature above 150°C. The slow pace of these processes has led to the conclusion that structural rearrangements rather than electronic transitions must be taking place, and, since hydrogen is known to be very mobile in the lattice, the assumption that hydrogen is involved in some way. This point of view has recently been reinforced through the emergence of a new preparation method, *hot-wire deposition* (HWCVD) [7], whereby SiH_4 is decomposed by a hot (~ 2000°C) tungsten filament and deposited on ~ 400°C substrates, and whereby a low hydrogen content of the amorphous layers and a low sensitivity to light-induced defect creation are jointly observed. HWCVD a-Si:H can be made with less than 1 at.% hydrogen while preserving the desirable characteristics of PECVD material. Since, in addition, the deposition rate can be kept much higher for hot-wire than for plasma-deposited material of comparable quality, HWCVD is now being investigated in many laboratories [8].

2.2. DEFECT EQUILIBRATION

Observation of the evolution in time, and at various ambient temperatures, of the a-Si:H defect density, either intrinsic or light-induced, has introduced the notion of a *thermal equilibrium state* into the discussion of amorphous silicon. This may be a bit surprising for a material which cannot even be obtained as a glass, and hence must be far from thermodynamic equilibrium. However, the thermal equilibrium meant is one between bonding and non-bonding configurations in a-Si:H [9], i.e. between Si-Si bonds and dangling bond defects.

The defect density is found to relax slowly to some equilibrium value, with the initial density being determined by the deposition conditions, and the relaxation being much faster at high than at low temperatures. The time dependence of the defect relaxation can be described by the stretched-exponential form

$$X = X_0 \exp[-(t/\tau)^\beta], \qquad (1)$$
$$\tau = \tau_0 \exp(E_B/kT), \qquad (2)$$

with order-of-magnitude values for β, τ_0, and E_B being $0.6, 10^{-11}$s, and 1eV. Stretched-exponential behaviour is characteristic for processes involving a wide distribution of event times, but generally fails to provide a microscopic model for those relaxations. The same functional form fits the relaxation of light-induced defects [10], as well as the dispersive diffusion of hydrogen in a-Si:H [11]. These observations have cemented the link between hydrogen motion and defect dynamics, and have in fact led to the introduction of the concept of a hydrogen glass to model the thermal equilibration of the silicon defects [11, 12].

442

A relationship between the relaxation and the phonon spectrum was suggested by Godet [13], who showed that for a number of amorphous substances the temperature dependence of the dispersion parameter β can be approximated as $\beta = T/T^*$, with kT^* being roughly equal to $\hbar\omega_0$, where ω_0 stands for the pseudo-transverse optical phonon frequency as deduced from Raman or infrared spectra. Interestingly, when the same is attempted for chalcogenide glasses, the point where the temperature equals T^*, i.e. where $\beta = 1$ and the process becomes monorelaxational, corresponds to the glass transition temperature T_g [14]. However, to use this as a basis for nominating T^* as the glass transition temperature of the hydrogen glass would be highly premature. Van de Walle [15] recently developed a model for hydrogen motion in a-Si:H which replaces the distributions of trapping energies and barrier heights that are invoked to justify the stretched-exponential formula, with just three energy levels: a ground state, a hydrogen trap above it, and a transport level (e.g. corresponding to the bond-center positions hydrogen would use in diffusing through the lattice). Instead of the stretched-exponential form of Eq. (1), Van de Walle derives a relation

$$t = -\tau' \ln \frac{X}{X_0} + \gamma \left(\frac{X}{X_0} - 1 \right), \tag{3}$$

which can be readily compared with Eq. (1) when the latter is rewritten as

$$t = \tau[-\ln(X/X_0)]^{1/\beta}. \tag{4}$$

It is shown in [15] that experimental data from Ref. [10] can be fitted equally successfully with Eq. (3) and Eq. (4). The parameter τ' was found to vary according to Eq. (2) in the same way as the stretched-exponential τ. Q. Zhang [16] subsequently found that only two levels – trapping and transport – are needed to obtain Eq. (3), and that τ' then simply becomes the free-hydrogen lifetime. He also showed that the agreement of Eq. (3) with experiment deteriorates as the temperature is lowered and the hydrogen motion becomes more dispersive. Fig.2 illustrates this by comparing the experimental data set from Kakalios *et al.* [11] with Eqs (3) and (4). Obviously, since the assumptions leading to the stretched exponential have only general plausibility and a good fit to the data in support, while the new model is based on known characteristics of hydrogen motion in silicon, the latter is to be preferred. The whole issue is an excellent example of incomplete knowledge leading to unjustified or unnecessary assumptions about an amorphous system.

2.3. THE DENSITY OF LOCALIZED STATES

We will have to return to the role of hydrogen in a-Si:H at a later stage, but should now first turn to an examination of the distribution of localized

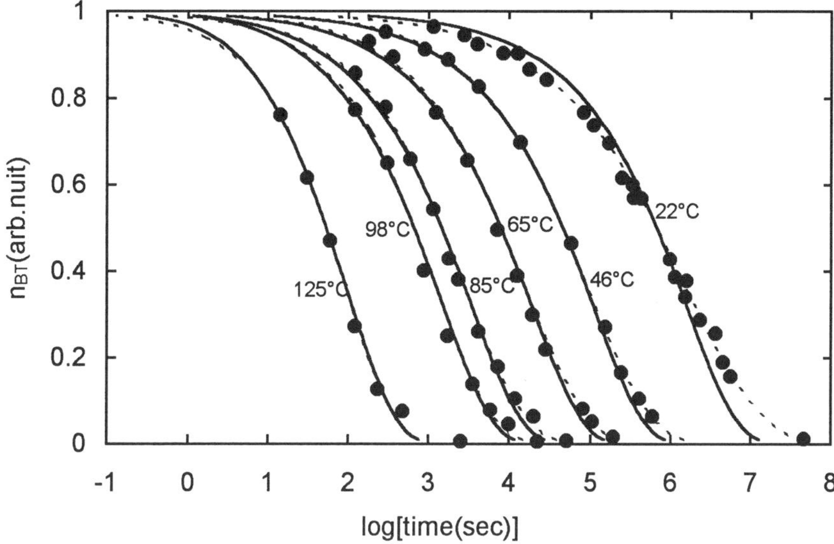

Figure 2. Fits with Eq. (3) (solid lines) and stretched exponentials (dashed lines) to the density of shallow occupied band-tail states as displayed in Ref. [11].

states in the gap, as it emerges from experimental observation. It is known from general principles that lattice disorder will produce tails of localized states extending from the band edges into the gap, and that the dangling-bond defects will generate another set of localized states at larger distances from the bands. However, the distinction between the two sets may not always be as clear-cut as this disorder or defect genesis might suggest. Indeed, the idea that there is an equilibration taking place between the most strained bonds and the broken bonds, after deposition, or whenever external conditions change, has now been generally accepted. This is the so-called weak to dangling bond conversion [17]. A further complication will be the occurrence of spatial (or indeed, temporal, as will be pointed out later) potential fluctuations that introduce a degree of inhomogeneity in the system.

Gap state spectroscopies should consequently avoid changing the weak *versus* dangling bond balance, whence small-signal excitation will be desirable. The resulting energy distributions will be averages over the sample. The latter means that not all resolved energy levels are necessarily available in all parts of the sample; when analyzing electronic transport properties, this fact may have to be taken into account. There seem to be three basic types of spectroscopy at hand to probe the energy distribution of localized gap states. They are based on, respectively, optical transitions, junction capacitance, or multiple-trapping transport.

2.3.1. *Optical Spectroscopy*

The optical spectroscopy is undoubtedly most widely used in a transmission/reflection mode to obtain estimates for the width of the band gap from the energy dependence of the absorption coefficient $\alpha(E)$, but it also has given rise to the one parameter that is now almost universally used to characterize the degree of lattice disorder in a-Si:H samples: the Urbach slope E_0. This quantity describes the exponential slope of the optical absorption

$$\alpha(E) = \alpha_0 \exp[(E^* - E)/E_0], \tag{5}$$

that is observed below the band gap in all amorphous semiconductors. Low values of E_0 indicate minimal disorder. For a-Si:H, values just below 50 meV have been measured for the best-quality material. It was shown by Stutzmann [18] that the value of E_0 also correlates with the dangling bond density of the material, and that it thus can be used as an overall quality indicator.

The straightforward optical transmission spectroscopy is not very sensitive in the region of low absorption caused by the states in the gap. It is therefore supplemented by techniques such as photo-thermal deflection spectroscopy (PDS) or the constant-photocurrent method (CPM) which measure the absorption coefficient indirectly. But whatever the technique used, absorption coefficients always contain information on both initial and final state of the optical transition, and hence lead to a so-called *joined* density of states (JDS). Deconvolution of the JDS is necessary to obtain the desired DOS, a process that requires that some assumptions be made about at least one of the bands. For example, it has become accepted that the exponential parts of $\alpha(E)$ curves represent the joint density of valence and conduction band tail states, and thus in fact the distribution $g_v(E)$ of valence band tail states since the conduction band tail is steeper[1] and the conduction band can hence be thought of as a step function.

One optical technique which is specifically suited for studying gap state energies is the optical modulation spectroscopy (OMS) [19], which measures the changes in optical transmission caused by across-gap illumination. The energies whereby photo-induced absorption or bleaching is resolved, can be related to energy levels in the gap, but it is difficult to deduce quantitative information on the density of gap states in this way.

2.3.2. *Capacitive Methods*

Deep-level transient spectroscopy (DLTS) is the best-known representative of this type of methods. The basic idea is that, at a metal-semiconductor

---

[1]This is known from transient photocurrent measurements, as will be discussed in a later section.

junction, a change in applied voltage will change the depletion region and force gap states to either give up or accept (depending on the direction of change) charge carriers. By using the sample as the dielectricum in a capacitor, induced changes can be monitored by measuring the capacitance. The method is rather indirect, and will not be discussed any further here.

2.3.3. *Transient Photocurrents*

Electrical transport in amorphous semiconductors is trap-limited. Due to the large number of localized states in the gap, free carriers will be very quickly trapped in this distribution of gap states, and remain trapped until they are thermally released after an average time of

$$t_r = \nu_0^{-1} \exp(E_t/kT), \tag{6}$$

where ν_0 is the attempt-to-escape frequency (on the order of 10^{12}s^{-1} for gap states of a-Si:H), E_t is the energy depth of the trap as measured from the band edge into the gap, k is the Boltzmann constant, and T is the absolute temperature. Once free, the carrier will move with free-carrier mobility μ_0 until trapped again. For electrons in a-Si:H, $\mu_0 \simeq 10\text{cm}^2\text{V}^{-1}\text{s}^{-1}$ has become the accepted value; earlier speculation that it might be as high as $500\text{cm}^2\text{V}^{-1}\text{s}^{-1}$ [20] proved to be unjustified [21].

As the average carrier will have been trapped many times before completing its path in an amorphous semiconductor, this current mode is called multiple-trapping (MT) transport. In MT mode each individual carrier will have interacted with a specific subset of the traps during its transit, and therefore stay in the sample for a long time if it encountered a deep trap, or exit quickly if it met shallow traps only, but a sufficiently large ensemble of carriers will contain information about a representative sample of localized states. Two types of photoconductive spectroscopy have been developed to extract that information from measured currents, one in the time and one in the frequency domain. When not specifically stated otherwise, these spectroscopies always assume that charge capture is instantaneous on the time scale of the experiment (such that only the release time from the traps need be considered), and that all traps have an equal capture probability, i.e. that the capture cross-section is energy-independent.

In the time-domain experiment, a laser flash will excite a large number of electrons from the valence band into the conduction band at $t = 0$, and the transient photocurrent is then monitored as a function of time. Relatively small, or large, currents at long times do mean relatively low, or high, densities of deep states, and so on. A full analysis of the transient current $I(t) \propto n_f(t)$, which will lead to the DOS $g(E)$, requires solving the rate equation for the free carrier density $n_f(t)$:

$$\frac{d}{dt} n_f(t) = n_t(t) f(t) - \frac{n_f(t)}{\tau}, \tag{7}$$

where $n_t(t) = n_0 - n_f(t)$ is the trapped charge density for an initial total excess density n_0, $f(t)$ the release probability of trapped charge, and τ the free carrier lifetime. The release probability $f(t)$ is the product of the capture probability at an earlier time $t - t'$, which is $(n_f(t-t')/n_0)(dt'/\tau)$, and $\Psi(t')$, the probability of a carrier staying trapped for the length of time t', summed over all times prior to t. $\Psi(t)$ is commonly known as the waiting-time distribution function. Introduction of

$$f(t) = \int_0^t dt' \left[\frac{n_f(t-t')}{n_0 \tau} \right] \Psi(t') \tag{8}$$

into Eq. (7), then leads a Volterra-type integral equation for the transient current

$$(1 + \tau \frac{d}{dt}) I(t) = \int_0^t dt' \left[(1 - \frac{I(t)}{I_0}) I(t - t') \right] \Psi(t') \tag{9}$$

where I_0 is the initial photo-generated current. Solving this equation for $\Psi(t)$ is required [22] as a first step towards $g(E)$. By using the approximation $p(t) = \delta[t - t_r]$ for the release probability per unit time out of a trap at depth E_t, where t_r is given by Eq. (6), one finally obtains

$$g(E) = (N/kT) t \Psi(t), \quad E = kT \ln(\nu_0 t), \tag{10}$$

whereby $N = \int g(E) dE$. The need to (numerically) solve the Volterra equation on the basis of the experimental $I(t)$ data has limited the attractiveness of this method.

The frequency-domain method, introduced by Oheda [23], makes use of a sinusoidal modulation of the light intensity, which gives rise to a modulated photocurrent (MPC) whose amplitude and phase are recorded. At any given modulation frequency ω, traps at energies less than $E_\omega = kT \ln(2\pi\nu_0/\omega)$ will be able to release trapped charge in phase with the changing light intensity, while release from deeper traps will be increasingly out of phase. This produces a phase shift, between the light modulation and the measured photocurrent, which will depend on – amongst other things – the relative numbers of traps with energies on either side of E_ω. Changing the frequency of the modulation will change this phase shift, and will provide the information input for a calculation [24] of the distribution of localized states. A combination of time- and frequency-domain methods was recently introduced by Main *et al.* [25]. They calculate the DOS by using the MPC formalism on a data set obtained by Fourier transforming standard transient photocurrent traces.

In turning eventually to a detailed discussion of the a-Si:H DOS, we will restrict ourselves largely to information obtained from time-of-flight (TOF) transient photoconductivity experiments. This restriction will allow us to pay sufficient attention to the premises of the method and the assumptions of the analysis, to permit a critical evaluation of the results.

3. The Time-of-Flight Experiment in a-Si:H

In a time-of-flight experiment [26], the a-Si:H layer is placed between two planar electrodes, blocking against injection of the charge carrier (electron or hole) to be studied, and with at least the top one being moderately transparent for across-gap light. An excess density of free carriers is then created in the sample immediately below the top electrode by a short pulse of highly-absorbed light, and either electrons or holes are moved to the other electrode by an applied field of appropriate polarity. A nitrogen-laser-pumped dye, typically lasing at 530nm for $\sim$3ns with some 10^{13}photons/pulse, is usually used as excitation source. The experimental arrangement is schematically shown in Fig. 3. Sufficiently blocking contacts

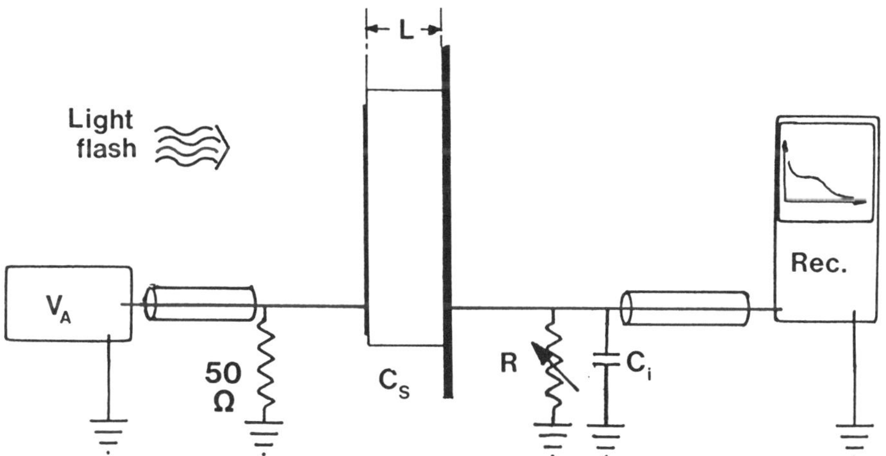

Figure 3. Schematic diagram of a basic TOF set-up. V_A is the voltage supply, L is the sample dimension, C_s and C_i are the capacitances of the sample and the input stage of the signal recorder (Rec.), and R is a variable resistance whereby the current is measured.

on a-Si:H can be provided by Schottky barriers with evaporated Cr layers, but it is preferable to have p-i-n structures where thin p^+ and n^+ surface layers provide higher barriers. The use of blocking contacts ensures that

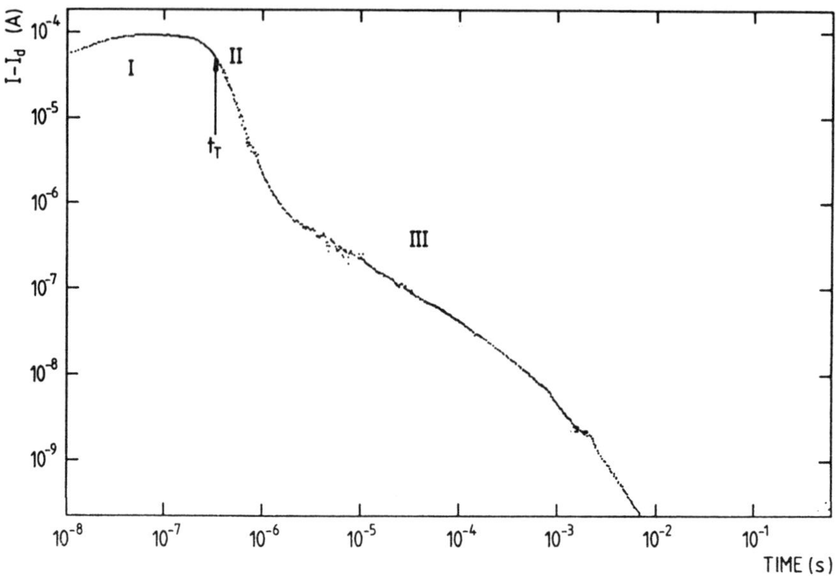

Figure 4.   Typical room temperature TOF transient photocurrent from a 5μm thick a-Si:H sample. The regions I, II and III are identified in the text.

only the primary photocurrent will be measured and, most importantly, that with electron-hole pairs being generated just inside the contact, electron and hole currents can be studied separately. The latter also means that recombination need not be taken into consideration. This feature is unique with TOF; in the other photoconductive methods both excess electrons and excess holes will always be present. In a-Si:H, where the electrons are much more mobile than the holes, most studies have been carried out with electrons.

A typical TOF photocurrent transient, measured at room temperature with 0.6V over a $\sim$5μm thick a-Si:H sample, can be seen in Fig. 4 [27]. Initially a fairly constant current is observed while the photo-generated carriers (most of which are trapped at any one time) move towards the back electrode[2]. As soon as a substantial number of carriers have been collected at the back electrode the current drops and a transit time t_T can be defined. This transit time will always roughly correspond to the time when the number of *free* carriers has been reduced by half [28]. Note that most of the carriers are then still trapped in the sample, but they do not contribute to the current of course – only *free* carriers do. The currents before and after t_T, designated as I and II in Fig. 4, are called the pre- en post-transit current. In both current regimes there will be significant trapping

---

[2]At lower temperatures electronic transport in a-Si:H becomes more dispersive and a decreasing current would be observed.

and release of charge going on throughout the DOS, in I without carriers being lost at the back contact, in II with such loss. In region III finally, the free carrier density has become so low that trapping has effectively ceased and only release of the trapped carrier distribution will define the current.

In the next section it will be shown how information about the tail states can be extracted from the short-time behaviour around the transit time. That is the domain of traditional TOF measurements. The long-time regime and the extraction of the density of *deep* states will be treated in a separate section. In both instances the results will be less than incontrovertible.

4. Distribution of Tail States

Interaction between the set of photo-excited carriers and the localized states in the tail of the transport band will dominate the initial stages of a transient photocurrent experiment due to the very short emission times – nanoseconds and less, see Eq. (6) – for charge trapped in shallow states. An analysis of the pre-transit current should therefore, in principle, be able to reveal the DOS in the tails. However, the electron mobility in a-Si:H is such that with the normally applied fields (on the order of 10^4V/cm; larger than the built-in, but smaller than the breakdown field) the transit time through the sample approaches the time-resolution of the set-up unless several μm thick samples are used. This does not leave much pre-transit current to be analysed.

Tail-state distributions have therefore traditionally been studied by modeling drift-mobility data sets. The drift mobility, which differs from the free-carrier mobility by the ratio of free to trapped charge, is for this purpose defined through the transit time t_T, the sample length L, and the applied field F as

$$\mu_d = L/t_T F, \tag{11}$$

and is determined experimentally as a function of temperature and electric field. Eq. (11) will only give meaningful values of μ_d if the electric field remains constant during the carrier transit. To achieve this, the field is applied only a short time before the laser is triggered while t_T has to be short with respect to the dielectric relaxation time of the material ($\sim$ 1ms for a-Si:H), and the photo-created charge Q_0 should be kept small compared to the charge $Q = CV$ on the sample electrodes. Fig. 5 shows a representative example [29] of an a-Si:H μ_d data set. The modeling is carried out with the transit-time expression [28]

$$t_T = t_0 + \sum_{i=1}^{m} (c_i t_0 / r_i), \tag{12}$$

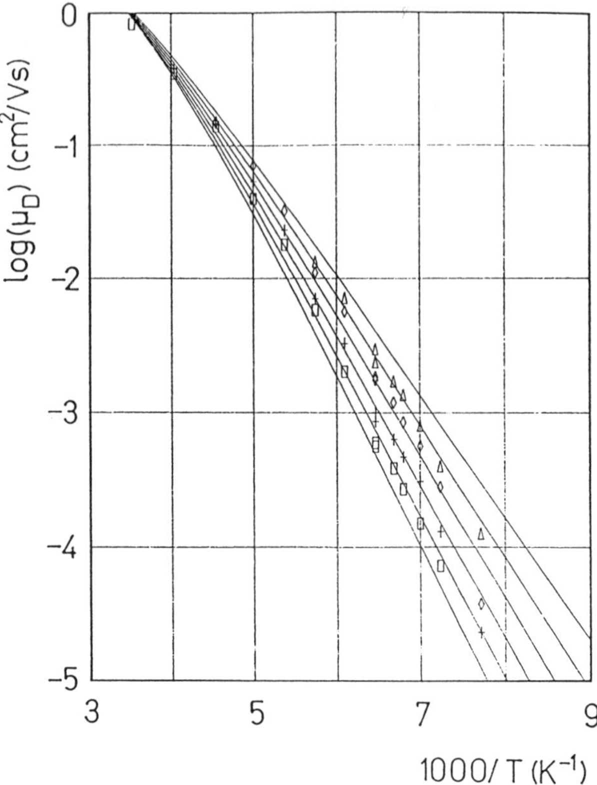

Figure 5. Drift mobility *vs* reciprocal temperature, measured at electric fields of $1.96(\square)$, $3.92(\dagger)$, $7.84(\diamond)$ and $15.7(\triangle)$kV/cm, for a 4.9μm thick Cr/a-Si:H/Cr sample. The full lines represent calculated values for the same fields, with in addition half the lowest and twice the highest field, for an exponential DOS, $T_0 = 250$K, $\nu_0 = 2.5 \times 10^{12}\mathrm{s}^{-1}$, and $\mu_0 = 7\mathrm{cm}^2\mathrm{V}^{-1}\mathrm{s}^{-1}$.

where $t_0 = L/\mu_0 F$ is the free carrier transit time, and c_i, r_i are the capture and release rates of the i-th trap. Since the traps do not form a discrete set in a-Si:H, a discretization of the continuous distribution $g(E)$ is used: $N_i = g(E_i)\Delta E$ will be the trap density at energy E_i, ΔE being the discretization step. Capture and release rates and the DOS are related by

$$c_i = bN_i \tag{13}$$

$$r_i = \nu_0 \exp(-E_i/kT), \tag{14}$$

where b is a capture constant. Detailed balance requires that

$$\nu_0 = bN_c, \tag{15}$$

with N_c the effective density of states in the conduction band which can be approximated as $g(0)kT$ for amorphous semiconductors. The actual modeling is done by comparing $\mu_d(T, F)$ values based on an assumed $g(E)$ and the Eqs. (11) - (15) with the experimental set. The major uncertainty in the results will arise from the need to make assumptions about some of the parameters, such as $g(0)$, or ν_0, or μ_0.

The summation in Eq. (12) should not run over the full range of traps, but just reach down to those that essentially control the current at t_T. As discussed in [28], qualitatively different t_T definitions may be obtained by making different choices for the limit m of the summation. However, while this choice is important when measured and calculated transit times are compared, it is less so when modeling $g(E)$ is the objective. In the latter instance, the energy or temperature dependence of the parameters may be more relevant, as will be shown below.

4.1. DIFFERENCES IN INTERPRETATION

It is a remarkable fact that, although there has been very little difference (with one exception) in the experimental drift mobility data sets that have been reported by various research groups, the interpretations of those data have been quite divergent. The exception to that uniformity deserves some special notice, since it is a high-profile one: Marshall, Street and Thompson [30] reported data from a sample which showed higher mobility values and smaller temperature dependences than observed in comparable samples by others. However, subsequent measurements at K.U.Leuven on similar samples from the same source and the same period, kindly supplied to us by Street and by Marshall, failed to show the divergent behaviour; they fitted nicely with all the rest of the data in the literature.

The first studies of a-Si:H drift mobilities were by Spear [31] and Tiedje [32] and their co-workers. In spite of the similarity of the data, it was concluded by Spear that the DOS must show a steep drop some 0.15eV below the conduction band, while Tiedje resolved a smooth exponential DOS

$$g(E) = g(0)\exp(-E/kT_0) \tag{16}$$

with characteristic temperature $T_0 = 312$K. Moreover, one of the data sets of [31] was used by Silver, Snow and Adler [33] to argue that the DOS should be represented by a double exponential, one with $T_{00} \simeq 700$K from the conduction band to ~ 0.15eV, and one with $T_0 = 300$K from there on.

While the initial data sets used only a small range of electric fields, subsequent sets which did span a much wider range did not result in any better agreement. It was easy to show by improved modeling that the double exponential proposed by Silver et $al.$ leads to a much stronger field dependence than observed experimentally, but double exponentials with different parameters were also being proposed by Vanderhaghen and Longeaud [34] and

452

Nebel and Bauer [35]. In the Nebel and Bauer analysis, an extensive study of the effect of energy-dependent capture constants (weak electron-phonon coupling) is included, but that does not alter their conclusion that "only a hybrid structure of $g(E)$ .. can fit $\mu_d(T, F)$". The Marshall *et al.* data mentioned earlier led to the proposal of a linearly varying DOS, at least between 0.085 and 0.145eV below E_c, a view later reinforced by the analysis of Marshall, Berkin and Main [36]. All these results are a bit surprising, since most experimental evidence seems to point to a simple exponential distribution of the tail states [5].

That a good description of TOF data sets can, nevertheless, be achieved with a single exponential, is shown by Fig. 5 where experimental data from Seynhaeve [29] are matched with theoretical curves calculated on the basis of such DOS. To achieve this agreement, the likely temperature dependence of some of the model parameters had to be introduced. What this means in practice, will be discussed in the next section, but already we can draw the conclusion that DOS models derived from $\mu_d(T, F)$ data sets need to be approached with caution.

4.2. TEMPERATURE DEPENDENT MODEL PARAMETERS

In the multiple-trapping model that is used for extracting DOS information from TOF results, temperature appears as an explicit parameter in the release rates r_i of Eq. (14), and through the $N_c = g(0)kT$ approximation. The latter means that in Eq. (15) either ν_0 or b will have some temperature dependence, or that the detailed-balance condition is set at one temperature only and the parameters are kept constant from there on. The difficulty of deciding which dependence is most credible for any of these parameters has led to the usage of neglecting any temperature dependence of the parameters with respect to the exponential one in the release time approximation. However, the differences between the various proposed models for the tail-state $g(E)$ are probably of the same order as the differences that would result from different assumptions about the temperature dependences. It is worthwhile therefore to examine the issue more closely.

A temperature dependence of the free-carrier mobility μ_0 has actually been used in [33]. As argued by Adler and Silver [37], scattering in extended states will give rise to a $T^{3/2}$ dependence for μ_0, arising from the presence of compensating charged centers and similar to what is found in compensated crystalline semiconductors. Longeaud *et al.* [34] on the other hand did use a T^{-2} dependence for μ_0 in analyzing their data in order to obtain a temperature-independent DOS, but they did so on an *ad hoc* basis. They further reported that a temperature dependence of ν_0 did not influence their results.

The temperature dependence of the mobility can also be examined from a more fundamental point of view. For instance, if hopping in tail states should play an important role in the transport process,

$$\mu = e\nu r^2/6kT \tag{17}$$

might be appropriate [38], with r the hopping distance and ν as discussed below. For the conduction band tail states, weak coupling of the electrons to the lattice may be assumed, whence the attempt-to-escape frequency ν may be written as [39]

$$\nu = \nu_0[1 + n(T)]^p, \tag{18}$$

where

$$n(T) = 1/[\exp(\hbar\omega/kT) - 1] \tag{19}$$

is the phonon occupation number and p is the number of phonons needed to reach the mobility edge. This relation yields a constant ν_0 at low temperatures and $\nu = \nu_0 T^p$ at high temperatures. According to Eq. (17) μ then varies as T^{p-1}. For purposes of DOS modeling these relationships are parametrized as

$$\nu = \nu_{300}(T/300\text{K})^p, \qquad \mu = \mu_{300}(T/300\text{K})^{p-1}. \tag{20}$$

Modeling of experimental data sets with the Eqs. (11)–(15) and (20) yield best results with $p = 3$ [29] although this value is not critical. It means that the temperature dependence for tail-state hopping mobility (T^2) is in practice indistinguishable from the one for extended-state transport ($T^{3/2}$). The calculated curves on Fig. 5 were obtained with the $p = 3$ value for an exponential DOS with $T_0 = 250$K, and with the ν_0 and μ_0 values shown with the figure in fact being the ν_{300} and μ_{300} values of Eq. (20). As pointed out before, the agreement between measured and calculated mobilities is good. The exponential distribution was found to give equally good agreement when fitting mobility data sets from other a-Si:H samples, either from our own laboratory, or from the literature. In all cases the value $p = 3$ could be used; T_0 values ranged between 230 and 250K. The only exception to that rule is the data set from [30].

5. Defect States

Classic TOF spectroscopy, based on observed transit times and calculated drift mobilities, can give information on the a-Si:H shallow states only (down to a few tenths of an eV) since emission from deeper states remains negligible on the t_T time scale. Conversely, since emission from deep-gap states will occur on a time scale which is long with respect to t_T, the time needed by a released carrier to leave the sample will – on the average – be negligible in comparison with the release time from a deep trap. An analysis of the current transient at much longer times will hence be necessary to gain insight into the distribution of defect states deeper in the gap. The data acquisition in a wider time domain that is needed for that purpose, is achieved by repeatedly carrying out the TOF experiment, while step-wise increasing the sampling time and the sensitivity, and then joining all the data sets on logarithmic scales.

5.1. POST-TRANSIT ANALYSIS

Extraction of the electronic DOS from the post-transit TOF signals was first worked out by Seynhaeve *et al.* [27, 29]. It is based on a number of assumptions which should not be forgotten when the results are interpreted. Specifically: the assumption that the capture cross-sections are energy-independent remains in effect, and it is further assumed that the current is strictly emission-limited. The latter assumption implies that all released carriers will actually leave the sample within a time on the order of t_T, rather than be retrapped in deep states. It can then be asserted that, for times $t > t_T$, the trapped charge distribution is frozen in until at $t_r \simeq \nu_0^{-1} \exp(E/kT)$ carriers trapped at energy depth E will be released and generate the current observed at $t = t_r$. Under the above circumstances, the post-transit current $I_{pt}(t)$ can be used to generate the DOS, $g(E)$.

5.1.1. *Practical Implementation*
Seynhaeve derived the relationship

$$I_{pt}(t).t = \frac{Q_0 t_0 \nu_0}{2g(0)} g(E) \quad , \quad E = kT \ln(\nu_0 t) \tag{21}$$

where good estimates for the quantities Q_0, t_0, ν_0 and $g(0)$ determine the quality of the result. While Q_0 and ν_0 can generally be obtained reasonably well from the total integrated current transient and from the temperature dependence of a characteristic feature on the transient (see e.g. [27] or [40]), t_0 has to be inferred from the analysis of t_T data sets (previous section), and literature data (e.g. [41]) are used for $g(0)$. A recent analysis by Yan *et al.* [42] resulted in an expression

$$I_{pt}(t).t = \frac{Q_0 kT}{\int_E^{E_f} g(E')dE'} g(E), \tag{22}$$

E_f being the dark Fermi level, which clearly indicates that only the *DOS profile* will be fixed by the post-transit analysis. To obtain absolute values

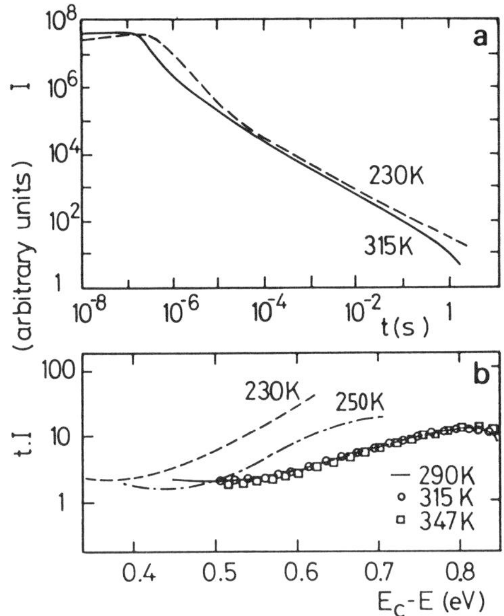

Figure 6. (a) TOF current transients from 5.6μm thick p-i-n sample with 1V applied voltage following a 530nm laser pulse at temperatures of 230 and 315K; (b) DOS profiles calculated from transients at the indicated temperatures.

for the density, information from other sources has to be added. This observation is especially relevant for studies of the light-induced defect density in a-Si:H; it is not *a priori* evident that light degradation will only affect the density profile, and not the parameters that set the calibration. This is, of course, also a problem with other techniques besides post-transit TOF, but it is a problem which is too often ignored.

An example of long-time TOF signals, and the a-Si:H deep-state density profile that is calculated from it according to Eq. (21), is shown in Fig. 6 [43]. The deep states that are revealed in this way are assumed to be mainly occupied dangling-bond states (D^-). Noteworthy in Fig. 6(b) is the fact that the resolved DOS profile becomes temperature dependent below room temperature. The observation is not specific to the sample used for Fig. 6, but is seen with all a-Si:H or silicon alloy samples. This feature will be discussed in the next few pages, and contains some interesting physics, but it is important to point out immediately that the high-temperature profile does represent the *"real"* DOS of the material under investigation, provided a temperature beyond which the curves will coalesce is actually observed. The latter will, for instance, not be the case when the sample under investigation contains such high density of deep traps that deep-retrapping will dominate everything. Amorphous silicon alloys readily fall

456

in this category; a TOF transit time can then not be observed.

5.1.2. *Analysis of the Temperature Dependence*

The comments above already indicate that the first questions to be answered in connection with the surprising temperature dependence of the resolved DOS profile, concern the validity of the post-transit analysis in those cases. The 'standard' TOF requirement that the experiment run its course before the applied electric field relaxes cannot be fulfilled for post-transit currents, but it is easy to show (and understand) that this does not influence the emission-limited currents; it merely retards the collection of the emitted charge a bit by lengthening the transit time. Whether this change could be sufficient to significantly increase the deep-retrapping probability, and thus invalidate the post-transit current analysis, is easily checked through the use of higher applied fields or thinner samples. Both tests come up negative [44], indicating that a more fundamental physical process must be involved.

In [40] the possible role of a shifting transport level or an energy dependence of the capture probability was investigated. It was shown by Monroe [45] that, since hopping between localized states will gain in importance relative to thermal emission to extended states when the temperature is lowered, a shift of the transport energy from the mobility edge into the tail states will occur. However, this shift is largely insufficient to account for the deviations seen in Fig. 6(b). Energy dependences of the capture probabilities, when ignored, may result in a distorted view of the underlying DOS (see [46] for an example), but uniting the various curves of Fig. 6(b) cannot be achieved in this way without compressing all data into the energy range between 0 and 0.5eV [40]. As evidence from other experiments such as optical modulation spectroscopy indicates that this is not the range where the D^- states should be located, this mechanism also fails to account for the observed shifts. It may be pointed out in passing that the apparent temperature dependence of the a-Si:H DOS function is not an artifact of the TOF technique. Indeed, the same pattern has been reported by Zhong and Cohen [47] for a DOS calculated on the basis of the MPC method, and by Bayley *et al.* [48] on the basis of Fourier-transformed transient photocurrent data.

Two further mechanisms have been proposed to account for the observed temperature dependence, one involving hopping-assisted release [44] that may be appropriate for some samples, and one, by Arkhipov, that involves random energy fluctuations [49] and looks most promising and general. It will be discussed in some detail below. Hopping-assisted release would mean that a carrier trapped at E_t would have a higher escape probability when hopping up to a nearby higher-energy state E_h and be thermally released from there. That carrier would then be 'prematurely' released and distort the DOS information. Comparing the thermal release rate $r_t = \nu_0 \exp(-E_t/kT)$ to the hop-and-release rate

$$r_h = \nu_0 \exp(-2\gamma R_h) \exp[-(E_t - E_h)/kT] + \nu_0 \exp(-E_h/kT), \qquad (23)$$

where $R_h \simeq g(E_h)^{-1/3}$ is the average intersite distance at depth E_h, and γ the localization length, it may be seen that r_h will be more favourable than r_t for

$$E_h > 2kT\gamma R_h. \qquad (24)$$

There obviously always will be some temperature low enough to satisfy the above relation, but realistic estimates for a-Si:H indicate that this process should be of minor importance in device-quality material. With lesser-quality material and with some of the alloys it might be a factor to be reckoned with.

5.2. RANDOM LOW-FREQUENCY ENERGY FLUCTUATIONS

In all discussions of carrier release up to here it has been implicitly understood that the energy distribution of the traps E_t is a stationary one, and that temporal energy fluctuations need hence not be introduced. However, this assumption need not necessarily be appropriate. There exists a substantial literature on the experimental observation of intrinsic random temporal fluctuations in a-Si:H, for instance random telegraph noise [50] or spontaneous oscillations [51]. And also theoretical analysis indicates [52] that dispersive transport, of which the MT transport in a-Si:H is one example, will lead to random energy fluctuations, tending to 1/f noise.

It is intuitively clear that fluctuating trap energies can lead to the type of apparent temperature dependence of the DOS shown in Fig. 6(b), if the fluctuations are sufficiently slow. At high temperatures, where the emission probability $\nu_0 \exp(-E_t/kT)$ is high, carriers will be re-emitted from essentially the same energy they were trapped at, and this will on the average be the average energy E_t at which the trap is located. At low temperatures on the other hand, the trapped carrier will be trapped for a long enough time to experience the full range of energies over which the trap fluctuates. Since release from the shallower end of that range is obviously more favourable than release from the deeper end, average carrier release will no longer occur from the average trap depth but from a shallower position. This process produces the progressively higher (apparent) densities at ever shallower energies as the temperature is lowered that are seen in Fig. 6(b). Experimentally, the activation energy of the release is observed, rather than the average depth of the trap.

458

5.2.1. *Theoretical Considerations*

The above arguments can be formalized in a rather simple model [49]. If we allow the energy level of the trap with respect to the mobility edge to be time-dependent, for instance such that

$$E(t) = E_t + \Delta \sin(2\pi f t) \tag{25}$$

with Δ and f being the amplitude and frequency of the variation, then the release probability will evidently also change in time, and each energy in the range $E_t \pm \Delta$ will contribute in accordance with the likelihood of its occurrence. The distribution in energy of such oscillating level will be given by

$$\gamma(E) = 2f \left[\frac{dE(t)}{dt}\right]^{-1} = \frac{1}{\pi\Delta\{1 - [(E - E_t)/\Delta]^2\}^{1/2}}, \tag{26}$$

for

$$E_t - \Delta < E < E_t + \Delta. \tag{27}$$

The most probable activation energy for a carrier jump from such state is then determined by the release probability averaged over the distribution $\gamma(E)$:

$$\exp\left(-\frac{E_a}{kT}\right) = \frac{1}{\pi\Delta} \int_{E_t-\Delta}^{E_t+\Delta} \frac{\exp(-E/kT)}{\{1 - [(E - E_t)/\Delta]^2\}^{1/2}} \, dE. \tag{28}$$

Since the integrand in the above expression has a pronounced maximum at the lower bound of integration, an activation energy $E_a \simeq E_t - \Delta$ will be observed in any process that involves thermal release of carriers trapped in a state of average energy E_t.

To proceed further, we now have to introduce a model for the fluctuation amplitude Δ. By assuming random temporal fluctuations of localized-state energies, characterized by the well-known $1/f$ law that has already been observed [53] in the noise spectrum of a-Si:H, we can provide sufficiently large values of Δ, to make the difference between E_t and E_a significant. A functional dependence

$$\Delta(f,T) = kT \, f_0/f, \tag{29}$$

where f_0 sets the characteristic time scale of the $1/f$ fluctuations, can then be used in Eq.(28). The assumed linear temperature dependence of Δ assures that the fluctuation amplitude will grow with temperature, but is not essential in the model. Since the length of time a carrier remains trapped at a given site will determine the lowest fluctuation frequency, and hence the largest fluctuation amplitude it can undergo there, we may use the reciprocal release time $1/t_r$ to approximate f in Eq.(29). The values of t_r and E_a are then related by the equation

$$t_r = \nu_0^{-1} \exp(E_a/kT) = \nu_0^{-1} \exp[(E_t - kT f_0 t_r)/kT]. \tag{30}$$

The above relationship allows us to obtain an implicit expression for the effective activation energy E_a:

$$E_a = E_t - (kT f_0/\nu_0) \exp(E_a/kT). \tag{31}$$

Up to here, we have only discussed thermal release from a single fluctuating trap. To examine now the influence of the $1/f$-type fluctuations on the release of carriers from a distribution of localized states $g(E)$ rather than from just one level E_t, we have to calculate the distribution of effective activation energies $g_{eff}(E_a)$ according to

$$g_{eff}(E_a) = g[E(E_a)]\frac{dE(E_a)}{dE_a} = g[E(E_a)]\left[1 + \frac{f_0}{\nu_0}\exp\left(\frac{E_a}{kT}\right)\right]. \tag{32}$$

The two distributions will coincide at sufficiently high temperatures where the relationship $(f_0/\nu_0)\exp(E_a/kT) \ll 1$ will be satisfied. But for lower temperatures the two distributions will only coincide down to the energy $E_b(T) \simeq kT\ln(\nu_0/f_0)$, and diverge for deeper energies. It also follows from Eq.(32) that the effective activation energies from all states with average energy $E > E_b$ will be compressed into a narrow energy range close to E_b.

5.2.2. *Comparison with Experiment*

The distribution $g_{eff}(E_a)$, calculated according to Eq.(32) for a model DOS based on standard a-Si:H parameters, and with the ratio ν_0/f_0 chosen as 10^{12}, is shown in Fig. 7 as a function of the temperature. The DOS parameters are shown on the figure, and with the accepted value for ν_0 being around 10^{12}s^{-1}, a characteristic frequency f_0 near 1Hz is implied. The latter value corresponds to the lowest frequencies, and hence largest intensities, where the previously-mentioned low-frequency fluctuations in a-Si:H could be followed [51, 53]. For comparison, a recent set of experimental a-Si:H DOS curves [54] is reproduced in Fig. 8. The similarity between Fig. 7 and Fig. 8 is remarkable. The abrupt cut-off of the distribution of activation energies that is shown in the calculated curves (e.g. just beyond 0.6eV for the 250K curve) does correspond to the experimental observations. If no such drop is actually shown on the 247K experimental curve of Fig. 8, it is only because the measured current drops to the noise level at that point.

The low-frequency random energy fluctuations, and the influence they exert on the thermal release of trapped carriers, will obviously have to be

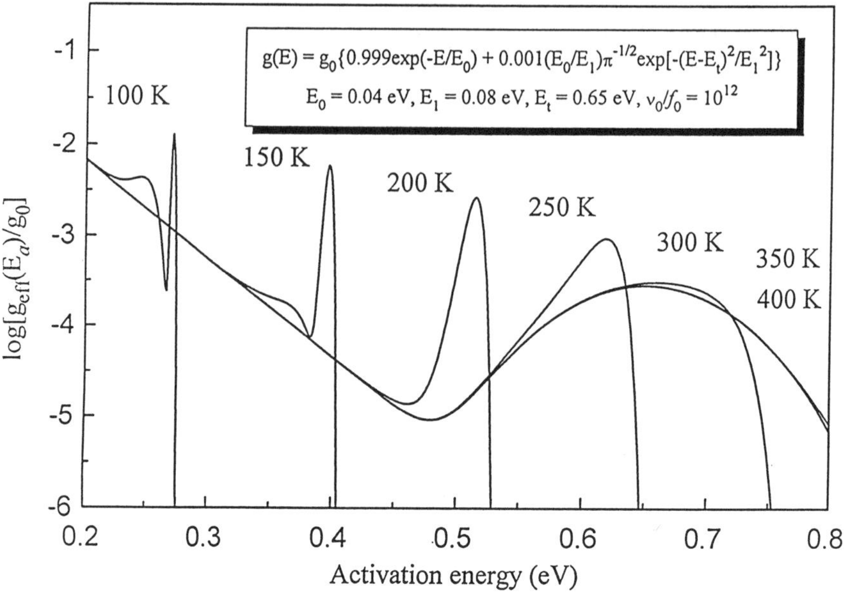

Figure 7. Temperature-dependent distributions of the effective activation energy predicted by the energy fluctuation model for the structured DOS defined in the inset.

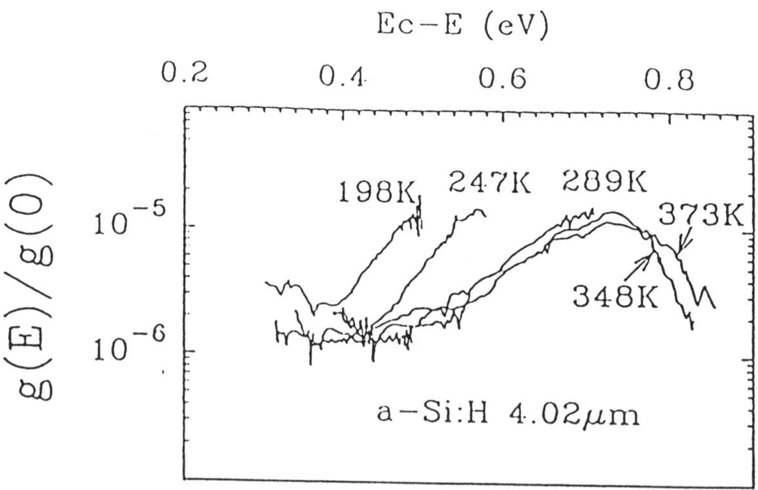

Figure 8. Relative density of localized states from an a-Si:H p-i-n sample, analyzed at indicated temperatures with the standard post-transit photocurrent analysis.

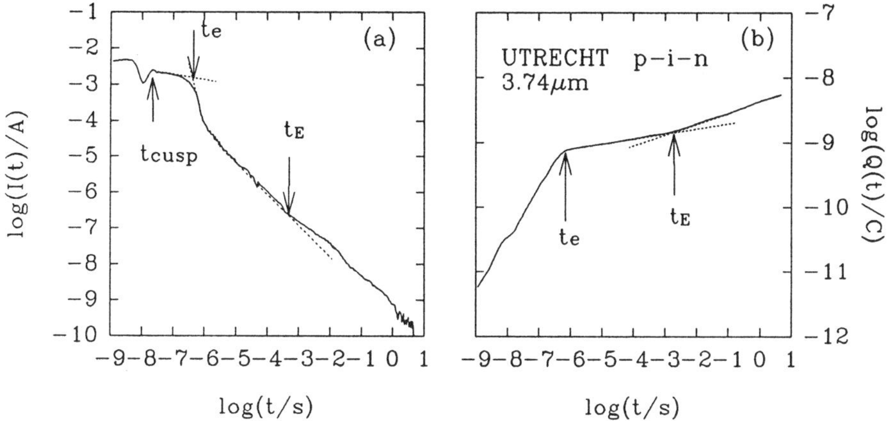

Figure 9. (a) Transient space-charge-limited current, and (b) collected charge measured after excitation with nitrogen-pumped dye laser (10^{13} photons in 3 ns pulse, λ=530nm) of a 3.74μm thick a-Si:H p-i-n diode, at 278K and with 3.0V reverse bias.

taken into account at all times, and not just when a spectroscopic evaluation of the DOS is attempted. The use of the energy-fluctuation model in equilibrium MT transport has already been discribed in [49], and other applications will undoubtedly follow. One area of particular relevance to a-Si:H and its light-induced defects, is the link that can be established between the motion of optically-excited electrons and holes, and the potential energy fluctuations that influence the release of trapped carriers [55].

5.3. SPACE-CHARGE LIMITED CURRENTS

To allow an investigation of the a-Si:H gap states, the TOF technique was successfully extended beyond its original restriction to times shorter than the dielectric relaxation time. The question may then also be asked whether the small-signal condition $Q_0 \ll CV$ could not likewise be circumvented. It was found that higher generation rates created space-charge limited current conditions, but that the situation could still be analyzed with some confidence [56, 57]. A typical space-charge limited TOF transient may be seen in Fig. 9(a), while the total collected charge in the experiment $Q(t) = \int_0^t I(t')dt'$ is shown in Fig. 9(b). The cusp and extraction times, t_{cusp} and t_e, indicated on the figure can – in principle – be used to obtain a value for the small-signal transit time t_T and hence the drift mobility, but as argued in [58], $t_{cusp} \simeq 0.7t_T$ is not easier to retrieve than t_T itself, and t_e (the time needed for the charge available for transport to fall below CV) is too much influenced by recombination processes to give more than an approximate indication of t_T.

However, the space-charge-limited transients can be useful for the study of the deep-state distribution. Due to the large amount of initial photo-generated carriers, all traps will have been filled to a larger extent than in the small-signal case, and the 'post-transit' signals will be correspondingly larger. The measurements can be used to identify an *emission time* t_E, as shown on the figures, which should correspond to charge emission from the top of the D^- band. It can be seen in Fig. 9(b) that at t_E most of the photo-generated charge is still in the sample (and hence in the deep traps). Measurements of the position of t_E as a function of temperature reveal the expected Arrhenius behaviour for thermal emission, and allow a determination of the attemt-to-escape frequency ν_0. For the sample of Fig. 9, $\nu_0 = 3.8 \times 10^{12}$Hz was obtained; for all a-Si:H samples investigated, values between 10^{11} and 10^{13}Hz were found.

6. Valence Band Side of the DOS

It is a widely-accepted fact, based on many experimental observations [5], that the valence band tail is much wider than the conduction band tail. This also means that there is a much higher density of localized states on the valence band side of the gap than there is on the conduction band side. Time-of-flight measurements on electrons and holes in the same samples confirm this notion in every possible way [59]. The room-temperature drift mobility values for holes, $\mu_{dh} < 10^{-2}$cm^{2}V^{-1}s^{-1}, are much lower than the $\mu_{de} \simeq 1$cm^{2}V^{-1}s^{-1} for electrons. Since Marshall *et al.* [60] showed that the free-hole mobility $\mu_{0h} \simeq 10$cm^{2}V^{-1}s^{-1}, which is essentially equal to the commonly accepted value for μ_{0e}, and since drift and free-carrier mobilities are related as $\mu_d = \mu_0 n_f/n_t$, it follows that the ratio n_f/n_t must be lower for holes and, with equal numbers of excess holes and electrons being generated, that therefore the number of traps must be correspondingly higher on the valence band side. Analysis of drift mobility data sets shows that the valence band tail may be approximated by an exponential distribution with characteristic temperature $T_{0v} \simeq 500$K.

Attempts to measure space-charge-limited current transients for the holes in a-Si:H proved unsuccessful.[3] While it was originally argued [61] that strong bimolecular recombination prevented a charge build-up in excess of CV, it turned out to be due to hole trapping in the high density of deep states above the valence band [59]. Calculations of collected charge – as illustrated by Fig.9(b) – easily show that the total charge transported by the hole transient in several times larger than CV in a-Si:H. However, most of that charge is trapped too deeply, and hence with too large emission

---

[3]This is strictly true for room temperature only. By raising the temperature to 377K, the onset of space-charge features may be observed.

times, to be able to sustain a space-charge-limited condition. Poor material quality does not adversely affect this result, since in that case a larger fraction of the generated excess carriers will be initially trapped and eventually recovered; measurements of electron transients from intentionally degraded samples have confirmed that observation. Moreover, the same inability to create space-charge-limited conditions is encountered with electron transients in a-Si:C:H samples [62] as soon as there is a few percent of carbon present. Again, it is known that alloying with carbon leads to a rapidly increasing density of localized states in the gap.

Based on the ratio of drift mobilities, which should be mostly affected by shallow tail-state densities, and the ratio of light intensities to produce the first signs of space-charge effects, which should be proportional to the deep-state densities, we may assert that the integrated localized-state density is two to three orders of magnitude larger on the valence than on the conduction band side of the gap.

7. The Defect Pool Model

An important point of current debate on a-Si:H, concerns the origin of the defect states. The traditional view (now often referred to as the *standard model*) sees the equilibrium distribution of deep states in undoped a-Si:H as mainly consisting of D^0, the neutral dangling bond, with D^- only appearing as an electron-occupied D^0 raised by the Hubbard interaction energy U. In the *defect pool model* (DPM) the charged states D^+ and D^- are part of the equilibrium distribution, and it will depend on system parameters whether they are more, or less prominent than the neutral ones. For doped materials, both models make use of ionized donor or acceptor states close to the bands, and deeper charge-compensating D^- or D^+ states [5], but the positioning of the latter in the gap differs.

The defect pool model has gradually grown out of a number of experimental observations. Adler [63] first introduced the notion that charged tetragonal sites T^+, T^-, with T^- actually having the lowest energy, should be part of the defect distribution in order to account for measured energy positions in the gap of D^- and the Fermi level E_F that were difficult to reconcile in the standard model. Bar-Yam and Joannopoulos [64] added the observation that all states should be part of one thermodynamical ensemble, with local disorder and the position of the Fermi level determining the charge state. When finally Stutzmann's weak-bond–dangling-bond conversion idea [17] was added by Smith and Wagner [65], a fairly comprehensive model for defect formation in a-Si:H was achieved. A good exposition of the essential elements of the DPM has been given by Winer [66].

The basic elements of the model are as follows: There is a distribution

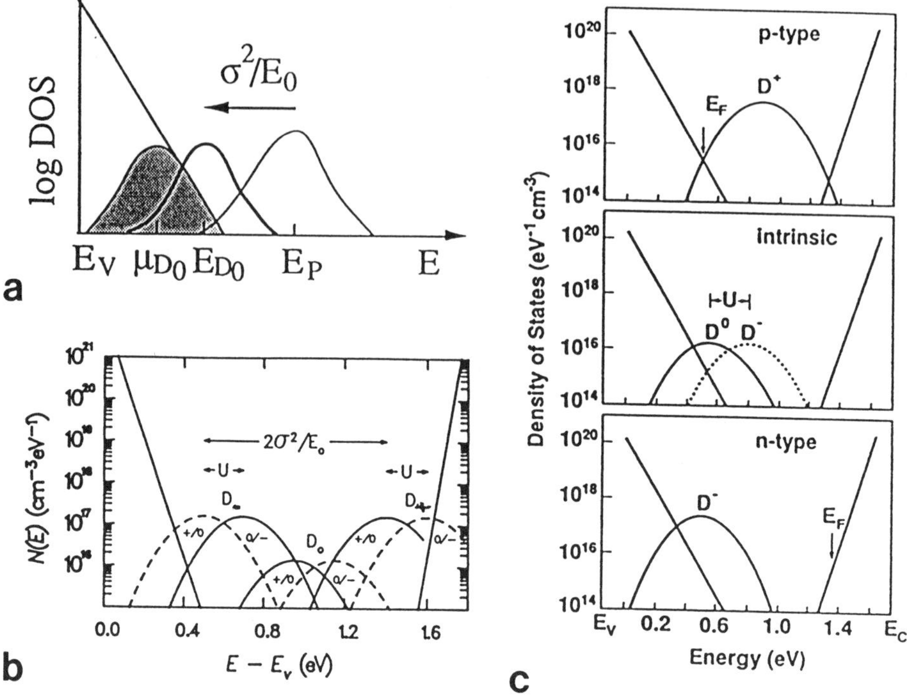

Figure 10. (a) Illustration of the defect pool concept; (b) density of localized states for undoped a-Si:H with charged defects dominating (from Powell *et al.* [67]); density of localized states in doped and undoped a-Si:H according to Winer [66]

of possible energies for coordination defects in a-Si:H (the *defect pool*, symbolized by the right-most distribution centered on E_p in Fig. 10(a)), but since there also will be equilibration between the weakest Si-Si bonds and the dangling bonds

$$\mathrm{Si\,Si} \;\rightleftharpoons\; \mathrm{D}^0, \qquad (33)$$

the defect distribution that actually forms will be displaced towards the valence band by σ^2/E_0, where σ is the Gaussian width of the defect pool and E_0 the characteristic exponential slope of the valence band. This shift results from a balance between the tendency to minimize the formation energy of the defect (valence band tail to E_{D^0}), and the elastic strain introduced by the formation of a defect away from the optimal position E_p.

However, this process will be modified whenever the defects can be charged. For instance, in n-type material the Fermi level E_F will be near the conduction band and an electron can easily drop onto the defect

$$\mathrm{D}^0 + \mathrm{e} \rightarrow \mathrm{D}^-, \qquad (34)$$

thus producing a negatively charged one. At the same time, the equilibrium of Eq. (33) will work to restore the D^0 concentration such that the net formation energy of the D^- defect is given by $2E_D - E_F + U$. Putting this energy in the chemical potential for the D^- state, and calculating the distribution of D^- states in the gap with it, Winer finds that the D^- band will dominate in the n-type material, and be shifted to $E_p + U - 2\sigma^2/E_0$ [66]. Analogous reasoning for the D^+ state in p-type a-Si:H shows that it will be located at the E_p energy. Fig. 10(c) illustrates the three cases. For a sufficiently large value of σ, the charged defects will also form in the undoped material, even to the extent of dominating the density of deep states as suggested in Fig. 10(b). Whether this situation occurs in standard a-Si:H is still being debated.

The DPM outline presented above differs from the discussion in the literature in one important aspect: the absence of hydrogen. From the genesis of the defect equilibration and weak-bond–dangling bond conversion ideas on, the role of hydrogen in catalyzing the transformations has been considered virtually self-evident. Eq. (33) is therefore often written in the form

$$\mathrm{Si\,Si} + \mathrm{Si\,H} \rightleftharpoons D^0 + \mathrm{Si\,H\,Si}, \tag{35}$$

i.e. a weak bond plus a Si-H bond convert into a dangling bond and a Si,H complex, and defect-to-hydrogen concentration ratios are routinely entered into the chemical potential calculations. An elaborate *hydrogen density of states* formalism has even been worked out for the DPM [68]. Unfortunately, whereas all these models involve some sort of joint configuration of a dangling bond and a Si-H bond, recent ESR evidence [69] indicates that no hydrogen is found within ~ 4.5Å of a dangling bond. Until these recent pulsed ESR experiments, inhomogeneous broadening of the D^0 ESR line made it difficult to be absolutely sure that hydrogen hyperfine structure was indeed absent. To counter this problem with line broadening, Stutzmann [70] newly measured ESR of the a-Si:H D^0 center at 400MHz instead of the standard 9.3GHz, and still could not resolve any hydrogen hyperfine structure. It is now quite sure therefore that there is no hydrogen close to an a-Si:H dangling bond.

In spite of this important flaw at the microscopic level, the DPM has gained credibility by providing a correct description for the fact that the charged dangling bonds that are found in doped amorphous silicon, do occur at energy positions in the gap which differ from the ones measured for the correspondingly charged defects in intrinsic a-Si:H. There is much experimental support for this observation, from the early results that troubled Adler, to more recent ones that led Kočka *et al.* [71] to formulate their (now superfluous) intimate-pair model, or to such current ones as, for instance, the OMS-resolved positions of the D^0 and D^- levels in intrinsic a-Si:H [72] and of the D^+ level in p-type material [73] with the latter one lying closest to the conduction band, in full agreement with the DPM.

Less universally accepted are claims, alluded to above, that intrinsic a-Si:H is likely to have a higher density of charged than of neutral dangling bonds. It is not clear whether the evidence presented for it, e.g. by Schumm *et al.* [74], could not also be explained by a DOS where the steep valence and conduction band tails become less steep as they penetrate into the gap. Such more slowly changing distributions have generally been needed in the analysis of experiments involving steady-state photoconductivity, where the DOS is being probed deeper in the gap, around the position of quasi-Fermi level [75].

8. Conclusions

While hydrogenated amorphous silicon is the one amorphous semiconductor whose semiconducting properties have found wide-spread application, our understanding of several important aspects of its behaviour remains poor. This is largely due to the fact that the electronic transport properties are strongly influenced by a defect density which involves less than 1 in 10^5 lattice sites, and that for many of the detailed models that are being proposed our analytical tools to test them lack the precision to deal with such small concentrations on a microscopic level. Equally unclear, beyond the passivation of most of the potential dangling-bond density, is the role of hydrogen in determining some of the a-Si:H properties and/or influencing their stability. Unsatisfactory as this situation may be, amorphous silicon will not be held back by it: it can be produced cheaply and it works!

Acknowlegments

Collaborations with Vladimir Arkhipov, Astrid Eliat, Miloš Nesládek, Geert Seynhaeve, Baojie Yan and Qing Zhang have been essential in developing the views on hydrogenated amorphous silicon expressed in these lecture notes. To all of them, my sincere thanks. Financial support from the Belgian *Nationaal Fonds voor Wetenschappelijk Onderzoek* is gratefully acknowledged.

References

1. R.C. Chittick, J.H. Alexander and H.F.·Sterling (1969) J. Electrochem. Soc. **117**, 77.
2. W.E. Spear and P.G. Le Comber (1975) Solid State Commun. **17**, 1193.
3. S. Roorda, W.C. Sinke, J.M. Poate, D.C. Jacobson, S. Dierker, B.S. Dennis, D.J. Eaglesham, F. Spaepen and P. Fuoss (1991) Phys. Rev. B **44**, 3702.

4. S. Acco, D.L. Williamson, P.A. Stolk, F.W. Saris, M.J. van den Boogaard, W.C. Sinke, W.F. van der Weg, S. Roorda and P.C. Zalm (1996) Phys. Rev. B **53**, 4415.
5. R.A. Street (1991) *Hydrogenated Amorphous Silicon* , Cambridge University Press, Cambridge, U.K.
6. D.L. Staebler and C.R. Wronski (1977) Appl. Phys. Lett. **31**, 292.
7. A.H. Mahan, J. Carapella, B.P. Nelson, R.S. Crandall and I. Balberg (1991) J. Appl. Phys. **69**, 6728.
8. See e.g. various contributions in *Amorphous Silicon Technology – 1996*, M. Hack, E.A. Schiff, S. Wagner, R. Schropp, A. Matsuda (eds), MRS Symp. Proc. **420**, Materials Research Society, Pittsburgh.
9. R.A. Street and K. Winer (1989) Phys. Rev. B **40**, 6236.
10. W.B. Jackson and J. Kakalios (1988) Phys. Rev. B **37**, 1020.
11. J. Kakalios, R.A. Street and W.B. Jackson (1987) Phys. Rev. Lett. **59**, 1037.
12. Ref. [5], p.209; J. Kakalios ans W.B. Jackson (1989) in *Amorphous Silicon and Related Materials*, H. Fritzsche (ed.), World Scientific, Singapore, p.207.
13. C. Godet (1994) Phil. Mag. B **70**, 1003.
14. V.K. Tikhomirov (1996) private communication.
15. C.G. Van de Walle (1996) Phys. Rev. B **53**, 11292.
16. Q. Zhang and G.J. Adriaenssens (1996) to be published.
17. M. Stutzmann (1987) Phil. Mag. B **56**, 63.
18. M. Stutzmann (1989) Phil. Mag. B **60**, 531.
19. Z. Vardeny, T.X. Zhou and J. Tauc (1989) in *Amorphous Silicon and Related Materials*, H. Fritzsche (ed.), World Scientific, Singapore, p.513.
20. M. Silver, E. Snow and D. Adler (1984) Solid State Commun. **51**, 581.
21. G.J. Adriaenssens and G. Seynhaeve (1987) J. Non-Cryst. Solids **97-98**, 133.
22. H. Michiel, J.M. Marshall and G.J. Adriaenssens (1983) Phil. Mag. B **48**, 187.
23. H. Oheda (1981) J. Appl. Phys. **52**, 6693.
24. C. Longeaud and J.P. Kleider (1992) Phys. Rev. B **45**, 11672; K. Hattori, Y. Adachi, M. Anzai, H. Okamoto and Y. Hamakawa (1994) J. Appl. Phys. **76**, 2841.
25. C. Main, R. Brüggeman, D.P. Webb and S. Reynolds (1992) Solid State Commun. **83**, 401.
26. A full desciption of the time-of-flight experiment for amorphous semiconductors will be found in W.E. Spear (1969) J. Non-Cryst. Solids **1**, 197.
27. G. Seynhaeve, R.P. Barclay, G.J. Adriaenssens and J.M. Marshall (1989) Phys. Rev. B **39**, 10196
28. G. Seynhaeve, G.J. Adriaenssens, H. Michiel and H. Overhof (1988) Phil. Mag. B **58**, 421.
29. G. Seynhaeve (1989) *Time-of-flight photocurrents in hydrogenated amorphous silicon*, Ph.D. thesis, Katholieke Universiteit Leuven.
30. J.M. Marshall, R.A. Street and M.J. Thompson (1986) Phil. Mag. B **54**, 51.
31. W.E. Spear (1983) J. Non-Cryst. Solids **59-60**, 1; A.C. Hourd and W.E. Spear (1985) Phil. Mag. B **51**, L13.
32. T. Tiedje (1984) in *Hydrogenated Amorphous Silicon, C*, J.I. Pankove (ed.), Semiconductors and Semimetals **21**, Academic Press, Orlando, p.207.
33. M. Silver, E. Snow and D. Adler (1986) J. Appl. Phys. **59**, 3506.
34. R. Vanderhaghen and C. Longeaud (1987) J. Non-Cryst. Solids **97-98**, 1059; C. Longeaud, G. Fournet and R. Vanderhaghen (1988) Phys. Rev. B **38**, 7493.
35. C.E. Nebel and G.H. Bauer (1989) Phil. Mag. B **59**, 463.
36. J.M. Marshall, J. Berkin and C. Main (1987) Phil. Mag. B **56**, 641.
37. D. Adler and M. Silver (1987) Phil. Mag. Lett. **56**, 113.
38. N.F. Mott and E.A. Davis (1979) *Electronic Processes in Non-Crystalline Materials*, 2nd ed., Clarendon Press, Oxford.
39. A.M. Stoneham (1981) Rep. Progr. Phys. **44**, 1251.
40. B. Yan and G.J. Adriaenssens (1995) J. Appl. Phys. **77**, 5661.
41. W.B. Jackson, S.M. Kelso, C.C. Tsai, J.W. Allen and S.-J. Oh (1985) Phys. Rev.

468

B **31**, 5187.

42. B. Yan, D. Han and G.J. Adriaenssens (1996) J. Appl. Phys. **79**, 3597.
43. M. Nesládek, A. Triska and G.J. Adriaenssens (1992) J. Non-Cryst. Solids **141**, 155.
44. M. Nesládek, G.J. Adriaenssens and A.S. Volkov (1991) J. Non-Cryst. Solids **137-138**, 443.
45. D. Monroe (1985) Phys. Rev. Lett. **54**, 146.
46. V.I. Arkhipov and G.J. Adriaenssens (1995) J. Non-Cryst. Solids **181**, 274.
47. F. Zhong and J.D. Cohen (1992) Mat. Res. Soc. Symp. Proc. **258**, 813.
48. P.A. Bayley, J.M. Marshall, C. Main, D.P. Webb, R. Van Swaaij and J. Bezemer (1996) J. Non-Cryst. Solids, in press.
49. V.I. Arkhipov and G.J. Adriaenssens (1996) Phil. Mag. Lett. **73**, 263.
50. C.E. Parman, N.E. Israeloff and J. Kakalios (1991) Phys. Rev. B **44**, 8391.
51. W. Eberle, W. Prettl and Y. Roizin (1995) J. Phys.: Condens. Matter **7**, 8165.
52. K.L. Ngai and F.-S. Liu (1981) Phys. Rev. B **24**, 1049.
53. C. Parman and J. Kakalios (1991) Phys. Rev. Lett. **67**, 2529; J. Fan and J. Kakalios (1994) Phil. Mag. B **69**, 595; M. Valenza *et al.* (1996) J. Appl. Phys. **79**, 923.
54. V.I. Arkhipov, G.J. Adriaenssens and B. Yan (1996) to be published.
55. V.I. Arkhipov and G.J. Adriaenssens (1996) to be published; G.J. Adriaenssens and V.I. Arkhipov (1996) Mat. Res. Soc. Symp. Proc. **420**, in press.
56. J. Kočka (1993) in *Proc. 7th Intern. School on Condens. Matter Phys.*, eds.: J.M. Marshall, N. Kirov and A. Vavrek (World Scientific, Singapore) p.129.
57. G. Juška, M. Viliunas, O. Klima, E. Šipek and J. Kočka (1994) Phil. Mag. B **69**, 227.
58. G.J. Adriaenssens, B. Yan and A. Eliat (1995) Mat. Res. Soc. Symp. Proc. **377**, 443.
59. B. Yan, G.J. Adriaenssens and A. Eliat (1996) Phil. Mag. B **73**, 543.
60. J.M. Marshall, R.A. Street, M.J. Thompson and W.B. Jackson (1987) J. Non-Cryst. Solids **97-98**, 563.
61. G. Juška, M. Viliunas, S. Vengris, A. Fejfar and J. Kočka (1995) in *Proc. 8th Intern. School on Condens. Matter Phys.*, eds.: J.M. Marshall, N. Kirov and A. Vavrek (Research Studies Press, Taunton UK) p.343.
62. A. Eliat, B. Yan, G.J. Adriaenssens and J. Bezemer (1996) J. Non-Cryst. Solids, in press.
63. D. Adler (1978) Phys. Rev. Lett. **41**, 1755.
64. Y. Bar-Yam, D. Adler and J.D. Joannopouls (1986) Phys. Rev. Lett. **57**, 467.
65. Z.E. Smith and S. Wagner (1987) Phys. Rev. Lett. **59**, 688.
66. K. Winer (1990) Phys. Rev. B **41**, 12150.
67. M.J. Powell, C. van Berkel, A.R. Franklin, S.C. Deane and W.I. Milne (1992) Phys. Rev. B **45**, 4160.
68. M.J. Powell and S.C. Deane (1996) Phys. Rev. B **53**, 10121.
69. S. Yamasaki and J. Isoya (1993) J. Non-Cryst. Solids **164-166**, 169.
70. M. Stutzmann (1996) presented at the *Chelsea Amorphous and Organic Semiconductors Meeting*, London, UK, 18-19 April.
71. J. Kočka, M. Vaněček and F. Schauer (1987) J. Non-Cryst. Solids **97-98**, 715.
72. W. Grevendonk, M. Verluyten, J. Dauwen, G.J. Adriaenssens and J. Bezemer (1990) Phil. Mag. B **61**, 393.
73. H. Herremans and W. Grevendonk (1994) Phys. Rev. B **50**, 7422.
74. G. Schumm, C.-D. Abel and G.H. Bauer (1991) J. Non-Cryst. Solids **137-138**, 351; G. Schumm, W.B. Jackson and R.A. Street (1993) Phys. Rev. B **48**, 14198.
75. G.J. Adriaenssens, S.D. Baranovskii, W. Fuhs, J. Jansen and Ö. Öktü (1995) Phys. Rev. B **51**, 9661, and references therein.

TUNNELING STATES

S. HUNKLINGER

Institut für Angewandte Physik der Universität Heidelberg

Albert-Ueberle-Str. 3-5

D-69120 Heidelberg, Germany

1. Introduction

The properties of real solids are not only determined by their crystalline lattice, or their random network in the case of amorphous solids, but also by the defects present. In particular at low temperatures, i.e., below the temperature of liquid helium, a specific type of defect, so-called Tunneling Systems (TSs) are responsible for many, very often unexpected properties. In this article we restrict ourselves to the discussion of the thermal, elastic and dielectric behaviour of insulators. We exclude metals from our considerations although TSs are present in these materials as well.

Studying dynamic properties we always have to keep in mind that the thermal energy $k_B T$ is the natural energy scale for the description of the phenomena. Energy differences and potential barriers have hardly any influence as long as they are small compared to $k_B T$. To be more specific let us consider a particle of mass m – e.g. an hydrogen atom in a hydrogen-bonded crystal or an impurity atom in a crystal – moving in a double-well potential with the parameters as specified in fig. 1. As long as $k_B T$ is larger than the barrier height V, the barrier has only a minor influence on the vibrational motion of the particle.

Lowering the temperature will lead to the situation $\Delta < k_B T < V$. Now the atom cannot vibrate freely in the double-well potential, but it may jump from one site to the other. In general these jumps will be thermally activated and the rate τ^{-1} is given by

$$\tau^{-1} = \tau_0^{-1} e^{-V/k_B T} , \tag{1}$$

where $\tau_0^{-1} = 2\pi\Omega$ is the frequency of the atom vibrating in one well. Cooling down further will lead to the condition $k_B T < \Delta < V$ and the atom will be localized in the lower well.

M. F. Thorpe and M. I. Mitkova (eds.), Amorphous Insulators and Semiconductors, 469–495.
© 1997 *Kluwer Academic Publishers. Printed in the Netherlands.*

470

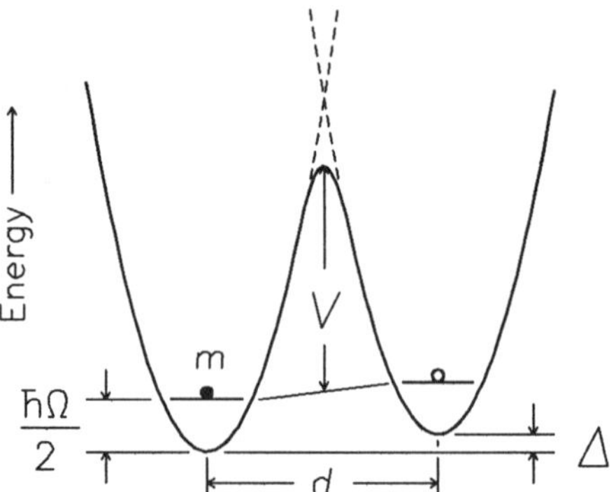

Figure 1. Double-well potential with barrier height V, asymmetry energy Δ and well distance d. $\hbar\Omega/2$ is the zero-point energy of the tunneling particle with mass m.

If temperatures are around or below 1 K a qualitatively new aspect comes into play. The jumping rate (1) becomes very small because the thermal energy is not sufficient any more to lift the particles over the barrier. But the particle always has an additional option to go from one well to the other: tunneling through the barrier. The tunneling rate and the resulting energy splitting Δ_0 is determined by the overlap of the wavefunctions of the particle in the left or right well. Since the tunneling frequency usually is rather small, $k_B T$ becomes comparable to Δ_0 only at low temperatures. If this condition is met, TSs have a decisive impact onto the properties of the solid.

At low temperatures were $k_B T \ll \hbar\Omega$, higher vibrational levels are not excited. Therefore only the behaviour of the ground state is of interest which is split into two levels due to tunneling. Therefore the TSs are often referred to as "two-level systems". In addition to the classical potential difference Δ, the quantum mechanical tunnel splitting Δ_0 contributes to the ground state splitting E which is given by

$$E = \sqrt{\Delta^2 + \Delta_0^2} \, . \tag{2}$$

For two identical harmonic wells one finds from elementary quantum mechanics [1]

$$\Delta_0 = 2\hbar\Omega\sqrt{\frac{2V}{\pi\hbar\Omega}}\, e^{-d\sqrt{2mV}/2\hbar} = 4E_0\sqrt{\frac{V}{\pi E_0}}\, e^{-V/E_0} \, , \tag{3}$$

where $E_0 = \hbar\Omega/2$ is the zero-point energy of the tunneling particle. For simplicity this expression is very often replaced by

$$\Delta_0 \approx \hbar\Omega e^{-V/E_0} = \hbar\Omega e^{-\lambda} \, . \tag{4}$$

Roughly speaking the tunnel splitting Δ_0 is given by the vibrational energy $\hbar\Omega$ of the particle multiplied by the probability $\exp(-\lambda)$ for tunneling. The so-called tunneling parameter $\lambda = d\sqrt{2mV}/2\hbar$ reflects the overlap of the wave functions for the particle on the two sides of the potential barrier.

The existence of TSs can be clearly demonstrated by their specific heat. The internal energy (and thus the specific heat) of a two-level system follows directly from partition function $Z = 1 + \exp(-E/k_\mathrm{B}T)$. For N independent TSs one obtains

$$C_V = N k_\mathrm{B} \left(\frac{E}{k_\mathrm{B}T}\right)^2 \frac{e^{E/k_\mathrm{B}T}}{(1 + e^{E/k_\mathrm{B}T})^2}\ . \tag{5}$$

As shown in fig. 2 this so-called Schottky anomaly exhibits a characteristic maximum at $T \approx 0.42\, E/k_\mathrm{B}$ which directly reflects the energy splitting.

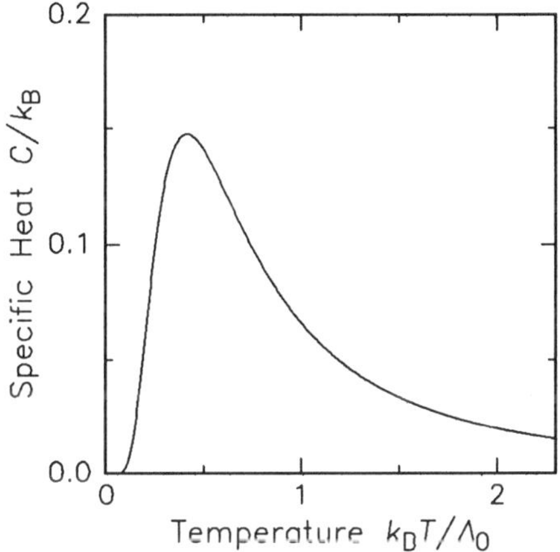

Figure 2.   Specific heat of a two-level system as a function of normalized temperature.

TSs couple to their environment via strain or electric fields which may alter Δ and/or Δ_0 changing the energy splitting E. For weak fields these changes vary linearly with field strength. Therefore we can introduce a perturbation potential $W_\mathrm{a} = \vec{\gamma}\cdot\vec{e}$ and $W_\mathrm{d} = \vec{p}\cdot\vec{F}$, where $\vec{\gamma}$ is the deformation potential, $\vec{p}$ the electrical dipole moment, $\vec{e}$ the amplitude of the strain field, and $\vec{F}$ the strength of the electric field. In the further discussion we neglect the tensorial character of $\vec{e}$ and $\vec{\gamma}$ in order to simplify the notation.

The macroscopic reaction of the TSs to external fields can be described with the help of the elastic or dielectric susceptibilities χ_a and χ_d, respectively. We do not reproduce here the derivations of the relevant equations but mention only the results that are needed for the discussion of the experimental data. The complex susceptibilities are related to the measured quantities through the following relations: In dielectric experiments

472

the dielectric function $\epsilon = \epsilon' + i\epsilon'' = 1 + \chi_{\rm d}' + i\chi_{\rm d}''$ is measured, where the imaginary parts reflect the losses. In acoustic experiments usually the change δv in the velocity of sound v and the attenuation coefficient α are measured. In this case the relevant quantities are linked by $\delta v = -\rho v^3 \chi_{\rm a}'/2$ and $\alpha \equiv l^{-1} = \omega \rho v \chi_{\rm a}''$.

Independent of the nature of the coupling field we can distinguish between two interaction processes: the resonant and the relaxation process. Resonance occurs if the frequency ω of the perturbing field fulfills the resonance condition $E = \hbar\omega$. This process causes losses only within the line width at the resonance. The temperature dependence of the susceptibility caused by this process is proportional to $\tanh(E/2k_{\rm B}T)$. This factor reflects the difference in the thermal population of the two levels. Therefore this process can only be observed up to temperatures $T \approx \hbar\omega/k_{\rm B}$. Resonant interaction also contributes to $\chi_{\rm res}'$ at low frequencies, i.e., it gives rise to an increase of the static susceptibilities.

The relaxation process has its origin in the modulation of the energy splitting by the applied field. The thermal population of the levels is disturbed and the TSs try to relax into the new equilibrium. Because of the finite relaxation time τ, the response of the TSs is delayed with respect to the applied field resulting in an energy dissipation. This leads to a contribution $\chi_{\rm rel}$ of the TSs to the susceptibility given by

$$\chi_{\rm rel} \propto \frac{1}{k_{\rm B}T} \left(\frac{\Delta}{E}\right)^2 {\rm sech}^2\left(\frac{E}{2k_{\rm B}T}\right) \frac{1}{1 + i\omega\tau} . \tag{6}$$

This process causes a loss maximum at the frequency $\omega \approx \tau^{-1}$. It also contributes to the static susceptibility. For $\omega\tau \gg 1$ the contribution to $\chi_{\rm rel}'$ vanishes. Because of the proportionality to Δ^2, symmetric TSs do not exhibit a relaxation process. This can be expressed in a more general manner: relaxation only occurs if the first derivative of the energy with respect to the perturbing field does not vanish. In a description of the phenomenon the relaxation time τ is of particular interest because it reflects the coupling of the TSs to their environment.

2. Tunneling Systems in Crystals

2.1. LEVEL SCHEME AND SPECIFIC HEAT

After this brief introduction we start with the discussion of well-defined TSs in crystals. In the simplest case the TSs are caused by substitutional atoms which do not properly fit into the lattice sites because they differ in shape and size from the atoms of the host lattice. These atoms experience a potential with the symmetry of the lattice and in many cases they do not sit at the exact lattice site but in an off-center potential minimum.

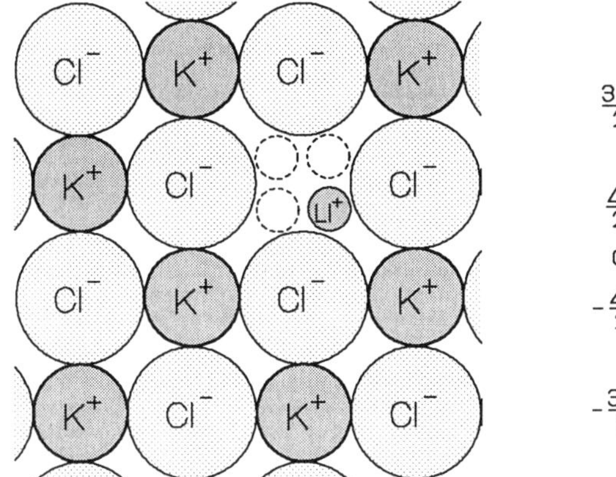
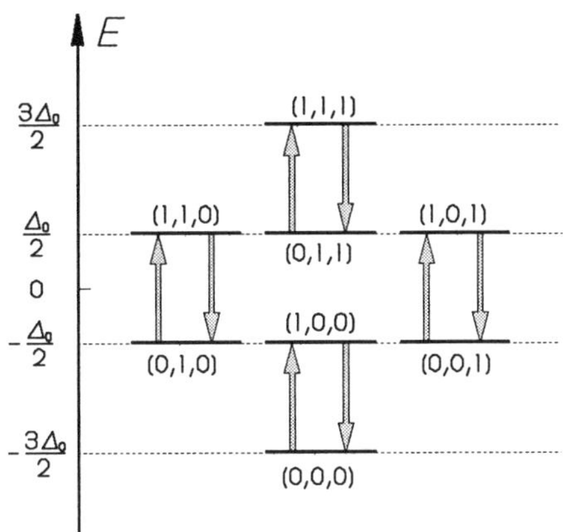

Figure 3. Schematic representation of a (100)-plane of KCl containing a Li⁺-ion. The eight off-center positions are located above and below the plane of drawing.

Figure 4. Level scheme of TSs formed by Li⁺-ions in KCl. The allowed transitions between the levels for an electric field applied in $\langle 100\rangle$-direction are indicated by arrows. The triples characterize the different levels.

In the following we discuss TSs formed by Li⁺-ions in KCl because this is a simple and intensively studied example. In this crystal Li⁺-ions with the radius $r_{\mathrm{Li}+} \approx 0.60$ Å replace K⁺-ions with $r_{\mathrm{K}+} \approx 1.33$ Å. In fig. 3 the ions in the (100)-plane of a KCl-crystal are drawn together with a substitutional Li⁺-ion. For the small Li⁺-ion there exist eight possible off-center positions in the $\langle 111\rangle$-directions, namely four above and four below the plane of drawing. The eight positions form a cubic cage with side length $d = 1.4$ Å. Tunneling occurs preferentially along the edges of this cube. In the simplest approximation we may use eqn. (4) to calculate the splitting of the ground state. Carrying out the calculation with the real potential leads to the level scheme shown in fig. 4 [2]. The levels are equally spaced whereby the inner two are threefold degenerate. From various experiments which we do not discuss here, the values $^6\Delta_0/k_{\mathrm{B}} = 1.6$ K and $^7\Delta_0/k_{\mathrm{B}} = 1.1$ K are known for the two isotopes ⁶Li and ⁷Li [3]. In other alkalihalides, six minima on the $\langle 100\rangle$-axes or twelve minima in $\langle 110\rangle$-directions are found depending on the host crystal and the dopant.

Eqn. (5) describes the specific heat of an ensemble of two-level systems. For KCl:Li all the eight levels of the TSs have to be taken into account. Because of the symmetry of the level scheme, the calculation leads to an expression identical with eqn. (5) except for the prefactor which differs by a factor 3. In fig. 5 data are presented for a KCl-crystal doped with 20 ppm ⁷Li [4]. After subtraction of the phonon contribution that is proportional to T^3, a typical Schottky anomaly is found. In fig. 6 this anomaly is shown for a sample doped with ⁶Li and for another doped with ⁷Li [4]. Note that

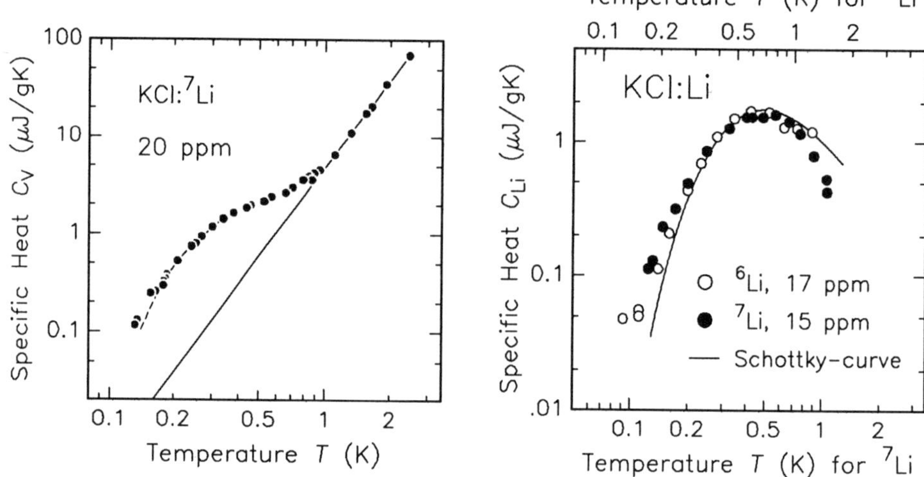

Figure 5. Specific heat of KCl doped with 20 ppm ^{7}Li. The Debye specific heat of the pure crystal is shown by the straight line. From [4].

Figure 6. Specific heat of ^{6}Li- and ^{7}Li-TSs. Note that two different temperature scales are used for the two isotopes. From [4].

two different temperature scales are used to account for the different tunnel splittings.

2.2. DYNAMIC BEHAVIOUR OF ISOLATED TUNNELING STATES

Now we turn to the response of TSs to electric and elastic fields. First we consider the reaction to quasi-static fields, i.e., we discuss the dielectric constant and the sound velocity at low frequencies. In fig. 7 and fig. 8 the contribution of the TSs to the dielectric susceptibility of KCl:Li [5] and LiF:OH [6] is plotted, respectively. The susceptibility of the TSs is deduced from the measured dielectric constant ϵ of the doped crystals. Since at low temperatures the dielectric constants ϵ_{KCl} and ϵ_{LiF} of the pure crystals are virtually temperature independent, the susceptibility χ'_{Li} or χ'_{OH} of the TSs can easily be deduced. It should be mentioned that OH$^-$-ions rotate in the cage formed by the neighboring ions and experience six potential minima in $\langle 100 \rangle$-direction.

How do Li-TSs couple to an electric field? Because of the symmetry of the wave function of the Li$^+$-ions their dipole moment vanishes in the absence of an external field. By weak electric fields a moment is induced which rises linearly with field strength. At strong fields this rise saturates because the ions become increasingly localized. If the strong field acts along the [111]-direction the ion will be trapped in the corresponding potential

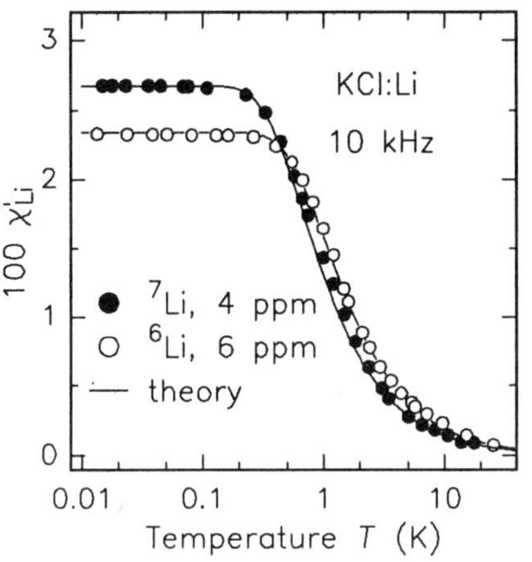

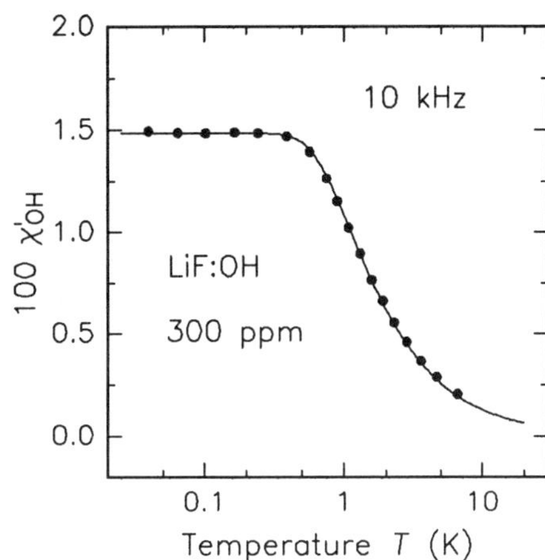

Figure 7. Susceptibility of ^{6}Li- and ^{7}Li-TSs in KCl measured at 10 kHz. The solid lines represent the theoretical curves obtained from eqn. (8). From [5].

Figure 8. Susceptibility of OH-TSs in LiF measured at 10 kHz. The solid line represents the theoretical curve calculated for the actual level scheme. Data from [6].

minimum resulting in dipole moment $p = ed\sqrt{3}/2$ for geometrical reasons. After correcting for the local field one obtains a dipole moment $p = 2.6$ D. An electric field in $\langle 100 \rangle$-direction shifts the energy levels in the manner shown in fig. 9. In analogy to atomic physics this phenomenon is called the Stark effect. For example the energy E_g of the lowest level follows the relation

$$E_\mathrm{g} = -\Delta_0 - \sqrt{\frac{\Delta_0^2}{4} + \frac{p^2 F^2}{3}} \, . \tag{7}$$

From the field dependence of the eigenvalues the partition function, and hence the field dependence of the free energy $\mathcal{F}$ can easily be calculated. Then using the real part of the susceptibility $\chi'_\mathrm{d} = -\partial^2 \mathcal{F}/\partial F^2|_{F=0}$ one immediately obtains

$$\chi'_\mathrm{d} = \frac{2}{3} \frac{np^2}{\epsilon_0 \Delta_0} \tanh\left(\frac{\Delta_0}{2k_\mathrm{B}T}\right) \, , \tag{8}$$

if the electric field is applied in $\langle 100 \rangle$-direction. It is worth mentioning that this expression is identical with that found for two-level systems. This surprising result becomes understandable if we look at fig. 4: the selection rules allow only transitions between two neighbored, equally spaced levels. A comparison between theory and experiment reveals an excellent agreement. In particular the $1/\Delta_0$-dependence of the effect can clearly be seen in fig. 7. Although the ^{6}Li-concentration is higher, the sample with ^{7}Li exhibits a larger susceptibility. Moreover, for the ^{6}Li-sample the point of

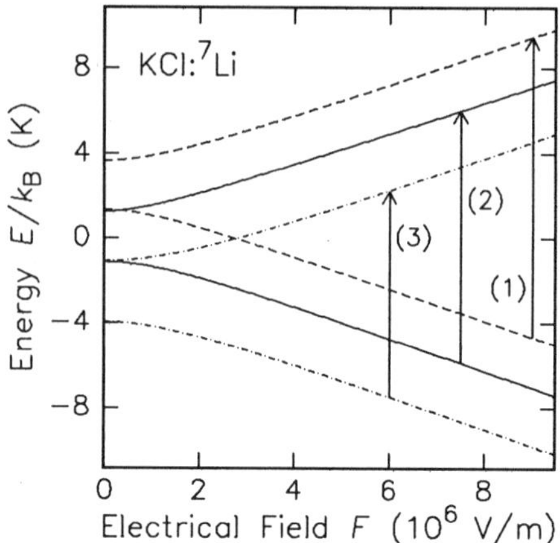

Figure 9. Field dependence of the level scheme for an electric field in $\langle 100 \rangle$-direction. The arrows mark the allowed transitions. From [3].

inflection in the susceptibility is found at higher temperatures which is due to the higher value of $^6\Delta_0$.

2.3. COUPLING TO ELASTIC FIELDS

Substitutional defects are surrounded by elastic distortions or strain fields which are changed by sound waves or thermal phonons. In general this coupling between sound wave and TSs is described with the help of a deformation potential γ. As already mentioned, both the elastic field and the deformation potential are tensors. Therefore the coupling strength depends on the orientation of the crystal relative to the propagation direction and polarization of the sound wave. It can be shown that Li-TSs in KCl do not couple to longitudinal waves propagating along a crystal axis. In fig. 10 the temperature dependence of the velocity of a shear wave propagating in a [110]-direction is shown. Whereas the [1$\bar{1}$0]-polarized wave is not influenced by the TSs, a strong variation of the velocity with temperature is observed if the polarization is in the [001]-direction. From such measurements it was concluded that the Li-TSs are oriented along the $\langle 111 \rangle$-directions [7].

Very recently the velocity of torsional waves propagating in [100]-direction has been investigated in KCl:Li [8]. The relative change of the velocity in a pure and in Li-doped KCl-crystals is shown in fig. 11. Apart from the sign (as mentioned $\delta v \propto -\chi'_a$) the behaviour is very similar to that of the dielectric susceptibility. As expected the effect is more pronounced for the heavier isotope, but there is also a qualitative difference. A minimum is found just below 1 K whereas the corresponding maximum is missing in the dielectric susceptibility. As in the dielectric case the elastic susceptibility

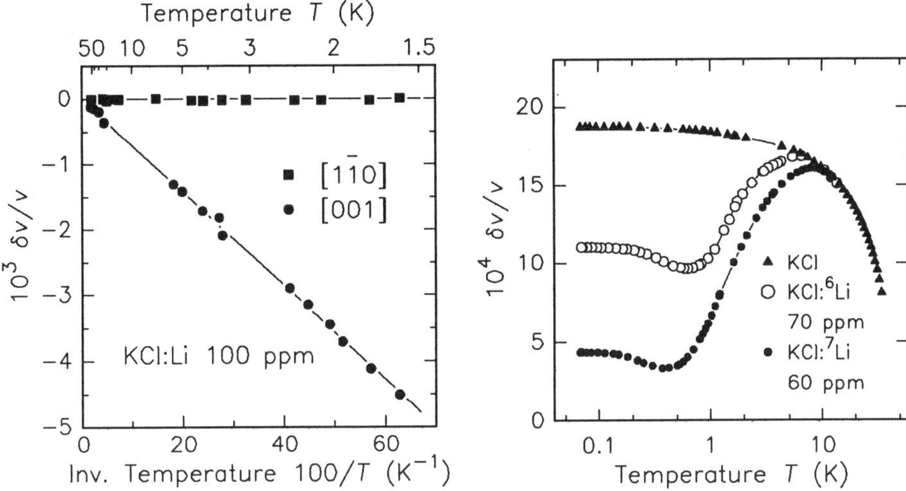

Figure 10. Relative change of the velocity of elastic shear waves propagating in [110]-direction as a function of the inverse temperature. The measurements have been carried out at 30 MHz with waves polarized in [1$\bar{1}$0]- and [001]-direction. From [7].

Figure 11. Relative change of the velocity of torsional waves propagating in [100]-direction in pure and Li-doped KCl as a function of temperature. The measurements have been carried out at 2 kHz. From [8].

is determined by the strain dependence of the energy levels. However, it turns out that the field dependence of the level scheme shown in fig. 9 does not hold for the elastic behaviour. Only four of the eight levels exhibit a quadratic field dependence at small fields. The energy of the other four levels varies linearly with the strain field σ. As a consequence the elastic susceptibility $\chi'_{\rm a} = \partial^2 \mathcal{F}/\partial\sigma^2|_{\sigma=0}$ differs from $\chi'_{\rm d}$. For the relative change of the sound velocity one finally obtains:

$$\frac{\delta v}{v} = -\frac{2n\gamma^2}{\rho v^2}\left[\frac{1}{\Delta_0}\tanh\left(\frac{\Delta_0}{2k_{\rm B}T}\right) + \frac{1}{2k_{\rm B}T}\mathrm{sech}^2\left(\frac{\Delta_0}{2k_{\rm B}T}\right)\right] . \qquad (9)$$

It is the second term which gives rise to the minimum in the temperature dependence of the velocity of sound.

2.4. COUPLED DEFECT PAIRS

In general point defects are randomly distributed on the lattice sites of the host crystal. With increasing concentration the mean distance between the defects decreases and the interaction between them becomes more and more important. For the defects discussed here there exist two types of interaction, namely via electrical dipoles and via elastic strain fields. From a theoretical point of view the treatment of dipolar interaction is much simpler because of the tensor character of the elastic fields. Since the elastic interaction can be neglected, the system KCl:Li can be treated in a simple way. For the energy W of interaction between two Li$^+$-defects with dipole moments $\vec{p}$ and $\vec{p}'$ we therefore write

$$W(\vec{p}, \vec{p}') = \frac{1}{4\pi\epsilon_0\epsilon_{KCl}}\left[\frac{\vec{p}\cdot\vec{p}'}{r^3} - \frac{3(\vec{r}\cdot\vec{p})(\vec{r}\cdot\vec{p}')}{r^5}\right]. \tag{10}$$

Here $\vec{r}$ is the vector connecting the two defects. Neglecting the orientation dependence of the interaction for simplicity we may define

$$J = \frac{1}{4\pi\epsilon_0\epsilon_{KCl}}\frac{p^2}{r^3} \tag{11}$$

as energy scale. If two defects are close enough they will form coupled pairs with properties completely different from those of isolated defect states. A two-dimensional visualization of this situation is shown in fig. 12.

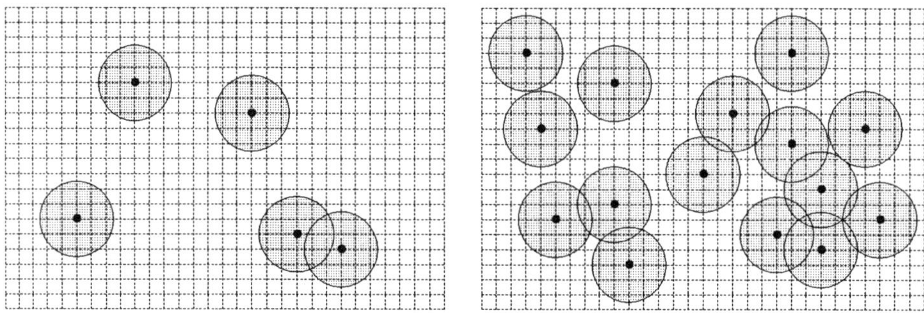

Figure 12. Schematic representation of the interaction of TSs. The square grid represents the crystal lattice, the dots the TSs. The dark areas indicate the region within which the dipolar forces are "noticeable". In going from the left to the right figure the concentration of the TSs has been increased. From [9].

The square grid represents the regular lattice of a crystal, the dots the impurities. A "radius of interaction" is drawn for which $J = \Delta_0$ holds. As long as the shaded areas do not overlap we may consider defects as "isolated". Overlapping areas indicate coupled TSs. At low concentration (left side) isolated TSs and one coupled pair are present. At higher concentration larger clusters come into play resulting in a complicated many-body problem. We will come back to this situation in the section on incoherent tunneling.

At a first glance the situation seems to be too complicated to be treated in a simple way. Even at such low concentrations that only isolated TSs and interacting pairs must be considered, the coupling of the two eight-level systems leads to a system with 64 levels. Fortunately, extensive simplification

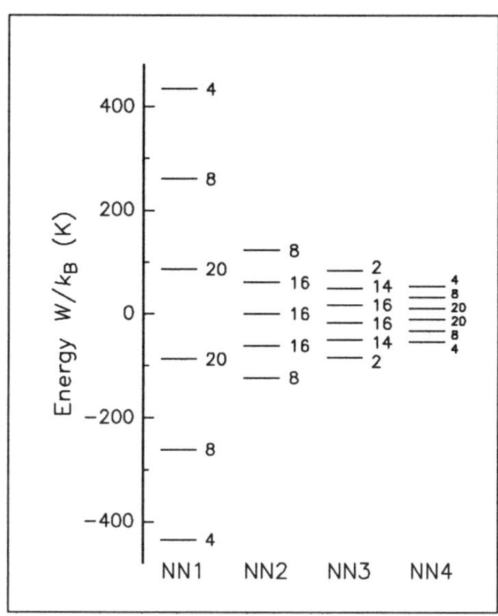

Figure 13. Energy levels of a classically interacting Li$^+$-pairs. The numbers next to the levels indicate the degree of degeneracy. From [10].

is possible. We start with the classical calculation of the interaction energy. Since J decreases with the separation of the TSs we have to specify the lattice sites of the interacting Li$^+$-defects. We consider here the neighbors and denote them by NN1, NN2, ...with increasing distance between the individual TSs. The interaction energy W depends on the relative orientation of the two dipoles. In our calculation we have to consider all possible orientations of the two dipole moments and compare the associated value of W. The classical electrostatic interaction energy is calculated for all the configurations and the level scheme shown in fig. 13 is found. Obviously, the levels are degenerate because of the high symmetry of the cubic crystal. The different configurations with the lowest potential energy are visualized in fig. 14 for the nearest and next nearest neighbors NN1 and NN2. For NN1 four possible configurations exist, for NN2 there are eight configurations.

The basic idea for further consideration is that the coupled Li$^+$-ions perform a coherent tunneling motion resulting in a tunnel splitting η that differs from Δ_0, the splitting of isolated defects. An interesting question is: can the existence of such pairs be demonstrated experimentally? As we will see, it are the NN2 pairs which can be seen in dielectric experiments. As indicated in fig. 13 the coupling energy of the classical ground state of NN2 pairs is $W/k_B = -123$ K. This state is eightfold degenerate but the degeneracy is lifted by the tunneling motion. A tedious calculation leads to a level scheme similar to that shown in fig. 4 for an isolated TS. The main

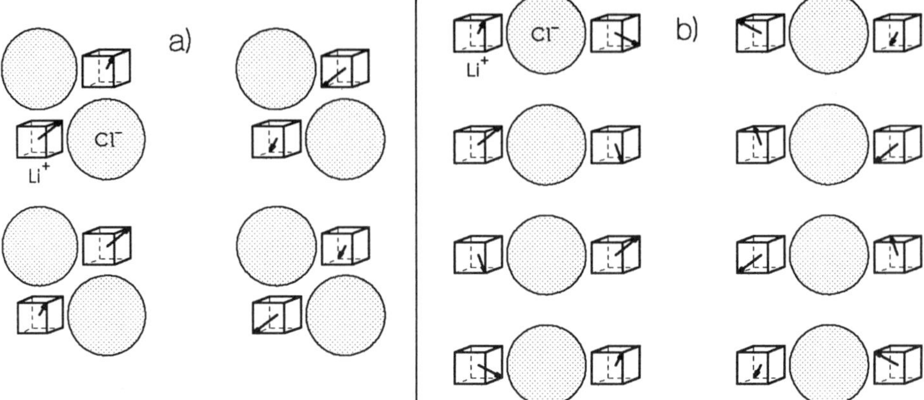

Figure 14. Schematic representation of those configurations of the pairs NN1 and NN2 that possess the lowest potential energy. From [10].

difference lies in the new level splitting η that is given by [11-13]

$$\eta = \frac{\Delta_0^2}{W} \, . \tag{12}$$

If we calculate the value of the tunnel splitting η of the different pair configurations we have to consider that only NN2 pairs (and again at larger distances the NN8 pairs) can change from one configuration (see fig. 13) to another simultaneously by tunneling along the edges of the cage. Since the Li^+-ions of the other pair configurations can only tunnel along other paths, Δ_0 has to be replaced in those cases by the appropriate tunnel splitting which is always considerably smaller. Therefore, among all configurations up to NN8, NN2 pairs exhibit the largest value of η although W is smaller for NN3 to NN7. For the tunnel splitting of pairs formed by the isotope ^7Li one finds the following values: $^7\eta/k_B = 0.001, 9.3, 0.007$ and 0.01 mK for NN1, NN2, NN3 and NN4 pairs, respectively.

In the discussion of the KCl:Li system we have assumed so far that all potential minima have the same depth or in other words that the asymmetry energy introduced in fig. 1 vanishes. In this case the tunneling splitting and energy splitting are identical, i.e. $E = \Delta_0$. From experiments it is known that such an assumption is meaningful for isolated defects because Δ_0 is relatively large. For pair states the situation is not as simple because $\eta \ll \Delta_0$. The asymmetry energy Δ caused by internal strain fields or uncontrolled impurities will be smeared out and can easily be comparable or even larger than η. Therefore we use eqn. (2) for coupled pairs and write for the energy splitting $E = \sqrt{\eta^2 + \Delta^2}$.

How can the coupled pairs be observed in an experiment? The most powerful tool for this purpose is through the generation and detection of rotary echoes using a technique very similar to NMR. At resonance a microwave field with frequency $\omega/2\pi$ induces transitions between the two levels. Because of the selection rules only transitions between two energetically neighbored levels are possible (see fig. 4) and in this sense we can consider the coupled pairs as two-level systems. The periodic change of the occupation of the two levels is accompanied by a harmonic oscillation of the electrical polarization on a macroscopic scale with the Rabi frequency

$$\Omega_{\rm R} = \frac{1}{\hbar}\,\frac{\eta}{E}\,\vec{p}_{\rm p}\cdot\vec{F}_0\,,\tag{13}$$

where $\vec{F}_0$ is the amplitude of the applied microwave field and $\vec{p}_{\rm p}$ the dipole moment of the coupled pairs. Because of the superposition of the contributions from TSs with different Rabi frequencies, – note that the asymmetry energy Δ is distributed – the macroscopic polarization vanishes rapidly after applying the microwave field. In the experiment mentioned here [14] the decaying Rabi oscillations could not be detected since the amplifiers were saturated after switching on the driving field. Therefore the phase of the microwave field was inverted after the time $t = t_{\rm p}$, causing a reversal of the time evolution of the polarization. Consequently, Rabi oscillations appear again and can easily be detected at $t = 2t_{\rm p}$, a phenomenon called "rotary echo". In the experiments frequencies around $\omega/2\pi = 1$ GHz – roughly corresponding to 50 mK – were applied and Rabi frequencies of the order of 1 MHz were observed.

Such experiments have been carried out at different frequencies on KCl doped with ^{6}Li, ^{7}Li and a mixture of both isotopes. Because $E = \hbar\omega$ in experiments based on resonant interaction a linear relationship between $\omega\cdot\Omega_{\rm R}$ and the field amplitude F_0 is expected from eqn. (13). Three clearly distinct straight lines are found experimentally for the three different kinds of coupled pairs (see fig. 15). Since the dipole moment $p_{\rm p}$ of NN2 pairs is known, the value of the tunnel splitting η can be calculated from the slopes of the lines. For $^6\eta$, $^{6/7}\eta$ and $^7\eta$ the values 20.5, 12.9, and 8.5 mK are found. These values compare favorably with those deduced from eqn. (12), namely 22, 14.4, and 9.3 mK. An even better agreement is obtained if the interaction energy is not calculated under the assumption of rigid dipoles. If the orientation of two individual charges is considered, the energy of interaction of NN2 pairs is found to be $W/k_{\rm B} = 139$ K [10].

2.5. PAIR-MODEL

With the help of rotary echoes only the well-defined defects NN2 are detectable. However, because of the random distribution of the Li^+-ions a

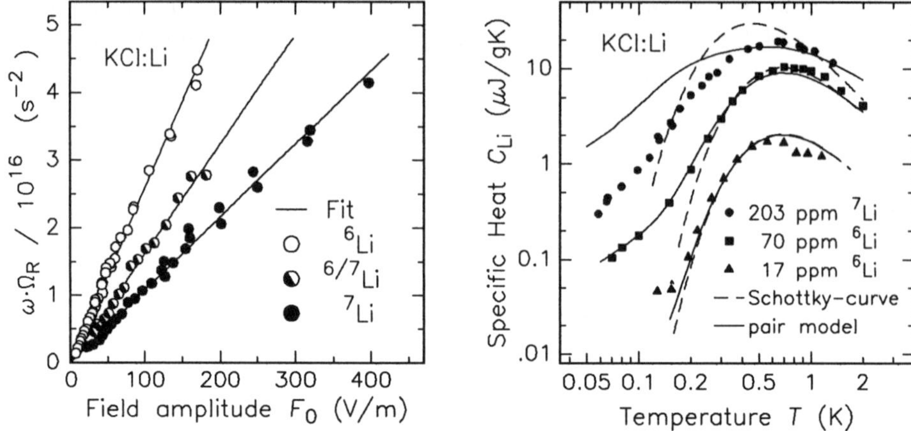

Figure 15. Product $\omega \cdot \Omega_R$ as function of the local field strength F_0 for samples containing different kinds of coupled pairs. From [10].

Figure 16. Specific heat of KCl:Li after subtraction of the contribution of the host lattice. The prediction of the pair-model [13] is shown by solid lines, the Schottky-anomaly by dashed lines. From [4] and [15].

broad spectrum of pairs must be present in the crystal. Coupled pairs of the type NN$\geq$8 are seen in measurements of the specific heat at higher concentrations. In fig. 16 the specific heat of KCl:Li is plotted for three different concentrations of lithium. Clearly, additional contributions are observed on the low temperature slope of the Schottky anomaly which cannot be understood on the basis of isolated TSs.

A qualitative explanation of this phenomenon is rather simple. The level splitting of the pairs depends on the interaction energy W which decreases with rising separation of the TSs forming the pairs. The problem lies in finding the appropriate distribution function for the energy splitting of the pairs. This problem is addressed in the pair-model [13] that is based on three simplifications. First it is assumed that the two interacting eight-level systems can be replaced by two two-level systems. This is possible because of the symmetry of the level scheme. Secondly, the distance between the interacting TSs is treated as a continuous variable. This assumption is obsolete for strongly coupled pairs. According to eqns. (11) and (12), however, their tunnel splitting is so small that they do not influence the specific heat above 50 mK. Finally, the appearance of the asymmetry energy Δ is ignored because it is expected to be rather small.

The theoretical curves drawn in fig. 16 demonstrate that this model correctly predicts the specific heat up to a concentration of about 100 ppm of lithium but fails at still higher concentrations for reasons we will discuss in the following section.

2.6. INCOHERENT TUNNELING

As visualized in fig. 12 the interaction between the TSs becomes more and more complicated with increasing concentration. The experimental results shown in the following two figures demonstrate that we are facing a many-body problem. In fig. 17 the temperature dependence of the dielectric susceptibility is plotted for samples doped with different amounts of lithium [5]. Three characteristic features are found. Most striking is the non-linear concentration dependence of the magnitude of the susceptibility. At the highest concentration of 1100 ppm the susceptibility becomes even smaller than at 210 ppm! Secondly, with increasing concentration the steep decrease of χ' shifts to higher temperatures. Finally a shallow maximum is observed indicating the occurrence of a relaxation process. The last phenomenon becomes more obvious in the imaginary part χ'' of the dielectric function shown in fig. 18 [5]. The loss cannot be caused by resonant interaction because $\hbar\omega$ is much smaller than Δ_0 or η. Therefore relaxation must occur although such a process does not take place in samples containing isolated TSs.

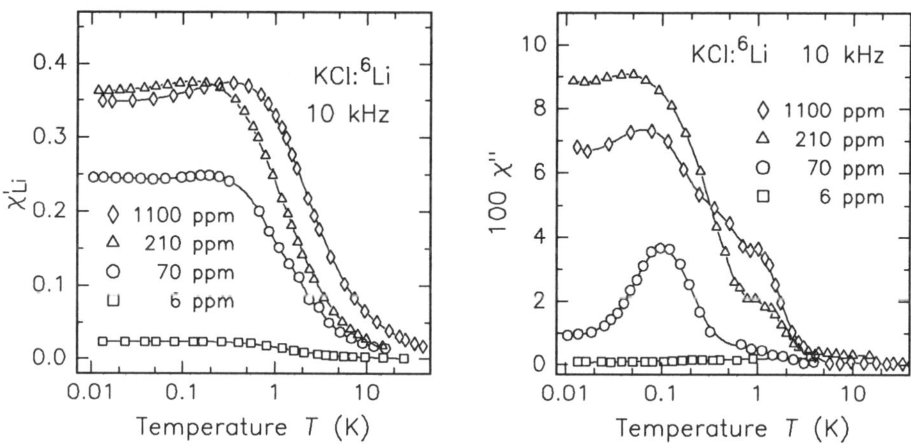

Figure 17.   Susceptibility of KCl:Li-samples containing different amounts of lithium measured at 10 kHz. The data of the sample containing 6 ppm lithium are also shown in fig. 7. From [5].

Figure 18.   Dielectric loss of KCl:Li-samples with different concentration of lithium. The measurements were performed at 10 kHz. From [5].

This complex theoretical problem has been solved only very recently [9]. The treatment is based on Mori's reduction method which permits a derivation of a continued fraction representation for the relevant two-times correlation functions [16]. We do not discuss here the rather sophisticated theoretical considerations but we need to use the results in order to explain

484

experimental results. In the theory the dimensionless factor

$$\mu = \frac{1}{3}\frac{np^2}{\epsilon\epsilon_0}\frac{1}{\Delta_0} \tag{14}$$

plays an important role. It reflects the ratio between the mean value of the interaction energy and the tunnel splitting.

For isolated TSs the phase of the wavefunction is preserved, i.e., the tunneling motion is coherent. Coupling to neighbored TSs disturbs the free tunneling motion and thus the phase of the tunneling particle. With increasing value of μ the number of interacting TSs rises and the disturbances become more and more serious. Finally, when $\mu \geq 1$ is reached the tunneling motion is virtually incoherent. At the same time the interaction of the TSs with external fields changes its character. The resonant interaction becomes more and more suppressed and relaxation phenomena set in. The relaxation time τ is not determined anymore by the interaction with the phonon bath but is due to the interaction amongst the TSs. For the dielectric susceptibility theory predicts [9]

$$\chi_{\text{res}} = \frac{2p^2}{3\epsilon_0 V_0}\sum_i (1 - R_i)\frac{1}{E_i}\tanh\left(\frac{E_i}{2k_{\text{B}}T}\right). \tag{15}$$

The summation has to be carried out over all TSs. V_0 is the volume of the sample and R_i the so-called relaxation amplitude. The energy splitting of the individual TSs is not a well-defined quantity anymore but depends on their local environment. Its mean value is given by

$$\overline{E_i} = \Delta_0\sqrt{1 + \mu^2}. \tag{16}$$

For the relaxational part of the susceptibility one finds

$$\chi_{\text{rel}} = \frac{p^2}{3\epsilon_0 V}\frac{1}{k_{\text{B}}T}\sum_i R_i\frac{1}{1 + i\omega\tau_i}, \tag{17}$$

where τ_i is the relaxation time of individual TSs. For a random distribution of TSs the mean value of the relaxation amplitude R can be calculated numerically. It increases steeply in the range $0.1 < \mu < 1$ and approaches unity for $\mu > 5$. Thus resonant interaction of isolated TSs determines the properties at low concentrations. Relaxational phenomena dominate the behaviour at high concentrations. There exists qualitative agreement between theory and experiment, in particular with respect to the three characteristic experimental features mentioned above. A quantitative comparison is not possible so far because the relaxation spectrum entering eqn. (17) is not known and can only be calculated using very rough approximations [9].

The dielectric loss has been measured at different temperatures in a wide frequency range [17]. From the observed loss peak the mean value of the relaxation time can be deduced. Surprisingly, it turned out that the heavier TSs consisting of ^{7}Li-ions relax faster than the lighter TSs. This clearly indicates that relaxation does not occur via coupling to phonons because in this case the opposite relation would be found. In fact this result shows that at high concentrations relaxation takes place via complicated collective motion within the ensemble of TSs. The surprising dependence on the mass of the TSs is also expected from the theoretical treatment of the incoherent tunneling phenomenon [9].

There exists an interesting possibility of a quantitative comparison. On the one hand for $T \to 0$ the relaxational contribution (17) to the dielectric susceptibility vanishes even at a high concentration of TSs since the relaxation times become extremely long. On the other hand the expression (15) for χ'_{res} can be approximated by [9]

$$\chi'_{\text{res}} = \frac{2np^2}{3\epsilon_0 \Delta_0} \frac{(\sqrt{1+\mu^2} - \mu)^2}{\sqrt{1+\mu^2}} \tanh\left(\frac{\Delta_0 \sqrt{1+\mu^2}}{2k_{\text{B}}T}\right) . \tag{18}$$

Using this equation together with eqn. (8) we obtain for $T \to 0$ the simple relation

$$\frac{\chi'}{\chi'_{\text{iso}}} = \frac{\left(\sqrt{1+\mu^2} - \mu\right)^2}{\sqrt{1+\mu^2}} . \tag{19}$$

Here we have written χ'_{iso} for the susceptibility given by eqn. (8) since this equation reflects the susceptibility due to isolated TSs. '

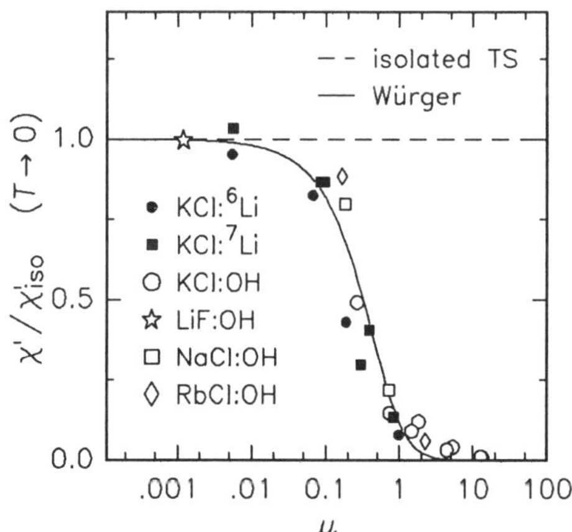

Figure 19. Ratio χ'/χ'_{iso} of different crystals with different concentrations of TSs as a function of μ. The theoretical curve (19) is shown as solid line [9]. From [18]

486

In fig. 19 this ratio is plotted as a function of μ together with experimental points not only from KCl:Li but also from measurements on systems containing OH^--ions. Although the agreement is impressive it should be noted that eqn. (19) is expected to hold only if the elastic interaction between the TSs can be neglected, i.e., if electrical dipolar forces couple the TSs.

3. Tunneling Systems in Amorphous Solids

3.1. TUNNELING MODEL

At low temperatures the thermal, acoustic and dielectric behaviour of amorphous solids differs significantly from that of crystalline solids. Most prominent examples are the almost linear temperature dependence of the specific heat C_V [19] and the surprising temperature variation $\delta v/v$ of the sound velocity [20].

As an example the specific heat of vitreous silica is plotted together with that of crystalline quartz in fig. 20. Clearly, at very low temperatures the specific heat of the glass exceeds that of the crystal by orders of magnitude. It is most remarkable that the magnitude of C_V is similar for all amorphous solids irrespective of their chemical composition. In fig. 21 the temperature variation of the sound velocity of cover glass is shown. The velocity first rises logarithmically, passes a maximum and then decreases with temperature again.

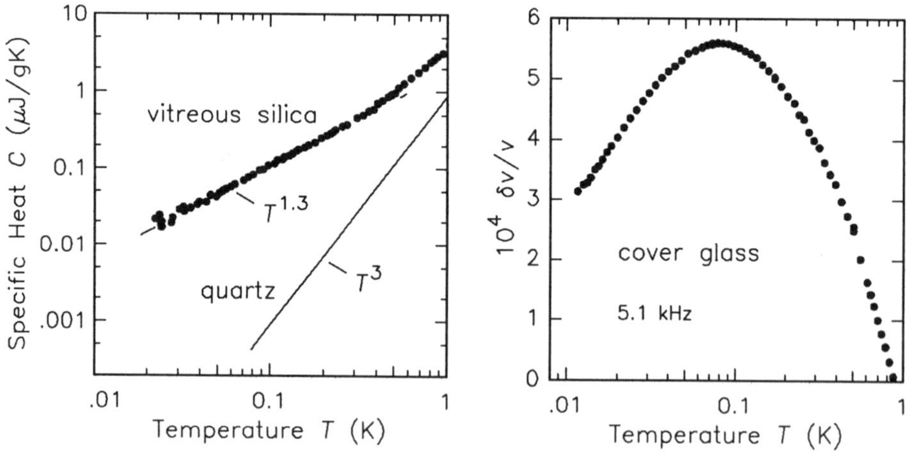

Figure 20. Specific heat of vitreous silica and of crystalline quartz as a function of temperature. From [19] and [21].

Figure 21. Relative variation of the velocity of sound of cover glass measured at 5.1 kHz. From [22].

In spite of intense efforts these and all the other "anomalies" can at present only be described on the basis of phenomenological models. We restrict ourselves in this discussion to the so-called Tunneling Model (TM) as it was originally introduced and is based on two simple assumptions [23]. First it is assumed that a small number of TSs exists in amorphous solids. Their microscopic nature is not well understood, the lack of long-range order prohibiting a proper characterization of the atomic potential energy landscape. Therefore the TSs do not correspond to well-defined atoms or molecules but rather involve structural rearrangements of more or less random local configurations. In such a case it is quite natural to assume that distribution functions of the relevant parameters to the TSs exist which are introduced on the basis of a second main assumption: Asymmetry energy Δ and tunneling parameter λ (see eqs. (2) and (4)) are assumed to be independent of each other and uniformly distributed:

$$P(\Delta, \lambda)\, \mathrm{d}\Delta\, \mathrm{d}\lambda = \overline{P}\, \mathrm{d}\Delta\, \mathrm{d}\lambda \,, \tag{20}$$

where $\overline{P}$ is a constant.

The alternative model, the so-called Soft Potential Model [24,25] involves a description of the anomalies in terms of a fourth order potential with appropriate distributions for its three parameters. In the description of the properties at very low temperatures it reduces essentially to the TM. At higher temperatures local quasi-harmonic potentials arising from quartic potentials with a vanishing or positive quadratic term, become important and determine the properties of glasses. Therefore this model is able to make predictions in a temperature range where tunneling is not important anymore.

From an experimental point of view the more relevant quantities are E and $r = (\Delta_0/E)^2$ instead of Δ and λ. Using these parameters eqn. (20) may be replaced by

$$P(E, r)\, \mathrm{d}E\, \mathrm{d}r = \frac{\overline{P}}{2r\sqrt{1 - r}}\, \mathrm{d}E\, \mathrm{d}r \,. \tag{21}$$

This distribution function tends to infinity at the boundaries of the allowed interval, i.e. for $r \to 0$ and $r \to 1$. Performing an integration over r provides us with the density of states of the TSs:

$$n(E) = \int_{r_{\min}}^{1} \frac{\overline{P}\, \mathrm{d}r}{2r\sqrt{1 - r}} = \overline{P} \ln r_{\min} \,. \tag{22}$$

We have introduced a lower limit $r_{\min}$ because otherwise the number of TSs would diverge. This also means that some kind of an upper cut-off must exist for the distribution of λ. Knowing the density of states allows a

488

calculation of the internal energy and the specific heat. Ignoring the weak logarithmic energy dependence of $n(E)$ we obtain the famous linear temperature variation of the specific heat, i.e. $C_V \propto T$. For the small deviations from the linear law, e.g. for the proportionality to $T^{1.3}$ in fig. 20, no generally accepted explanation exists so far.

3.2. ACOUSTIC PROPERTIES

Acoustic experiments have been a very important tool in the investigation of low temperature properties of amorphous solids. As discussed in the introductional section, TSs give rise to resonant and relaxation interaction. While TSs in crystals are well-defined units, the parameters of TSs in amorphous solids are smeared out according to eqn. (20). As a consequence we have to calculate average values of the relevant quantities with the help of distribution functions in order to be able to compare the experiments with theory. Carrying out the corresponding averaging procedure for the acoustic susceptibility one finds for the absorption coefficient α or for the internal friction $Q^{-1} = \alpha v / \omega$ at ultrasonic frequencies

$$\frac{\alpha v}{\omega} = Q_{\mathrm{res}}^{-1} = \frac{\pi C}{\sqrt{1 + I/I_{\mathrm{c}}}} \tanh \left(\frac{\hbar \omega}{2 k_{\mathrm{B}} T} \right) , \qquad (23)$$

if we consider the resonant interaction only. Here $C = \overline{P} \gamma^2 / \rho v^2$ reflects the coupling between a sound wave and TSs and γ represents the deformation potential as before. If the acoustic intensity I is well below its critical value I_{c} the absorption process is linear. At higher intensities the upper level becomes populated due to the absorption mechanism itself. This effect is called saturation and causes a reduction of the resonant acoustic loss. In fig. 22 data on the temperature variation of the attenuation of vitreous silica is shown at a frequency of about 1 GHz and at two different intensities [26]. The saturation effect and the increase of the absorption with decreasing temperature at low intensities can clearly be seen. It should be added that the corresponding behaviour is also found in measurements of the dielectric loss [27].

The rise of the absorption at higher temperatures is caused by the relaxation process. This effect is shown in more detail in fig. 23 where the internal friction at a low frequency is plotted for the same glass [22]. Measurements at low frequencies are especially suitable to obtain information on the distribution functions because the contribution of the resonant process can be neglected and the relaxation effect becomes more pronounced. Before the we can express the internal friction Q^{-1} in an equation we need to consider the relaxation mechanism. It is known that below 1 K the simple one-phonon process dominates where a single thermal phonon is absorbed

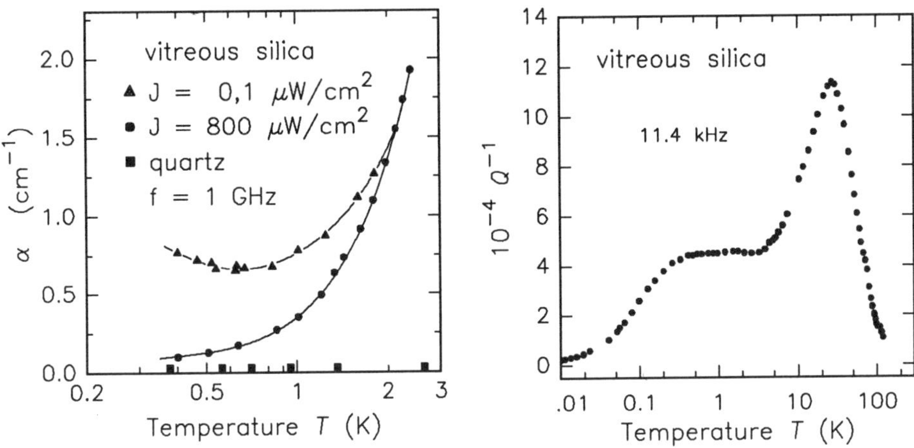

Figure 22. Temperature variation of the ultrasonic absorption in vitreous silica and crystalline quartz at about 1 GHz. At the lowest temperature the magnitude of the absorption depends strongly on the acoustic intensity. From [26].

Figure 23. Internal friction of vitreous silica at 11.4 kHz. The strong peak at higher temperatures is attributed to thermally activated processes, not discussed here. From [22].

or emitted by the relaxing TS. In this case the relaxation rate is given by

$$\tau^{-1} = ArE^3 \coth(E/2k_\mathrm{B}T) \,, \tag{24}$$

where A is a constant, characteristic of the amorphous solid. Because the parameter r varies in the range $0 < r < 1$ there exists a wide distribution of relaxation rates for a given energy splitting E even at a fixed temperature. From ultrasonic measurements it is known that at temperatures of about 1 K TSs with an energy splitting $E/k_\mathrm{B} = 1$ K exhibit a minimum relaxation time $\tau_{\min}(r = 1) \approx 1$ ns. On the other hand at the same temperature relaxation times $\tau > 10^4$ s have been observed in so-called heat release experiments [28]. In this type of experiment the temperature of the sample is reduced in such a short time that a noticeable part of the TSs is not able to reach thermal equilibrium. Because of the wide distribution of relaxation times these TSs will successively relax and deliver energy to the phonon bath. This means that there exists a continuous heat release. From eqs. (21) and (24) the heat release is expected to be inversely proportional to the time of measurement.

The main contribution to the acoustic (or dielectric) relaxational losses arises from those TSs having an energy splitting $E \approx k_\mathrm{B}T$ and relaxing on a time scale comparable with the period of the sound wave, i.e., from those TSs for which $\omega\tau \approx 1$. Analytic solutions can be obtained for $\omega\tau_{\min} \gg 1$

490

and $\omega\tau_{\min} \ll 1$. One finds

$$Q^{-1} = \frac{\pi^4 A C k_{\mathrm{B}}^3 T^3}{12\omega} \qquad \text{and} \qquad Q^{-1} = \frac{\pi C}{2} \tag{25}$$

for low and relatively high temperatures, respectively. The plateau at high temperatures directly reflects the uniform distribution given by eqn. (22).

At first glance the measurements shown in fig. 23 seem to be in agreement with the predictions of the TM. It turns out, however, that it is extremely difficult to prove the existence of the expected T^3-increase. At present it seems that the rise is less steep [22, 29]. Recently it has been argued that at very low temperature the relaxation in glasses does not occur via one-phonon processes. In theoretical attempts to explain the temperature dependence it has been proposed that in this temperature range the TSs move collectively [30]. In this case relaxation would take place within the ensemble of TSs and a rise linear in T is predicted.

As is the case in crystals the motion of the tunneling particles is disturbed at higher temperatures due to thermal motion of the environment. Thus a transition from coherent to incoherent tunneling is expected to occur also in amorphous solids. In theoretical considerations it has been shown that the onset of this phenomenon causes an increase in the absorption in the temperature range between 5 K and 10 K in agreement with observation [31].

Intimately related to the absorption process is a change in the velocity of sound reflecting the variation of χ_{a} with temperature. For the contribution of the resonant process one finds a logarithmic variation for $\hbar\omega \ll k_{\mathrm{B}}T$ of the form

$$\left.\frac{\delta v}{v}\right|_{\mathrm{res}} = C \ln\left(\frac{T}{T_0}\right), \tag{26}$$

where T_0 is an arbitrary reference temperature [20]. The validity of this equation has been verified for a large number of amorphous solids. Recently it has been observed that the slope of this rise (see fig. 21) depends on the amplitude of the applied strain [22, 29]. So far a generally accepted explanation of the strong nonlinearity does not exist.

The contribution of the relaxation process to δv is negligible as long as $\omega\tau_{\min} \ll 1$. At high temperatures, when $\omega\tau_{\min} \approx 1$, the contribution of the relaxation process becomes dominant and the velocity starts to decrease. In this regime the TM predicts a logarithmic decrease of the velocity reflecting the type of relaxation process, i.e., the slope should depend on the relaxation mechanism. As long as the one-phonon process (24) is dominant, the relation

$$\left.\frac{\delta v}{v}\right|_{\mathrm{rel}} = -\frac{3}{2} C \ln\left(\frac{T}{T_0'}\right) \tag{27}$$

should hold, where T_0' is another reference temperature. Above 1 K the decrease due to relaxation should become much steeper because relaxation via multi-phonon processes sets in. The total change of the velocity is the sum of the two eqns. (26) and (27).

As can be seen in fig. 21 and as is known from many other experiments, the ratio of the slopes given by eqn. (26) at low temperatures and by the sum of the contributions of eqns. (26) and (27) near and above the maximum is not -1/2 but rather -1 [22]. The reason for this discrepancy is not yet understood.

As mentioned above, measurements of the thermal properties gave first evidence for the anomalous behaviour of amorphous solid at low temperatures [19]. The thermal conductivity Λ in fig. 24 increases proportional to T^2 up to about 3 K and levels off above that temperature. At least at low temperatures, heat is transported by phonons as in crystals but phonon propagation is hindered by the TSs. Using the kinetic formula

$$\Lambda = \frac{1}{3} C_{\mathrm{D}} \, v \, l \,, \tag{28}$$

where C_{D} represents the T^3-dependent Debye specific heat, the thermal conductivity can be described in a simple way. In order to obtain the correct temperature dependence $l \propto 1/T$ must hold. We may apply eqn. (23) to thermal phonons by putting $I = 0$ and $\tanh(\hbar\omega/2k_{\mathrm{B}}T) \approx 1$ and find immediately the correct relationship. Thus we can conclude that TSs limit the heat transport in amorphous solids at low temperatures.

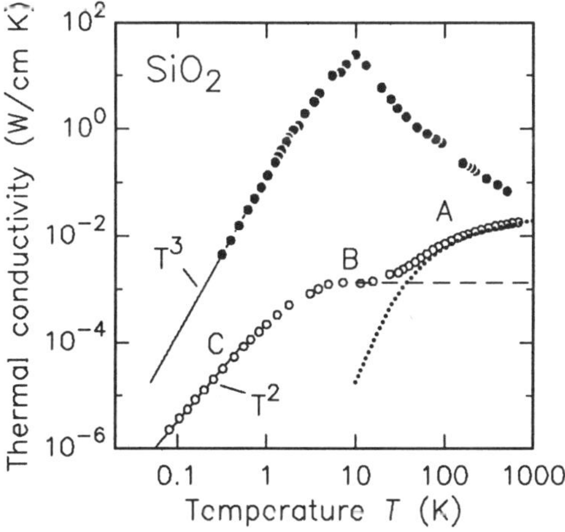

Figure 24. Thermal conductivity of vitreous silica and crystalline quartz as a function of temperature. From [19].

492

The conductivity in the plateau region between 1 K and 10 K is reduced by additional scattering. In the Soft Potential Model this scattering is attributed in a natural way to the quasi-harmonic potentials that have the same origin as the TSs. It should be noted that alternative explanations for the heat transport at higher temperatures are under discussion.

3.3. INTERACTION BETWEEN TUNNELING SYSTEMS

So far we have only discussed the interaction of TSs with external elastic (or electric) fields. Many experimental results indicate that interaction between the TSs is also of importance in amorphous systems. Its is very likely that the elastic interaction predominates here. Experimentally there are two more or less direct ways of observing this phenomenon. Also in other experiments, e.g. in the surprising memory effects discovered recently in dielectric experiments at very low temperatures [32], information on the interaction between the TSs can be obtained.

As we have mentioned, the resonant absorption of acoustic or electric waves depends on the difference in the populations of the two levels. An intense pulse leads to an equipartition and burns a "hole" of width $\Delta\omega$ into the spectral distribution at the corresponding energy. If a second, weaker probing pulse of slightly different frequency is sent through the sample, its attenuation will reflect the population difference at that frequency. As an example the result of such a measurement carried out in a borosilicate glass is shown in fig. 25. Saturation due to the first pulse can still be seen by the probing pulse if its frequency is displaced by as much as 50 MHz. Furthermore it was observed that the width of the hole is much wider than expected from lifetime broadening, depends linearly on temperature, and increases with the time of measurement.

This remarkable effect is ascribed to temporal fluctuations of the level splitting of the TSs at resonance. The fluctuations are caused by neighboring TSs which are thermally excited and de-excited and couple elastically to the resonating systems. Therefore TSs having an energy close to the phonon energy of the intense pulse will temporarily be in resonance with the sound wave, although at $T = 0$ an energy difference would exist. As a consequence TSs can be excited within a wider energy range and a broad hole is observed by the probing pulse. The excursion of the energy splitting of a given TS will depend on time since the contributions of the neighboring TSs add statistically. After very long times the hole will reach its largest value when all the TSs in the neighborhood have undergone a thermal transition, i.e. when half of the systems have changed their state an odd number of times. With increasing temperature the number of thermally activated TSs rises and results in an increase of the hole width. This mechanism of hole broad-

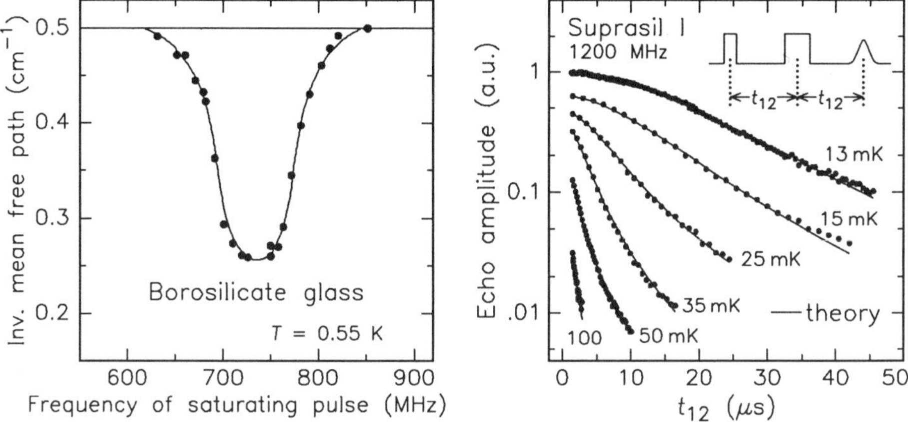

Figure 25. Resonant attenuation of a weak probing pulse at 735 MHz as a function of the frequency of the intense saturating pulse. From [33].

Figure 26. Decay of the amplitude of the spontaneous echo in vitreous silica as a function of the pulse separation t_{12} at 1200 MHz and different temperatures. The curves are displaced vertically for clarity. The lines are fits with the model of spectral diffusion. The insert shows the sequence of the exciting pulses and the echo. From [35].

ening is called "spectral diffusion" by analogy with the mechanism known from magnetic resonance experiments. It has been discussed in detail and agreement between theory and experiment has been obtained [34].

With decreasing temperature, both the relaxation time τ and the phase memory time τ_2, also known as "transverse" relaxation time, become longer and longer. Therefore at low enough temperatures coherent echoes can be generated similar to the rotary echo discussed above. As an example we mention here new measurements of the decay time of spontaneous echoes in vitreous silica [35]. The echo decay is governed by the spectral diffusion process just discussed in conjunction with the hole burning experiment. Thus the echo decay reflects the interaction between the TSs. In fig. 26 the decay of the amplitude is shown as a function of pulse separation time t_{12}. In the same figure the applied pulse sequence and the observed echo at time $2t_{12}$ are drawn schematically. As expected from the theory of spectral diffusion the decay is nonexponential and much faster at higher temperatures. In fact the solid lines in fig. 26 represent fits obtained from theory [36]. For short times a Gaussian-like decay curve proportional to $\exp[-(2t_{12}/\tau_2)^2]$ is predicted, for long times the echo is expected to decay proportional to $\exp(-\sqrt{2t_{12}}/m)$, where m is a constant. In the experiment $\tau_2 \propto T^{-2}$ and $m \propto T^{-1}$ was found in agreement with theory. It is worth mentioning that the "classical" one-phonon process has been used in the fitting procedure and not a relaxation time based on collective motions of the TSs as proposed in a recent theory on the relaxation at low temperatures [30].

494

4. Conclusions

The investigations of TSs has a long tradition in low temperature physics; nevertheless many open questions still remain. Isolated TSs in model systems are well understood whereas the interaction between TSs has been studied for a long time and is still not fully understood. For the interaction via electrical dipole moments an extensive theory has been worked out recently and leads to good agreement with experimental results. The situation is less clear if the TSs interact elastically because of the complexity of the problem. Although there has been much effort in the past to clarify the situation, we have not addressed these attempts here because many open questions still remain. In particular the interesting theoretical and experimental results on highly concentrated TSs, for example in the case of KBr:KCN, have not been discussed here although they have very often been considered as crystalline models for TSs in glasses [37].

TSs in amorphous materials are presently only understood on the basis of phenomenological models. The experimental and theoretical problems are closely related to the exact shape of the relevant distribution functions. A more fundamental question is that of the atomic nature of the TSs which has not been addressed at all in this article. No generally accepted answers can be given at present. Attempts have been made to attribute TSs to specific local atomic configurations which are different for different systems. If such an approach leads to a good description of experimental results the universality of the observed phenomena is difficult to explain. At present it seems that computer simulations of simple glasses are the most promising attempts to come to an understanding of the structure of the TSs [38]. A further open problem is still the interaction between TSs. There is no doubt that TSs interact but it is not clear whether the interaction is of great importance or not. It could even be that the interaction is responsible for the creation of the experimentally observed TSs and the distribution of their parameters [39].

5. Acknowledgments

Helpful discussions with Simon Bandler, Johannes Classen, Christian Enss, and Manfred von Schickfus are gratefully acknowledged.

References

1. See for example: Merzbacher, E. (1970) *Quantum Mechanics*, John Wiley & Sons, New York
2. Gomez, M., Bowen, S.P. and Krumhansl, J.A. (1967) *Phys. Rev.* **153**, 1009
3. Wang, X. and Bridges, F. (1992) *Phys. Rev.* **B46**, 5122
4. Harrison, J.P., Peressini, P.P. and Pohl, R.O. (1968) *Phys. Rev.* **171**, 1037

5. Enss, C., Gaukler, M., Hunklinger, S., Tornow, M., Weis, R. and Würger, A. (1996) *Phys. Rev.* **B53**, 12094

6. Moy, D., Potter, R.C. and Anderson, A.C. (1983) *J. Low Temp.* **52**, 115

7. Byer, N.E. and Sack, H.S. (1966) *Phys. Rev. Lett.* **17**, 72; (1968) *J. Phys. Chem. Solids* **29**, 677

8. Hübner, M., Enss, C. and Weiss, G., to be published

9. Würger, A. (1996) *From Coherent Tunneling to Relaxation*, Springer-Verlag, Heidelberg

10. Weis, R. (1995) *Messung der dielektrischen Eigenschaften wechselwirkenden Tunnelsysteme bei ticfcn Temperaturen am Beispiel von KCl:Li*, PhD Thesis, Heidelberg

11. Baur, M. E. and Salzman, W.R. (1966) *Phys. Rev.* **151**, 710

12. Klein, M.W. (1984) *Phys. Rev. B.* **29**, 5825

13. Terzidis, D. and Würger, A. (1994) *Z. Physik* **94**, 341

14. Weis, R., Enss, C., Leinböck, B., Weiss, G. and Hunklinger, S. (1995) *Phys. Rev. Lett.* **75**, 2220

15. Dobbs, J.N. and Andersson, A.C. (1986) *Phys. Rev.* **B33**, 4172

16. Mori, H. (1965) *Theor. Phys.* **33**, 423 and (1965) *Theor. Phys.* **34**, 3909

17. Enss, C., Gaukler, M., Nullmeier, M., Weis, R. and Würger, A. (1996) submitted to *Phys. Rev. Lett.*

18. Würger, A., Weis, R., Gaukler, M. and Enss, C. (1996) *Europhys. Lett.* **33**, 533

19. Zeller, R.C. and Pohl, R.O. (1971) *Phys. Rev.* **B4**, 2029

20. Piché, L., Maynard, R., Hunklinger, S. and Jäckle. J (1974) *Phys. Rev. Lett.* **32**, 1426

21. Lasjaunias, J.C., Ravex, A., Vandorpe, M. and Hunklinger, S. (1975) *Sol. State Commun.* **17**, 1045

22. Classen, J., Enss, C., Bechinger, C., Weiss, G. and Hunklinger, S. (1994) *Ann. Physik* **3**, 315

23. Phillips, W.A. (1972) *J. Low. Temp. Phys.* **7**, 351
    Anderson, P.W., Halperin, B.I. and Varma, C. (1972) *Phil. Mag.* **25**, 1

24. Karpov, V.G., Klinger, M.I. and Ignatiev, F.N. (1983) *Soviet JETP* **57**, 439

25. Parshin, D.A. (1993) *Physica Scripta* **T49**, 180

26. Hunklinger, S., Arnold, W. and Stein, S. (1974) *Phys. Lett.* **45A**, 311

27. von Schickfus, M. and Hunklinger, S. (1977) *Phys. Lett.* **64A**, 144

28. See for example: Parshin, D.A. and Sahling, S. (1993) *Phys. Rev.* **47**, 5677

29. Esquinazi, P., König, R. and Pobell, F. (1992) *Z. Phys.* **B87**, 305

30. Burin, A.L. and Kagan, Yu. (1994) *JETP* **79**, 347

31. Rau, S., Enss, C., Hunklinger, S., Neu, P. and Würger, A. (1995) *Phys. Rev.* **B52**, 7179

32. Salvino, D.J., Rogge, S., Tigner, B. and Osheroff, D.D. (1994) *Phys. Rev. Lett.* **73**, 268
    Rogge, S., Natelson, D. and Osheroff, D.D. (1996) *Phys. Rev. Lett.* **76**, 3136

33. Arnold, W. and Hunklinger, S. (1975) *Sol. State Commun.* **17**, 833

34. Black, J.L. and Halperin, B.I. (1977) *Phys. Rev.* **B16**, 2879

35. Enss, C., Ludwig, S., Weis, R. and Hunklinger, S. (1996) *Czech. J. Phys.* **46**, 2247

36. Hu, P. and Hartmann, S.R. (1974) *Phys. Rev.* **B9**, 1

37. Sethna, J.P. and Chow, K.S. (1985) *Phase Trans.* **5**, 317

38. Heuer, A. and Silbey, R.J. (1993) *Phys. Rev. Lett.* **70**, 3911

39. Yu, C.C. and Leggett, A.J. (1988) *Comments. Cond. Mat. Phys.* **14**, 231

ELECTRONIC STRUCTURE OF AMORPHOUS INSULATORS AND SEMICONDUCTORS BY X-RAY PHOTOELECTRON AND SOFT X-RAY SPECTROSCOPIES

Christiane SÉNÉMAUD

Laboratoire de Chimie-Physique, 11 rue Pierre et Marie Curie

75231 PARIS cedex 05, FRANCE

Abstract

X-Ray Photoelectron Spectroscopy (XPS) and Soft X-Ray Spectroscopy (SXS) are attractive experimental methods which give complementary information on the electronic structure of amorphous insulators and semiconductors. By XPS, both core levels and valence band (VB) distributions are obtained. By soft X-Ray emission (XES) and absorption spectroscopies (XAS), local and symetry selected VB and CB states are determined. Consequently the XPS and SXS methods give quite complementary information concerning the electronic structure of materials. In this paper, the XPS and SXS methods are described and their possibilities are discussed. Results concerning the electronic structure of insulators and semiconductors specially in the case of amorphous silicon-based systems are discussed.

1. Introduction

Due to their large range of applications, amorphous insulators and semiconductors are important materials in electronics and optoelectronics. To understand the physical properties of these systems, it is essential to have information on their electronic structure. The theoretical determination of densities of states (DOS) of amorphous systems are based on different models of atomic arrangement and consequently experimental data are quite helpful. Among the experimental methods which inform on the electronic structure of materials, X-ray photoelectron (XPS) and soft X-ray (SXS) spectroscopies have the advantage to give direct information on either the valence band (VB) or the conduction band (CB) states. Consequently they are quite complementary to methods such as optical absorption spectroscopy widely used in the range of semiconductors, which inform on combined VB and CB states. In these methods it is possible to relate the modifications of the energy gap to the shift of either the VB or the CB edges.

M. F. Thorpe and M. I. Mitkova (eds.), Amorphous Insulators and Semiconductors, 497–506.
© *1997 Kluwer Academic Publishers. Printed in the Netherlands.*

The aim of this paper is to discuss the possibilities of XPS and SXS methods and to show their complementarity in the determination of electronic structure of materials. We illustrate these possibilities by presenting some results obtained for various amorphous semiconductors and insulators. Starting from amorphous silicon (a-Si) as a typical example, we show the evolution of the VB and CB as another element in incorporated in the a-Si matrix. Among Si alloys, either low gap or wide gap system can be obtained according to the incorporated element and to its concentration. Examples are given in each case. Before showing some experimental results obtained recently, some details concerning the experimental methods are first described.

2. Description of experimental methods

2.1 X-RAY PHOTOELECTRON SPECTROSCOPY (XPS)

XPS is based on the absorption of a photon $h\nu$ and the ejection of a photoelectron, whose kinetic energy E_{kin} is related to the binding energy E_B of an electron in the sample atom.

In a first approximation, $E_B = h\nu - E_{kin}$. The measurement of E_{kin} provides information on the bonding state of the atomic electron. Due to the small escape depth of the emitted photoelectron, only a superficial layer of the sample is analysed, typically 0.5 to 10 nm; as a consequence, a carefull sample preparation and clean vacuum systems are required.

Both core level, with energy lower than $h\nu$, and valence band distributions are investigated. As concerns the VB, energy distribution curves (EDC) are obtained which correspond to densities of states (DOS) modulated by photoionization cross sections.

$$I (E) \propto \sigma_s N_s (E) + \sigma_p N_p (E) + ...$$

σ_s, σ_p are the photoionisation cross sections; $N_s(E)$, $N_p(E)$ ... are the partial DOS of s, p... states. Consequently, information concerning particular symetry states can be enhanced. In the case of a compound, the DOS of the total VB is investigated. When synchrotron radiation as exciting beam, EDC's at various photon energies can be measured; they concern various thicknesses of the sample.

The energy resolution of XPS data is limited by the energy distribution of the incident photon beam. The use of a monochromator associated to the X-ray source give significant gain of resolution though involving a loss of intensity.

2.2 X-RAY EMISSION SPECTROSCOPY (XES)

In XES, radiative electronic transitions between the VB and a localized core level are analysed, (let us recall that non-radiative decay of inner hole give rise to Auger processes). Initial core holes are created by means of either charged particles or x-ray photons, possibly synchrotron beam. Dipole selection rules govern the radiative decay of the inner vacancy and imply that only VB states with particular angular momentum are investigated. Thus K emission spectra reflect the p-like VB states; L emission spectra correspond to s, d-like states... Moreover, as the core level concerns a specific atomic site, the DOS around each atomic species of a compound are investigated separately. The thickness of sample participating to the emission is determined mainly by the penetration of incident particles; it is in most cases a few 100 nm and these measurements concern the bulk sample.

The spectral intensity of an X-ray emission corresponds to the convolution product of the VB DOS by the core level Lorentzian profile L. In the independant electron approximation :

$$I\,(h\nu) \propto \nu^3 \int M_{if}^{\,2}\,(E)\; N_{OCC}^{\,l\pm1}\,(E)\; L\,(h\nu\text{-}E)\; dE$$

M_{if} (E) is the matrix element of the transition; it does not affect the main features of an XES spectrum over a limited spectral range;

$N_{OCC}^{\,l\pm1}$ is the occupied states DOS with $l\pm1$ symetry

$L(h\nu\text{-}E)$ is the core hole lorentzian profile.

2.3   X-RAY ABSORPTION SPECTROSCOPY (XAS)

In XAS, transitions between a core level and the conduction band states are analysed. In the spectral range close to an absorption edge, the variation of the photoabsorption coefficient μ is in first approximation proportionnal to the density of empty states, N_{emp} above E_F, with particular angular momentum around a specific site:

$$\mu\,(h\nu) \propto \nu \int M_{if}^{\,2}\,(E)\; N_{emp}^{\,l\pm1}\,(E)\; L\,(h\nu\text{-}E)\; dE$$

The spectral range close to the absorption edge (about 50 eV from E_F) corresponds to XANES (X-ray Near Edge Absorption Structure). Beyond a few tens of eV, the variations of μ exhibit oscillations which are related to the local atomic structure around the absorbing atom : This spectral range which corresponds to EXAFS (extended X-ray Absorption Fine Structure) is not considered in the following.

By XANES, either the volume or the surface of the sample are concerned, according to the energy range analysed and the technique chosen. XANES and EXAFS techniques have been widely developped in the recent years with the use of synchrotron radiation.

For both XES and XAS spectra, the life time of the core hole induces a Lorentzian broadening which limits the resolution of the method and implies to study only transitions involving a sharp core level.

Thus, XPS and SXS are complementary methods to study the electronic structure of materials. For a valuable comparison of experimental data, it is particularly interesting to set all the experimental curves in a common energy scale referring to the Fermi level. This is directly achieved in XPS spectra which are referred to E_F. For SXS data, the spectral distributions are measured in a transition energy scale; in this case, the determination of the binding energy of the core level involved in the emission or absorption transition is necessary. It can be obtained directly by XPS measurement of the core hole energy or by combining XPS and XES data. In this case the relaxation effects are neglected.

3.   Experimental results

3.1   AMORPHOUS Si (a-Si)

In figure 1, we report the XPS valence band, the X-ray emission band Si Kβ (3p--->1s) and the photoabsorption spectrum close to the Si K edge (1s--->np) for a-Si (1,2). The spectra are plotted in a common energy scale with E_F as origin. The total VB distribution given by the XPS spectrum exhibits two peaks. The lower BE one, rather sharp, is at the same BE as XES maximum which corresponds to Si 3p states.The higher BE XPS peak corresponds to Si 3s states; however the small shoulder observed on Si Kβ at about 7 eV BE, reveals that Si 3s are partly hybridized to Si 3p in this range. Let us recall that the crystalline silicon (c-Si) VB spectrum exhibits three well-resolved peaks associated to Si p, sp and s states which are well corroborrated by theoretical DOS of pure c-Si see (ref 3). In amorphous silicon, sp and s states merge into one hump and the leading edge becomes more abrupt. These effects of the loss of long range order are well supported by calculations taking into account the existence of odd-membered rings of bonds and fluctuations of bond angles and bond lengths. The Si K photoabsorption spectrum reported in figure 1 gives the unoccupied Si p states. Let us note that X-ray emission and absorption curves are directy measured in the same energy scale and that no energy adjustment is needed to obtain the relative energy position of emission and absorption edges. The comparison of spectra from a-Si, hydrogenated or not, have shown

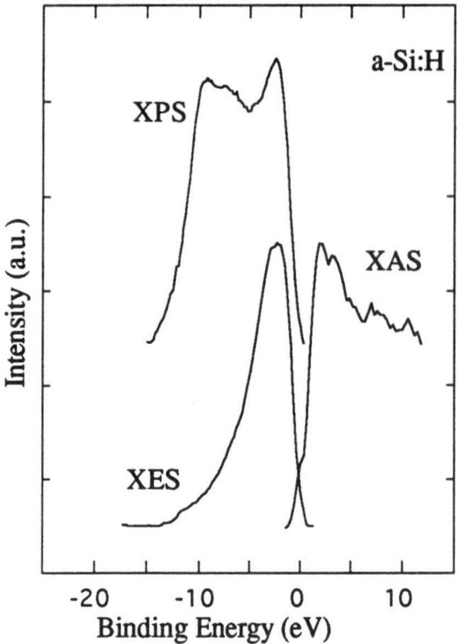

Figure 1 XPS VB, XES Si Kβ and XAS Si K for a-Si:H

that the presence of hydrogen involves a small receeding of the VB edge which accounts well for the increase of the optical gap observed from optical spectra (4).

The incorporation of another atom in a-Si network induces noticeable modifications of its optical properties. Low gap, wide gap systems or insulators can be obtained, according to the element incorporated and to its concentration. In all cases, it is particularly interesting to investigate separately the VB and CB states near the edges, which are responsible for the optical properties of the systems.

3.2   a-Si-Ge$_x$:H, a-Si-Ni$_y$:H (LOW GAP ALLOYS)

The incorporation of Ge in a-Si :H is accompanied by a decrease of the optical gap E_G as revealed by optical spectroscopy. The variation of E_G follows the relation:
$$E_{04} = 1.92 - 0.67 \ x(Ge) \qquad (5)$$

The study of the Si Kβ emission and Si K absorption spectra from a-Si$_{1-x}$Ge$_x$:H alloys have shown that the decrease of E_G is only due to a progressive shift of the top of the VB towards E_F as x increases, whereas the bottom of the CB remains roughly at the same energy position (6,7) (fig.2). XAS measurements from Si K and Ge L$_3$ edges which give respectively Si p and Ge s,d states show that Si p and Ge s,d states are mixed at the bottom of the CB edge of the alloys (fig. 3).The slope of the VB edge does not change under alloying and consequently the spreading of the optical absorption edge is only due to a spreading of the CB edge as revealed by the study of the Si K and Ge L$_3$ edges . This spreading of the CB edge is interpreted by an increase of disorder in the alloy; thus these results show that mixed sp antibonding states located at the bottom of

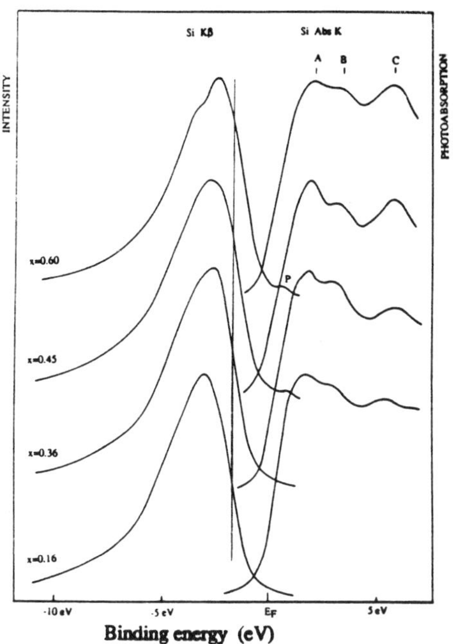

Figure 2 - Si Kβ and Si Abs K for a-Si$_{1-x}$Ge$_x$:H with x= 0.16, 0.36, 0.45 and 0.60

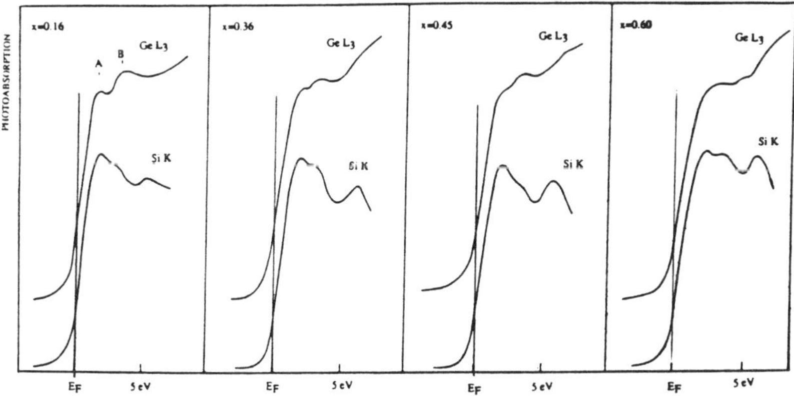

Figure 3 - Si K and Ge L$_3$ XAS for a-Si$_{1-x}$Ge$_x$:H with x= 0.16, 0.36, 0.45 and 0.60

502

the CB are more sensitive to disorder than pure p states that are present at the top of the VB.

By incorporating a transition metal such as Ni in a-Si network, a larger decrease of the optical gap is obtained and a transition to metallic character is observed at about y(Ni) = 0.26 (8). In these alloys, XES Si Kβ (Si 3p ---> Si 1s) and Ni Lα (Ni 3d ---> Ni 2p) can be analysed with good energy resolution. They provide respectively the Si 3p and Ni 3d VB states; the total VB distribution is given by XPS spectra from the same alloys. In figure 4, XES (Si Kβ, Ni Lα), XAS (Si K) and XPS VB spectra are plotted comparatively for a-Si$_{1-y}$ Ni$_y$:H alloys with y = 0.21, 0.44 and 0.71 (9). The results show that Si 3p states are strongly modified as Ni is incorporated. Si 3p maximum is significantly shifted towards higher BE; simultaneously, at the top of the VB, Si 3p states are progressively replaced by Ni d states which are responsible for the metallic character of the alloy. The XPS for the same systems show dramatic changes between 21% and 44% or 71% alloys. This result is mainly due to the influence of photoionisation cross sections. For low Ni content, the first two low BE XPS features are in coïncidence with the maxima of Ni Lα and Si Kβ states and hence are associated to the Ni 3d and Si 3p states respectively. The peak at about 10 eV BE can be assigned to Si 3s states. the XPS VB spectra are dominated by the Ni d states. It is worth noting that in these cases, XES data are quite helpfull to interpret the XPS data. The experimental data are in good agreement with DOS calculation performed recently (10).

3.3   a-Si-N$_X$:H, a-Si-C$_X$:H ALLOYS

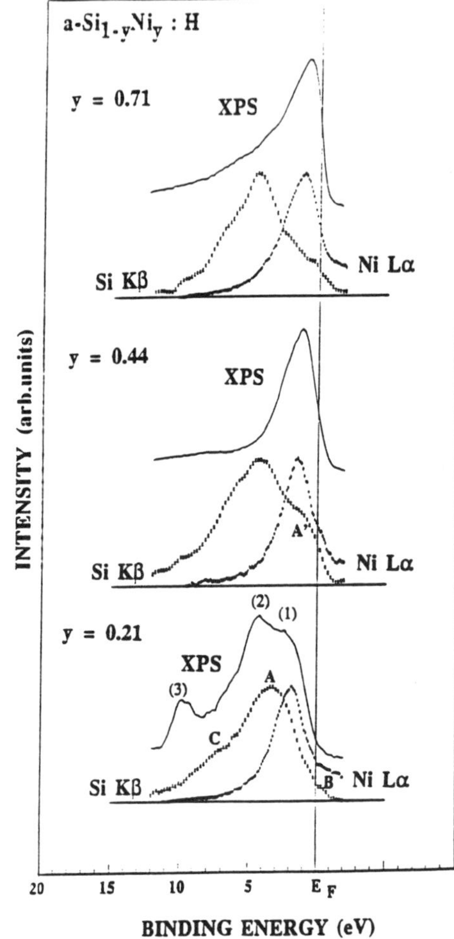

Figure 4 - XPS and XES Si Kβ, Ni Lα for a-Si$_{1-y}$Ni$_y$:H with x= 0.21, 0.44 and 0.71

By addition of nitrogen or carbon in the a-Si matrix, a progressive widening of the gap is observed and insulators (silicon nitride and carbide) can be obtained.

In a-Si-N$_x$:H compounds, only slight modifications are observed from x = 0 to 0.45, (fig.5-a). The maximum of Si 3p states and the low BE edge of the VB are shifted towards high BE. In this case the VB distribution is mainly determined by Si-Si bonds, as confirmed by the study of Si 2p core levels.

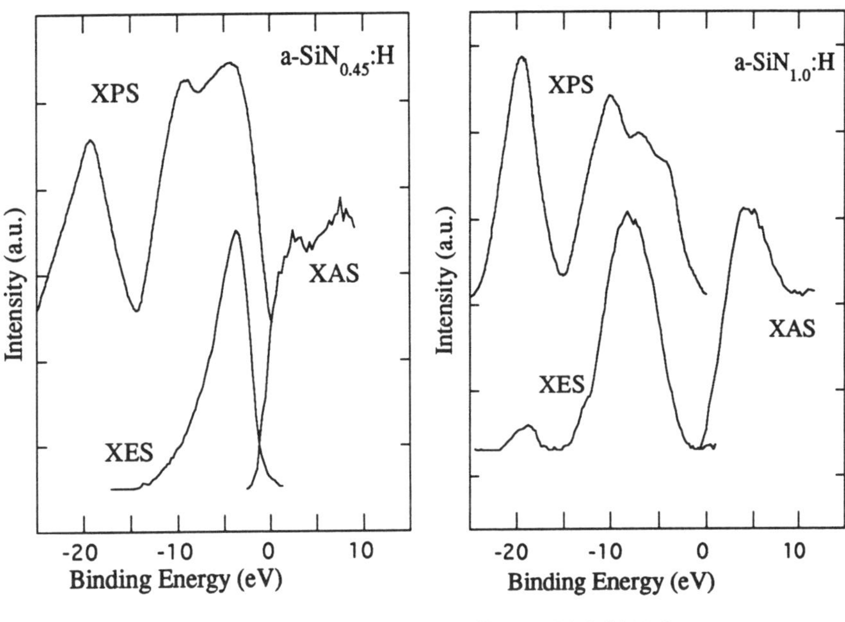

Figure 5 - XPS VB, XES Si Kβ and XAS Si K for
a) a-SiN$_{0.45}$:H   b) a-SiN$_{1.0}$:H

For x = 1.0 (fig 5-b), the shape of XPS VB and Si Kβ emission are completely changed as compared to lower N content. The VB exhibits three peaks and its shape is close to that of silicon nitride. The maximum of Si 3p coïncides with the intermediate XPS curve (2,11). Consequently, the first shoulder of XPS can be assigned to pure N states. A DOS calculation by J. Robertson for Si$_3$N$_4$ attributes this peak to the N 2pπ lone pair (13). In this alloy, Si-N bonds are dominant and the electronic structure is noticeably modified as compared to pure a-Si. At higher BE, about 20 eV, a strong peak is observed on XPS spectrum and a small feature is present at the same energy on Si Kβ. It results from N 2s states, slightly hybridized with Si 3p states, in agreement with theoretical predictions. The edge of Si K photoabsorption edge is not significantly modified in x= 0.45 alloy; it is shifted by about 2 eV in x=1.0 alloy, showing that in this case, the increase of the optical gap is due to shifts of both VB and CB edges. These results confirm well those from optical spectra which exhibit dramatic changes of the optical spectra of a-Si N$_x$:H alloys beyond x = 0.4 (14).

Similarly the addition of carbon in a-Si matrix involves an increase of the optical gap. At low C content, typically x(C) = 0.2, the distributions of the valence and conduction bands are practically identical to pure a-Si one; only a small shift of the VB

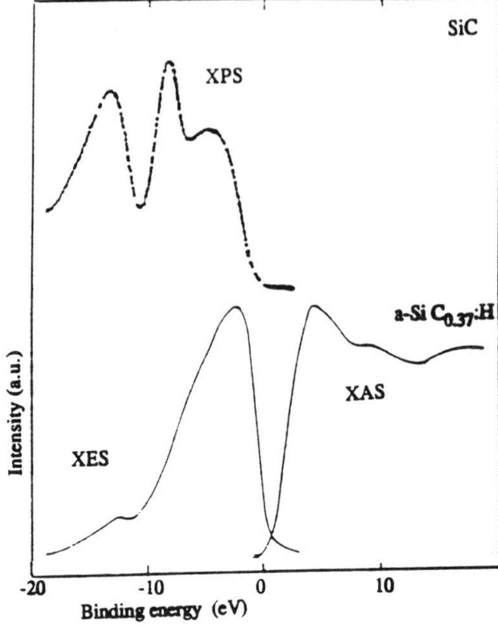

Figure 6 - XES Si Kβ and XAS Si Abs K
for a-SiC₀.₃₇:H ; XPS from SiC

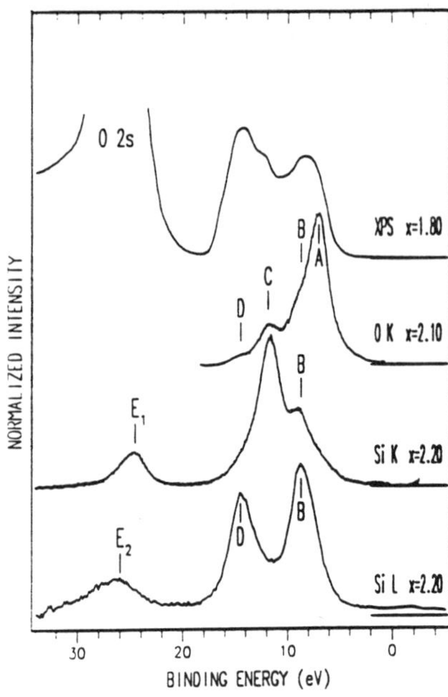

Figure 7 - XPS, XES Si K and L,
O K from a-Si O$_x$ (ref. 18)

edge is observed which explains the small increase of the optical gap. In figure 6, the XES and XAS spectra from a-Si C$_{0.37}$:H are reported. Pure SiC XPS spectrum is also reported as a reference (15,16). The Si 3p states given by Si Kβ are located at the top of the VB; their distribution is noticeably broadened as compared to Si, due to the existence of Si-C bonds. The XPS spectrum exhibits a sharp peak at about 9.5 eV which is in

coïncidence with a feature on Si Kβ , showing that Si 3p states are mixed to C 2p states; the higher BE peak is due to C 2s states which are mixed to Si 3s (and 3s states). The experimental results are very well supported by theoretical DOS calculations (17).

Consequently though similar behaviour of optical spectra is observed for a-Si N$_x$:H and a-Si C$_x$:H alloys, SXS and XPS reveal that important differences are observed in the electronic structure of these alloys.

3.4   Si O$_x$:H

The XPS VB, XES O K, Si Kβ and L from SiO$_x$:H (x= 1.8 to 2.2) are reported in fig. 7 from reference (18). These spectral curves give respectively the total VB, the O 2p VB, and the Si 3p and 3s,d states. They show that O 2p states are located at the top of the

VB; they are mixed to Si 3p states whose maximum is at C (12 eV) and probably to Si 3d states (peak B). The Si 3s states are located at about 14 BE, they are mixed to O 2p states. At higher BE (about 26 eV), O 2s states are mixed to Si 3p and 3s states. There is a very good agreement between the experimental data and it is worth noting that XES data are necessary to identify the XPS peaks.

4.   Conclusion

Valuable information on the electronic structure of amorphous insulators and semiconductors can be obtained from combined measurements performed by XPS and SXS methods. The data are quite complementary to those deduced from optical spectroscopy; in particular, changes in the gap widths of a compound as a function of its composition can be associated to the shift of either the VB or the CB edges. Examples are given for amorphous Si-based systems. Other amorphous semiconductors such as III-V and chalcogenides systems have been studied by using these techniques. In the case of the Sb_2S_3-Tl_2S system, the XPS and XES data give respectively the total VB distribution and the sulfur 3p states. They reveal that in most cases, the S 3p VB states are located at the top of the VB (19).

References:

1. Sénémaud, C. and Lima, M.T. (1979) K X-Ray spectra of amorphous and crystalline silicon, *J. Non Cryst. Solids,* 33, 141-148.
2. Driss Khodja, M., (1993) Thèse, *Université Pierre & Marie Curie, Paris.*
3. Ley, L., Cardona, M. and Pollak, R.A., (1979) Photoemission in Semiconductors, *Topics in Applied Physics* Vol. 27, 10-172. *Springer Verlag.*
4. Drévillon, B., Sénémaud, C., Cardinaud, C. and Driss Khodja, M., (1986) Electronic structure studies of plasma-deposited amorphous silicon, *Philos. Mag.* B 54, 335-342.
5. Chahed, L., Gheorghiu, A., Thèye, M.L., Ardelean, I., Sénémaud, C. and Godet, C. (1989) Studies of the density of states at the band edges and in the pseudo-gap in a-SiGe:H alloys by combined photothermal deflection spectroscopy and X-rays spectroscopy, *J. Non Cryst. Solids,* 114, 471-473.
6. Sénémaud, C. and Ardelean, I. (1990) Electronic structure of hydrogenated amorphous silicon-germanium alloys studied by x-ray photoelectron spectroscopy and soft-x-ray spectroscopy, *J.Phys.: Condens. Matter,* 2, 8741-8750.
7. Sénémaud, C., Cardinaud, C.and Villela, G., (1984) Direct observation of occupied gap states in amorphous $Si_{1-x}Ge_x$:H alloys by SXS, *Solid State Comm.* Vol.50, 643-645.
8. Davis, E.A., Bayliss, S.C., Asal, R. and Manssor, M. (1989) Free-carrier behaviour in a-$Si_{1-y}Ni_y$:H, *J. Non Cryst. Solids,* 114, 465-467.
9. Gheorghiu, A., Sénémaud, C., Asal, R, and Davis,E.A., (1995) Study of the electronic structure of a-$Si_{1-y}Ni_y$:H alloys by XPS and XES,*J. Non Cryst. Solids,* 182, 293-301.
10. Gheorghiu, A., Sénémaud, C., Belin-Ferré, E., Dankhazi, Z., Magaud-Martinage, L.and Papaconstantopoulos, D.A., (1996) Comparison of theoretical and experimental electronic distributions of Si-Ni and Si-Er alloys, *J. Phys. Condens. Matter* 8, 719- 728.

11. Kärcher, R., Ley, L. and Johnson, R.L., (1984) Electronic structure of hydrogenated and unhydrogenated amorphous SiN_x ($0 \leq x \leq 1.6$): A photoemission study, *Phys. Rev.* B 30, 1896-1910.
12. Sénémaud, C., Driss Khodja, M., Gheorghiu, A., Harel, S., Dufour, G., and Roulet, H., (1993) Electronic structure of silicon nitride studied by both soft x-ray spectroscopy and photoelectron spectroscopy, *J. Appl. Phys.*, 74, 5042-5046.
13. Robertson, J., (1991) Electronic structure of silicon nitride, *Philos.Mag.*, B 63, 47-77.
14. Sotiropoulos, J., Fuhs, W. and Nickel, N., (1993) Structure and optical properties of a-$Si_{1-x}N_x$ alloys.*J. Non Cryst. Solids* 164, 881-884.
15. Solomon, I., Schmidt, P., Sénémaud, C. and Driss Khodja, M., Band structure of carbonated amorphous silicon studied by optical, photoelectron and x-ray spectroscopies (1988) *Phys. Rev.* B 38, 13263-13270.
16. Driss Khodja, M., Dufour, G., Gheorghiu, A., Roulet, H., Sénémaud, C., Cauchetier, M., Croix, O. and Luce, M., (1992) Electronic structure of laser-synthetized Si-C by photoelectron and soft x-ray spectroscopy, *Materials Science & Engineering,* B 11, 97-101.
17. Robertson, J., (1992) The electronic and atomic structure of hydrogenated amorphous Si-C alloys, *Philos. Mag.* B 66, 615-638.
18. Simunek, A. and Wiech, G., (1991) Partial densities of statyes and chemical shifts in the system a-SiO_x:H ($O \leq x \leq 2.2$) *J. Non Cryst. Solids* 137-138, 903-906.
19. Gheorghiu, A., Lampre, I., Dupont, S., Sénémaud, C., El Idrissi Raghni, M.A., Lippens, P.E. and Olivier-Fourcade, J., (1995) Electronic structure of chalcogenide compounds from the system $Tl_2S-Sb_2S_3$ studied by XPS and XES, *J. Alloys and Comp.*, 228, 143-147.

AMORPHOUS INSULATORS AND SEMICONDUCTORS
May 26 - June 8, 1996
Sozopol, Bulgaria

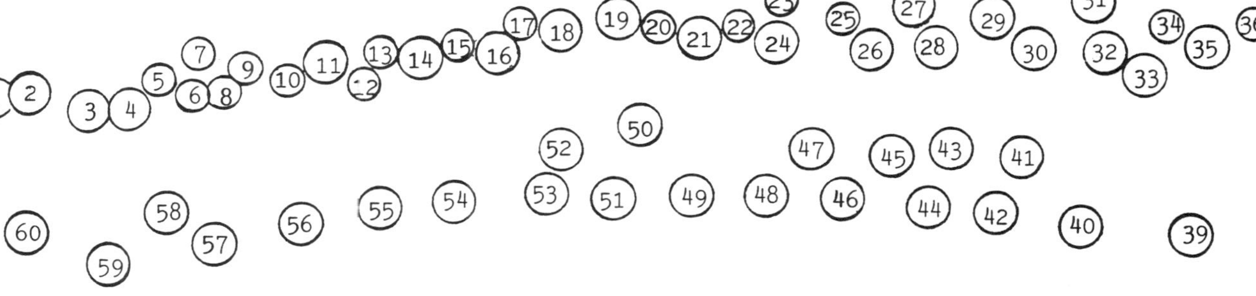

[1] I. Kasik, [2] N. Mousseau, [3] A. Pryde, [4] I. Pocsik, [5] M. Soltwisch, [6] M. Dove, [7] W. Gerike, [8] N. Tomozeiu, [9] G.J. Adriaenssens, [10] L. Aldon, [11] G. Kasper, [12] M.T. Mora, [13] V. Arkhipov, [14] S. Yannopoulos, [15] A.R. Day, [16] R. Zallen, [17] J. Classen, [18] D. Drabold, [19] M.F. Thorpe, [20] M.S. Marinov, [21] M.I. Mitkova, [22] M. El Souaidi, [23] J. Schmelzer, [24] S. Hunklinger, [25] S. Borick, [26] Z.G. Ivanova, [27] W. Bresser, [28] E. A. Smorgonskaya, [29] M. Gambhir, [30] D. Ilieva, [31] N. Zotov, [32] T. Petkova, [33] E. Courtens, [34] P. Boolchand, [35] V. Boev, [36] A. Bakai, [37] Y. Dimitriev, [38] E. Emelianova, [39] V. Tikhomirov, [40] I. Iliev, [41] M. Sendova-Vassileva, [42] C.E. Jesurum, [43] D. Nesheva, [44] T. Lobkovskaya, [45] J. Vaklidova, [46] M. Foret, [47] C. Sénémaud, [48] R. Popescu, [49] Y. Wang, [50] L. Pusztai, [51] O. Matsuda, [52] M. Coeck, [53] N. Bollé, [54] A. Kodolbas, [55] E. Cernoskova, [56] E. Mytilineou, [57] I. Kotsalas, [58] P. Vlaimirov, [59] A. Krylov, [60] M. Tzolov.

Not in photograph: A. Alexopoulos, C.A. Angell, M. Arai, E. Brousse, J. Dong, R. Elliott, I. Gutzow, R. Iordanova, M. Koos, C. Levelut, J-L. Prat, A. Sapelkin, O. Shpotyuk, F. Terki, P. Vashishta, T. Vassilev, N. Vedishcheva, A. Veleva-Dobreva, A.C. Wright, J.W. Zwanziger.

PARTICIPANTS

G.J. Adriaenssens
Laboratorium voor Halfgeleiderfysica
Katholieke Universiteit Leuven
Celestijnenlaan 200 D
B-3001 Haverlee-Leuven
BELGIUM

Laurent Aldon
Laboratoire de Physicochimie des
Matériaux Solides
Université Montpellier II
Place Eugène Bataillon
34095 Montpellier Cedex 05
FRANCE

Aris Alexopoulos
Department of Physics
Monash University
Clayton, Victoria 3168
AUSTRALIA

C.A. Angell
Department of Chemistry
Arizona State University
Box 871604
Tempe, AZ 85287-1604
U.S.A.

Masatoshi Arai
BSF
National Lab. for High Energy Physics
1-1 Oho
Tsukuba 305
JAPAN

Vladimir Arkhipov
Lab. voor Halfgeleiderfysica
Katholieke Universiteit Leuven
Celestijnenlaan 200 D
B-3001 Heverlee-Leuven
BELGIUM

Alexander Bakai
National Scientific Center
Kharkov Institute of Physics &
Technology
Akademicheskaja St., 1
310108 Kharkov
UKRAINE

Victor Boev
Central Lab. of Electrochemical Power
Sources
Bulgarian Academy of Sciences
Acad. G. Bonchev Str. - Bl. 10
1113 Sofia
BULGARIA

Nadia Bollé
Katholieke Universiteit Leuven
Lab. voor Halfgeleiderfysica
Celestijnenlaan 200D
3001 Heverlee
BELGIUM

P. Boolchand
Dept. of Electrical and Computer
Engineering
Mail Location 30
University of Cincinnati
Cincinnati, OH 45221
U.S.A.

512

Steven S. Borick
Department of Chemistry
Arizona State University
Box 871604
Tempe, AZ 85287-1604
U.S.A.

Wayne Bresser
Dept. of Electrical & Computer
Engineering
PO Box 210030
University of Cincinnati
Cincinnati, OH 45221-0030
U.S.A.

Elodie Brousse
Lab. de Physicochimie des Matériaux
Solides
Université Montpellier II
Place E. Bataillon c.c. 03
34095 Montpellier Cedex 05
FRANCE

Eva Cernoskova
Joint Lab. of Solid State Chemistry
Czech. Acad. of Sci. & Univ.
Pardubice
Studentska 84
530 09 Pardubice
CZECH REPUBLIC

Johannes Classen
Institut für Angewandte Physik
Ruprecht-Karls-Universität Heidelberg
Albert-Ueberle-Str. 3-5
69120 Heidelberg
GERMANY

Michèle Coeck
Katholieke Universiteit Leuven
Lab. voor Halfgeleiderfysica
Celestijnenlaan 200 D
B-3001, Heverlee-Leuven
BELGIUM

E.L. Courtens
LSMV, case 069
Univ. de Montpellier II
F-34095 Montpellier Cedex 5
FRANCE

A. Roy Day
Department of Physics
Marquette University
Milwaukee, WI 53201-1881
U.S.A.

Yanko Dimitriev
University of Chemical
   Technology and Metallurgy
Department of Silicates
Boul. Kliment Ohridski 8
1756 Sofia, BULGARIA
BULGARIA

Jianjun Dong
Dept. of Physics and Astronomy
Ohio University
Athens, OH 45701-2979
U.S.A.

M. Dove
Department of Earth Sciences
University of Cambridge
Downing Street
Cambridge CB2 3EQ
UNITED KINGDOM

David Drabold
Dept. of Physics and Astronomy
Ohio University
Athens, OH 45701-2979
U.S.A.

Roger Elliott
University of Oxford
Theoretical Physics
1 Keble Road
Oxford OX1 3NP
UNITED KINGDOM

Mohamed El Souaidi
Universitat Autónoma de Barcelona
Departament de Física
Grup Física Materials I
08193 Bellaterra (Barcelona)
SPAIN

Evguenia Emelianova
Moscow Institute of Physics &
Engineering
Physics Department
Kashirskoye shosse 31
Moscow 115409
RUSSIA

Marie Foret
Université de Montpellier II
Laboratoire des Verres
Place E. Bataillon
F.34095 Montpellier
FRANCE

Manoj Gambhir
Cambridge University
Cavendish Laboratory
Madingley Road
Cambridge CB3 0EH
UNITED KINGDOM

Walter Gerike
Universität Rostock
Department of Physics
Universitätsplatz 3
18051 Rostock
GERMANY

I. Gutzow
Institute of Physical Chemistry
Bulgarian Academy of Sciences
1113 Sofia
BULGARIA

S. Hunklinger
Institute für Angewandte Physik
Ruprecht-Karls-Universität Heidelberg
Albert-Ueberle-Str. 3-5
69120 Heidelberg
GERMANY

Ilian D. Iliev
Central Lab. of Electrochemical Power
Sources
Bulgarian Academy of Sciences
Bl. 10 - Acad. G. Bonchev Str.
1113 Sofia
BULGARIA

Dora P. Ilieva
University of Chemical
   Technology and Metallurgy
8 Kliment Ohridski blvd.
1756 Sofia
BULGARIA

Reni S. Iordanova
Institute of General and Inorganic
Chemistry
Bulgarian Academy of Sciences
Acad. G. Bonchev Str. - Bl. 11
1113 Sofia
BULGARIA

Zoja G. Ivanova
Institute of Solid State Physics
Bulgarian Academy of Sciences
72 Tzarigradsko Chaussee Blvd.
1784 Sofia
BULGARIA

514

C. Esther Jesurum
M.I.T. Lab for Computer Science
545 Technology Square, Rm. 338
Massachusetts Institute of Technology
Cambridge, MA 02139
U.S.A.

Ivan Kasik
Inst. of Radio Engineering &
Electronics
Czech Academy of Sciences
Chaberska 57
Prague 8 (182 51)
CZECH REPUBLIC

Gernot Kasper
Institut für Angewandte Physik
Ruprecht-Karls-Universität Heidelberg
Albert-Ueberle-Str. 5
D-69120 Heidelberg
GERMANY

A. Osman Kodolbas
Dept. of Physics Engineering
Hacettepe University
06532 Beytepe-Ankara
TURKEY

Margit Koos
Research Institute for Solid State
Physics
Hungarian Academy of Sciences
Budapest XII. Konkoly Thege M. u.
29-33
H-1525 Budapest P.O. Box 49
HUNGARY

Ioannis P. Kotsalas
National Technical University of
Athens
Department of Physics
Iroon Politechniou 9 Str. (Zografou
Campus)
GR-157 80 Athens
GREECE

Andrey S. Krylov
Faculty of Computational Mathematics
& Cybernetics
Moscow State University
Leninskie Gory
119899 Moscow
RUSSIA

Claire Levelut
Laboratoire des Verres
Université Montpellier II
place Eugène Bataillon, case 69
34095 Montpellier Cedex 5
FRANCE

Tatiana A. Lobkovskaya
Lab. of Low Temperature, Physical
Department
4 Svobody Square
Kharkov State University
Kharkov 310077
UKRAINE

Marin S. Marinov
Central Lab. of Mineralogy &
Crystallography
Bulgarian Academy of Sciences
Rakovski Street 92
1000 Sofia
BULGARIA

Osamu Matsuda
Dept. of Physics - Graduate School of
Science
Osaka University
1-1 Machikaneyama
Toyonaka 560
JAPAN

Maria I. Mitkova
Central Lab. of Electrochemical Power
Sources
Bulgarian Academy of Sciences
Acad. G. Bonchev str. - Bl. 10
1113 Sofia, BULGARIA
BULGARIA

Maria T. Mora
Universitat Autonoma de Barcelona
Department de Fisica
Grup Fisica Materials I
08913 Bellaterra (Barcelona)
SPAIN

Normand Mousseau
Département de physique
Université de Montréal
C.P. 6128, succ. Centre-ville
Montréal, (Québec) H3C 3J7
CANADA

Eugenia Mytilineou
Physics Department
University of Patras
GR-265 00 Patra
GREECE

Diana Nesheva
Institute of Solid State Physics
Bulgarian Academy of Sciences
72 Tzarigradsko Chaussee Blvd.
1784 Sofia
BULGARIA

Tamara Petkova
Central Lab. of Electrochemical Power
Sources
Bulgarian Academy of Sciences
Acad. G. Bonchev Str. - Bl. 10
1113 Sofia
BULGARIA

Istvan Pocsik
Research Institute for Solid State
Physics
Hungarian Academy of Sciences
Budapest XII. Konkoly Thege M. u.
29-33
H-1525 Budapest P.O. Box 49
HUNGARY

Radian Popescu
Institute of Physical Chemistry
Romanian Academy
Spl. Independentei 202
77208 Bucharest
ROMANIA

Jean-Luc Prat
Laboratoire des Verres, UMR 5587
Université Montpellier II, C-C-069
Place Eugène Bataillon
34095 Montpellier Cedex 5
FRANCE

Alexandra Pryde
Department of Earth Sciences
University of Cambridge
Downing Street
Cambridge CB2 3EQ
UNITED KINGDOM

Laszlo Pusztai
Studsvik Neutron Research Laboratory
University of Uppsala
S-611 82 Nykoping
SWEDEN

516

Andrei V. Sapelkin
Dept. of Applied Physics
De Montfort University
The Gateway
Leicester LE1 9BH
UNITED KINGDOM

Juern Schmelzer
Department of Physics
University of Rostock
Universitätsplatz
18051 Rostock
GERMANY

Marushka Sendova-Vassileva
Central Lab. for Solar Energy & New
Energy Sources
Bulgarian Academy of Sciences
72 Tzarigradsko Chaussee
1784 Sofia
BULGARIA

Christiane Sénémaud
Laboratoire de Chimie-Physique, URA
CNRS 176
Université Pierre et Marie Curie
11 rue Pierre et Marie Curie
75231 - Paris cedex 05, FRANCE
FRANCE

O.I. Shpotyuk
Dept. of New Perspective Technologies
Lviv Scientific Research Institute of
Materials
Stryjska str. 202
Lviv 290031
UKRAINE

Emilia A. Smorgonskaya
A.F. Ioffe Physico-Technical Institute
Russian Academy of Sciences
Polytechnicheskaya 26
St. Petersburg 194021
RUSSIA

Manfred Soltwisch
Institut fuer Experimentalphysik
Freie Universitaet Berlin
Arnimalle 14
D-14195 Berlin
GERMANY

Ferial Terki
Laboratoire des Verres, UMR 5587
Université Montpellier II, C-C-069
place Eugène Bataillon
34095 Montpellier Cedex 5
FRANCE

M.F. Thorpe
Dept. of Physics and Astronomy
Michigan State University
East Lansing, MI 48824
U.S.A.

Victor Tikhomirov
Katholieke Universiteit, Leuven
Laboratorium voor Halfgeleiderfysica
Celestijnenlaan 200D
B-3001 Heverlee
BELGIUM

Nicolae Tomozeiu
Univ. of Bucharest - Fac. of Physics
Str. Fizicienilor Nr. 1, Cod 76900
P.O. Box 11-Mg
Sector 5 Bucharest
ROMANIA

Marian B. Tzolov
Central Lab. for Solar Energy & New
Energy Sources
Bulgarian Academy of Sciences
72 Tzarigradsko Chaussee
1784 Sofia
BULGARIA

P. Vashishta
Department of Physics & Astronomy
Nicholson Hall
Louisiana State University
Baton Rouge, LA 70803-4001
U.S.A.

Tsvetan Vassilev
Institute of Physical Chemistry
Bulgarian Academy of Sciences
Acad. G. Bonchev Str. - Bl. 11
1113 Sofia
BULGARIA

Natalia M. Vedishcheva
Institute of Silicate Chemistry
Russian Academy of Sciences
Ul. Odoevskogo, 24, Korp. 2
St. Petersburg 199155, RUSSIA
RUSSIA

Anka Veleva-Dobreva
Institute of Physical Chemistry
Bulgarian Academy of Sciences
Acad. G. Bontchev Str. - Bl. 11
Sofia 1113
BULGARIA

Pavel V. Vladimirov
RRC "Kurchatov Institute"
Kurchatov sq. 1
123182 Moscow
RUSSIA

Yong Wang
Dept. of Physics - K. Murase Lab.
Osaka University - Fac. of Sci.
Machikaneyama 1-1
Toyonaka, Osaka 560
JAPAN

A.C. Wright
Department of Physics
J.J. Thomson Physical Laboratory
University of Reading, Whiteknights
Reading RG6 6AF
UNITED KINGDOM

S.N. Yannopoulos
Foundation for Research &
Technology-Hellas
Institute of Chemical Engineering
    and High Temperature Chemical
Processes
P.O. Box 1414
GR -265 00 Patras
GREECE

Richard Zallen
Department of Physics
Virginia Tech
Blacksburg, VA 24061
U.S.A.

Nikolay Zotov
Central Lab. of Mineralogy &
Crystallography
Bulgarian Academy of Sciences
Rakovski Street 92
Sofia 1000
BULGARIA

Joseph W. Zwanziger
Department of Chemistry
Indiana University
Bloomington, IN 47405
U.S.A.

INDEX

Absorption spectra of chalcogenides, 77
Acoustic properties, 488
Acoustic waves, 255
Acoustical modes, 392
Adam-Gibbs equation, 9
Aerogels, 270, 276
Ag-Te-I, 75
Alkali chloride glasses, 5
Alloys, 502
Amorphous carbon, 151, 202, 321
Amorphous diamond, 291, 423
Amorphous germanium, 293
Amorphous silicon, 215, 437, 499
Angular constraints, 305
Anharmonicity, 2, 54
Anisotropy, 245
Anomalous dispersion, 124
As-S, 86
Average coordination, 230, 340

Banana graphs, 316
Bandtails, 410
Barium crown glasses, 388
Bethe lattice, 424
Binding energy, 500
Boltzmann factor, 138, 301
Bond angle distributions, 233
Bond bending, 257, 339
Bond stretching, 257, 339
Borate glasses, 250
Boson peak, 385
Brillouin scattering, 277, 389

Canonical ensemble, 155
Car-Parrinello methods, 13, 147, 416
Chain structure, 73
Chain terminator, 76
Chalcogenides, 72, 86, 227
Charged dangling bonds, 465
Chemical bond, 245

Chemical potential, 237
Chemical short-range ordering, 230
Chemical structure of glasses, 236
Chemical structure, 238
Chemically ordered network, 50
Coefficient of thermal expansion, 238
Cole-Davidson relaxation, 392
Colloidal aggregates, 266
Compton scattering, 96
Computer modeling, 133, 296
Computer simulations, 329
Conjugate gradient algorithm, 333
Connectivity percolation, 309
Constraint counting, 217, 289, 304, 342, 365
Continuous random network, 134, 292
Cooling rates, 72
Coordination constraints, 230
Coordination number distributions, 232
Coordination of the vitreous tellurium, 75
Coordination parameter, 340
Corner-sharing, 192
Correlation functions, 105
Coulomb potential, 152, 180
Covalent amorphous solids, 293
Covalent glasses, 226
Covalent network, 50
Covalently bonded, 46
Crack propagation, 151
Cristobalite, 349
Crystalline silicon, 500
Crystallization, 56
Crystallography, 351
Cu-As-Te glass, 231
Current degradation, 440

Dangling bond, 321, 438, 465
Data correction, 98
Debye specific heat, 491

520

Deep states, 455
Deep-retrapping, 456
Defect energies, 78
Defect equilibration, 441, 465
Defect pairs, 477
Defect pool model, 463
Defect structures, 405
Defects in amorphous chalcogenides, 78
Defects, 80
Degrees of coordination, 78
Dense random packing, 134
Density functional theory, 412
Density of states, 280, 320
Diamond cubic cell, 290
Diamond films, 323
Diamond lattice, 221
Diamond-like carbon, 302, 344
Dielectric experiments, 471
Diffraction patterns, 93, 215
Diffuse scattering, 370
Dipolar interaction, 477
Dipole selection rules, 498
Dissolving agent, 80
Domain decomposition, 170
Doping, 440
DOS profile, 454
Drift mobility, 449
Dynamic fracture, 151
Dynamical matrix, 406
Dynamical scaling, 256

Eden growth, 266
Effective activation energy, 459
Elastic constants, 320
Elastic deformation, 345
Elastic fields, 476
Elastic properties, 306
Elasticity tensor, 329
Elasticity, 257, 329
Electron diffraction, 370
Electron gas, 413
Electronic structure, 497

Electronic structure of chalcogenide glasses, 77
Electronic structure, 405
Emission time, 462
Emission-limited, 454
Empirical potentials, 141, 407
Empirical theory of glass formation, 49
Empirical tight-binding approximation, 408
Energy gap, 78
Energy level diagram, 78
Entropy, 29, 54
Equation for the molar volume, 238
ERS experiments, 465
Escape depth, 498
Ewald summation, 152, 162, 180
Exponential bandtails, 405
Exponential DOS, 451
Extended electronic states, 77

Fictive temperature, 387, 241
Finite-size effects, 400
First diffraction peak, 126, 385
Fitting parameters and constraints, 226
Floppy modes, 1, 151, 194, 289, 319, 341, 405
Floppy networks, 341
Floppy to rigid transition, 405
Fluctuation amplitude, 458
Form factors, 262
Fourier transform, 92, 99, 226
Fractal dimension, 309
Fractal networks, 258
Fractal structure, 266, 272
Fractal, 255
Fraction of 4-fold coordinated boron atoms, 240
Fractional coordination number constraints, 233
Fracture surface, 193
Fracture, 201
Fragile liquids, 1, 5, 37, 39
Fragility, 2
Free ends, 80

521

Free volume theory, 54
Free-carrier mobility, 445
Free-hydrogen lifetime, 442

GaAs, 146
Gaussian fitting, 230
Gels, 270
GeO$_2$, 86
Ge-Se, 86
Gibbs free energy, 238
Glass formation, 21, 27, 45, 57
Glass forming ability, 45
Glass forming regions, 46
Glass structure, 245
Glass transition, 1, 52, 385
Glassformation, 1, 72, 74
Glassy alloys, 133
Glow-discharge, 439
Graphite, 151, 202

Hardness, 323, 341, 344
Harmonic wells, 470
High pressure, 191
High temperature, 151
High-temperature phases, 374
Hooke springs, 306
Hopping, 453, 456
Hopping-assisted release, 456
Hot-wire deposition, 441
Hydrogen, 439, 465
Hydrogenated amorphous silicon, 342,
439
Hydrostatic pressure, 158
Hypersonic attenuation, 389

Ideal solution, 236
Impurities, 387
Inelastic neutron scattering, 370
Inelastic scattering, 276
Infrared reflectivity, 396
Interatomic potential, 184
Interlayered bond, 79
Inverse Monte Carlo, 296
Ioffe-Regel limit, 256, 278, 385

IR spectroscopies, 245
Irreversible processes, 23
Isotopic substitution, 124

Jumping rate, 470

KAF models, 78
Kauzmann paradox, 10, 27, 52
KBr:KCN, 494
KCl, 473
Keating potential, 136, 295, 331
Kinetics, 21
Kohlrausch stretched exponent, 41

Lanczos method, 424
Landau free energy, 361
Landau order parameter, 28
Landau theory, 356, 361
Leached glasses, 256, 267
Length scales, 245
Lennard-Jones potential, 162
LiF, 475
Light scattering, 281
Light-induced defect, 440, 455
Lindemann criteria, 1, 17
Linear problem, 332
Liouville operator, 160
Liquid-glass transition, 386
Local density approximation, 418
Local minimum, 337
Localized excitations, 256
Lone pair electrons, 73, 76
Long-range order, 4
Long-range potentials, 163

Magnon dynamics, 266
Markov chain, 216
Mass balance of components, 236
Maxwell relation, 5
Maxwell-Boltzmann distribution, 159
MDS, 78
Mechanical properties, 289
Medium range order, 79, 240
Melt quenching, 83

Microcanonical ensemble, 155
Mode coupling theory, 386
Model networks, 319
Model of associated solutions, 235
Model of ideal associated solutions,
  239
Model size, 227
Modeling of amorphous materials, 225
Modulated photocurrent, 446
Molecular dynamics, 119, 140, 151,
  154, 215, 405
Molecular-like units, 232
Monovalent iodine atoms, 76
Monochromator, 498
Monte Carlo, 117, 152, 215
Multiple-trapping, 445, 452

Nanoclusters, 199
Nano-indentation, 323
Negative effect, 80
Networks, 289
Neutron diffraction, 83, 245, 257, 292,
  354
Newton-Raphson method, 333
NIH, 345
Non-equilibrium crystallization, 60
Nonlinear problems, 333
Nonplanar triangles, 73
Nuclear magnetic resonance, 245

One-fold coordinated, 339
Optical spectroscopy, 444
Order N methods, 421
Oxidation of the sulphur, 80
Oxy-reduction nature, 80

Pair correlation functions, 220, 225,
  261
Parallel algorithms, 172
Parallel computer, 169
Parrinello-Rahman approach, 156
Partial pair distribution function, 226,
  230
Particle size, 261

Passivation, 440
Pebble game, 306
PECVD, 439
Percolation threshold, 8, 256, 258, 309,
  313
Periodic boundary conditions, 112
Persistence length, 261
Phase diagrams, 12
Phase transformations, 21, 398
Phase transition, 349, 356, 362, 375
Phonon, 359, 473
Phonon-fracton crossover, 256
Photoinduced changes, 79
Photoinduced effects, 79
Photoinduced processes, 78
Photovoltaic material, 342
Pivot sites, 312
Plucked network, 339, 343
Polymers, 5
Porod law, 264
Porous silica, 151, 192, 255
Post-transit analysis, 454
Potential minimum, 474
Powder patterns, 246
Preparation conditions, 387

Quadrupolar spins, 249
Quartz, 349, 486

Radial distribution function, 134, 292
Radiative decay, 498
Raman scattering, 252, 257, 351, 386,
  398
Random bonding, 230
Random ensemble of atoms, 227
Random moves, 228
Random networks, 46, 83, 115, 289,
  306, 329
Randomization, 302
Rayleigh scattering, 256
Reflectivity spectra, 396
Refractive index of zinc phosphate
  glasses, 238
Regular associated solutions, 236

Relaxation processes, 56
Relaxation time, 484
Relaxational phenomena, 484
Relaxational processes, 386
Reverse Monte Carlo, 215, 225
Rigid bodies, 312
Rigid unit mode, 324, 350, 363
Rigidity percolation, 1, 289, 343
Rings, 500
RMC simulations, 227
Rutile, 398

Salt-like products, 235
Scaling laws, 274
Scaling relations, 263
Schottky anomaly, 473
Selective solubility, 80
Selenium rings, 76
Self-similarity, 257
Semi-empirical interactions, 144
Semiconductors, 505
Sharpening function, 92
Shell model, 354
Shift towards longer wavelengths, 79
Short-range order, 240
Short-range potentials, 163
Silica aerogels, 255
Silica gels, 272
Silica, 181, 255
Silicate networks, 323
Silicates, 349, 375
Silicon diselenide, 151, 181
Silicon nitride, 151, 181
Silicon, 411, 437
Silver halides, 73
Single-bond dissociation energies, 342
Sintering, 199
SiO_2, 86, 292, 395, 411
SiO_4, 120
Soft mode, 350, 357, 373
Soft X-ray spectroscopies, 497
Softening of the network, 344
Sol-gel silica, 387
Sol-gel synthesis, 396

Sound velocity, 387
Sound wave, 476
Space-charge limited currents, 461
Specific heat, 472, 486
Spectral shape, 278
Staebler-Wronski effect, 440
Steric effects, 74, 108
Stiffness threshold, 341, 346
Strain, 329
Stress carrying backbone, 309
Stress, 329
Stretched-exponential, 441
Strong glasses, 390
Strong liquids, 1, 5, 37, 39
Strongly ordered network, 50
Structural correlations functions, 166
Structural disorder, 215, 278
Structural models, 109, 115
Structural motifs, 237
Structure factors, 219, 262
Structure of the chalcogenide atoms, 73
Structure, 289
Super-cell, 297, 409
Superionic conductor glasses, 216
Surface floppy modes, 317

Tail states, 449
Tellurite glasses, 250
Temporal energy fluctuations, 457
Te-Te bonds, 76
Tetrahedral building blocks, 345
Tetrahedral, 221
Tetrahedron, 73
Thermal evaporation, 439
Thermal expansion, 378
Thermodynamic potentials, 237
Thermodynamic properties, 235
Thermodynamics, 21
Three dimensional network, 73
Three-center integrals, 411
Three-fold coordination, 76
Tight binding, 145, 406, 415
Time correlation functions, 167
Time-of-flight, 447, 462

Titania, 395
Topological criterion, 48
Topological disorder, 47
Torsion angles, 108
Torsional waves, 476
Total pair distribution function, 226
Total radial distribution function, 230
Transient photocurrents, 445
Transit time, 448
Transport energy, 456
Tunneling model, 486
Tunneling modes, 405
Tunneling particle, 470
Tunneling rate, 470
Tunneling states, 469, 474
Tunneling systems in crystals, 472
Two-dimensional nonperiodic network, 73

Uncooled melts, 37
Uncoordinated networks, 341
Uniqueness, 218
Unpaired dangling electron, 78
Urbach slope, 444

Valence band tail, 462
Variable angle correlation spectroscopy, 246
Variation of bond angles, 73
Velocity-Verlet algorithm, 160
Vibrational dynamics, 385
Vibrations, 2, 276
Vicosity, 4, 5, 55
Vitreous silica, 85, 105, 486
Vitreous state, 22
Vitrification, 22, 72
Vogel-Fulcher, 53
Vycor, 267, 269

Waiting-time, 446
Wannier functions, 407
Weak to dangling bond conversion, 443, 463

X-ray, 498
X-ray diffraction, 83, 398

Zachariasen rules, 5
Zeolites, 369, 378
Zero-point energy, 470